Sustainability Sciences in Asia and Africa

Agroecosystem Dynamics and Sustainable Practices

Sustainability as a scientific endeavor requires broad understanding of the interconnections in our global environment. It is therefore positioned at the crossroads of research in natural sciences, technology and human sciences. This book series supports the global efforts by providing timely coverage of the progress, opportunities, and challenges towards achieving sustainable development goals (SDGs) in Asia and Africa. It fosters the research in transdisciplinary academic fields spanning across agriculture and global food security, health and diseases, access to clean water and affordable energy, climate resilience and effects of global climate change on biodiversity, water and land. This series is planned to narrate the stories of scientific development and progression towards SDGs from the regions of Asia and Africa. The overall aim of the series is to provide a comprehensive publication platform and act as a knowledge engine in the growth of sustainability sciences with a special focus on developing nations. The series constitute of four sub-series each focusing on a specific theme closely linked to a set of SDGs. Each sub-series is leaders and experts in agricultural sciences, environmental sciences, biomedicine and medical sciences respectively. Books in this series are useful for environmentalists, biotechnologists, plant scientists, medical practitioners, conservation biologists, policy makers, NGOs and scientific bodies working in the areas of sustainability sciences in the developing regions.

This book series publishes research associated with the prevailing opportunities and challenges faced by agroecosystems in Asia and Africa. It also explores the future scope of agroecological techniques and offers guidance on sustainable management of these systems. The series discusses important areas such as sustainable agricultural production, soil health, water resource management, biodiversity conservation, animal health and productivity, insect management strategies, genome editing and climate change mitigation. Further, it includes the challenges faced by African agriculture in arid and semi-arid regions, particularly in response to climate change. It highlights issues such as water shortages, soil degradation, desertification, and the need for sustainable farming techniques.

Sustainable agroecosystem maintenance entails strategies that improve overall environmental health and economic viability while limiting negative environmental impacts. The series takes a multidisciplinary approach to investigate creative ways for boosting sustainable agriculture, increasing food security, and improving lives for people in global south. This book series acts as a global publishing platform to collate and disseminate knowledge on fostering resilient and productive agroecosystems, thus enhancing accessibility, and fostering informed decision-making.

Kamel A. Abd-Elsalam • Aftab Ahmad
Editors

Gene Editing in Cereals

Approaches and Applications

Springer

Editors
Kamel A. Abd-Elsalam
Plant Pathology Research Institute
Agricultural Research Center
Giza, Egypt

Aftab Ahmad
Natural and Medical Sciences
Research Center
University of Nizwa
Nizwa, Oman

ISSN 2730-6771 ISSN 2730-678X (electronic)
Sustainability Sciences in Asia and Africa
ISSN 3059-247X ISSN 3059-2488 (electronic)
Agroecosystem Dynamics and Sustainable Practices
ISBN 978-981-95-6721-8 ISBN 978-981-95-6722-5 (eBook)
https://doi.org/10.1007/978-981-95-6722-5

This Springer imprint is published by the registered company Springer Nature Singapore Pte Ltd.
The registered company address is: 152 Beach Road, #21-01/04 Gateway East, Singapore 189721, Singapore

Preface

In an era marked by unprecedented threats to global food security, growing population, the wild cards of global warming, and the ever-present threat of disease and pests, the industry of agriculture stands at a fork in the road. Never have food systems been so critical to being sustainable and resilient, and at the heart of that revolution lies the revolutionary power of gene editing. Among the many crops that will determine our collective fate, cereals, the source of the world's food, are the central theme.

This publication, *Gene Editing in Cereals: Approaches and Applications*, is the result of this crucial period. It is a comprehensive and current assessment of how gene editing, and indeed the versatile CRISPR/Cas systems, is revolutionizing essentially the cereal crop improvement landscape. Since science itself has evolved dramatically over the recent years, a clear pathway from cutting-edge research to practical application has been missing. That is the intention of this book.

Pioneered by the finest plant geneticists, molecular biologists, and agronomists, the book gives in-depth insight into precision reconfiguring the genomic landscape of cereals. CRISPR/Cas systems have transcended their role as incremental genetic tools, emerging as powerful platforms for targeted redesign of cereal genomes and enhanced disease resistance through epigenome editing. We explain how gene editing may be applied to combat the destructive effects of abiotic and biotic stresses, with an in-depth review of issues and solutions for maize, rice, and wheat crops.

Along with fundamental science, this book also discusses applied and future applications of gene editing. There are extensive discussions on CRISPR-Cas-based gene editing application in triticale, barley, and sorghum, and multi-step gene editing process for improvement of quality and nutritional traits in cereals. The interaction of technology is a leitmotif that recurs, with specialist articles on Bioinformatics tools applied in Cereal Genome Editing and the novel use of CRISPR-based diagnostics for the diagnosis of pathogens, which mirrors the multidisciplinary nature of the topic. We also highlight the great strides being made within individual crops, with such titles as "Harnessing CRISPR/Cas Gene Editing for Wheat Improvement" and "Gene Editing in Maize: Progress and Challenges," reflecting the actual achievements already being realized.

It is written for general readers, from scholar researchers and students who want the most up-to-date facts to business practitioners and policymakers attempting to chart the regulatory and ethical limits of this technology. It delights in the success even-handedly but documents the ongoing challenges and the milestone significance of establishing public trust and awareness. By providing a brief and wide-ranging overview, *Gene Editing in Cereals* seeks to encourage future generations of researchers and farmers and be an influencing force for a world where everyone is food-secure.

We would also like to express our warmest appreciation to the Springer team and specifically Sanchi Bhimrajka, our indefatigable Publishing Editor. Your ongoing support and professionalism at every step of the publication process have been of invaluable assistance.

We are immensely thankful to the Series Editors, Prof. Kamel A. Abd-Elsalam, Agricultural Research Center, Giza, Egypt, and Prof. Olubukola O. Babalola, Vice-President of The World Academy of Sciences (TWAS) and eminent Research Director at North-West University, South Africa. Your creativity and brilliant ideas have greatly enriched the Agroecosystem Dynamics and Sustainable Practices series, a worth-reading book in the field.

We also appreciate the efforts of Prof. Adel Abdel Azeem, President of the Agricultural Research Center, Giza, Egypt, and members of the series editorial board for their unwavering dedication to dissemination of knowledge and innovation in agroecosystem dynamics and sustainable practices. Your dedication has played a remarkable role in making this gigantic achievement possible.

Giza, Egypt Kamel A. Abd-Elsalam
Nizwa, Oman Aftab Ahmad

Contents

Part I CRISPR Gene Editing Tools

Editors and Contributors

About the Editors

Kamel A. Abd-Elsalam, PhD, is currently a Research Professor and Deputy Director for Research at the Plant Pathology Research Institute, Agricultural Research Center, Giza, Egypt. His research interests include developing and improving plant biosecurity diagnostic tools, understanding fungal pathogen genomes, and creating eco-friendly hybrid nanomaterials for controlling toxigenic fungi and plant diseases as well as applications in agroecosystems. He earned his PhD in Molecular Plant Pathology from Christian Albrechts University of Kiel (Germany) and Suez Canal University (Egypt). In 2008, he was awarded a postdoctoral fellowship from the same institution. He served as a Visiting Associate Professor at Mae Fah Luang University in Thailand, the Institute of Microbiology at TUM in Germany, the Laboratory of Phytopathology at Wageningen University in the Netherlands, and the Plant Protection Department at Sassari University in Italy.

He has published 40 books related to nanobiotechnology applications in agriculture and plant protection with major publishing houses such as Royal Society of Chemistry (RSC), Springer, Taylor & Francis, and Elsevier. Since 2019, he has served as the Editor-in-Chief of the Elsevier book series Nanobiotechnology for Plant Protection and as the Series Editor for Applications of Genome Modified Plants and Microbes in Food and Agriculture. Additionally, he has been an active member of the Elsevier Advisory Panel since 2020, providing feedback and suggestions to improve Elsevier's products and services. He has published more than 327 scientific research articles in international and regional specialized journals with high impact factors, and has an H-index of 55, an i-10 index of 170, and over 11,346 citations. He served as a Guest Editor for the *Journal of Fungi, Plants, and Microorganisms*, and as a Review Editor for *Frontiers in Genomic Assay Technology*, *Agrochemicals*, and *Plant Nano Biology*. He was ranked in the top 2% of the most influential scientists in the world in nanobiotechnology for 2021, 2022, 2023, and 2024 by Elsevier and Stanford University. In 2014, he was awarded the Federation of Arab Scientific Research Councils Prize for excellent scientific research in biotechnology (fungal genomics), ranking first. Additionally, he received the Egyptian State Award for Agricultural Scientific Excellence from ASRT, Egypt, in 2022 with a silver medal.

Aftab Ahmad, PhD, is currently working as an Associate Professor at the Natural and Medical Sciences Research Center (NMSRC), University of Nizwa, Oman. His research interests include developing biotic stress-tolerant gene-edited plants using CRISPR/Cas technology. He has published 4 international books on CRISPR and more than 40 research articles in international peer-reviewed journals including *Frontiers in Plant Science*, *Journal of Biomedical Sciences*, *Scientific Reports-Nature*, *PLOS One*, *Journal of Translational Medicine*, *Plant Molecular Biology Reporter*, *Molecules*, *Journal of Cereal Science*, and *Genetic Resources and Crop Evolution*. He pursued his PhD and postdocs in Plant Molecular Biology from the University of Shizuoka, Japan, under the Center of Excellence (COE, 21st Century) program. He works as a guest editor in *Frontiers in Genome Editing* and *Frontiers in Plant Sciences*. He also served as a visiting scientist in the Department of Plant Sciences, University of California, Davis, USA. He has more than ten years of teaching experience in biochemistry and molecular biology, and his research focuses on CRISPR/Cas-based genome editing in plants.

Contributors

Naglaa A. Abdallah National Biotechnology Network of Expertise (NBNE), Academy of Scientific Research (ASRT), Cairo, Egypt

Genetics Department, Faculty of Agriculture, Cairo University, Giza, Egypt

Toka E. Abdelkhalek Biotechnology English Program, Faculty of Agriculture, Cairo University, Giza, Egypt

Kamel A. Abd-Elsalam Plant Pathology Research Institute, Agricultural Research Center (ARC), Giza, Egypt

Hussien Abodalam Plant Pathology Department, Faculty of Agriculture, Cairo University, Giza, Egypt

Aftab Ahmad Natural and Medical Sciences Research Center (NMSRC), University of Nizwa, Nizwa, Oman

Biochemistry/Center for Advanced Studies in Agriculture and Food Security (CASAFS), University of Agriculture, Faisalabad, Pakistan

Farah K. Ahmed Biotechnology English Program, Faculty of Agriculture, Cairo University, Giza, Egypt

T. P. Ajeesh Krishna Division of Plant Molecular Biology and Biotechnology, Department of Biosciences, Rajagiri College of Social Sciences, Cochin, Kerala, India

Ahmed Al-Harrasi Natural and Medical Sciences Research Center (NMSRC), University of Nizwa, Nizwa, Oman

Hassan Almoammar Wellness and Preventive Medicine Institute, Health Sector, King Abdulaziz City for Science and Technology (KACST), Riyadh, Saudi Arabia

Fatma Nur Basturk Institute of Graduate Studies in Sciences, Istanbul University, Istanbul, Turkey

Juan Manuel Bello-López División de Investigación, Hospital Juárez de México, Ciudad de México, Mexico

Yaroslav B. Blume Institute of Food Biotechnology and Genomics, National Academy of Sciences of Ukraine, Kyiv, Ukraine

Kadir Boztas Department of Biology, Faculty of Science, Istanbul University, Istanbul, Turkey

Emirhan Çağla Institute of Graduate Studies in Sciences, Istanbul University, Istanbul, Turkey

Luis Alberto Castillo-Díaz Departamento de Medicina y Ciencias de la Salud, Universidad de Sonora, Hermosillo, Mexico

Graciela Castro-Escarpulli Laboratorio de Investigación Clínica y Ambiental, Departamento de Microbiología, Escuela Nacional de Ciencias Biológicas, Instituto Politécnico Nacional, Ciudad de México, Mexico

Stanislaus Antony Ceasar Division of Plant Molecular Biology and Biotechnology, Department of Biosciences, Rajagiri College of Social Sciences, Cochin, Kerala, India

Gül Cevahir Öz Department of Biology, Faculty of Science, Istanbul University, Istanbul, Turkey

Berke Çevik Institute of Graduate Studies in Sciences, Istanbul University, Istanbul, Turkey

Ansu Chetri Department of Chemistry and Biochemistry, Sharda School of Engineering and Sciences, Sharda University, Greater Noida, India

Yusuf Can Gerçek Department of Biology, Faculty of Science, Istanbul University, Istanbul, Turkey

Siddhesh B. Ghag School of Biological Sciences, UM-DAE Centre for Excellence in Basic Sciences, University of Mumbai campus, Kalina, Santacruz (East), Mumbai, India

Ian Godwin Centre for Crop Science, Queensland Alliance for Agriculture and Food Innovation, The University of Queensland, St Lucia, Australia

Luis Uriel Gonzalez-Avila Laboratorio de Investigación Clínica y Ambiental, Departamento de Microbiología, Escuela Nacional de Ciencias Biológicas, Instituto Politécnico Nacional, Ciudad de México, Mexico

Rawan K. Hassan Biotechnology English Program, Faculty of Agriculture, Cairo University, Giza, Egypt

Samia Hassan Department of Biochemistry, The University of Agriculture, Faisalabad, Pakistan

Swati Hazra Department of Life Science, Sharda School of Bio-Science and Technology, Sharda University, Greater Noida, India

Cecilia Hernández-Cortez Laboratorio de Bioquímica Microbiana, Departamento de Microbiología, Escuela Nacional de Ciencias Biológicas, Instituto Politécnico Nacional, Ciudad de México, Mexico

Annena Ijaz Department of Biochemistry, Government College University, Faisalabad, Pakistan

Sohini Jash JMS School of Life Sciences, University of Sussex, Falmer, Brighton, UK

Zhiyin Jiao Institute of Millet Crops, Hebei Academy of Agriculture & Forestry Sciences/Hebei Branch of China National Sorghum Improvement Centre, Shijiazhuang, China

Anatolii V. Karelov Institute of Food Biotechnology and Genomics, National Academy of Sciences of Ukraine, Kyiv, Ukraine

Abeer A. Khalaf Genome Engineering Laboratory, Agricultural Genetic Engineering Research Institute (AGERI), Agricultural Research Center (ARC), Giza, Egypt

Muhammad Naeem Khan School of Pharmaceutical Science and Technology, Tianjin University, Tianjin, China

Arshiba Khawar Department of Biochemistry, The University of Agriculture, Faisalabad, Pakistan

Seran Kırkıncı Institute of Graduate Studies in Sciences, Istanbul University, Istanbul, Turkey

Natalia O. Kozub Institute of Food Biotechnology and Genomics, National Academy of Sciences of Ukraine, Kyiv, Ukraine

Institute of Plant Protection, National Academy of Agrarian Sciences, Kyiv, Ukraine

Guoquan Liu Centre for Crop Science, Queensland Alliance for Agriculture and Food Innovation, The University of Queensland, St Lucia, Australia

Peng Lv Institute of Millet Crops, Hebei Academy of Agriculture & Forestry Sciences/Hebei Branch of China National Sorghum Improvement Centre, Shijiazhuang, China

Theivanayagam Maharajan Division of Plant Molecular Biology and Biotechnology, Department of Biosciences, Rajagiri College of Social Sciences, Cochin, Kerala, India

Sayanti Mandal Department of Chemistry and Biochemistry, Sharda School of Engineering and Sciences, Sharda University, Greater Noida, India

Karen Massel Centre for Crop Science, Queensland Alliance for Agriculture and Food Innovation, The University of Queensland, St Lucia, Australia

Mirza Abid Mehmood Plant Pathology, Institute of Plant Protection, Muhammad Nawaz Shareef University of Agriculture, Multan, Pakistan

Hande Morgil Department of Biology, Faculty of Science, Istanbul University, Istanbul, Turkey

Rana Ali Haider Department of Biochemistry, University of Agriculture, Faisalabad, Pakistan

Nayla Munawar Department of Chemistry, College of Science, United Arab Emirates University, Al-Ain, United Arab Emirates

Luis Fernando Muñoz-Mateo Laboratorio de Investigación Clínica y Ambiental, Departamento de Microbiología, Escuela Nacional de Ciencias Biológicas, Instituto Politécnico Nacional, Ciudad de México, Mexico

Komal Pujare School of Biological Sciences, UM-DAE Centre for Excellence in Basic Sciences, University of Mumbai campus, Kalina, Santacruz (East), Mumbai, India

Sameer H. Qari Department of Biology, AlJumum University College, Umm Al-Qura University, Makkah, Saudi Arabia

Eman H. Radwan National Biotechnology Network of Expertise (NBNE), Academy of Scientific Research (ASRT), Cairo, Egypt

Zoology Department, Faculty of Science, Damanhour University, Damanhour, Egypt

Khaled H. Radwan Genome Engineering Laboratory, Agricultural Genetic Engineering Research Institute (AGERI), Agricultural Research Center (ARC), Giza, Egypt

National Biotechnology Network of Expertise (NBNE), Academy of Scientific Research (ASRT), Cairo, Egypt

Mythili Ravichandran Department of Microbiology, Vivekanandha Arts and Science College for Women, Sankagiri, Tamil Nadu, India

Andres Saldaña-Padilla Laboratorio de Investigación Clínica y Ambiental, Departamento de Microbiología, Escuela Nacional de Ciencias Biológicas, Instituto Politécnico Nacional, Ciudad de México, Mexico

Laboratorio de Bioquímica Microbiana, Departamento de Microbiología, Escuela Nacional de Ciencias Biológicas, Instituto Politécnico Nacional, Ciudad de México, Mexico

Enakshee Samanta Department of Life Science, Sharda School of Bio-Science and Technology, Sharda University, Greater Noida, India

Ihor O. Sozinov Institute of Plant Protection, National Academy of Agrarian Sciences, Kyiv, Ukraine

Oksana I. Sozinova Institute of Food Biotechnology and Genomics, National Academy of Sciences of Ukraine, Kyiv, Ukraine

Institute of Plant Protection, National Academy of Agrarian Sciences, Kyiv, Ukraine

Veysel Süzerer Bingöl University, Vocational School of Health Services, Department of Pharmacy Services, Bingöl, Türkiye

Ramsha Syed Department of Chemistry and Biochemistry, Sharda School of Engineering and Sciences, Sharda University, Greater Noida, India

Himani Thakur Department of Life Science, Sharda School of Bio-Science and Technology, Sharda University, Greater Noida, India

Engin Tilkat Batman University, Faculty of Science and Literature, Department of Biology, Batman, Türkiye

Yiğit Sabri Unlu Institute of Graduate Studies in Sciences, Istanbul University, Istanbul, Turkey

Selin Gül Ünsal Istanbul Technical University, Faculty of Science and Letters, Department of Molecular Biology and Genetics, Istanbul, Türkiye

Deepak Kumar Upadhyay Department of Chemistry and Biochemistry, Sharda School of Engineering and Sciences, Sharda University, Greater Noida, India

Kashish Vats Department of Life Science, Sharda School of Bio-Science and Technology, Sharda University, Greater Noida, India

Jinping Wang Institute of Millet Crops, Hebei Academy of Agriculture & Forestry Sciences/Hebei Branch of China National Sorghum Improvement Centre, Shijiazhuang, China

Zhifang Wang Institute of Millet Crops, Hebei Academy of Agriculture & Forestry Sciences/Hebei Branch of China National Sorghum Improvement Centre, Shijiazhuang, China

Nazlıcan Yıldırım Institute of Graduate Studies in Sciences, Istanbul University, Istanbul, Turkey

Alaa Youssef Genome Engineering Laboratory, Agricultural Genetic Engineering Research Institute (AGERI), Agricultural Research Center (ARC), Giza, Egypt

Paulo Zaini Department of Plant Sciences, University of California, Davis, CA, USA

Baohong Zhang Department of Biology, East Carolina University, Greenville, NC, USA

Iram Zovia Department of Chemistry, College of Science, United Arab Emirates University, Al-Ain, United Arab Emirates

Chapter 1
CRISPR-Modified Cereal Crops: Mitigating the Effects of Climate Change on Agricultural Production

Khaled H. Radwan, Abeer A. Khalaf, Alaa Youssef, Eman H. Radwan, and Naglaa A. Abdallah

Abstract Climate change threatens the productivity or stability of global agriculture, particularly the major cereals that serve as global staples (i.e., wheat, rice, maize, barley, sorghum, and millet). The combination of higher temperatures, fluctuating rainfall, salinity, drought, and pest and pathogen pressure significantly undermines agricultural yields and food security. Genome editing technologies, particularly CRISPR/Cas systems, have proven to be transformative tools for developing climate-resilient crops with increased tolerance to abiotic and biotic stresses, improved nutritional quality, and sustainable performance in variable environments.

This chapter discusses genome-editing methods and their role in improving cereal crops against the main climatic stress factors, including drought, heat, salinity, elevated CO_2 levels, and pest infestation. Relevant case studies are used to illustrate impressive applications of CRISPR, with a focus on increasing crop yields, improving tolerance to environmental stressors, and enhancing nutritional

K. H. Radwan (✉)
Genome Engineering Laboratory, Agricultural Genetic Engineering Research Institute (AGERI), Agricultural Research Center (ARC), Giza, Egypt

National Biotechnology Network of Expertise (NBNE), Academy of Scientific Research (ASRT), Cairo, Egypt

A. A. Khalaf · A. Youssef
Genome Engineering Laboratory, Agricultural Genetic Engineering Research Institute (AGERI), Agricultural Research Center (ARC), Giza, Egypt

E. H. Radwan
National Biotechnology Network of Expertise (NBNE), Academy of Scientific Research (ASRT), Cairo, Egypt

Zoology Department, Faculty of Science, Damanhour University, Damanhour, Egypt

N. A. Abdallah
National Biotechnology Network of Expertise (NBNE), Academy of Scientific Research (ASRT), Cairo, Egypt

Genetics Department, Faculty of Agriculture, Cairo University, Giza, Egypt

K. A. Abd-Elsalam, A. Ahmad (eds.), *Gene Editing in Cereals*, Sustainability Sciences in Asia and Africa, https://doi.org/10.1007/978-981-95-6722-5_1

biochemistry. This chapter focuses on the introduction of the omics/bioinformatics-assisted process of localizing target genes and implementing genome editing in breeding pipelines. This includes off-target effects, delivery methods, regeneration protocols, and regulatory frameworks. The application of genome editing in climate-smart agriculture is therefore key to sustainable grain production and global food security in the future.

Keywords Cereal crops · Omics and bioinformatics · Climate-smart agriculture · Prime editing · Base editing · CRISPR/Cas

1.1 Introduction

Climate change is a global challenge for agriculture, impacting crop productivity, food security, and the livelihoods of farmers. These challenges include shifts in rainfall patterns, intensification of extreme weather events, and an increase in temperature. The impacts have been extensively documented and represent an existential threat to agriculture worldwide (Yeleliere et al. 2023).

In areas with low food security, like Africa and South Asia, any change in temperature or precipitation has an immediate response to crop yield. The increasing incidences of droughts and floods also contribute to global yield losses (Chhogyel et al. 2020; Rusmayandi et al. 2023). This makes rural areas that depend on rain-fed agriculture especially vulnerable to climate change, as fluctuations in climate have been found to encourage the emergence of previously unseen pests and pathogens (Thomas et al. 2013; Waiba 2023).

Adaptation measures can be integrated into food systems through cross-sectoral resilience-building strategies, coupled with investments in agricultural research, which can significantly lower climate vulnerabilities (Favas et al. 2024). However, the application of such policies is variable and varies largely due to socio-economic conditions.

Cereal crops are one of the cornerstones of global food systems as they provide staple food rich in carbohydrates to billions of people around the world. The susceptibility of crops to abiotic and biotic stresses aggravated by climate change creates major challenges for food security and sustainable agriculture (Zhou et al. 2022).

Drought, salinity, and nutrient deficiencies are among the most significant abiotic stresses that limit cereal production (Morris and Jacquet 2024; Wang et al. 2018). Increasing temperatures and erratic rainfall are projected to reduce cereal yields substantially, with an average reduction of 17% (Nelson et al. 2014). Rice, a staple food that requires a lot of water, is especially prone to drought stress (Packirisamy et al. 2023).

Under thermal stress, wheat exhibits decreased growth parameters, which makes better breeding programs more necessary (Dunaiska et al. 2023). In addition, biotic stresses caused by pathogens and pests may bring additional constraints to the production of cereals. Higher temperatures make conditions more favorable for pests and diseases, making stress responses in plants more complex (Saijo and Loo 2020).

Minimizing abiotic stress effects through genome editing, such as CRISPR/Cas9, is at the heart of cereal resilience. Designing stress-tolerant varieties has become imperative (Hillary and Ceasar 2019; Mathivanan et al. 2023). Genome editing holds great promise for the optimization of metabolic pathways that we are beginning to understand in plants and that show potential for improving drought and heat tolerance (Krishna et al. 2022; Liu et al. 2023).

Applications of genome editing in crops enhance yield, nutritional quality, and disease resistance traits (Ansari et al. 2020; Numan et al. 2021; Zafar et al. 2020). CRISPR allows for faster and more accurate alterations to the genome than conventional breeding approaches. Newer genome-editing technologies, like base editing and prime editing, boost the accuracy of these improvements in cereals (Basu et al. 2023).

In addition, CRISPR/Cas9 technology makes it possible to create non-genetically modified organisms (non-GMOs) while maintaining such desirable traits. The public skepticism about GMOs drives the need for alternative approaches, and genome editing offers a way to produce more regulatory-friendly genetically modified (Jaganathan et al. 2018). It increases the likelihood and acceptability of new technologies that can be applied to guarantee food production and ameliorate the consequences of climate change.

This chapter aims to explore innovative strategies to improve the resilience of cereal crops. It focuses on the potential of genome-editing technologies, particularly CRISPR/Cas systems, as well as emerging tools such as base and prime editing, as transformative solutions to address challenges in major cereal crops. Also, it highlights the integration of these tools with omics and bioinformatics approaches, discusses practical applications and successful case studies in wheat, rice, maize, and other cereals, and addresses key challenges related to technical limitations, regulatory frameworks, and public acceptance. Ultimately, it aims to provide a roadmap for the use of genome editing in breeding programs to develop cereal varieties that can ensure global food security under changing environmental conditions.

1.2 Cereal Crops Under Climate Stress

1.2.1 Drought and Heat Stress

Drought stress is one of the major abiotic stress factors that causes the highest annual yield reduction for cereals worldwide (Numan et al. 2021). Hence, there is a diversity of strategies that crops developed to tolerate drought stresses, including physiological adaptations such as stomatal closure, modification of root structures, and synthesis of protective molecules (Gupta et al. 2020; Kim et al. 2022).

Nevertheless, the sensitivity of different plant species to heat stress is observed in the form of loss of reproductive functions and oxidative damage, causing considerable harm to yield and quality of crops (Devi et al. 2022; Jha et al. 2023). At the mechanistic level, different pathways have been suggested to control heat superstability. The heat shock proteins (HSPs) family is most important to protect plants from heat stress by refolding the denatured proteins.

Genes involved in the synthesis of HSPs may provide targets for both natural and engineered alleles that enhance tolerance to higher temperatures, allowing plants to perform better under heatwaves (Havrlentová et al. 2021). HSP pathways are amenable to modification using gene editing strategies, which could be employed to address HSP expression in cereal crops.

1.2.2 *Salinity Stress in Cereal Crops*

Salinity consequently causes osmotic stress, ionic toxicity, and oxidative stress in plants, disrupting essential physiological processes of growth and development (Kakar et al. 2019; Syed et al. 2021). As a response, it adversely affects seed germination, root development, nutrient uptake, and growth of cereal plants. To sustain turgor pressure and cellular function in plants exposed to salinized conditions, plants accumulate compatible solutes, or osmolytes, including proline and glycine betaine (Ayoub et al. 2021). This osmotic adjustment is important for plants' salinity tolerance (Kakar et al. 2019).

A cornerstone feature of plant development and adaptation is the transcriptional factors, such as DREB (dehydration-responsive element-binding proteins), and NAC (NAM, ATAF1/2, and CUC2), that are key players in the regulation of networks of salt-responsive genes. Genome editing and manipulation of these TFs can increase the expression of stress-induced genes for the resistance traits against drought and salinity. As an example, it has been reported that activation of the NAC gene family is essential for enhancing plant response against salinity stress (Thilakarathne et al. 2025; Xiong et al. 2025).

Overexpression of genes associated with the synthesis and accumulation of osmolytes can be targeted to represent an alternate route to saline tolerance. Elevating proline biosynthesis gene expression can help to optimize proline accumulation in cereal crops to improve tolerance against saline stress through osmotic adjustment (Ayoub et al. 2021).

1.2.3 *CO_2 Enrichment and Its Mixed Effects on Grain Quality*

The effects of atmospheric carbon dioxide (CO_2) on crop production, particularly for cereals, are becoming increasingly evident as its concentration continues to rise. It is known that photosynthesis will change at elevated levels of CO_2, so biomass production will increase; grain quality is often harder to predict and can differ between cereal species according to the environmental growing conditions (Balla et al. 2014).

The effect of heat stress with CO_2 enrichment is particularly important during reproductive stages of cereal growth, when the crops are most sensitive to excessive temperatures (Zenda et al. 2022). Moreover, they emphasized that this can further aggravate adverse effects on grain yield and quality through altered accumulation of nutrients associated with an increased rate of grain filling due to high CO_2 at higher temperatures (Liu 2025; Patil and Kaliwal 2017).

Nevertheless, wild relatives of the cultivated cereals can provide a genetic resource that can help to counteract the adverse effects of the heat stress and CO_2 enrichment, possibly leading to improved grain quality by introducing the advantageous traits into breeding programs (Bahrami et al. 2020, 2021).

1.2.4 Pest and Disease Dynamics in a Warming World

The metabolic rates of pests are typically increasing with higher temperatures, which are associated with increased reproductive rates and geographical distribution range expansions (Broberg et al. 2018; Kayastha et al. 2023).

Pests like the maize weevil (*Sitophilus zeamais*) are a threat to cereal crops in the field as well as in storage. High infestation by this pest results in low seed quality, leading to serious economic losses in grain production (Macaluso et al. 2022). In storage, pests such as *Rhyzopertha dominica* and *Trogoderma granarium* also represent a major threat that leads to quantitative and qualitative loss of wheat and other cereals (Ali and Baloch 2023). The influence of these pests highlights the need for strong pest management strategies that include pre-harvest and post-harvest phases.

Cereals defend themselves by employing different defense mechanisms against pests and diseases. Thicker cuticles and fibrous tissues can inhibit insect feeding and lower disease incidence by limiting pathogen penetration (Sinha et al. 2013). Some cereals have also developed biochemical defenses, such as the production of secondary metabolites that are toxic to herbivorous pests or inhibit pathogen growth (Kitonde et al. 2019). Interestingly, crop rotation and diversification have been promoted to manage pest lifecycles and disease occurrence. (Nainggolan et al. 2023).

1.3 Evolution of Cereal Improvement from Conventional Breeding, and Genetic Engineering to Genome Editing

Traditional breeding strategies have established the basis for cereal crop enhancement for centuries. These include selection based on phenotypic traits, hybridization, and backcrossing, which include the combination of beneficial traits from different parent plants to produce better varieties (Haroon et al. 2022; Ji et al. 2014).

As an example, traits like disease resistance and yield can easily be improved over successive generations through selection.

Nonetheless, due to the restrictive nature of the gene pool of parent plants, traditional breeding is usually slow in progress and does not allow fast introgression of novel traits (Haroon et al. 2022; Raj and Nadarajah 2022). Moreover, these approaches often necessitate laborious field trials and thorough assessments before newly developed cultivars can be commercialized (Haroon et al. 2022).

By the end of the twentieth century, however, genetic engineering led to a new era extremely different from these traditional approaches. Through *Agrobacterium*-mediated transformation and biolistic (gene gun) methods, genetic traits were inserted directly into a plant's genome to express specific desired characteristics (Ji et al. 2014; Varshney et al. 2018).

Therefore, genetic modification of maize has produced varieties that express desirable traits, including pest resistance and herbicide tolerance, resulting in easier plant husbandry and potential yield augmentations (Wang et al. 2019). Nevertheless, the prevailing use of GM crops, having gained negative opinions on safety and environment from the public, has resulted in restricted acceptance in certain regions (Huang et al. 2022).

The most recent advances in plant biotechnology focus increasingly on genome-editing technologies like CRISPR/Cas9, which allow higher precision than ever for specific and quantifiable targeting and editing of genomic sequences (Guo et al. 2023; Radchuk et al. 2012). This technology enables the rapid generation of crops with improved tolerance (Guo et al. 2023) (Fig. 1.1).

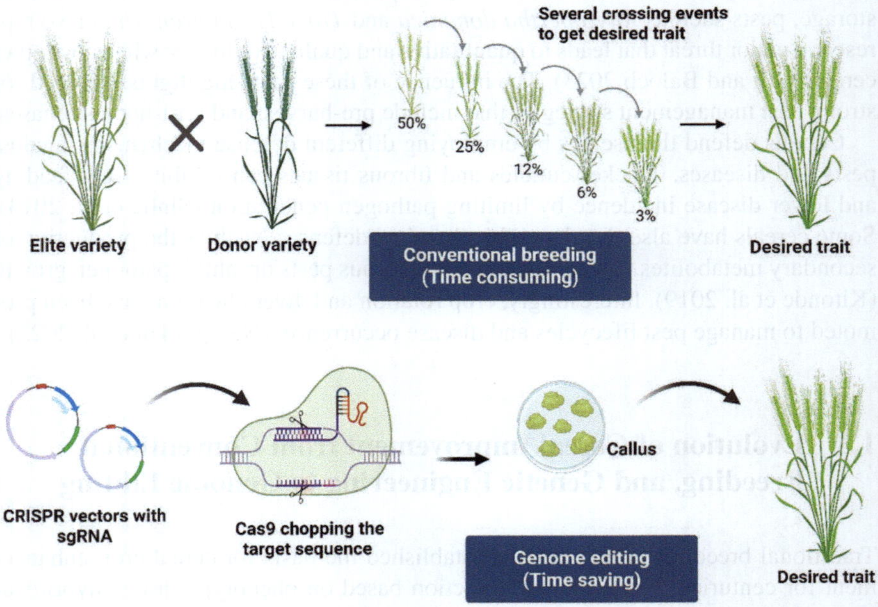

Fig. 1.1 Genome-editing accelerates the development of climate-resilient cereal varieties. Unlike conventional breeding, which involves several cycles of crossing and backcrossing, CRISPR enables the rapid introduction of targeted traits into elite backgrounds through precise gene editing

1.4 Emergence of CRISPR–Cas Systems in Cereals

CRISPR–Cas systems were a game changer for plant biotechnology where it can perform precise site-specific genome editing. The CRISPR–Cas9 system employs a guide RNA that targets the Cas9 endonuclease to sequences of interest to create double-strand breaks (DSBs) in the genome at user-defined sites. This action allows the cell repair to conduct changes via non-homologous end joining or homology-directed (Tang and Fu 2018).

The versatility of the CRISPR platform has been used to create essential traits in major food cereals. For instance, targeted mutations of the *OsMLO* gene in wheat showed broad-spectrum resistance to powdery mildew, demonstrating the potential of CRISPR–Cas9 technology for pathogen resistance management (Knott and Doudna 2018). Moreover (Z. Zhang et al. 2019b), reported successful editing of the *OsRR22* gene in rice that significantly improved salt tolerance. In addition, herbicide resistance in cereals has been achieved by targeted gene editing (Zafar et al. 2020) which has allowed the development of more efficient weed management strategies. Moreover, the CRISPR system allowed for the breeding of cereals that can tolerate abiotic stresses, including drought and salt (Cram et al. 2019).

1.4.1 Delivery Systems in Cereal Crops

Different routes of delivery have been devised for the entry of CRISPR components into plant cells, with each method having its own unique benefits and drawbacks (Ahmar et al. 2024). Nevertheless, a new strategy is to introduce preassembled CRISPR/Cas ribonucleoprotein (RNP) complexes into plant cells. By going through this method, there would not be a need for a delivery mechanism of DNA, and it also enhances editing efficiency and decreases the off-target effect (Sathiyabalan et al. 2024). In this method, RNPs are delivered into immature embryo cells using particle bombardment techniques that reduce the reliance on plant transformation protocols (Liang et al. 2017).

Biolistic delivery, also known as “gene gun” technology, is a non-covalent approach that involves physically transferring CRISPR components into plant tissues. This method uses microprojectiles encapsulated with either DNA or RNPs, which are fired into plant cells, allowing for successful delivery of the microprojectiles into the nucleus and efficient gene modification (Svitashev et al. 2016).

Agrobacterium tumefaciens is a widely used vehicle to transfer foreign DNA to the plant genomes. In this method, the natural infection mechanism of *Agrobacterium* is utilized, where T-DNA is transferred to the plants, and the CRISPR/Cas components get stably integrated (Basu et al. 2023; Dalla Costa et al. 2020). This approach has been utilized in a broad range of crops, especially in cereals, and it frequently leads to non-targeted integration of bacterial DNA.

In addition, virus-induced genome editing (VIGE) is a useful strategy for directed mutagenesis in cereal crops by using effective viral vectors for delivering the

CRISPR/Cas system. Particularly clear successes have been achieved in various cereal crops using certain viral vectors (e.g., Barley stripe mosaic virus, BSMV). It has been demonstrated that BSMV can mediate heritable gene editing with editing efficiencies as high as 100% in some cases (Baysal et al. 2025; Tamilselvan-Nattar-Amutha et al. 2023). Such high efficiency makes VIGE more ideal for multiplexed gene editing, which is important for simultaneously improving multiple agronomic traits (Uranga and Daròs 2023; Wang et al. 2022).

Also, protoplast transformation mediated by polyethylene glycol (PEG) has been well established for the direct delivery of CRISPR/Cas9 RNPs into plant cells. This method enables the simultaneous delivery of the RNP (Ribonucleoprotein) complexes with the Cas9 protein and guide RNA into the cytoplasm to reach the target DNA (Nakajima et al. 2017; Sant'Ana et al. 2020). PEG treatment also increases the uptake efficiency of these complexes, as it makes the permeability of the protoplast membrane more effective (Z. Zhang et al. 2019b).

Such efficacy is especially appreciated because it excludes the incorporation of foreign DNA, providing a transgene-free plant, which is more and more relevant from a regulatory perspective (Brandt et al. 2020). Aspects like PEG concentration, incubation time, and temperature during transformation can affect PEG-mediated transformation efficiency. The optimization of these parameters needs to be precise, as they have been observed to significantly affect the outcomes of editing (Jiang et al. 2024; Poddar et al. 2023) (Fig. 1.2).

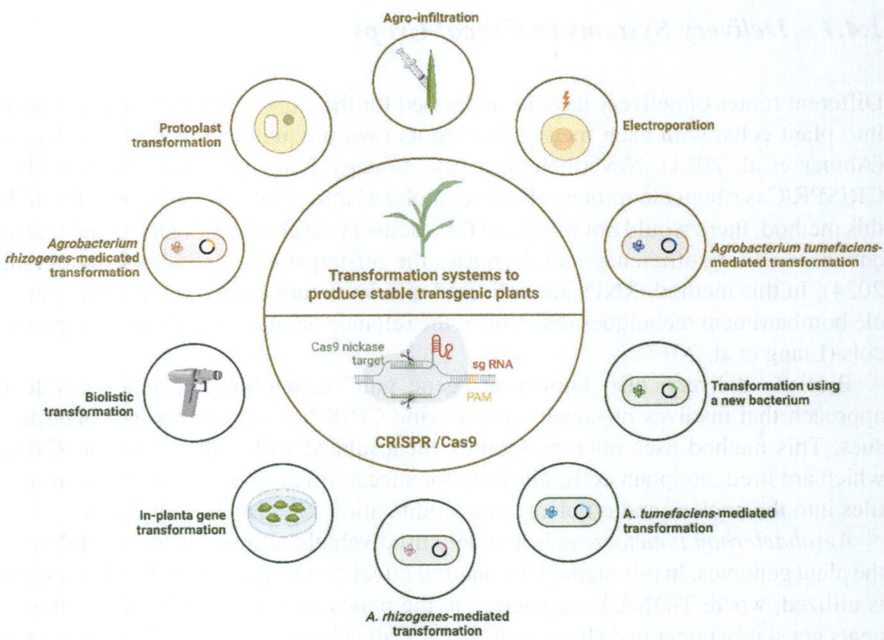

Fig. 1.2 Delivery systems used for stable genetic transformation in plants with CRISPR-Cas9 constructs. Methods include Agrobacterium-mediated transformation; tumefaciens and rhizogenes biolistics, electroporation, and protoplast transformation; among others. Efficient delivery is critical for successful genome editing in cereal crops

1.5 CRISPR Implementation in Key Cereal Crops

1.5.1 Wheat (**Triticum aestivum**)*: Hexaploid Complexity*

The modulation of stress-responsive genes, involved in osmotic adjustment and root development, can enhance performance when water is limited and thus contribute to yield stability under prolonged drought (Abdallah et al. 2025; Potlapalli et al. 2020). Moreover, knockdown of the *TaCSN5* gene through CRISPR results in effective and durable resistance to *Puccinia striiformis f.* sp. *Tritici* (Bai et al. 2021).

CRISPR systems also enable multiplex editing, or the targeting of multiple loci in the genome at the same time. (X. Huang et al. 2023b) made use of this feature with CRISPR/Cas9 to modify the TaUAM3 gene, greatly improving the resistance of wheat plants to stripe rust fungi. These methods not only increase disease resistance but also allow for the improvement of multiple traits at the same time.

*1.5.2 Maize (***Zea mays***)*

Targeting of *ZmGA20ox3* gene (involved in gibberellin biosynthesis) enhanced growth architecture and drought tolerance of maize by editing this gene with both of the above functions in addressing behavioral expertise and implications to common environmental stressors that may cause disease susceptibility to worsen (Liu et al. 2023).

CRISPR/Cas9 was extensively employed to knock out genes involved in susceptibility to fungal pathogens; one such promising target is the *ZmACD6* gene, conferring resistance to the plant pathogen, *Ustilago maydis*. The editing of this gene resulted in the development of maize that was more resistant to the pathogen, demonstrating the ability of CRISPR as a tool for plant protection (Guo et al. 2023; Z. Zhang et al. 2019b). Moreover, the transcription factor ZmNF–YA1 has been reported to improve tolerance to drought and salinity (Yang et al. 2022). Moreover, the successful editing of several genes in maize to improve traits such as increased stress tolerance and nutritional quality was successfully reported (Zhu et al. 2021).

*1.5.3 Rice (***Oryza sativa***)*

A remarkable example of the improvement of rice resistance against diseases was through using CRISPR/Cas9 to edit genes, including *Bsr-d1* and *Pi21*, all of which play a critical role in the resistance of rice to blast and bacterial blight pathogens, respectively. CRISPR/Cas9-stimulated targeted mutagenesis has allowed for rice plants with greatly increased resistance to these diseases, demonstrating the utility of CRISPR/Cas9 in improving agricultural resilience (Zhou et al. 2022).

Moreover, the *BADH2* gene has been edited using CRISPR/Cas9 to enhance the production of 2-acetyl-1-pyrroline (2-AP), a compound known to contribute to the aromatic rice flavor in several aromatic varieties (Hui et al. 2022). In addition, the utilization of CRISPR editing of rice callus tissue to mimic the cauliflower orange mutation has been successful, which could lead to golden rice types that are enriched in vitamin A precursor β-carotene (Endo et al. 2019), important for overcoming vitamin A deficiency through carotenoid accumulation.

*1.5.4 Barley (***Hordeum vulgare***)*

Genome-editing approaches in barley have focused on ionic transport and osmotic regulation genes. Editing the signaling pathway *HvGSK1*, which is related to salt tolerance and drought resistance, has also shown potential for the development of more resilient barley cultivars (Kloc et al. 2024).

Enhancing disease resistance is one of the most important fields of CRISPR in barley. *Fusarium graminearum* is one of the most important pathogens of barley, and hence, research has focused on developing Fusarium-resistant barley lines to minimize crop loss. CRISPR technology has been used by researchers to knock out susceptibility genes, such as the *MLO* gene, resulting in resistant barley plants to powdery mildew. This method shows the potential for better disease control with reduced dependence on chemical fungicides (Gasparis et al. 2018; Kershanskaya et al. 2019).

Moreover, CRISPR technology is being used to alter physical attributes that improve barley agricultural performance. More specifically, quick release of naked barley, which is more favorable for malting because it has less husk, has been developed through targeted mutagenesis of the Nudum (NUD) gene (Hisano et al. 2025). The development of genome editing since the advent of CRISPR technology has enabled progress towards genetic improvement of barely allowing precise mutations to be generated in a short time.

*1.5.5 Sorghum (***Sorghum bicolor***) and Millets (***Pennisetum glaucum, Eleusine coracana***, etc.)*

A major use of CRISPR in sorghum includes improving insect and disease resistance. A comparable study indicated that disruption of the gene encoding *SbCCD8* involved in strigolactone biosynthesis could lead to a significant reduction of

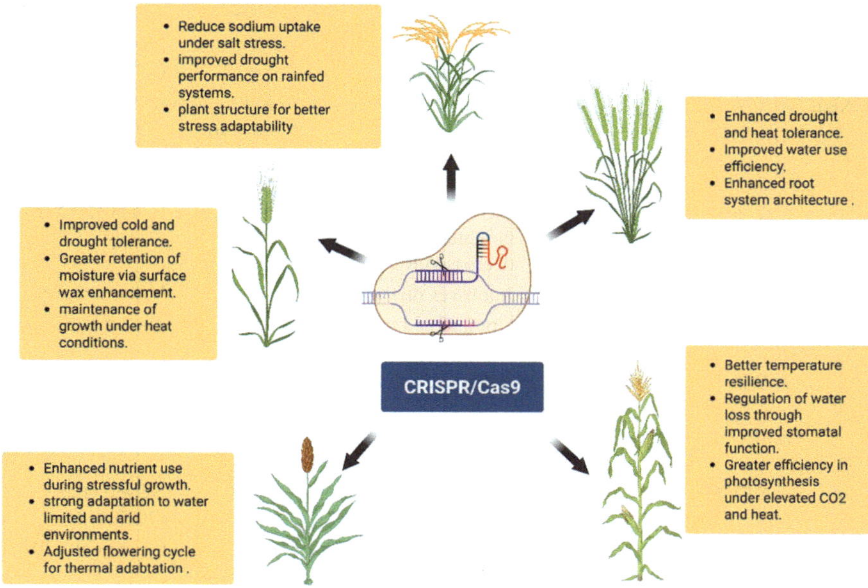

Fig. 1.3 Examples of successful CRISPR/Cas edits in cereals aimed at improving resilience to climate-related stresses such as drought, salinity, and heat

orobanchol production and thus affect the germination of the parasitic plant considerably (Hao et al. 2023).

Studies have focused on genes contributing to salt stress tolerance, such as *SbbHLH85*. This was highlighted by the work of (Song et al. 2022), which found that knockouts of this gene using CRISPR-/Cas9-regulated root architecture development through inhibiting root development, and thus the overexpression of this gene promoted root hair growth and Na + absorption in saline conditions. These observations reveal the contribution of *SbbHLH85* to salt tolerance, and genetic editing offers opportunities to develop sorghums more adapted to cope with environmental stress challenges (Fig. 1.3 and Table 1.1).

Table 1.1 Functional gene edits performed using CRISPR/Cas systems in cereals to enhance resilience to climate-related stressors

Crop	Gene	Trait improvement	References
Wheat	*TaRPK1*	Receptor-like protein kinase 1 (RPK1) regulates root architecture system (RAS), abiotic stress, and yield	Rahim et al. (2024)
	TaAT1	Increased salt–alkaline tolerance, heightened plant growth, and decreased ROS accumulation	Sun et al. (2023)
	TaSAL1	Improves drought tolerance	Mohr et al. (2022)
	TaMBF1c	Upregulated in response to heat stress	Tian et al. (2022)
	TaHAG1	Enhanced salt tolerance	Zhang et al. (2021)
	TaMKP1	Enhanced resistance to rust and powdery mildew in wheat	Liu et al. (2023)
	TaGRF4	Resistance against leaf rust and powdery mildew	Biswal et al. (2023)
	TaHRC	Suppresses calcium-mediated immune response, triggers wheat Fusarium head blight susceptibility	Chen et al. (2024)
	TaeIF4E	Viral resistance, improvement in plant height, and grain length	Kan et al. (2023)
	TaCIPK14	Increased wheat resistance to *Puccinia striiformis*	He et al. (2023)
	TaPPD-1	Increase spike architecture and grain	Errum et al. (2023)
	TaTRXH9	Early heading	Fan et al. (2023)
	TaSPL13	Increased grain size and number	Gupta et al. (2023)
	TaARF12	Reduce plant height with larger spikes shows higher grain yield	Kong et al. (2023)
	TaBAK1-2	Increased immunity toward virus	Hahn et al. (2021)
	TaCKX2-D1	Increased grain number per spikelet	A. Zhang et al. (2019a)
	TaGLW7	Increased wheat grain size and weight	Z. Zhang et al. (2019b)
	TaGW2	Increased weight and protein	Z. Zhang et al. (2019b)
	TaGW8	Increased grain size and weight	Z. Zhang et al. (2019b)
Rice	*OsDSG1*	Enhanced salt tolerance in rice	Ly et al. (2024)
	OslKP2	Increased leaf size improves the tolerance against drought	Shim et al. (2023)
	ObHLIH024	Enhanced salt tolerance	Alam et al. (2022)

(continued)

Table 1.1 (continued)

Crop	Gene	Trait improvement	References
	OxMAD526	Enhanced drought tolerance	Anjala and Augustine (2022)
	OsPIN9	Chilling tolerance	Xu et al. (2020)
	OsPQT3	Enhanced resistance to abiotic stresses increases grain yield	Alfatih et al. (2020)
	OsERA1	Enhanced drought tolerance	Ogata et al. (2020)
	OsPUB67	Reduced tolerance to drought	Qin et al. (2020)
	OsDST	Enhanced leaf water retention ability, lower stomatal density, and under drought	Santosh Kumar et al. (2020)
	OsPYL9	Enhance grain yield under drought	Usman et al. (2020)
	OsLOGLS	Increased grain yield under abiotic conditions	C. Wang et al. (2020b); Wang et al. (2020c)
	OswiR535	Increased tolerance against abiotic stresses	Yue et al. (2020)
	OsNAC006	Drought and heat tolerance	B. Wang et al. (2020a)
	OsPINS6OsGS3, and OsMYB30	Improvement of Rice yield and cold tolerance	Zeng et al. (2020)
	OsSRL1/2	Drought tolerance (higher grain filling under stress)	Liao et al. (2023)
	OxRR22	Enhance rice salinity tolerance	A. Zhang et al. (2019a)
	OxLRR2	Reduced BPH (Brown planthopper) infestation and enhanced natural biological control by attracting natural enemies	Kuai et al. (2024)
	OxNAC	Enhanced innate immunity	Son et al. (2024)
	OxBSR-d1	Enhances the blast resistance	Zhang et al. (2024)
	OsHPP04	Enhanced resistance to rice root-knot nematode	Q. Huang et al. (2023a)
	OsSH	Broad-spectrum disease resistance	Li et al. (2023)
	OsV-ATPase d	Increased resistance against Southern rice black-streaked dwarf virus (SRBSDV), but it decreased resistance against Rice stripe virus (RSV) in rice	Lu et al. (2023)
	OsCPK18/OsCPK4	Enhanced disease resistance and yield in rice	Li et al. (2022)
	OsDJA2 and OsERF104	Resistance to Pyricularia oryzae	Távora et al. (2022)
	OxXa13	Transgene-free bacterial blight-resistant rice with retained fertility	Li et al. (2019)

(continued)

Table 1.1 (continued)

Crop	Gene	Trait improvement	References
	OsPFT1	Sheath blight resistance	Shah et al. (2019)
	OxNRAMP5	Reduction in cadmium accumulation	Songmei et al. (2019)
	OsWDR48, and OsbZIP10	Improved rice grain quality	Jiang et al. (2024)
	OsCPF735A3/4, OsIPT1-10, OsLOG1, OsLOG1-10, OsPUP1-13, OsENT1-4, OsCXX1-11	Improved yield	Wu et al. (2024)
	OsPUB33	Improved grain size and weight	Xie et al. (2025)
	OsCpSRP43, OsCpSRP54a, and OsCpSRP54b	Increased photosynthesis	Caddell et al. (2023)
	OsMADS17	Increase in both grain number and grain weight	Li et al. (2023)
	OsAP2-39	Improved grain weight and yield	Liu et al. (2023)
	OsSPL16 and GW8	Improves grain yield	Usman et al. (2020)
	OsIPA1	Enhanced grain yield	Song et al. (2022)
Maize	*ZmU6*	Enhanced drought tolerance	Qi et al. (2018)
	ZmHKT1	Enhanced salt stress tolerance	Jiang et al. (2018)
	ZmBET5L1	More tolerance to drought stress and primary root elongation rate under osmotic stress	Zhao et al. (2023)
	ZmbZIP68	Enhanced cold tolerance	Li et al. (2022)
	ZmCST1	Reduce stomatal opening, carbon starvation, and early leaves senescence	Wang et al. (2019)
	ZmGDIa	Resistance to maize rough dwarf disease	Liu et al. (2022)
	ZmLOX3	Enhance smut resistance	Pathi et al. (2020)
	ZmSRL5	Enhanced drought resistance	Pan et al. (2020)
	ZmACO2	Increase in grain yield	Ning et al. (2021)
	ZmCLE7, ZmFCP1	Increase yield-related traits	Liu et al. (2022)
	ZmCENH3	High yield	Zong et al. (2017)
Barley	*HvMORC1*	Resistance to powdery mildew	Kumar et al. (2018)
	HvMORC1, HvMORC6a	Resistance to plant pathogenic fungi	Galli et al. (2022)
	HvdsRED	Resistance against wheat dwarf virus	Kis et al. (2019)
	HveIF4E	Resistance to virus infection	Kershanskaya et al. (2019)
	HvCMF7, HvMUS81 HvZYP1,	Resistance to virus infection	Tamilselvan-Nattar-Amutha et al. (2023)
	HvCKX1	Increased in grain yield	Holubová et al. (2018)

(continued)

Table 1.1 (continued)

Crop	Gene	Trait improvement	References
	HvHPT and HvHGGT	Influence of barley grain development	Zeng et al. (2020)
	HvPDS	Increased numbers of spike per plant	Han et al. (2021)
	HvARE1	Higher yield and plant height	Karunarathne et al. (2022)

Traits include abiotic stress tolerance, disease resistance, and productivity under suboptimal conditions

1.6 Omics and Bioinformatics Integration for Target Identification in Cereal Crops

The integration of omics technologies and bioinformatics contributes to cereal target identification through the provision of information that has a profound impact on breeding programs aimed at replacing unfavorable genetic material with superior germplasm for the improvement of various traits such as abiotic stress tolerance, disease resistance, and nutritional quality. Different omics technologies, genomics, transcriptomics, proteomics, and metabolomics, are used to generate high-throughput biological data, and bioinformatics methods are used to discover genes as possible targets (Kaur et al. 2021; Sarfraz et al. 2025).

Kaur et al. (2021) used omics data to enhance traits in several important cereal crops and reported that they provide considerable benefits in pre-breeding and breeding stages. Moreover, they emphasized that combining different types of omics information enabled the speedy identification of genes associated with positive or negative traits and to investigate their expression in multiple environmental situations.

The evolution of sequencing technologies allows researchers to access whole-genome data from many cereal species, providing insight into their genetic architecture. Whole-genome sequencing (WGS) is a potential tool to identify the genetic variation linked with the desired traits such as abiotic stresses, biotic stress, and nutritional contents (Jeyasri et al. 2021). These genomic resources support the development of marker-assisted selection and targeted genome-editing approaches.

Improved high-throughput sequencing technologies, including RNA-seq, have also made comprehensive profiling of gene expression to abiotic and biotic stresses possible in cereal crops. Such data can be useful in identifying the temporal and spatial expression of genes behind target traits, thereby helping the selection of candidates for genome editing (Ashraf et al. 2022; Jeyasri et al. 2021).

MicroRNAs (miRNAs) are important post-transcriptional regulators of gene expression and have been involved in the response of plants to environmental stressors. The probable miRNA target genes are predicted using bioinformatics approaches, and their roles in stress regulation are studied and validated. Specifically, by identifying important miRNA–mRNA interactions, regulatory networks that

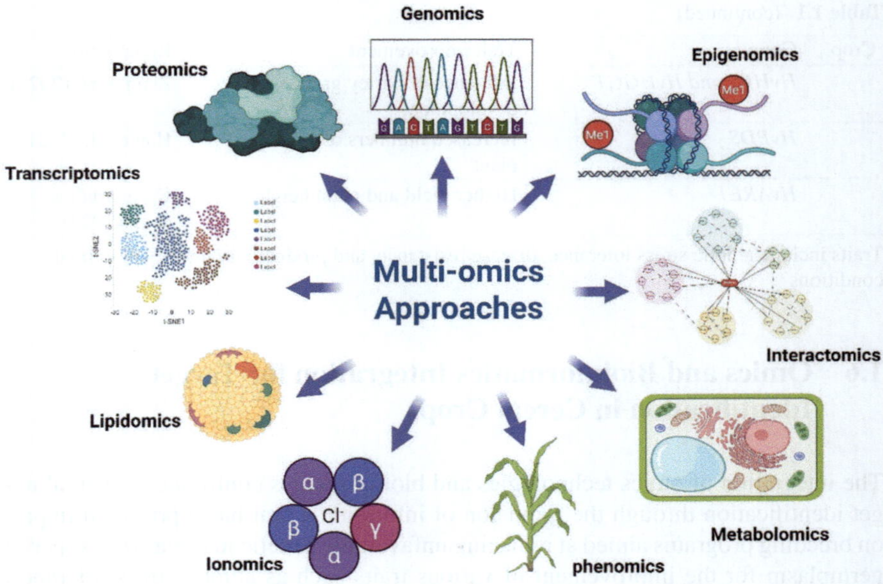

Fig. 1.4 Integration of multi-omics approaches (genomics, transcriptomics, proteomics, metabolomics) with CRISPR-based genome editing for target identification and functional validation in cereal crops under climate stress conditions

affect essential traits could be explained and contribute to further developing resilient cereals (Ashraf et al. 2022).

Nevertheless, the expansion of omics technologies is leading to the generation of massive amounts of data, which requires bioinformatics tools to encapsulate and analyze this data. Many bioinformatic platforms provide an opportunity for broad integration of genomic and transcriptomic data to identify putative gene targets from expression profile or molecular interaction and metabolic pathway data (Misra et al. 2019).

The systematic incorporation of these integrated omics and bioinformatic approaches not only fast tracks the target identification timeline but also complements the ongoing breeding programs, by adding a more concise and predictive component for cereal crop improvement. This integrated approach keeps potential to evolve resilient and high-yielding cereal genotypes against elapsing challenges of agriculture (Kaur et al. 2021; Nayak et al. 2021) using data driven indications (Fig. 1.4).

1.7 Challenges and Limitations of CRISPR in Cereal Crops

CRISPR, being a disruptive plant-breeding technology particularly in cereals, its successful deployment in agricultural applications is still hindered by many challenges and limitations that need further attention. A major drawback of the CRISPR/

Cas9 technology is the possibility of off-target effects at unintended sites in the genome with sequence similarity to the target site (Rasheed et al. 2021; Ricroch et al. 2017). These off-target changes can result in negative phenotypes, and such unwanted traits can slow the adoption of CRISPR crops. Variants of Cas with improved targeting specificity, e.g., Cas12, have been engineered (Sampath et al. 2023).

Efficient plant regeneration and transformation continue to be a major limiting factor restricting CRISPR application in many cereal crops. Plant transformation protocols differ widely among species and typically involve lengthy, laborious techniques that make obtaining high numbers of edited plants for evaluation cumbersome. In some cereal species, for example, it was difficult to regenerate stable transgenic lines (Basu et al. 2023), which limits the more widely application of CRISPR in those crops. Generating such tissue culture-independent transformation protocols is necessary to maximize regeneration efficiency for particular germplasm and expand the available crops for CRISPR-enhanced trait development base.

GM regulatory frameworks can limit the use of CRISPR/Cas systems in agriculture. In some countries, CRISPR-derived products will come under the GMO umbrella, leading to rigorous risk assessments that can and will delay commercialization (Saini et al. 2025). The regulatory environment is also evolving over time with continuing discussions around genome-edited plants, generating uncertainty for researchers and breeders.

1.8 Conclusion

Climate change is threatening global cereal production and food security through rising temperatures, more frequent droughts and floods, changes in soil salinity, and changing pest and disease pressure. Such challenges require precise and novel tactics to increase the resilience and productivity of crops. Genome editing, particularly CRISPR/Cas systems, has redefined the concept of cereal improvement by allowing researchers to make changes in genes associated with stress tolerance, disease resistance, yield, and nutritional quality at unprecedented resolution.

In combination with omics and bioinformatics, CRISPR can accelerate the discovery of candidate genes that improve transfer and subsequent transformation protocols for genome editing, so that they can be used in less manageable crops such as wheat. However, there are still concerns regarding off-target effects, regulatory uncertainty, transformation efficiency, equity, and acceptability.

To exploit the possibilities of advanced genome-editing, breeding pipelines, regulations, and a cooperative technological or infrastructural setup are needed to implement breeding solutions that will be successful in the future. As climate change progresses, cereals that can be maintained with high-yielding and nutrient-rich varieties will be of great importance to global food systems and farmers' livelihoods to find sustainable agricultural solutions for the twenty-first century.

References

Abdallah NA, Elsharawy H, Abulela HA, Thilmony R, Abdelhadi AA, Elarabi NI (2025) Multiplex CRISPR/Cas9-mediated genome editing to address drought tolerance in wheat. GM Crops Food 16(1):1–17

Ahmar S, Usman B, Hensel G, Jung K-H, Gruszka D (2024) CRISPR enables sustainable cereal production for a greener future. Trends Plant Sci 29(2):179–195

Alam MS, Kong J, Tao R, Ahmed T, Alamin M, Alotaibi SS, Abdelsalam NR, Xu J-H (2022) CRISPR/Cas9 mediated knockout of the OsbHLH024 transcription factor improves salt stress resistance in rice (Oryza sativa L.). Plants 11(9):1184

Alfatih A, Wu J, Jan SU, Zhang ZS, Xia JQ, Xiang CB (2020) Loss of rice PARAQUAT TOLERANCE 3 confers enhanced resistance to abiotic stresses and increases grain yield in field. Plant Cell Environ 43(11):2743–2754

Ali S, Baloch N (2023) Efficacy of essential oils of two local weed plants in controlling major stored wheat insect pests Rhyzopertha dominica (Fabricius) and Trogoderma granarium (Everts) under laboratory conditions. Sarhad J Agric 39(4):995–1002

Anjala K, Augustine R (2022) Designing of guide rna constructs for crispr/cas9-mediated editing of rice transcription factor osmads26 for enhancing drought tolerance. J Appl Biology Biotechnol 11(1):176–182

Ansari WA, Chandanshive SU, Bhatt V, Nadaf AB, Vats S, Katara JL, Sonah H, Deshmukh R (2020) Genome editing in cereals: approaches, applications and challenges. Int J Mol Sci 21(11):4040

Ashraf U, Mahmood S, Shahid N, Imran M, Siddique M, Abrar M (2022) Multi-omics approaches for strategic improvements of crops under changing climatic conditions. In: Principles and practices of OMICS and genome editing for crop improvement. Springer, pp 57–92

Ayoub N, Ali Z, Sher MA, Saddique MAB (2021) Insights into the mechanisms of susceptibility and tolerance against salinity stress in rice plant. Pak J Biochem Biotechnol 2(1):1–23

Bahrami F, Arzani A, Rahimmalek M (2020) A novel tolerance index to identify heat tolerance in cultivated and wild barley genotypes. *bioRxiv*:2020.2005. 2031.125971

Bahrami F, Arzani A, Rahimmalek M (2021) Tolerance to high temperature at reproductive stage: trade-offs between phenology, grain yield and yield-related traits in wild and cultivated barleys. Plant Breed 140(5):812–826

Bai X, Zhan G, Tian S, Peng H, Cui X, Islam MA, Goher F, Ma Y, Kang Z, Xu Z-S (2021) Transcription factor BZR2 activates chitinase Cht20. 2 transcription to confer resistance to wheat stripe rust. Plant Physiol 187(4):2749–2762

Balla K, Bencze S, Bónis P, Árendás T, Veisz O (2014) Changes in the photosynthetic efficiency of winter wheat in response to abiotic stress. Cent Eur J Biol 9:519–530

Basu U, Riaz Ahmed S, Bhat BA, Anwar Z, Ali A, Ijaz A, Gulzar A, Bibi A, Tyagi A, Nebapure SM (2023) A CRISPR way for accelerating cereal crop improvement: progress and challenges. Front Genet 13:866976

Baysal C, Kausch AP, Cody JP, Altpeter F, Voytas DF (2025) Rapid and efficient in planta genome editing in sorghum using foxtail mosaic virus-mediated sgRNA delivery. Plant J 121(2):e17196

Biswal AK, Hernandez LRB, Castillo AI, Debernardi JM, Dhugga KS (2023) An efficient transformation method for genome editing of elite bread wheat cultivars. Front Plant Sci 14:1135047

Brandt KM, Gunn H, Moretti N, Zemetra RS (2020) A streamlined protocol for wheat (Triticum aestivum) protoplast isolation and transformation with CRISPR-Cas ribonucleoprotein complexes. Front Plant Sci 11:769

Broberg M, Dubey M, Sun M-H, Ihrmark K, Schroers H-J, Li S-D, Jensen DF, Brandström Durling M, Karlsson M (2018) Out in the cold: identification of genomic regions associated with cold tolerance in the biocontrol fungus Clonostachys rosea through genome-wide association mapping. Front Microbiol 9:2844

Caddell D, Langenfeld NJ, Eckels MJ, Zhen S, Klaras R, Mishra L, Bugbee B, Coleman-Derr D (2023) Photosynthesis in rice is increased by CRISPR/Cas9-mediated transformation of two truncated light-harvesting antenna. Front Plant Sci 14:1050483

Chen F, Chen L, Yan Z, Xu J, Feng L, He N, Guo M, Zhao J, Chen Z, Chen H (2024) Recent advances of CRISPR-based genome editing for enhancing staple crops. Front Plant Sci 15:1478398

Chhogyel N, Kumar L, Bajgai Y (2020) Consequences of climate change impacts and incidences of extreme weather events in relation to crop production in Bhutan. Sustainability 12(10):4319

Cram D, Kulkarni M, Buchwaldt M, Rajagopalan N, Bhowmik P, Rozwadowski K, Parkin IA, Sharpe AG, Kagale S (2019) WheatCRISPR: a web-based guide RNA design tool for CRISPR/Cas9-mediated genome editing in wheat. BMC Plant Biol 19:1–8

Dalla Costa L, Piazza S, Pompili V, Salvagnin U, Cestaro A, Moffa L, Vittani L, Moser C, Malnoy M (2020) Strategies to produce T-DNA free CRISPRed fruit trees via Agrobacterium tumefaciens stable gene transfer. Sci Rep 10(1):20155

Devi P, Jha UC, Prakash V, Kumar S, Parida SK, Paul PJ, Prasad PV, Sharma KD, Siddique KH, Nayyar H (2022) Response of physiological, reproductive function and yield traits in cultivated chickpea (Cicer arietinum L.) under heat stress. Front Plant Sci 13:880519

Dunaiska T, Batjuka A, Petjukevičs A, Škute N (2023) Physiological response of some wheat cultivars to thermal stress. ENVIRONMENT. TECHNOLOGIES. RESOURCES. Proceedings of the International Scientific and Practical Conference 1:60

Endo A, Saika H, Takemura M, Misawa N, Toki S (2019) A novel approach to carotenoid accumulation in rice callus by mimicking the cauliflower Orange mutation via genome editing. Rice 12:1–5

Errum A, Rehman N, Uzair M, Inam S, Ali GM, Khan MR (2023) CRISPR/Cas9 editing of wheat Ppd-1 gene homoeologs alters spike architecture and grain morphometric traits. Funct Integr Genomics 23(1):66

Fan Y, Li M, Wu Y, Wang X, Wang P, Zhang L, Meng X, Meng F, Li Y (2023) Characterization of thioredoxin gene TaTrxh9 associated with heading-time regulation in wheat. Plant Physiol Biochem 201:107903

Favas C, Cresta C, Whelan E, Smith K, Manger MS, Chandrasenage D, Singhkumarwong A, Kawasaki J, Moreno S, Goudet S (2024) Exploring food system resilience to the global polycrisis in six Asian countries. Front Nutr 11:1347186

Galli M, Martiny E, Imani J, Kumar N, Koch A, Steinbrenner J, Kogel KH (2022) CRISPR/Sp Cas9-mediated double knockout of barley Microrchidia MORC1 and MORC6a reveals their strong involvement in plant immunity, transcriptional gene silencing and plant growth. Plant Biotechnol J 20(1):89–102

Gasparis S, Kała M, Przyborowski M, Łyżnik LA, Orczyk W, Nadolska-Orczyk A (2018) A simple and efficient CRISPR/Cas9 platform for induction of single and multiple, heritable mutations in barley (Hordeum vulgare L.). Plant Methods 14:1–14

Guo Y, Zhao G, Gao X, Zhang L, Zhang Y, Cai X, Yuan X, Guo X (2023) CRISPR/Cas9 gene editing technology: a precise and efficient tool for crop quality improvement. Planta 258(2):36

Gupta A, Hua L, Zhang Z, Yang B, Li W (2023) CRISPR-induced miRNA156-recognition element mutations in TaSPL13 improve multiple agronomic traits in wheat. Plant Biotechnol J 21(3):536–548

Gupta A, Rico-Medina A, Caño-Delgado AI (2020) The physiology of plant responses to drought. Science 368(6488):266–269

Hahn F, Sanjurjo Loures L, Sparks CA, Kanyuka K, Nekrasov V (2021) Efficient CRISPR/Cas-mediated targeted mutagenesis in spring and winter wheat varieties. Plants 10(7):1481

Han Y, Broughton S, Liu L, Zhang X-Q, Zeng J, He X, Li C (2021) Highly efficient and genotype-independent barley gene editing based on another culture. Plant Commun 2(2):100082

Hao J, Yang Y, Futrell S, Kelly EA, Lorts CM, Nebie B, Runo S, Yang J, Alvarez S, Lasky JR (2023) CRISPR/Cas9-mediated mutagenesis of carotenoid cleavage dioxygenase (CCD)

genes in sorghum alters strigolactone biosynthesis and plant biotic interactions. Phytobiomes J 7(3):339–351
Haroon M, Wang X, Afzal R, Zafar MM, Idrees F, Batool M, Khan AS, Imran M (2022) Novel plant breeding techniques shake hands with cereals to increase production. Plants 11(8):1052
Havrlentová M, Kraic J, Gregusová V, Kovácsová B (2021) Drought stress in cereals–a review. Agriculture 67(2):47–60
He F, Wang C, Sun H, Tian S, Zhao G, Liu C, Wan C, Guo J, Huang X, Zhan G (2023) Simultaneous editing of three homoeologues of TaCIPK14 confers broad-spectrum resistance to stripe rust in wheat. Plant Biotechnol J 21(2):354–368
Hillary VE, Ceasar SA (2019) Application of CRISPR/Cas9 genome editing system in cereal crops. Open Biotechnol J 13(1):173
Hisano H, Sakai H, Hamaoka M, Munemori H, Abe F, Meints B, Sato K, Hayes PM (2025) Rapid development of naked malting barley germplasm through targeted mutagenesis. Mol Breed 45(3):1–16
Holubová K, Hensel G, Vojta P, Tarkowski P, Bergougnoux V, Galuszka P (2018) Modification of barley plant productivity through regulation of cytokinin content by reverse-genetics approaches. Front Plant Sci 9:1676
Huang J, Gao L, Luo S, Liu K, Qing D, Pan Y, Dai G, Deng G, Zhu C (2022) The genetic editing of GS3 via CRISPR/Cas9 accelerates the breeding of three-line hybrid rice with superior yield and grain quality. Mol Breed 42(4):22
Huang Q, Lin B, Cao Y, Zhang Y, Song H, Huang C, Sun T, Long C, Liao J, Zhuo K (2023a) CRISPR/Cas9-mediated mutagenesis of the susceptibility gene OsHPP04 in rice confers enhanced resistance to rice root-knot nematode. Front Plant Sci 14:1134653
Huang X, Bai X, Qian C, Liu S, Goher F, He F, Zhao G, Pei G, Zhao H, Wang J (2023b) TaUAM3, a UDP-Ara mutases protein, positively regulates wheat resistance to the stripe rust fungus. Food Energy Secur 12(3):e456
Hui S, Li H, Mawia AM, Zhou L, Cai J, Ahmad S, Lai C, Wang J, Jiao G, Xie L (2022) Production of aromatic three-line hybrid rice using novel alleles of BADH2. Plant Biotechnol J 20(1):59–74
Jaganathan D, Ramasamy K, Sellamuthu G, Jayabalan S, Venkataraman G (2018) CRISPR for crop improvement: an update review. Front Plant Sci 9:985
Jeyasri R, Muthuramalingam P, Satish L, Pandian SK, Chen J-T, Ahmar S, Wang X, Mora-Poblete F, Ramesh M (2021) An overview of abiotic stress in cereal crops: negative impacts, regulation, biotechnology and integrated omics. Plants 10(7):1472
Jha PK, Beebe S, Alvarez-Toro P, Mukankusi C, Ramirez-Villegas J (2023) Characterizing patterns of seasonal drought stress for use in common bean breeding in East Africa under present and future climates. Agric For Meteorol 342:109735
Ji Z, Shi J, Zeng Y, Qian Q, Yang C (2014) Application of a simplified marker-assisted backcross technique for hybrid breeding in rice. Biologia 69:463–468
Jiang M, Zhang H, Song Y, Chen J, Bai J, Tang J, Wang Q, Fotopoulos V, Zhu QH, Yang R (2024) Transcription factor OsbZIP10 modulates rice grain quality by regulating OsGIF1. Plant J 119(5):2181–2198
Jiang Z, Song G, Shan X, Wei Z, Liu Y, Jiang C, Jiang Y, Jin F, Li Y (2018) Association analysis and identification of ZmHKT1; 5 variation with salt-stress tolerance. Front Plant Sci 9:1485
Kakar N, Jumaa SH, Redoña ED, Warburton ML, Reddy KR (2019) Evaluating rice for salinity using pot-culture provides a systematic tolerance assessment at the seedling stage. Rice 12:1–14
Kan J, Cai Y, Cheng C, Chen S, Jiang C, He Z, Yang P (2023) CRISPR/Cas9-guided knockout of eIF4E improves wheat yellow mosaic virus resistance without yield penalty. Plant Biotechnol J 21(5):893
Karunarathne SD, Han Y, Zhang XQ, Li C (2022) CRISPR/Cas9 gene editing and natural variation analysis demonstrate the potential for HvARE1 in improvement of nitrogen use efficiency in barley. J Integr Plant Biol 64(3):756–770
Kaur B, Sandhu KS, Kamal R, Kaur K, Singh J, Röder MS, Muqaddasi QH (2021) Omics for the improvement of abiotic, biotic, and agronomic traits in major cereal crops: applications, challenges, and prospects. Plants 10(10):1989

Kayastha S, Behera A, Sahoo J, Mahapatra M (2023) Growing green: sustainable agriculture meets precision farming: a review. Bhartiya Krishi Anusandhan Patrika 38(4):349–355

Kershanskaya O, Nelidova D, Esenbaeva G, Mukiyanova G, Nelidov S (2019) Modern CRISPR/Cas9 biotechnology is a current challenge to create elite disease resistant crops cultivars. In: Current challenges in plant genetics, genomics, bioinformatics, and biotechnology, vol 24, p 113

Kim JH, Sung JH, Shahid S, Chung E-S (2022) Future hydrological drought analysis considering agricultural water withdrawal under SSP scenarios. Water Resour Manag 36(9):2913–2930

Kis A, Hamar É, Tholt G, Bán R, Havelda Z (2019) Creating highly efficient resistance against wheat dwarf virus in barley by employing CRISPR/Cas9 system. Plant Biotechnol J 17(6):1004

Kitonde C, Dossaji SF, Lukhoba CW, Wagacha J, Xiong Q (2019) In vitro studies of 3-O-Methylquercetin against phytopathogenic fungi of major cereals. J Agric Sci Pract 4(4):102–112

Kloc Y, Dmochowska-Boguta M, Żebrowska-Różańska P, Łaczmański Ł, Nadolska-Orczyk A, Orczyk W (2024) HvGSK1. 1 controls salt tolerance and yield through the brassinosteroid signaling pathway in barley. Int J Mol Sci 25(2):998

Knott GJ, Doudna JA (2018) CRISPR-Cas guides the future of genetic engineering. Science 361(6405):866–869

Kong X, Wang F, Wang Z, Gao X, Geng S, Deng Z, Zhang S, Fu M, Cui D, Liu S (2023) Grain yield improvement by genome editing of TaARF12 that decoupled peduncle and rachis development trajectories via differential regulation of gibberellin signalling in wheat. Plant Biotechnol J 21(10):1990–2001

Krishna TPA, Maharajan T, Ceasar SA (2022) Application of CRISPR/Cas9 genome editing system to reduce the pre-and post-harvest yield losses in cereals. Open Biotechnol J 16(1)

Kuai P, Lin N, Ye M, Ye M, Chen L, Chen S, Zu H, Hu L, Gatehouse AM, Lou Y (2024) Identification and knockout of a herbivore susceptibility gene enhances planthopper resistance and increases rice yield. Nat Food 5(10):846–859

Kumar N, Galli M, Ordon J, Stuttmann J, Kogel KH, Imani J (2018) Further analysis of barley MORC 1 using a highly efficient RNA-guided Cas9 gene-editing system. Plant Biotechnol J 16(11):1892–1903

Li C, Li W, Zhou Z, Chen H, Xie C, Lin Y (2019) A new rice breeding method: CRISPR/Cas9 system editing of the Xa13 promoter to cultivate transgene-free bacterial blight-resistant rice. Plant Biotechnol J 18(2):313

Li Y, Wu S, Huang Y, Ma X, Tan L, Liu F, Lv Q, Zhu Z, Hu M, Fu Y (2023) OsMADS17 simultaneously increases grain number and grain weight in rice. Nat Commun 14(1):3098

Li Z, Fu D, Wang X, Zeng R, Zhang X, Tian J, Zhang S, Yang X, Tian F, Lai J (2022) The transcription factor bZIP68 negatively regulates cold tolerance in maize. Plant Cell 34(8):2833–2851

Liang Z, Chen K, Li T, Zhang Y, Wang Y, Zhao Q, Liu J, Zhang H, Liu C, Ran Y (2017) Efficient DNA-free genome editing of bread wheat using CRISPR/Cas9 ribonucleoprotein complexes. Nat Commun 8(1):14261

Liao X, Yang X, He Q, Yao Y, Xiao J, Shi W (2023) Afforestation resists the drought disturbance on soil carbon sink in subtropical agricultural ecosystems. Agric Ecosyst Environ 357:108670

Liu C, Kong M, Yang F, Zhu J, Qi X, Weng J, Di D, Xie C (2022) Targeted generation of Null Mutants in ZmGDIα confers resistance against maize rough dwarf disease without agronomic penalty. Plant Biotechnol J 20(5):803

Liu X, Yu Y, Yao W, Yin Z, Wang Y, Huang Z, Zhou JQ, Liu J, Lu X, Wang F (2023) CRISPR/Cas9-mediated simultaneous mutation of three salicylic acid 5-hydroxylase (OsS5H) genes confers broad-spectrum disease resistance in rice. Plant Biotechnol J 21(9):1873–1886

Liu Z (2025) The behavioral dimension of CO2 fertilization effect: evidence from US harvesting records. Agric Econ 56(1):5–26

Lu Q, Luo X, Yang X, Zhou T, Zhang Y, Lan Y, Zhang D, Zheng L, Li Y, Li L (2023) CRISPR/Cas9-mediated gene editing of vacuolar ATPase subunit d mediates phytohormone biosynthesis and virus resistance in rice. Front Plant Sci 14:1122978

Ly LK, Ho TM, Bui TP, Nguyen LT, Phan Q, Le NT, Khuat LTM, Le LH, Chu HH, Pham NB (2024) CRISPR/Cas9 targeted mutations of OsDSG1 gene enhanced salt tolerance in rice. Funct Integr Genomics 24(2):70

Macaluso M, Farina P, Rossi L, Bianchi A, Venturi F, Chiriboga Ortega RD, Bedini S, Conti B, Guidi L, Zinnai A (2022) Chemical vs. enzymatic refining to produce peanut oil for edible use or to obtain a sustainable and cost-effective protector for stored grains against Sitophilus zeamais (Coleoptera: Curculionidae). Foods 11(9):1224

Mathivanan K, Uthaya Chandirika J, Mathimani T, Rajaram R, Zhang R (2023) A study on exopolysaccharides production, biosorption, and detoxification properties of Lysinibacillus fusiformis KMNTT-10 in growth media treated with Pb2+ ions. Emerg Mater 6(5):1491–1502

Misra BB, Langefeld C, Olivier M, Cox LA (2019) Integrated omics: tools, advances and future approaches. J Mol Endocrinol 62(1):R21–R45

Mohr T, Horstman J, Gu YQ, Elarabi NI, Abdallah NA, Thilmony R (2022) CRISPR-Cas9 gene editing of the Sal1 gene family in wheat. Plants 11(17):2259

Morris V, Jacquet J (2024) The animal agriculture industry, US universities, and the obstruction of climate understanding and policy. Clim Chang 177(3):41

Nainggolan D, Abay AT, Christensen JH, Termansen M (2023) The impact of climate change on crop mix shift in the Nordic region. Sci Rep 13(1):2962

Nakajima I, Ban Y, Azuma A, Onoue N, Moriguchi T, Yamamoto T, Toki S, Endo M (2017) CRISPR/Cas9-mediated targeted mutagenesis in grape. PLoS One 12(5):e0177966

Nayak SN, Aravind B, Malavalli SS, Sukanth B, Poornima R, Bharati P, Hefferon K, Kole C, Puppala N (2021) Omics technologies to enhance plant based functional foods: an overview. Front Genet 12:742095

Nelson GC, Van Der Mensbrugghe D, Ahammad H, Blanc E, Calvin K, Hasegawa T, Havlik P, Heyhoe E, Kyle P, Lotze-Campen H (2014) Agriculture and climate change in global scenarios: why don't the models agree. Agric Econ 45(1):85–101

Ning Q, Jian Y, Du Y, Li Y, Shen X, Jia H, Zhao R, Zhan J, Yang F, Jackson D (2021) An ethylene biosynthesis enzyme controls quantitative variation in maize ear length and kernel yield. Nat Commun 12(1):5832

Numan M, Serba DD, Ligaba-Osena A (2021) Alternative strategies for multi-stress tolerance and yield improvement in millets. Genes 12(5):739

Ogata T, Ishizaki T, Fujita M, Fujita Y (2020) CRISPR/Cas9-targeted mutagenesis of OsERA1 confers enhanced responses to abscisic acid and drought stress and increased primary root growth under nonstressed conditions in rice. PLoS One 15(12):e0243376

Packirisamy J, Subiramaniyan A, Perumal C, Arumugam R, Natarajan A, Ramalingam K, Chinnaraju NK (2023) Physiological and morphological changes in rice during drought stress: a review. Plant Arch (09725210) 23(2)

Pan Z, Liu M, Zhao H, Tan Z, Liang K, Sun Q, Gong D, He H, Zhou W, Qiu F (2020) ZmSRL5 is involved in drought tolerance by maintaining cuticular wax structure in maize. J Integr Plant Biol 62(12):1895–1909

Pathi KM, Rink P, Budhagatapalli N, Betz R, Saado I, Hiekel S, Becker M, Djamei A, Kumlehn J (2020) Engineering smut resistance in maize by site-directed mutagenesis of LIPOXYGENASE 3. Front Plant Sci 11:543895

Patil L, Kaliwal B (2017) Effect of CO2 concentration on growth and biochemical composition of newly isolated indigenous microalga Scenedesmus bajacalifornicus BBKLP-07. Appl Biochem Biotechnol 182(1):335–348

Poddar S, Tanaka J, Running KL, Kariyawasam GK, Faris JD, Friesen TL, Cho M-J, Cate JH, Staskawicz B (2023) Optimization of highly efficient exogenous-DNA-free Cas9-ribonucleoprotein mediated gene editing in disease susceptibility loci in wheat (Triticum aestivum L.). Front Plant Sci 13:1084700

Potlapalli BP, Schubert V, Metje-Sprink J, Liehr T, Houben A (2020) Application of Tris-HCl allows the specific labeling of regularly prepared chromosomes by CRISPR-FISH. Cytogenet Genome Res 160(3):156–165

Qi X, Dong L, Liu C, Mao L, Liu F, Zhang X, Cheng B, Xie C (2018) Systematic identification of endogenous RNA polymerase III promoters for efficient RNA guide-based genome editing technologies in maize. Crop J 6(3):314–320

Qin Q, Wang Y, Huang L, Du F, Zhao X, Li Z, Wang W, Fu B (2020) A U-box E3 ubiquitin ligase OsPUB67 is positively involved in drought tolerance in rice. Plant Mol Biol 102:89–107

Radchuk V, Radchuk R, Pirko Y, Vankova R, Gaudinova A, Korkhovoy V, Yemets A, Weber H, Weschke W, Blume YB (2012) A somaclonal line SE7 of finger millet (Eleusine coracana) exhibits modified cytokinin homeostasis and increased grain yield. J Exp Bot 63(15):5497–5506

Rahim AA, Uzair M, Rehman N, Fiaz S, Attia KA, Abushady AM, Yang SH, Khan MR (2024) CRISPR/Cas9 mediated TaRPK1 root architecture gene mutagenesis confers enhanced wheat yield. J King Saud Univ Sci 36(2):103063

Raj SRG, Nadarajah K (2022) QTL and candidate genes: techniques and advancement in abiotic stress resistance breeding of major cereals. Int J Mol Sci 24(1):6

Rasheed A, Gill RA, Hassan MU, Mahmood A, Qari S, Zaman QU, Ilyas M, Aamer M, Batool M, Li H (2021) A critical review: recent advancements in the use of CRISPR/Cas9 technology to enhance crops and alleviate global food crises. Curr Issues Mol Biol 43(3):1950–1976

Ricroch A, Clairand P, Harwood W (2017) Use of CRISPR systems in plant genome editing: toward new opportunities in agriculture. Emerg Top Life Sci 1(2):169–182

Rusmayandi G, Hertini ES, Harahap MAK, Nugroho RJ (2023) Climate change, agricultural production, food security, and livelihoods of rural communities in rural districts: a qualitative approach. West Sci Interdiscip Stud 1(03):102–110

Saijo Y, Loo EP i (2020) Plant immunity in signal integration between biotic and abiotic stress responses. New Phytol 225(1):87–104

Saini H, Devrani A, Synrem G, Priyanka. (2025) Application of CRISPR technology in plant improvement: an update review. Adv Agric 2025(1):4578877

Sampath V, Rangarajan N, Sharanappa C, Deori M, Veeraragavan M, Ghodake B, Kaushal K (2023) Advancing crop improvement through CRISPR technology in precision agriculture trends-a review. Int J Environ Clim Chang 13(11):4683–4694

Sant'Ana RRA, Caprestano CA, Nodari RO, Agapito-Tenfen SZ (2020) PEG-delivered CRISPR-Cas9 ribonucleoproteins system for gene-editing screening of maize protoplasts. Genes 11(9):1029

Santosh Kumar V, Verma RK, Yadav SK, Yadav P, Watts A, Rao M, Chinnusamy V (2020) CRISPR-Cas9 mediated genome editing of drought and salt tolerance (OsDST) gene in indica mega rice cultivar MTU1010. Physiol Mol Biol Plants 26:1099–1110

Sarfraz Z, Zarlashat Y, Ambreen A, Mujahid M, Iqbal MS, Fatima SA, Iqbal MS, Iqbal R, Fiaz S (2025) Plant biochemistry in the era of omics: Integrated omics approaches to unravel the genetic basis of plant stress tolerance. Plant Breed

Sathiyabalan A, Amir M, Sandhu R, Bhattacharya S (2024) Evaluating CRISPR/Cas9's progressive role in crop enhancement and sustainable agriculture. J Adv Biol Biotechnol 27:855–867

Shah PR, Varanavasiappan S, Kokiladevi E, Ramanathan A, Kumar K (2019) Genome editing of rice PFT1 gene to study its role in rice sheath blight disease resistance. Int J Curr Microbiol App Sci 8(6):2356–2364

Shim Y, Seong G, Choi Y, Lim C, Baek SA, Park YJ, Kim JK, An G, Kang K, Paek NC (2023) Suppression of cuticular wax biosynthesis mediated by rice LOV KELCH REPEAT PROTEIN 2 supports a negative role in drought stress tolerance. Plant Cell Environ 46(5):1504–1520

Sinha R, Bhattacharyya D, Majumdar AB, Datta R, Hazra S, Chattopadhyay S (2013) Leaf proteome profiling of transgenic mint infected with Alternaria alternata. J Proteome 93:117–132

Son S, Song G, Nam S, Lee G, Im J, Lee KS, Park YJ, Suh E-J, Park SR (2024) CRISPR/Cas9-mediated mutagenesis of rice NAC transcription factor genes results in altered innate immunity. Plant Physiol 195(2):1138–1142

Song Y, Li S, Sui Y, Zheng H, Han G, Sun X, Yang W, Wang H, Zhuang K, Kong F (2022) SbbHLH85, a bHLH member, modulates resilience to salt stress by regulating root hair growth in sorghum. Theor Appl Genet 135:201

Songmei L, Jie J, Yang L, Jun M, Shouling X, Yuanyuan T, Youfa L, Qingyao S, Jianzhong H (2019) Characterization and evaluation of OsLCT1 and OsNramp5 mutants generated through CRISPR/Cas9-mediated mutagenesis for breeding low Cd rice. Rice Sci 26(2):88–97

Sun W, Zhang H, Yang S, Liu L, Xie P, Li J, Zhu Y, Ouyang Y, Xie Q, Zhang H (2023) Genetic modification of Gγ subunit AT1 enhances salt-alkali tolerance in main graminaceous crops. Natl Sci Rev 10(6):nwad075

Svitashev S, Schwartz C, Lenderts B, Young JK, Mark Cigan A (2016) Genome editing in maize directed by CRISPR–Cas9 ribonucleoprotein complexes. Nat Commun 7(1):13274

Syed A, Sarwar G, Shah SH, Muhammad S (2021) Soil salinity research in 21st century in Pakistan: its impact on availability of plant nutrients, growth and yield of crops. Commun Soil Sci Plant Anal 52(3):183–200

Tamilselvan-Nattar-Amutha S, Hiekel S, Hartmann F, Lorenz J, Dabhi RV, Dreissig S, Hensel G, Kumlehn J, Heckmann S (2023) Barley stripe mosaic virus-mediated somatic and heritable gene editing in barley (Hordeum vulgare L.). Front Plant Sci 14:1201446

Tang Y, Fu Y (2018) Class 2 CRISPR/Cas: an expanding biotechnology toolbox for and beyond genome editing. Cell Biosci 8(1):59

Távora FT, Meunier AC, Vernet A, Portefaix M, Milazzo J, Adreit H, Tharreau D, Franco OL, Mehta A (2022) CRISPR/Cas9-Targeted knockout of rice susceptibility genes OsDjA2 and OsERF104 reveals alternative sources of resistance to Pyricularia oryzae. Rice Sci 29(6):535–544

Thilakarathne AS, Liu F, Zou Z (2025) Plant signaling hormones and transcription factors: key regulators of plant responses to growth, development, and stress. Plants 14(7):1070

Thomas TS, Mainuddin K, Chiang C, Rahman A, Haque A, Islam N, Quasem S, Sun Y (2013) Agriculture and adaptation in Bangladesh: current and projected impacts of climate change. Intl Food Policy Res Inst 1281

Tian X, Qin Z, Zhao Y, Wen J, Lan T, Zhang L, Wang F, Qin D, Yu K, Zhao A (2022) Stress granule-associated TaMBF1c confers thermotolerance through regulating specific mRNA translation in wheat (Triticum aestivum). New Phytol 233(4):1719–1731

Uranga M, Daròs JA (2023) Tools and targets: the dual role of plant viruses in CRISPR–Cas genome editing. Plant Genom 16(2):e20220

Usman B, Nawaz G, Zhao N, Liao S, Qin B, Liu F, Liu Y, Li R (2020) Programmed editing of rice (Oryza sativa L.) OsSPL16 gene using CRISPR/Cas9 improves grain yield by modulating the expression of pyruvate enzymes and cell cycle proteins. Int J Mol Sci 22(1):249

Varshney RK, Singh VK, Kumar A, Powell W, Sorrells ME (2018) Can genomics deliver climate-change ready crops? Curr Opin Plant Biol 45:205–211

Waiba SP (2023) Economics impact of climate change on agriculture GDP in Bhutan: empirical evidence from ARDL approach. Int J Econ (IJEC) 2(2):706

Wang B, Zhong Z, Wang X, Han X, Yu D, Wang C, Song W, Zheng X, Chen C, Zhang Y (2020a) Knockout of the OsNAC006 transcription factor causes drought and heat sensitivity in rice. Int J Mol Sci 21(7):2288

Wang B, Zhu L, Zhao B, Zhao Y, Xie Y, Zheng Z, Li Y, Sun J, Wang H (2019) Development of a haploid-inducer mediated genome editing system for accelerating maize breeding. Mol Plant 12(4):597–602

Wang C, Wang G, Gao Y, Lu G, Habben JE, Mao G, Chen G, Wang J, Yang F, Zhao X (2020b) A cytokinin-activation enzyme-like gene improves grain yield under various field conditions in rice. Plant Mol Biol 102(4):373–388

Wang J, Vanga SK, Saxena R, Orsat V, Raghavan V (2018) Effect of climate change on the yield of cereal crops: a review. Climate 6(2):41

Wang K, Gong Q, Ye X (2020c) Recent developments and applications of genetic transformation and genome editing technologies in wheat. Theor Appl Genet 133(5):1603–1622

Wang Y, Tang Q, Pu L, Zhang H, Li X (2022) CRISPR-Cas technology opens a new era for the creation of novel maize germplasms. Front Plant Sci 13:1049803

Wu J, Wang R, Zhao W, Zhao K, Wu S, Zhang J, Wang H, Fu G, Huang M, Li Y (2024) Combined subsoiling and ridge–furrow rainfall harvesting during the summer fallow season improves

wheat yield, water and nutrient use efficiency, and quality and reduces soil nitrate-N residue in the dryland summer fallow–winter wheat rotation. Front Plant Sci 15:1401287
Xie Z, Sun Y, Zhan C, Qu C, Jin N, Gu X, Huang J (2025) The E3 ligase OsPUB33 controls rice grain size and weight by regulating the OsNAC120–BG1 module. Plant Cell 37(1):koae297
Xiong H, He H, Chang Y, Miao B, Liu Z, Wang Q, Dong F, Xiong L (2025) Multiple roles of NAC transcription factors in plant development and stress responses. J Integr Plant Biol 67:510
Xu R, Li J, Liu X, Shan T, Qin R, Wei P (2020) Development of plant prime-editing systems for precise genome editing. Plant Commun 1(3):100043
Yang Y, Wang B, Wang J, He C, Zhang D, Li P, Zhang J, Li Z (2022) Transcription factors ZmNF-YA1 and ZmNF-YB16 regulate plant growth and drought tolerance in maize. Plant Physiol 190(2):1506–1525
Yeleliere E, Antwi-Agyei P, Guodaar L (2023) Farmers response to climate variability and change in rainfed farming systems: insight from lived experiences of farmers. Heliyon 9(9):e19656
Yue E, Cao H, Liu B (2020) OsmiR535, a potential genetic editing target for drought and salinity stress tolerance in Oryza sativa. Plants 9(10):1337
Zafar K, Sedeek KE, Rao GS, Khan MZ, Amin I, Kamel R, Mukhtar Z, Zafar M, Mansoor S, Mahfouz MM (2020) Genome editing technologies for rice improvement: progress, prospects, and safety concerns. Front Genom Edit 2:5
Zenda T, Wang N, Dong A, Zhou Y, Duan H (2022) Reproductive-stage heat stress in cereals: impact, plant responses and strategies for tolerance improvement. Int J Mol Sci 23(13):6929
Zeng Y, Wen J, Zhao W, Wang Q, Huang W (2020) Rational improvement of rice yield and cold tolerance by editing the three genes OsPIN5b, GS3, and OsMYB30 with the CRISPR–Cas9 system. Front Plant Sci 10:1663
Zhang A, Liu Y, Wang F, Li T, Chen Z, Kong D, Bi J, Zhang F, Luo X, Wang J (2019a) Enhanced rice salinity tolerance via CRISPR/Cas9-targeted mutagenesis of the OsRR22 gene. Mol Breed 39:1–10
Zhang Y, Lin X-F, Li L, Piao R-H, Wu S, Song A, Gao M, Jin Y-M (2024) CRISPR/Cas9-mediated knockout of Bsr-d1 enhances the blast resistance of rice in Northeast China. Plant Cell Rep 43(4):100
Zhang Y, Ren Q, Tang X, Liu S, Malzahn AA, Zhou J, Wang J, Yin D, Pan C, Yuan M (2021) Expanding the scope of plant genome engineering with Cas12a orthologs and highly multiplexable editing systems. Nat Commun 12(1):1944
Zhang Z, Hua L, Gupta A, Tricoli D, Edwards KJ, Yang B, Li W (2019b) Development of an Agrobacterium-delivered CRISPR/Cas9 system for wheat genome editing. Plant Biotechnol J 17(8):1623–1635
Zhao R, Li N, Lin Q, Li M, Shen X, Peng Y, Du Y, Ning Q, Li Y, Zhan J (2023) ZmBET5L1 inhibits primary root growth and decreases osmotic stress tolerance by mediating vesicle aggregation and tethering in maize. Plant Cell Environ 46(3):975–990
Zhou Y, Xu S, Jiang N, Zhao X, Bai Z, Liu J, Yao W, Tang Q, Xiao G, Lv C (2022) Engineering of rice varieties with enhanced resistances to both blast and bacterial blight diseases via CRISPR/Cas9. Plant Biotechnol J 20(5):876–885
Zhu X, Xu K, Liu Y, Guo R, Chen L (2021) Assessing the vulnerability and risk of maize to drought in China based on the AquaCrop model. Agric Syst 189:103040
Zong Y, Wang Y, Li C, Zhang R, Chen K, Ran Y, Qiu J-L, Wang D, Gao C (2017) Precise base editing in rice, wheat and maize with a Cas9-cytidine deaminase fusion. Nat Biotechnol 35(5):438–440

Part I
CRISPR Gene Editing Tools

Chapter 2
Next-Generation Plant Gene-Editing Tools Empowered by Artificial Intelligence

Kamel A. Abd-Elsalam, Rawan K. Hassan, and Toka E. Abdelkhalek

Abstract The innovations in artificial intelligence (AI) are transforming genetic editing by making it more accurate and effective. Some technologies associated with AI supplement the current practices, and others provide a solution to previously existing problems. AI-powered assistants help popularize guide RNA (gRNA) editing, making it more specific. AI can also reduce off-target edits by foreseeing editing and repairing pathways to enhance the safety and familiarity of genetic applications. Advancing methodologies such as base and prime editing have shown that more complex edits are possible and allow for single-base fixes without creating double-strand fusion. Capabilities of AI-based gene editing in agriculture range widely, with potential returns of increased crop yields, disease-protected crops, and crops able to adapt to changing environments. The developments are vital toward dealing with food security in a world affected by climate change and an increase in population. However, there is a technical gap in implementing AI in genetic editing, which includes the problem of good datasets and opaque algorithms. In the future, we are likely to see an even more intimate association between AI and gene editing, especially as they incorporate multi-omics data and give an overall picture of genetic interactions. Also, it is important that automation in the field of gene editing is bound to make the work more efficient in research and labs, so that the respective field proceeds at a fast pace.

Keywords Artificial intelligence (AI) · Gene editing · CRISPR · Guide RNA (gRNA) · Off-target effects · Base editing · Prime editing · Genome engineering · Agricultural biotechnology · Crop improvement · Multi-omics data · Precision genome editing · Bioinformatics · Machine learning · Data challenges

K. A. Abd-Elsalam (✉)
Plant Pathology Research Institute, Agricultural Research Center, Giza, Egypt

R. K. Hassan · T. E. Abdelkhalek
Biotechnology English Program, Faculty of Agriculture, Cairo University, Giza, Egypt

K. A. Abd-Elsalam, A. Ahmad (eds.), *Gene Editing in Cereals*, Sustainability Sciences in Asia and Africa, https://doi.org/10.1007/978-981-95-6722-5_2

2.1 Introduction

The recent development of gene editing, especially using such technology as CRISPR–Cas9 (Wang and Doudna 2023), has transformed the arena of genetic engineering inasmuch as it now allows the precise editing of DNA. Artificial intelligence (AI) and machine learning (ML) integration are boosting the development of a more straightforward process to analyze complicated genomic data, support in locating genetic susceptibilities, and predict how diseases change and how they can be addressed (Zafar et al. 2021; Liu et al. 2022a, b).

Although CRISPR–Cas9 can achieve a precision far beyond any other tool, there are issues of clinical translation, i.e., increasing the efficiency and selectivity of editing or decreasing immunogenicity (Villiger et al. 2024). Some of these solutions using AI/ML are innovative, including the faster discovery and design of new gene-editing enzymes, as well as better mutagenesis screens (Chia et al. 2025; Dai et al. 2024). AI/ML models can also forecast the efficiency and specificity of guide RNAs (gRNAs), which makes gene-editing technologies safer.

Newer modalities (such as base editing and prime editing) are less likely to have off-target edits than most CRISPR modalities and are more flexible in what they can be used to edit (Liu et al. 2022a, b; Dai et al. 2024). The AI-based models also enhance safety levels by forecasting guide RNA effectiveness, including particularities. Gene editing and AI are reshaping the sphere of genetics, opening up the quest for agricultural activity (Zafar et al. 2021; Liu et al. 2022a, b; Chia et al. 2025).

This chapter aims to examine the effectiveness of CRISPR–Cas9 and other emerging gene-editing technologies in the field of agriculture. It will see how AI/ML can advance the research in the field of GM crops, and evaluate the problems in the translation to the agricultural sector. It will also gauge AI/ML models in the prediction of the results of crop performance and the agricultural impact on sustainability and food security.

2.2 Foundations and Evolution of Gene-Editing Technologies

The development of so-called genome-editing tools has transformed the scope of biological research since it allows us to make precise changes to organisms. Major technologies include meganucleases, zinc finger nucleases (ZFNs), transcription activator-like effector nucleases (TALENs), and CRISPR in their ability to perform precise mutations in the genome (Gaj et al. 2016). A lot of these basic inventions are developed in universities and patented to foster subsequent commercial development, creating a great variety of technology transfer models (and conflicts there between). One of the primary concepts of gene editing is the use of the targeted DNA double-strand breaks (DSBs) to induce the repair cascades of the cell (Szostak et al. 1983). The repair of these DSBs occurs via two primary pathways, namely, homology-directed repair (HDR) and nonhomologous end-joining (NHEJ).

Whereas the latter is an imprecise process allowing the insertion and deletion of nucleotides (indels) to+ occur (Bibikova et al. 2002). The speed at which these CRISPR, as well as other genome-editing, technologies have developed has raised a variety of commercialization strategies, such as deposits in the public domain, open access agreements, material transfer deals, etc., and option licensing arrangements. This heterogeneity encourages experimentation and rivalry, bringing up a model of federalism in technology transfer (Takata et al. 1998).

2.2.1 Historical Overview of Gene-Editing Technologies

The historical background of the respective gene-editing technologies signifies a massive transformation of the old techniques of gene therapy that were limited to introducing a single gene into human cells toward specific manipulation of the human genome to treat individuals. Important nuclease-based platforms have played a central role in the transformation (Maeder and Gersbach 2016). The topic of gene editing came into the limelight since scientists were keen to find practical ways of editing genomes (Khalil 2020).

The history of genetics could be traced to the works of Gregor Mendel in the 1860s, when his studies with the use of pea plants predetermined the current knowledge in the genetic field. The discovery of the molecular basis of heredity waited, however, a century. Classical breeding techniques had limitations in terms of their application because they could only be used in closely related species, and some processes that were involved were cumbersome, such as mutagenesis (Maeder and Gersbach 2016).

In the late twentieth century, more refined editing methods were discovered: ZFNs and Agrobacterium-mediated transformation. The invention of the TALENs and CRISPR systems in the twenty-first century became a revolution in the area, and CRISPR, especially, has become commonly used as it is efficient and easy to apply. Although promising, gene editing in humans has also been associated with a number of ethical issues, an aspect that has spilled over to the formulation of policies that govern the use of this technology in different nations, with the resultant delay in the commercialization of genetically modified organisms. However, the pace toward accepting the CRISPR technology seems to increase, and this makes one think that the same can happen to genome-edited crops (Ahmad 2025). With genome editing still evolving, its potential uses in agriculture will focus on seeding more food production and environmental stress tolerance, which are challenging needs in how to support food security (Kale et al. 2021). The current development and debate on the ethical consequences show the necessity of striking the right balance between innovation and social considerations of gene editing.

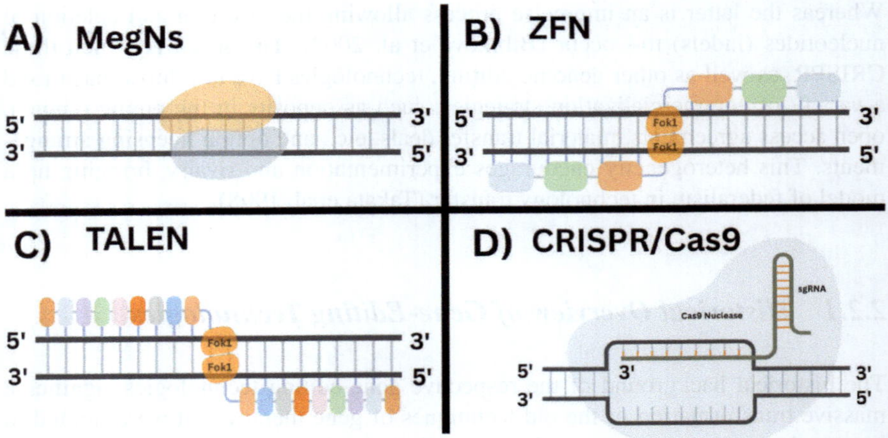

Fig. 2.1 Symbolic presentation of four endonuclease used in the methods of gene editing: (**a**) Meganuclease (MegN) (**b**) Zinc finger nuclease (ZFN) (**c**) Transcription activator-like effector nuclease (TALEN) (**d**) CRISPR/Cas9 system

2.2.2 Next-Generation Gene-Editing Tools

The development of biotechnologies allows editing of the genome in a precise manner, which creates new opportunities to correct a mutation and regulate the expression of various genes. Editors are constantly refining four endonucleases employed in gene-editing technologies (Fig. 2.1) such as meganucleases, ZFNs, TALENs, and CRISPR–Cas platforms (Gaj et al. 2016). The task of programming CRISPR–Cas systems is simpler than older ones; however, different forms of technology can have diverse mechanisms, advantages, limitations, and ethical issues (Khan 2019).

2.2.2.1 Meganucleases (MegNs)

Homing endonucleases, or meganucleases (MegNs), are among the oldest developed genome-editing tools. These are naturally produced enzymes that can specifically cleave long DNA. The most investigated family is the LAGLIDADG family, which comprises I-CreI and I-SceI enzymes (Silva et al. 2011). Meganucleases work by generating a breakage in both strands of the DNA at a specific place in the genome and activating the cellular DNA repair pathways that can be used to edit the genome (Stoddard 2014).

Meganuclease technology has as its fundamental concept the re-engineering of the DNA-binding specificity of these enzymes to novel sequences. This has been realized by an effort of rational protein design with the realization technique. But the main drawback to this is that the meganucleases being structurally coupled between their cleavage, as well as the binding sites, makes it challenging and technically involved to reprogram them to new targets (Mandip and Steer 2019). To deal with these issues, researchers have introduced chimeric proteins combining the exquisite specificity of binding of the zinc fingers (ZFs) or transcription activator-like effectors (TALEs) with the cleavage specificity of meganucleases (Walker-Daniels 2013). Engineered monomeric enzymes are modular tools to take advantage of the benefits of both platforms to increase the specificity and performance of targets and cleavages.

A major advantage of the meganucleases has been that they create DSBs with 3219Genev overhangs, believed to enhance homologous recombination over that seen with the 5219Genev generation of FokI-based nucleases (Chevalier and Stoddard 2001). Moreover, meganucleases represent the smallest category of genome-editing enzymes and are thus extremely suitable to employ in generating gene delivery. The small size also permits convenient incorporation into viral vectors and permits multiple editing events of genomic sites using multiple meganuclease monomers delivered in a single construct (Redondo et al. 2008).

2.2.2.2 Zinc Finger Nucleases (ZFNs)

Zinc finger nucleases (ZFNs) are designed restriction enzymes that are novel, effective, and specific genome-editing instruments. ZFNs are a complex of several zinc finger domains forming the specific binding sites to DNA sequences and a FokI endonuclease domain, inducing targeted double-strand DNA breaks (Bibikova et al. 2001; Carroll 2011). This peculiar form enables ZFNs to enable accurate editing of the genomes of different organisms. Among the primary advantages of ZFNs is the possibility to create DSBs at targeted sites and introduce gene editing. ZFNs are comparatively easy to design because they use a combinatorial assembly of zinc fingers having known recognition motifs (Kim et al. 1996). Such simplicity in design has also rendered ZFNs attractive to scientists wanting to use them to manipulate genes to accomplish different purposes.

Nevertheless, ZFNs also have great challenges. Another problem that stands out is "context-dependent specificity," the phenomenon that allows the nuclease's effectiveness to depend on neighboring stretches of DNA. Such fluctuations may cause unintended cleavages, which make the genome fragmented and unstable (Stoddard 2011). Furthermore, although ZFNs work quite well on focusing on a single site, they cannot conduct more complicated operations like the so-called knock-in or genetic manipulation that necessitates the use of genes with more complex structure.

The possibility of off-target effects is another concern whereby, ZFNs do not precisely cut DNA at its intensive points. Those off-target cleavages may be due to single-nucleotide mutations or the presence of ill-concerted protein domain interactions posing cellular toxicity. Nevertheless, in spite of these disadvantages, ZFNS have been quite promising ever since they came to be (Stoddard 2011). They have already found application in deliberate gene editing of multiple model organisms. The capability of being able to edit genes with precision presents a lot of potential uses in research activities and treatment purposes; therefore, ZFNs play an influential role in the context of genome editing (Urnov et al. 2010).

2.2.2.3 Transcription Activator-Like Effector Nucleases (TALENs)

TALENs, Transcription Activator-Like Effector Nucleases, form part of the latest research in the development of genome editing technology that aims to overcome the drawbacks of the preceding nucleases, such as meganucleases and zinc finger nucleases (ZFNs). They are complex proteins involving distinct FokI endonucleases engineered with DNA-binding domains of TALEs. TALE domains have tandem 34 amino acid repeats, and their variable amounts make up di-residues that determine nucleotide affinity (Huo et al. 2019).

TALENs are also more modular and base-specific compared to ZFNs; therefore, they enable broader sequence-specific targeting of the DNA genome in diverse organisms. They dimerize to form the double-strand break to achieve the genome editing knock-in/knock-out (Khalil 2020). In spite of their benefits, TALENs have impediments such as large proteins hindering their delivery through viral vectors and the complex design process, which can be discouraging to researchers not versed in molecular biology. Nevertheless, TALENs have a future since they express a high specificity, minimum off-target effect, and efficacy in editing cells. They have found success in plant genome engineering (Jansen et al. 2002; Khalil 2020).

2.2.2.4 CRISPR/Cas Systems

CRISPR/Cas is a revolutionary genome editing service and was presented as a third-generation instrument in 2013. First discovered in the 1980s in the genomes of bacteria and archaea as clustered regularly interspaced short palindromic repeats (CRISPR), it is an adaptive immune system against phage infection. Cas genes and CRISPR loci encode proteins that are essential to degradation and DNA specificity (Mojica et al. 2005).

In contrast to ZFNs and TALENs, where protein-DNA complementarity is engineered, CRISPR/Cas takes advantage of RNA to serve as a recognition component against DNA, conferring benefits, including simpler design of target and easier off-target prediction and editing many sites on a genome at once (multiplexing). CRISPR systems are grouped into Class 1 (multiple protein complex systems) and Class 2 (single-effectors with proteins, such as Cas9, Cas12, and Cas13) (Doudna and Charpentier 2014).

The extremely popular Cas9 nuclease is complemented by a single-guide RNA (sgRNA) to bind and cut the target DNA, with the specificity necessitating a protospacer adjacent motif (PAM). This is also relatively cheap and very efficient because the only entity that should be designed to target new ailments is the sgRNA (Vasebi and Khakvar 2014). CRISPR/Cas9 is simple and versatile, with a wide application in research and biotechnology, so despite the possible off-target effects, the system is used everywhere (Hille et al. 2018).

2.3 Artificial Intelligence Fundamentals and Their Role in Gene Editing

CRISPR-based genome editing (GED) has brought promising prospects into the study of genes and the improvement of agricultural products (Dixit et al. 2024; Abadi et al. 2017). AI systems aid in the development of guide RNAs (gRNAs) within CRISPR-Cas systems, such as DeepCRISPR and CRISTA, that predict the best gRNAs using genomic context and off-target implications (Abdar et al. 2023). Meanwhile, these models have improved base editing and prime editing, so it is possible to make precise changes without using sequences of donor DNA to provide templates (Adli 2018). Zinc-finger nucleases (ZFNs), TALENs, and CRISPR/Cas9 are among the highest technology GEDs, with CRISPR/Cas9 being the most preferred due to its convenience and flexibility. AI-based tools also enhance GEDs to perform more accurately, efficiently, and affordably by targeting as many diseases as possible (Abdar et al. 2021).

2.3.1 AI in Biological Data Analysis

The analytical capabilities of deep learning (DL) and machine learning (ML) have been intensively studied in the analysis of complex biological data, which would help build algorithms and trend prediction models in different fields of biology (Anzalone et al. 2019). Bioinformatics algorithms have employed Artificial Intelligence (AI) as a way of resolving biological problems by analyzing and interpreting information (Arbab et al. 2020). ML has been applied to simulating protein and gene behavior as well as analyzing the DNA-expression arrays, and DL has been used in sustainable agriculture and genomics. These technologies have played an important role in the diagnosis and forecasting of diseases based on the data set of symptoms and medical images, and help healthcare specialists (Bock et al. 2022).

The occurrences of AI development in the biomedical field also bring up thoughts on difficulties and the future. It is evident that revolutionary changes can occur in the sphere of agriculture (Yan et al. 2021). Food and Agriculture Organization, which is an international body working under the United Nations, forecasts that the world population will exceed more than 9 billion individuals by the year 2050, which is a pointer to the potentiality of the agricultural sector to meet the nutritional needs of this population. Economic development depends on agriculture because it consumes approximately 38% of the earth's surface (Callif et al. 2017). Most agricultural activities are only manual, and their automation would contribute greatly to their productivity as well as effectiveness (Nierode et al. 2016).

The use of AI in agriculture can alter the industry by changing how to better manage the soil, better water estimation, better fertilizer utilization, and crop yield forecasting (Tsubouchi et al. 2024). Real-time crop monitoring and harvesting are being improved by such technologies as drones and robots, and analyzed with the help of deep learning and computer vision (Vojnits et al. 2022). AI products are more powerful when it comes to efficiency via crop production, soil mapping, irrigation, and weeding (Lazzarotto et al. 2020).

2.3.1.1 AI and Next-Generation Gene-Editing Tools in Cereals

High-quality seeds have good germination and vigor, hence the resulting strong plant growth and development. Genetic engineering is transforming agriculture through the improved quality traits in seeds, such as increasing resistance to stress, nutritive value, and productivity (Sun et al. 2024). The gene-editing methods are CRISPR-Cas9 and transgenic. These inventions enable the production of seeds with a better nutrient content, disease resistance, and abiotic stresses such as drought and salinity (Ahmar et al. 2020). One of the examples of such projects is Bt cotton and herbicide-tolerant soybeans, which taught that genetically modified seeds may lead to higher yields and decreased use of chemicals. Nevertheless, there are moral, environmental, and socioeconomic concerns that it is important to take into consideration to meet the concerns of the people with a view to promoting responsible usage (Hallauer and Darrah 1985). The combination of genetic engineering with old breeding techniques allows the development of more resilient and stronger seed varieties. Genetic engineering has great promise in increasing seed attributes of nature such as vigor, nutritive content, and stress tolerance, as illustrated in Fig. 2.2 (Ashraf et al. 2024).

Such advancements will cause increased and even more consistent harvests to be grown, which will remove the food security and responsible farming issues. Breeding of rice is not easy, and it takes a long period, and there is also pressure to come up with more rice to feed the fast-growing global population (Seck et al. 2012). To overcome these barriers, a novel breeding technique, Rapid Generation Advance (RGA), has been implemented on inbred rice varieties, which consists of planting the higher-density breeding populations in a greenhouse to induce a phenotype of early flowering and seed filling, hence leading to a rapid generation advancement and economical cost savings (Wang et al. 2022). Furthermore, floral traits, including stigma exsertion, which is very important in the production of hybrid rice, can be enhanced by special breeding strategies. There is a massive development that has taken place in hybrid rice breeding technology in recent decades. It explained how hybrid rice is traditionally bred, modern breeding modes, recent breakthroughs, molecular markers and genomics techniques, and how new breeding techniques of hybrid rice were developed (Lenaerts et al. 2018). Stages driven by AI assist in the exploration of large genomics data by researchers to determine significant

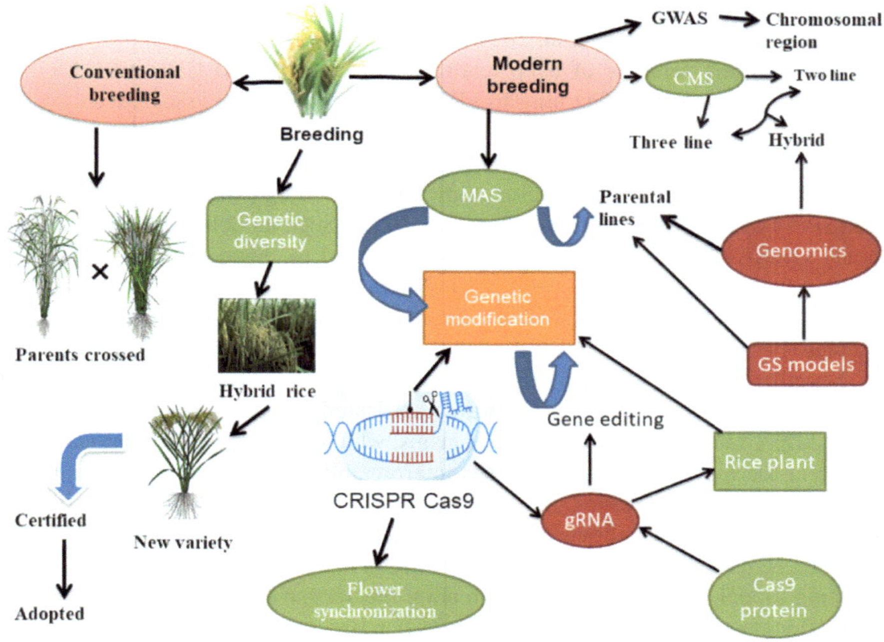

Fig. 2.2 Conventional and modern breeding strategies for hybrid rice production (Ashraf et al. 2024)

characteristics associated with such qualities as drought tolerance, pest resistance, and high yield. Such knowledge assists in the application of highly sophisticated gene-editing technologies such as CRISPR/Cas9 that can specifically modify the genetic makeup of cereal crops to improve their productivity and resistance (Wang et al. 2023a, b). Also, AI is used to design gene-editing experiments, predicting the effects of certain changes in terms of plant traits and streamlining the process of breeding and increasing its efficiency (Horgan and Crisol 2013). This mixture of AI and quality-modifying fast-forwarding together with high-throughput phenotyping and information analytics accelerates the development of next-generation cereals tailor-made to address the needs of a growing global population within the assortment of natural circumstances (Lin et al. 2020).

2.3.2 *DeepCRISPR*

The successful genome editing in CRISPR–Cas systems depends on accurate predictions of gRNA activity, bearing in mind on-target and off-target effects (Chuai et al. 2018). Neural networks, such as deep learning, play a role in modeling the intricate patterns in big data to tell gRNA activities as characterized by their sequences (Toufikuzzaman et al. 2024).

The previous versions of deep learning were attentive to CRISPR-Cas9 and applied CNNs and RNN learning to gRNA sequence-activity associations (Wang et al. 2019). One can predict the on-target and off-target activity using these models, thus the best gRNAs can be selected. The CRISPR–Cas12 system was adopted for deep learning and performed precise predictions according to the peculiarities of the CRISPR-Cas12 system (Konstantakos et al. 2022). Such models as CRISPR-Net or novel sgRNA-DNA sequence encoding can be used to enhance off-target activity prediction in CRISPR-Cas9 (Wang et al. 2020). Other experiments ameliorate the accuracy of quantifying off-target effects and high on-target activity in agronomic species (Lee 2023).

Nevertheless, it is still difficult to predict the editing effect of CRISPR–Cas9 editing properly, and this is also important to design a good working gRNA (Ding et al. 2025). The use of attention mechanisms, which emphasize the significant elements in the input sequence, increases interpretability. To refine Cas9 variants to more specific genome editing, new methods that combine high-throughput sequencing with deep learning validate and predict the off-target sites of variant Cas9, which can then be used to predict new off-target sites (Zhang et al. 2021). Apindel, a deep learning model with BiLSTM and Attention mechanisms, has better results in the prediction detail and accuracy (Liu et al. 2022a, b).

2.3.3 CRISTA

CRISTA has made a major improvement in an optimized sgRNA design by introducing and handling bulge events as well as data imbalance in off-target prediction. This average was biased since the amount of uncleaved site samples was so vast compared to cleaved sites that, in the long run, it could create a bias when conjoined (Hu et al. 2020). To address this, CRISTA used leave-one-sgRNA-out and leave-study-out testing using bootstrapping that undermined both majority and oversampled minority classes to generate a balanced training dataset.

The actual performance of a machine learning model is most appropriate to assess when using a full set of testing data, regardless of this balance (Omotezako et al. 2025). Thus, every testing experiment was kept in consideration. Using benchmark tests in the design of sgRNA, CRISTA, and DeepCRISPR showed better results supporting the idea of using learning-based solutions as the foundation of our models (Naha et al. 2024).

The problems of CRISPR analysis are the inconsistencies in experiments, the lack of information, and non-sequence characteristics. In response to them, we suggest a new sgRNA design strategy to make no distinction between on- and off-target activity (Desai et al. 2024). We have come up with two complementary tools, Navitas, which predicts guide RNA activity, and Optimus, which identifies optimal sgRNAs.

First, Navitas employed a linear regression where mean squared error (MSE) is used as a loss function to predict the frequency of cleavages after pooling together on- and off-target data. Nevertheless, it was inconsistent, especially with the minority range values. We improved Navitas by adding XGBoost, Random Forest, and Neural networks-feedForward to it, and used the model with the best prediction outcomes. Optimus used an Evolutionary Algorithm (EA) to define its optimization techniques, including the Navitas model in its fitness module (Crista et al. 2024).

2.3.4 DeepHF

The past years have been characterized by the evolution of machine learning, and especially bioinformatics, due to the supply of large datasets and the advancement of computational resources and deep neural networks. These networks have had success in forecasting the CRISPR/Cas9 on-target editing intensities. The utilization of deep learning for this purpose was first applied to Seq-deepCpf1, using the CRISPR high-throughput data (Wang et al. 2019). It was however outperformed by other models such as DeepHF and CRISPRon that although are high performance in terms of cross-validation (e.g., Spearman correlation of 0.87), encounter issues when dealing with endogenous and functional data sets (Spearman correlation <0.5) (Chen et al. 2020).

Researchers have identified one of the biggest disadvantages of deep learning models as their reliance on large quantities of data to be accurate. These models are subject to overfitting should lower volumes of data be used to train them. The solution of using transfer learning (TL) is one of the potential solutions, in which a model trained using target data is less significant in size when compared to its source data, and the initial model is trained (Fan et al. 2014). To work in an enhanced way and maximize prediction performance, source data have to be relevant to target data (Elkayam and Orenstein 2022).

The final method is the last-layer method, where, during fine-tuning, it is only the weights of the final hidden layer that are re-trained. This mechanism was applied in the DeepHF study to acquire performance improvements in the prediction of the functionality datasets (Xiao et al. 2024). It should be noted that while being well established in the realm of computer vision, the last-layer method might not be relevant in the context of bioinformatics due to reduced source datasets and less parameterized networks (OMdl et al. 2023).

2.4 Integration of AI with Gene-Editing Technologies

In the digital environment today, artificial intelligence is developing fast with the aspiration to create intelligent machines that seem human in their thought process (Abideen 2013). It includes machine learning (ML) applications, such as deep neural networks (DNN), artificial neural networks (ANN), support vector machines (SVM), and technologies, such as the Internet of Things (IoT) (Keen and Moore 2015). The potential of AI in agriculture should not be underestimated, as scientists are capable of comprehending plant behavior in a different climate setting with its help (Schaly et al. 2021).

The release of Summit, which can be considered the most powerful supercomputer in the world, with its 27,000 GPUs, represents a considerable advance in the use of AI in agriculture and in solving such problems as abiotic stresses and crop yield (Tarim et al. 2021). The next-generation applications of AI can combine high throughput and multi-omics data to improve crop performance predictive modeling (Lu et al. 2022).

A potential AI tool called machine learning will enable the automation of the learning process and increase the efficiency of the interpretation of data, which will subsequently enhance the performance of breeding and crop productivity (Shedden-Mora et al. 2020). The statistical analysis of multimodal data can be performed using such tools as MATLAB and Python, and helps plant biologists obtain patterns of stresses and expand the knowledge base of how plants interact with pathogens (Bagabir et al. 2022). This leads to a better understanding of agricultural data, which opens up the possibility of breeding methods that will be more successful (Selvakumar et al. 2022).

2.4.1 AI-Driven Guide RNA (gRNA) Design and Optimization

The success of GED depends on the gRNA choice of sequence. Some gRNA would be able to disrupt almost every target allele in a cell population, and some gRNA would have little to no observed activity (Grigorian Shamagian et al. 2019). Thus, an assortment of gRNA design tools is created, and they mostly use ML and DL algorithms to cope with this difficulty (Ewens and Farre-Mensa 2022). CRISPR/Cas technology consists of various elements, and the complexity grows in advanced editing methods, such as epigenome editing, where the selection of the effector and the appropriate effector combination varies according to specific needs (Fig. 2.3).

Recently, there has been ongoing progress in the development of off-target prediction algorithms. These algorithms are typically trained using data obtained from in vitro cleavage assays, which involve using immortalized cell lines (Hua et al. 2019). Immortalized cell lines provide a controlled environment for conducting experiments related to GED, which makes them a valuable resource for training predictive models. It has been observed that these off-target prediction algorithms tend to rely heavily on sequence-based information, which demonstrates a strong connection with the actual cleavage activity of CRISPR/Cas9 or similar gene-editing systems (Nedelea et al. 2018). Such physical properties of the genome as chromatin access and features of DNA methylation patterns are currently poorly used in AI models, but they offer insights into the architecture of how DNA is three-dimensionally sequenced in cells and can alter the accessibility of that DNA to gene

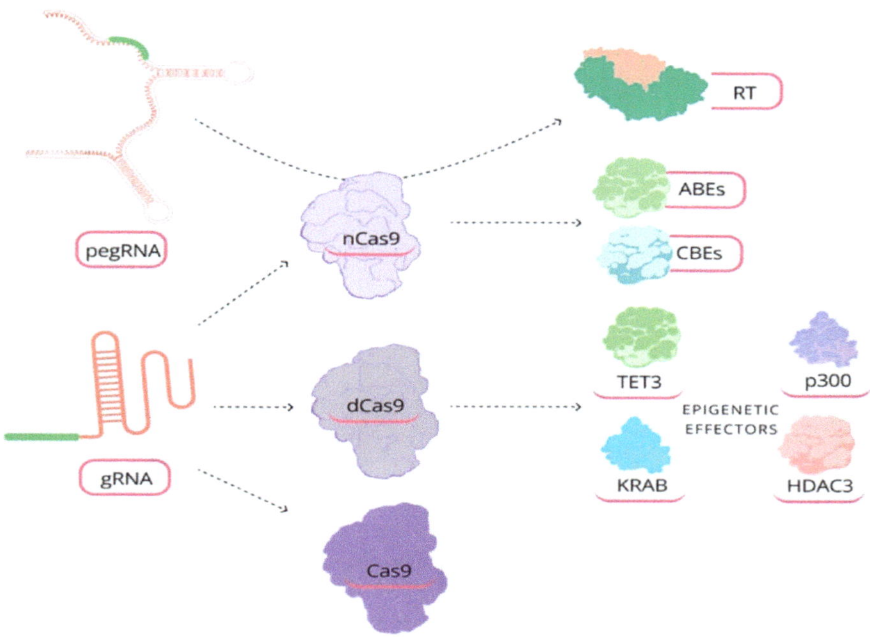

Fig. 2.3 Modular nature of CRISPR-based editing technologies (Dixit et al. 2024)

editing. Designed an approach called piCRISPR, which takes into account a mixture of both sequence and physically informed features, such as chromatin accessibility and DNA methylation, among others (Stalidzans and Dace 2021). By an in-depth evaluation based on a large record of CRISPR/Cas9 editing events, piCRISPR performed better than all other currently available prediction tools for off-target cutting activity. The CNN models presented the most successful benchmarks of performance, and CNN S5E2 scored the top summation of precision (AU—ROC = 0.998) (Chempak Kumar and Mubarak 2024). piCRISPR has the potency to identify new off-target cutting locales and prime the formation of CRISPR/Cas9 experiments with diminished perspective of off-target cutting (Abd El-Aziz et al. 2024).

2.4.2 *AI in Predicting Editing Outcomes and Repair Pathways*

DNA double-strand breaks are, primarily, repaired by one of two pathways: non-homologous end-joining (NHEJ) and homology-directed repair (HDR). In addition to this, there is evidence for microhomology-mediated end-joining (MMEJ). HDR is complex and precise, can only occur in the G2 and S phase of the cells, and

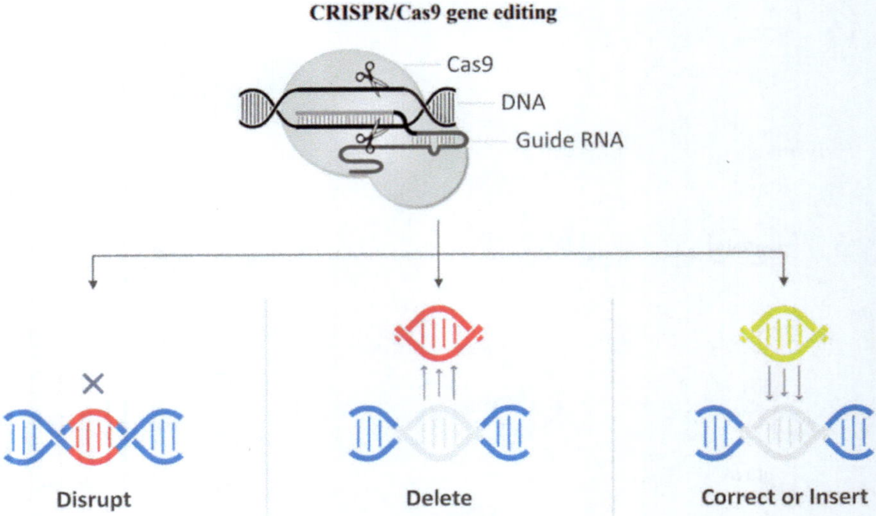

Fig. 2.4 Mechanisms of CRISPR gene editing (Salomonsson and Clelland 2024)

repairs double-strand breaks (DSBs), using homologous template sequences (Fig. 2.4) (Zhang et al. 2014).

The CRISPR–Cas9 procedure entails generating a double-strand break (DSB) in the DNA that, in turn, is fixed by the cells through a non-homologous end-joining (NHEJ) repair process, which is precise but inaccurate and tends to create indels (Lemos et al. 2018). As the other process, microhomology-mediated end joining (MMEJ) does exploit small segments of microhomology, although it, too, is imprecise, often adding deletions or changing the positions of genes (Allen et al. 2019). Details of which repair activity NHEJ and MMEJ will use depend on the sequence properties of the target DNA (Zhang et al. 2018).

Interestingly, most of the template-free CRISPR/Cas9 editing variants have proven to be an intricate bioinformatics task to predict, as they differ across cell lines and mutated Cas9. The developments in this area have been enabled through machine learning (Lemos et al. 2018). Different tools based on machine learning have been established to predict these outcomes. The first is the training of deep neural networks and k-nearest Neighbor algorithms to predict deletions of microhomology (MH), MH-less deletions, and 1 bp insertions (Zhang et al. 2018). Regarding another mode, Gradient Boosting Decision Trees are used with spacer sequences and PAM as inputs to estimate several statistics, such as average deletion length (Zhang et al. 2018).

Based on Convolutional Neural Networks (CNN) and Neural Architecture Search (NAS), CROTON is a deep learning platform that predicts the likelihood of deletions and insertions as well as the frequency of frameshifting using DNA sequence data (Hu et al. 2019). FORECAST produces candidate mutations and predicts whether a repair is going to be successful or not with a multi-class Logistic

Regression model (Liu et al. 2022a, b). Some models have binary features defining the target sequences and predicting with Logistic Regression the ratios of insertions to deletions, as well as allocation of deletions and insertions. Nevertheless, the explicit nature of CRISPR-based gene editing is unknown yet, and synthetically made properties or special one-hot encoding can impair the outcomes of prediction (Wang et al. 2023a, b).

2.4.3 AI for Next-Generation Editors

The phenotyping of plant traits has played a crucial role in the domestication of crops, and this involves the total phenotypic character or phenome of a plant, which represents genomic expression under environmental conditions (Chia et al. 2025; Dai et al. 2024). Plant phenomics entails the use of sophisticated technology that collects and processes profound data about plant development, architecture, and behavior (Spanjol and Noble 2023). The domain is key with respect to plant-environment relationships and the production of elite crop types. Computer vision and computer analysis allow high-throughput phenomics to make biological response predictions, leading to digital agricultural practices that use AI and ML for crop management (Nye 2023; Gill et al. 2022).

2.4.3.1 Base Editing and AI

Base editing is a highly specific genome-editing technology that allows highly specific changes in individual nucleotides without generating double-stranded breaks. It has a modular design that enables it to have different configurations, thus providing flexibility in its application. The most recent developments in ML and DL have been oriented at enhancing base-editing outcomes (Petrušić et al. 2024).

Among them, consider a development, BE-Hive ML model, which uses a deep conditional autoregressive model to predict tactical sequences and performance in disease-related single-nucleotide variants (SNHs) (Yang and Chen 2023) at a level exceeding 90%. Another ML model was found to generalize better across datasets (the range of precision was 0.49 and 0.72 with various base editors) (Li et al. 2024). One of the important discoveries is that the efficacy of editing is also significantly determined by surrounding nucleotides, which confuses estimations of efficiency and possible bystander mutations.

To overcome these difficulties, a DL model was trained to consider the information not only of the sequence but also of the target nucleotide position, and it reported superior performance compared to existing ones with a reported Pearson correlation of 0.50 to 0.95 on adenine and cytosine base editors (Zhang et al. 2023). Furthermore, BE–DICT, a DL model based on an attention algorithm, was used to make editing probabilities of diverse target nucleotides and demonstrated great AUC values of different base editors.

Prediction tools of efficiencies of adenine and cytosine base editors are also possible, such as DeepABE and DeepCBE, respectively. Additionally, models such as DeepCas9variants and DeepBE have been established to predict the performing functions of particular Cas9 variants and base editors, showing high correlation coefficients (Wang et al. 2023a, b).

2.4.3.2 Prime Editing and AI

Prime editing is an emerging technique that utilizes reverse transcription to insert programmed sequence modifications into DNA sequences. It is an adaptable GED tool, capable of making a wide range of genetic changes, but achieving high editing efficiency and product purity necessitates PED guide RNA (pegRNA) experimental optimization. It comprises three key components: a reverse transcriptase, a pegRNA, and a Cas9 nickase. The pegRNA has the edit sequence as well as the target sequence to program the intended change in the DNA (Rottner et al. 2025). Designing the pegRNAs is more difficult following the generation of guides in other CRISPR-based editing techniques. Luckily, there are tools like Easy-Prime and Prime Design to aid in this complicated process of design. Easy-Prime has been trained on PED datasets that were published earlier. Easy-Prime uses the familiar and newly identified characteristics and predicted secondary structure and RNA folding to rank pegRNA candidates and predict their editing efficiency. They have shown optimization of prime editor guides for correcting mutations in 89.5% of the 152,351 Genome-Wide Association Studies (GWAS) variants (Antoniades and Chan 2025). Easy-Prime can also generate optimum pegRNAs for many genetic variations associated with diseases. PrimeDesign is another tool for pegRNA design, designed for a number of different editing tasks, containing single nucleotide substitutions, additions, deletions, and inversions (Yu et al. 2023).

2.4.3.3 Epigenome Editing and AI

Epi-gene editing (epi-GED) allows localized editing of gene control without DNA sequence modification and modifies DNA methylation, histone modifications, and RNA editing (Tran and Han 2024). The method can be used in treating diseases, functional genomics, and stem cell therapies. CRISPR/Cas-based epi-GED applies particular DNA motives exhibiting CRISPR/Cas nucleases fused to epigenetic modifiers.

Epigenomic data are increasingly analyzed using machine learning (ML) and deep learning (DL) techniques to facilitate an improved understanding of the processes of epigenetics. The DL tool EpiCas-DL identifies the sgRNA activity of CRISPR-mediated epi-GED with the incorporation of relative gene expression, methylation, and chromatin accessibility characteristics with an AUC of 0.87 (Levy and Ruohola-Baker 2023). It is promising, but AI is less well-studied about the epi-GED tool than the traditional editing approaches because of the dynamic character of epigenomic data (Akatsuka et al. 2020).

2.5 Applications of AI-Enhanced Gene Editing in Agriculture

The combination of artificial intelligence (AI) and genetics is transforming crop breeding and improving the features of plants to produce higher agricultural yields and nourishment. The promising advancements in genetic empowerment of crops are catalyzed by AI-based models, genomic data, and such technologies as CRISPR–Cas9 (Patil 2023; Kumar et al. 2025). AI is of great importance in accurate forecasting and alteration of crop characteristics, increasing disease tolerance, and harvests. The recent crop breeding is tedious, but due to AI, there is genomic selection and trait prediction. AI algorithms use genetic markers in large genomic data sets to build predictive models that evaluate a plant at its early stages of development faster and with the elimination of time-consuming field trials and repetitive selection testing (Pawar et al. 2023). AI also integrates environmental and historical data to increase predictions of crop performance. The technology both hastens breeding and boosts genotype-to-phenotype profiling and contributes to the creation of resilient crops and eventually resolving food security, hunger, malnutrition, and challenges of climate change (Ampatzidis and Vougioukas 2009; Patil 2023).

2.5.1 Crop Yield and Quality Improvement

The creation of new varieties of crops with interesting characteristics normally took a long time to complete (years or even decades), the selection, crossing, and testing indexes. But, with the incorporation of artificial intelligence (AI) in crop breeding, this science is transforming through the capacity of artificial intelligence (AI) to shortcut invasions and produce high-yield and high-quality crops in a short period (Ma et al. 2024; Chen et al. 2025).

2.5.1.1 Selection Based on Data

Algorithms are fundamentally changing the breeding of plants, making it easier to analyze the large amount of genomic information to develop genetic markers in desirable traits such as yield, nutritional value, and the ability to fight diseases (Ma et al. 2024). The field can be further improved with predictive modeling and machine learning, as it enables breeders to predict plant reactions to any condition and avoid excessive and expensive field trials (Chen et al. 2025).

2.5.1.2 Efficient Trait Scanning

The breeders can also scale in their recording of various phenotypic traits, albeit within a reasonable amount of time, with the phenotyping tricks that are enabled by AI (drones, remote sensing, computer vision). The automatic process of recording the data of the traits is objective and bias-free, and the results are consistent and repeatable across the generations of breeding (Patil 2023). This reduces the time of selecting the plants with the best attributes to give to their further reproduction.

2.5.1.3 Creating a Simulated Breeding Facility

AI allows simulation of the virtual breeding environment in which thousands of breeding scenarios may be modeled and tested. The nature of these simulated environments makes this kind of breeding process cost less in terms of physical resources and is environmentally friendly (Chen et al. 2025). Virtual breeding also assists in estimating the performance of certain combinations of traits under various climatic and soil conditions.

2.5.2 Disease and Pest Resistance

Genome editing brings original approaches to pest and disease resistance, supplementing the existing strategies based on traditional breeding and transgenic plants (Sun et al. 2017). In contrast to the traditional approaches, both plants and insect pests can be accurately managed by genome editing. Editing can result in the desired mutations in pests to reduce the destruction. Knocking out cadherin receptors in Helicoverpa armigera using CRISPR/Cas9 cuts resistance to Bt toxins as an example. In the same way, the insect susceptibility to plant defenses has been enhanced by targeting midgut receptors and other detoxification genes (CYP6AE cluster) (Sun et al. 2017).

Another method that is promising is disruption of the insect communication. The altered sense of odors by olfactory gene mutations (Or83b in Drosophila or Orco in Spodoptera litura) causes the hosts and mates to be detected more poorly and results in lower pest success (Larsson et al. 2004; Sun et al. 2017). Moreover, CRISPR-edited odorant receptor genes such as OR16 in every H. armigera male confuse them into searching for the mature females and hence produce sterile eggs (Sun et al. 2017).

The abd-A expression has been targeted in pests to create deformations and lethality in *Spodoptera* species and *Plutella xylostella* (McFarlane et al. 2018). These gene-editing tools have a big potential for sustainable targeted pest control and cutting down dependence on chemical pesticides.

2.5.3 Stress Tolerance and Environmental Adaptation

Utilization of gene editing and artificial intelligence (AI) technologies provides revolutionary alternatives to enhancing crop resistance to environmental adversities. One of the most prominent is the improvement of wheat's ability to tolerate stresses using AI-mediated gene editing. The workers applied AI to massive wheat genomic data sets, and they performed machine learning to discover essential genes related to drought, heat, and saline–alkali resistance (Chen et al. 2025; Patil 2023). This data-related methodology allowed the selection of target genes in CRISPR–Cas9 editing with precision (Cao et al. 2021).

Copy-paste gene editing involved the usage of CRISPR–Cas9 to edit genes related to osmotic regulation and antioxidant pathways, increasing wheat's capacity to conserve water and limit the damage of cells caused by heat. It was revealed in the laboratory trial that the physiological traits were alleviated, including restoration of the water balance and drought-induced photosynthesis, as well as enhanced antioxidant response during heat stress. The field tests also indicated that the gene-modification technology of wheat was better and more resistant to the unfavorable conditions of the environment as compared to control by their survival and better yielding (Chen et al. 2025). These results portray how AI and gene editing can address agricultural challenges through climate change. Further development of these technologies would probably bring varieties of wheat even more resistant to stress and help support the idea of global food security and sustainability of agriculture (Kumar et al. 2025).

2.6 Challenges and Ethical Considerations

The use of gene-editing technologies, especially CRISPR–Cas9, in agriculture raises many challenges and ethical questions that need to be addressed to ensure responsible adoption. The regulatory environment is one of the great challenges. The levels of control over genetically modified organisms (GMOs) vary in different countries, and the process of maneuvering through the systems may prove to be embroiled (Niemiec and Howard 2020; Singh 2021). Disunity might limit innovation and deter the adoption of technologies that can be used to create positive developments in the field of agriculture.

The other issue that concerns me is the acceptance by the people. Several consumers still fear GMOs because of the safety, environmental, and ethical concerns. Communication and openness regarding the advantages and dangers of gene editing are key to gaining popular confidence (Karagyaur et al. 2019). There is a possibility of unforeseen effects that pose ethically whether our intervention on ecosystems has longer impacts due to gene editing. As an example, releasing genetically modified crops may upset the local biodiversity or spawn resistant pests. Therefore, risk measures strongly need to be assessed and monitored (Karunarathna et al. 2024).

There is also concern about equity, especially in access to gene-editing technologies. The African continent, smallholder farmers, and less developed nations can be unable to afford these innovations and enhance the gap between the rich agricultural sector and the poor one. These ethical issues and challenges are crucial to the ethical and sustainable use of gene editing in agriculture to make sure that emerging technologies play a positive role in food security and environmental sustainability (Dixit et al. 2024).

2.7 Future Perspectives and Emerging Trends

2.7.1 Integration of AI with Multi-Omics Data

The omics disciplines offer a comprehensive method of biological systems study, especially by combining genomics, transcriptomics, proteomics, and metabolomics under the thematic-based term of multi-omics (Lin and Lane 2017). Genomics deals with a complete set of genomes in an organism, involving its structure, functioning, and evolution. It facilitates the discernment of complicated hereditary attributes and their effect on vegetative nature, supplementing part assorted sleeve assists in advancement (Yang et al. 2021). WGS helps to discover new genes and regulatory elements, whereas the most effective technique in identifying single-nucleotide polymorphisms (SNPs) and small insertions and deletions (INDELs) is genotyping-by-sequencing (Morganti et al. 2019).

Transcriptomics, especially by using RNA sequencing (RNA-seq), entails an investigation of the entire accumulated RNA conversion in the genome in a certain condition. The field throws light into patterns of gene expression, and researchers can identify which genes are expressed in different stages of development or certain environments. As an example, RNA-seq may be used to explain the transcriptomic alterations in fruit maturation or plant responses to disease in horticulture (Chen et al. 2019). In sum, transcriptomics forms an essential intermediary between the genome and phenotype, and most importantly, it will strongly affect our insights into the molecular processes in plants and become a key source of information on how crops can be improved (Lee et al. 2023).

2.7.2 AI-Driven Automation in Gene-Editing Labs

The suggested model will combine multi-omics artificial intelligence (AI) and machine learning (ML) methods to improve understanding of plant phenotypes and plant predictive modeling in horticultural studies. It is started by a thorough multi-level data collection of various layers of the omics gene, transcriptome, proteome, and metabolome data, with phenotypic data collected by high-throughput methods like the features of imaging and sensors to get the environmental data (Cheng et al. 2019). Then, the gathered data is normalized and preprocessed to provide its similarity through quality control, normalization, and extraction of features. Next, the analysis of data collected at different layers of the omics is performed by multivariate statistical approaches, as well as network-based approaches. The machine learning algorithms are then used to develop the model of predictability, and we use supervised/unsupervised learning or reinforcement learning. The models are tested, and these are insightful into biology. Lastly, they are used in feeding horticultural practices, including fertilization approaches, disease identification, and crop production forecasts (Pitt 2021).

2.8 Conclusion

The combination of synthetic methods and future-era, quality-modifying tools such as CRISPR–Cas systems is transitioning the landscape of cutting-edge agriculture. AI accelerates the discovery and research of hereditary traits, which allows for more precise and efficient progression processes. It is used together with gene editing to produce crops that yield better, are pest-resistant, and are resistant to climate change, which are global concerns such as food security and sustainability. As these improvements continue to make headway, their skills and ethical usage will be fundamental in ensuring that the future experiences modernized rural growth that satisfies the emerging needs of the increasingly rising global population, ensuring conservation of natural sensitivity.

References

Abadi S, Yan WX, Amar D, Mayrose I (2017) A machine learning approach for predicting CRISPR-Cas9 cleavage efficiencies and patterns underlying its mechanism of action. PLoS Comput Biol 13(10):e1005807

Abd El-Aziz AA, Mahmood MA, Abd El-Ghany S (2024) A robust EfficientNetV2-S classifier for predicting acute lymphoblastic leukemia based on cross validation. Symmetry 17(1):24

Abdar M, Samami M, Mahmoodabad SD, Doan T, Mazoure B, Hashemifesharaki R et al (2021) Uncertainty quantification in skin cancer classification using three-way decision-based Bayesian deep learning. Comput Biol Med 135:104418

Abdar M, Salari S, Qahremani S, Lam HK, Karray F, Hussain S et al (2023) UncertaintyFuseNet: robust uncertainty-aware hierarchical feature fusion model with ensemble Monte Carlo dropout for COVID-19 detection. Inf Fusion 90:364–381

Abideen AS (2013) Constraints in the application of biotechnology to cassava production in Nigeria. Int Lett Nat Sci 4:26

Adli M (2018) The CRISPR tool kit for genome editing and beyond. Nat Commun 9:1911

Ahmad F (2025) Historical perspectives of cutting-edge genome editing techniques. In: Genome editing for crop improvement: theory and methodology. CABI, GB, pp 1–16

Ahmar S, Gill RA, Jung K-H, Faheem A, Qasim MU, Mubeen M, Zhou W (2020) Conventional and molecular techniques from simple breeding to speed breeding in crop plants: recent advances and future outlook. Int J Mol Sci 21(7):2590. https://doi.org/10.3390/ijms21072590

Akatsuka K, Sasaki-Honda M, Sawai T (2020) Ethical framework for next-generation genome and epigenome editing. Am J Bioeth 20(8):32–36

Allen F, Crepaldi L, Alsinet C, Strong AJ, Kleshchevnikov V, De Angeli P et al (2019) Predicting the mutations generated by repair of Cas9-induced double-strand breaks. Nat Biotechnol 37(1):64–72

Ampatzidis YG, Vougioukas SG (2009) Field experiments for evaluating the incorporation of RFID and barcode registration and digital weighing technologies in manual fruit harvesting. Comput Electron Agric 66(2):166–172

Antoniades C, Chan K (2025) Using artificial intelligence to spot heart failure from ECGs: is it prime time? Eur Heart J 46:ehae906

Anzalone AV, Randolph PB, Davis JR, Sousa AA, Koblan LW, Levy JM et al (2019) Search-and-replace genome editing without double-strand breaks or donor DNA. Nature 576(7785):149–157

Arbab M, Shen MW, Mok B, Wilson C, Matuszek Ż, Cassa CA, Liu DR (2020) Determinants of base editing outcomes from target library analysis and machine learning. Cell 182(2):463–480

Ashraf H, Ghouri F, Baloch FS, Nadeem MA, Fu X, Shahid MQ (2024) Hybrid rice production: a worldwide review of floral traits and breeding technology, with special emphasis on China. Plants 13(5):578. https://doi.org/10.3390/plants13050578

Bagabir SA, Ibrahim NK, Bagabir HA, Ateeq RH (2022) Covid-19 and artificial intelligence: genome sequencing, drug development and vaccine discovery. J Infect Public Health 15(2):289–296

Bibikova M, Carroll D, Segal DJ, Trautman JK, Smith J, Kim YG, Chandrasegaran S (2001) Stimulation of homologous recombination through targeted cleavage by chimeric nucleases. Mol Cell Biol 21(1):289–297

Bibikova M, Golic M, Golic KG, Carroll D (2002) Targeted chromosomal cleavage and mutagenesis in drosophila using zinc-finger nucleases. Genetics 161(3):1169–1175

Bock C, Datlinger P, Chardon F, Coelho MA, Dong MB, Lawson KA et al (2022) High-content CRISPR screening. Nat Rev Methods Primers 2(1):1–23

Callif BL, Maunze B, Krueger NL, Simpson MT, Blackmore MG (2017) The application of CRISPR technology to high content screening in primary neurons. Mol Cell Neurosci 80:170–179

Cao Q, Shi Z, Zhang G et al (2021) Progress in the application of CRISPR/Cas9 technology in wheat breeding. Adv Biotechnol 11(6):661–667

Carroll D (2011) Genome engineering with zinc-finger nucleases. Genetics 188(4):773–782

Chempak Kumar A, Mubarak DMN (2024) Ensembled CNN with artificial bee colony optimization method for esophageal cancer stage classification using SVM classifier. J Xray Sci Technol 32(1):31–51

Chen G, Ning B, Shi T (2019) Single-cell RNA-seq technologies and related computational data analysis. Front Genet 10:317

Chen Y, Zhang L, Wang H, Weinan E (2020) Ground state energy functional with Hartree–Fock efficiency and chemical accuracy. J Phys Chem A 124(35):7155–7165

Chen X, Cui X, Ma H (2025) The application of artificial intelligence and gene editing technologies in high-yield crop breeding: innovations and prospects for future agriculture. Adv Resour Res 5(1):146–166

Cheng JC, Chen W, Chen K, Wang Q (2020) Data-driven predictive maintenance planning framework for MEP components based on BIM and IoT using machine learning algorithms. Autom Constr 112:103087

Chevalier BS, Stoddard BL (2001) Homing endonucleases: structural and functional insight into the catalysts of intron/intein mobility. Nucleic Acids Res 29(18):3757–3774

Chia BS, Seah YFS, Wang B, Shen K, Srivastava D, Chew WL (2025) Engineering a new generation of gene editors: integrating synthetic biology and AI innovations. ACS Synth Biol 14(3):636–647

Chuai G, Ma H, Yan J, Chen M, Hong N, Xue D et al (2018) DeepCRISPR: optimized CRISPR guide RNA design by deep learning. Genome Biol 19:1–18
Crista L, Radulov I, Crista F, Imbrea F, Manea DN, Boldea M et al (2024) Utilizing principal component analysis to assess the effects of complex foliar fertilizers regarding maize (Zea mays L.) productivity. Agriculture 14(8):1428
Dai M, Li X, Zhang Q, Liang T, Huang X, Fu Q (2024) Health research in the era of artificial intelligence: advances in gene-editing study. Med Plus 1:100027
Desai PS, Parmar BB, Desai FP, Patel AM (2024) Caesalpinia Crista (Kanchaki) as green corrosion inhibitor for zinc in hydrochloric acid solutions. Chem Afr 7(4):2173–2187
Ding S, Zheng J, Jia C (2025) DeepMEns: an ensemble model for predicting sgRNA on-target activity based on multiple features. Brief Funct Genomics 24:elae043
Dixit S, Kumar A, Srinivasan K, Vincent PDR, Ramu Krishnan N (2024) Advancing genome editing with artificial intelligence: opportunities, challenges, and future directions. Front Bioeng Biotechnol 11:1335901
Doudna JA, Charpentier E (2014) The new frontier of genome engineering with CRISPR-Cas9. Science 346(6213):1258096
Elkayam S, Orenstein Y (2022) DeepCRISTL: deep transfer learning to predict CRISPR/Cas9 functional and endogenous on-target editing efficiency. Bioinformatics 38:i161
Ewens M, Farre-Mensa J (2022) Private or public equity? The evolving entrepreneurial finance landscape. Annu Rev Financ Econ 14(1):271–293
Fan Q, Feng H, Sun W, Xie Y, Wu CH, Allen WD, Schaefer HF III (2014) The Li··· HF van der Waals minimum and the barrier to the deep HF–Li potential well. Mol Phys 112(5–6):770–773
Gaj T, Sirk SJ, Shui SL, Liu J (2016) Genome-editing technologies: principles and applications. Cold Spring Harb Perspect Biol 8(12):a023754
Gill SS, Xu M, Ottaviani C, Patros P, Bahsoon R, Shaghaghi A et al (2022) AI for next generation computing: emerging trends and future directions. Internet of Things 19:100514
Grigorian Shamagian L, Madonna R, Taylor D, Climent AM, Prosper F, Bras-Rosario L et al (2019) Perspectives on directions and priorities for future preclinical studies in regenerative medicine. Circ Res 124(6):938–951
Hallauer AR, Darrah LL (1985) Compendium of recurrent selection methods and their application. Crit Rev Plant Sci 3(1):1–33. https://doi.org/10.1080/07352688509382202
Hille F, Richter H, Wong SP, Bratovič M, Ressel S, Charpentier E (2018) The biology of CRISPR-Cas: backward and forward. Cell 172(6):1239–1259
Horgan FG, Crisol E (2013) Hybrid rice and insect herbivores in Asia. Entomol Exp Appl 148(1):1–19. https://doi.org/10.1111/eea.12080
Hu H, Xiao A, Zhang S, Li Y, Shi X, Jiang T et al (2019) DeepHINT: understanding HIV-1 integration via deep learning with attention. Bioinformatics 35(10):1660–1667
Hu C, Shu L, Huang X, Yu J, Li L, Gong L et al (2020) OPA1 and MICOS regulate mitochondrial crista dynamics and formation. Cell Death Dis 11(10):940
Hua K, Zhang J, Botella JR, Ma C, Kong F, Liu B, Zhu JK (2019) Perspectives on the application of genome-editing technologies in crop breeding. Mol Plant 12(8):1047–1059
Huo Z, Tu J, Xu A, Li Y, Wang D, Liu M et al (2019) Generation of a heterozygous p53 R249S mutant human embryonic stem cell line by TALEN-mediated genome editing. Stem Cell Res 34:101360
Jansen R, Embden JDV, Gaastra W, Schouls LM (2002) Identification of genes that are associated with DNA repeats in prokaryotes. Mol Microbiol 43(6):1565–1575
Kale RR, Shingote PR, Wasule DL, Mirajkar SJ, Rathod DR, Moharil MP (2021) Historical developments of genome editing in plants. In: Genome editing in plants. CRC Press, pp 1–11
Karagyaur MN, Efimenko AY, Makarevich PI, Vasiluev PA, Akopyan ZA, Bryzgalina EV, Tkachuk VA (2019) Ethical and legal aspects of using genome editing technologies in medicine. Современные технологии в медицине 11(3 (eng)):117–132
Karunarathna I, Hapuarachchi T, Ekanayake U, Gunathilake S (2024) The ethics of genetic editing: navigating the future of science. Proceedings of the Uva Clinical Research

Keen JC, Moore HM (2015) The genotype-tissue expression (GTEx) project: linking clinical data with molecular analysis to advance personalized medicine. J Pers Med 5(1):22–29
Khalil AM (2020) The genome editing revolution. J Genet Eng Biotechnol 18(1):68
Khan SH (2019) Genome-editing technologies: concept, pros, and cons of various genome-editing techniques and bioethical concerns for clinical application. Mol Ther Nucleic Acids 16:326–334
Kim YG, Cha J, Chandrasegaran S (1996) Hybrid restriction enzymes: zinc finger fusions to Fok I cleavage domain. Proc Natl Acad Sci 93(3):1156–1160
Konstantakos V, Nentidis A, Krithara A, Paliouras G (2022) CRISPR–Cas9 gRNA efficiency prediction: an overview of predictive tools and the role of deep learning. Nucleic Acids Res 50(7):3616–3637
Kumar K, Sharma R, Ranjan K, Pandey DK (2025) Artificial intelligence in biotechnology. In: Artificial intelligence and biological sciences. CRC Press, pp 192–218
Larsson MC, Domingos AI, Jones WD, Chiappe ME, Amrein H, Vosshall LB (2004) Or83b encodes a broadly expressed odorant receptor essential for Drosophila olfaction. Neuron 43(5):703–714
Lazzarotto CR, Malinin NL, Li Y, Zhang R, Yang Y, Lee G et al (2020) CHANGE-seq reveals genetic and epigenetic effects on CRISPR–Cas9 genome-wide activity. Nat Biotechnol 38(11):1317–1327
Lee M (2023) Deep learning in CRISPR-Cas systems: a review of recent studies. Front Bioeng Biotechnol 11:1226182
Lee JS, Jang HK, Kang WH, Yeom SI (2023) Transcriptome analysis of pepper-Phytophthora infestans interaction based on a pipeline of a simplified and effective RNA-seq analysis (PoRAS). Hortic Sci Technol 41:100
Lemos BR, Kaplan AC, Bae JE, Ferrazzoli AE, Kuo J, Anand RP et al (2018) CRISPR/Cas9 cleavages in budding yeast reveal templated insertions and strand-specific insertion/deletion profiles. Proc Natl Acad Sci 115(9):E2040–E2047
Lenaerts B, Collard BCY, Demont M (2018) Global survey of rice breeders to investigate characteristics and willingness to adopt alternative breeding methods. Agric Food Secur 7(1):40. https://doi.org/10.1186/s40066-018-0191-3
Levy S, Ruohola-Baker H (2023) AI-driven designed protein epigenomics. Clin Res Oncol 1(1):1
Li B, Zhu X, Zhao D, Li Y, Yang Y, Li J et al (2024) igRNA prediction and selection AI models (igRNA-PS) for bystander-less ABE Base editing. J Mol Biol 436(18):168714
Lin E, Lane HY (2017) Machine learning and systems genomics approaches for multi-omics data. Biomark Res 5:1–6
Lin Z, Qin P, Zhang X, Fu C, Deng H, Fu X, Huang Z, Jiang S, Li C, Tang X, Wang X, He G, Yang Y, He H, Deng XW (2020) Divergent selection and genetic introgression shape the genome landscape of heterosis in hybrid rice. Proc Natl Acad Sci 117(9):4623–4631. https://doi.org/10.1073/pnas.1919086117
Liu G, Lin Q, Jin S, Gao C (2022a) The CRISPR-Cas toolbox and gene editing technologies. Mol Cell 82(2):333–347
Liu X, Wang S, Ai D (2022b) Predicting CRISPR/Cas9 repair outcomes by attention-based deep learning framework. Cells 11(11):1847
Lu J, Ding J, Liu Z et al (2022) Retrospective analysis of the preparation and application of immunotherapy in cancer treatment. Int J Oncol 60(2):1–23
Maeder ML, Gersbach CA (2016) Genome-editing technologies for gene and cell therapy. Mol Ther 24(3):430–446
Mandip KC, Steer CJ (2019) A new era of gene editing for the treatment of human diseases. Swiss Med Wkly 149(0304):w20021
McFarlane GR, Whitelaw CBA, Lillico SG (2018) CRISPR-based gene drives for pest control. Trends Biotechnol 36(2):130–133
Mojica FJ, Díez-Villaseñor CS, García-Martínez J, Soria E (2005) Intervening sequences of regularly spaced prokaryotic repeats derive from foreign genetic elements. J Mol Evol 60:174–182

Morganti S, Tarantino P, Ferraro E, D'Amico P, Duso BA, Curigliano G (2019) Next generation sequencing (NGS): a revolutionary technology in pharmacogenomics and personalized medicine in cancer. In: Translational research and onco-omics applications in the era of cancer personal genomics, pp 9–30

Naha R, Strohm R, Schaumkessel Y, Urbach J, Wittig I, Reichert AS et al (2024) SLP2 and MIC13 synergistically coordinate MICOS assembly and crista junction formation. iScience 27(12)

Nedelea AM, Mironiuc M, Huian MC, Bîrsan M, Bedrule-Grigoruţă MV (2018) Modeled interdependencies between intellectual capital, circular economy and economic growth in the context of bioeconomy. Amfiteatru Economic 20(49):616–630

Niemiec E, Howard HC (2020) Ethical issues related to research on genome editing in human embryos. Comput Struct Biotechnol J 18:887–896

Nierode G, Kwon PS, Dordick JS, Kwon SJ (2016) Cell-based assay design for high-content screening of drug candidates. J Microbiol Biotechnol 26(2):213

Nye BENJAMIN (2023) Authoring by editing and revising: considering generative AI tools. In: Critical thinking and ethics in the age of generative AI in education. A critical look into the future of learning, pp 34–41

OMdl T, Vázquez JC, Redolar-Ripoll D (2023) Improvement of social cognition in a down's syndrome patient after bilateral PFC deep HF-rTMS: a clinical case report

Omotezako T, Rodrigues M, Eleanor N, Wang Y, Wee G, Chong B, Matsubara A (2025) Subvisible microscale texture is present on the crista cutis of the skin and interacts with incident light to create a soft and radiant 'shitsukan'appearance. Int J Cosmet Sci 47:691

Patil A (2023) Use of artificial intelligence to hasten progress in plant genetics. Intelligence (AI) 1:3

Pawar V, Patil A, Tamboli F, Gaikwad D, Mali D, Shinde A (2023) Harnessing the power of AI in pharmacokinetics and pharmacodynamics: a comprehensive review. Int J Pharma Qual Assur 14(2):426–439

Petrušić I, Ha WS, Labastida-Ramirez A, Messina R, Onan D, Tana C, Wang W (2024) Influence of next-generation artificial intelligence on headache research, diagnosis and treatment: the junior editorial board members' vision–part 1. J Headache Pain 25(1):151

Pitt H (2021) What knowledge is required to grow food? A framework for understanding horticulture's skills 'crisis'. J Rural Stud 85:59–67

Redondo P, Prieto J, Munoz IG, Alibés A, Stricher F, Serrano L et al (2008) Molecular basis of xeroderma pigmentosum group C DNA recognition by engineered meganucleases. Nature 456(7218):107–111

Rottner AK, Lundin A, Li S, Firth M, Maresca M, Sienski G (2025) Optimized prime editing of the Alzheimer's disease-associated APOE4 mutation. Stem Cell Rep 20(1):102372

Salomonsson SE, Clelland CD (2024) Building CRISPR gene therapies for the central nervous system: a review. JAMA Neurol 81(3):283–290

Schaly S, Ghebretatios M, Prakash S (2021) Baculoviruses in gene therapy and personalized medicine. Biologics Targets Ther 15:115–132

Seck PA, Diagne A, Mohanty S, Wopereis MCS (2012) Crops that feed the world 7: rice. Food Secur 4(1):7–24. https://doi.org/10.1007/s12571-012-0168-1

Selvakumar SC, Preethi KA, Ross K, Tusubira D, Khan MWA, Mani P et al (2022) CRISPR/Cas9 and next generation sequencing in the personalized treatment of cancer. Mol Cancer 21(1):83

Shedden-Mora MC, Pan Y, Heisig SR, von Blanckenburg P, Rief W, Witzel I et al (2020) Optimizing expectations about endocrine treatment for breast cancer: results of the randomized controlled psy-breast trial. Clin Psychol Eur 2(1):e2695

Silva G, Poirot L, Galetto R, Smith J, Montoya G, Duchateau P, Pâques F (2011) Meganucleases and other tools for targeted genome engineering: perspectives and challenges for gene therapy. Curr Gene Ther 11(1):11–27

Singh AN (2021) Ethical controversies and challenges in human genome editing. Int Med J 28(2):134–137

Spanjol J, Noble CH (2023) From the editors: engaging with generative artificial intelligence technologies in innovation management research—some answers and more questions. J Prod Innov Manag 40(4):383
Stalidzans E, Dace E (2021) Sustainable metabolic engineering for sustainability optimisation of industrial biotechnology. Comput Struct Biotechnol J 19:4770–4776
Stoddard BL (2011) Homing endonucleases: from microbial genetic invaders to reagents for targeted DNA modification. Structure 19(1):7–15
Stoddard BL (2014) Homing endonucleases from mobile group I introns: discovery to genome engineering. Mob DNA 5:1–16
Sun D, Guo Z, Liu Y, Zhang Y (2017) Progress and prospects of CRISPR/Cas systems in insects and other arthropods. Front Physiol 8:608
Sun L, Lai M, Ghouri F, Nawaz MA, Ali F, Baloch FS, Nadeem MA, Aasim M, Shahid MQ (2024) Modern plant breeding techniques in crop improvement and genetic diversity: from molecular markers and gene editing to artificial intelligence—a critical review. Plants 13(19):2676. https://doi.org/10.3390/plants13192676
Szostak JW, Orr-Weaver TL, Rothstein RJ, Stahl FW (1983) The double-strand-break repair model for recombination. Cell 33(1):25–35
Takata M, Sasaki MS, Sonoda E, Morrison C, Hashimoto M, Utsumi H et al (1998) Homologous recombination and non-homologous end-joining pathways of DNA double-strand break repair have overlapping roles in the maintenance of chromosomal integrity in vertebrate cells. EMBO J 17:5497
Tarim EA, Karakuzu B, Oksuz C, Sarigil O, Kizilkaya M, Al-Ruweidi MKA et al (2021) Microfluidic-based virus detection methods for respiratory diseases. Emerg Mater 4:143–168
Toufikuzzaman M, Hassan Samee MA, Sohel Rahman M (2024) CRISPR-DIPOFF: an interpretable deep learning approach for CRISPR Cas-9 off-target prediction. Brief Bioinform 25(2):bbad530
Tran NT, Han R (2024) Rapidly evolving genome and epigenome editing technologies. Mol Ther 32(9):2803–2806
Tsubouchi A, An Y, Kawamura Y, Yanagihashi Y, Nakayama H, Murata Y et al (2024) Pooled CRISPR screening of high-content cellular phenotypes using ghost cytometry. Cell Rep Methods 4(3)
Urnov FD, Rebar EJ, Holmes MC, Zhang HS, Gregory PD (2010) Genome editing with engineered zinc finger nucleases. Nat Rev Genet 11(9):636–646
Vasebi Y, Khakvar R (2014) CRISPR-Cas: the effective immune systems in the prokaryotes. pp. 334–344
Villiger L, Joung J, Koblan L, Weissman J, Abudayyeh OO, Gootenberg JS (2024) CRISPR technologies for genome, epigenome and transcriptome editing. Nat Rev Mol Cell Biol 25(6):464–487
Vojnits K, Nakanishi M, Porras D, Kim Y, Feng Z, Golubeva D, Bhatia M (2022) Developing CRISPR/Cas9-mediated fluorescent reporter human pluripotent stem-cell lines for high-content screening. Molecules 27(8):2434
Walker-Daniels J (2013) CRISPR and genomic engineering. Mater Methods 3:164
Wang JY, Doudna JA (2023) CRISPR technology: a decade of genome editing is only the beginning. Science 379(6629):eadd8643
Wang D, Zhang C, Wang B, Li B, Wang Q, Liu D et al (2019) Optimized CRISPR guide RNA design for two high-fidelity Cas9 variants by deep learning. Nat Commun 10(1):4284
Wang J, Zhang X, Cheng L, Luo Y (2020) An overview and metanalysis of machine and deep learning-based CRISPR gRNA design tools. RNA Biol 17(1):13–22
Wang F, Liu Y, Zhang A, Kong D, Bi J, Liu G, Yu X, Luo L (2022) Breeding an early maturing, blast resistance water-saving and drought-resistance rice (WDR) cultivar using marker-assisted selection coupled with rapid generation advance. Mol Breed 42(8):46. https://doi.org/10.1007/s11032-022-01319-3

Wang B, Tian P, Sun Q, Zhang H, Han L, Zhu B (2023a) A novel, effective machine learning-based RNA editing profile for predicting the prognosis of lower-grade gliomas. Heliyon 9(7):e18075

Wang X, Chang X, Ma L, Bai J, Liang M, Yan S (2023b) Global and regional trends in greenhouse gas emissions from rice production, trade, and consumption. Environ Impact Assess Rev 101:107141. https://doi.org/10.1016/j.eiar.2023.107141

Xiao J, Zhang Y, Li B, Wang H, Zhang JZ, Zhu T (2024) Machine learning based density functional method for chemical reactions I: organic reactions

Yan X, Stuurman N, Ribeiro SA, Tanenbaum ME, Horlbeck MA, Liem CR et al (2021) High-content imaging-based pooled CRISPR screens in mammalian cells. J Cell Biol 220(2):e202008158

Yang L, Chen J (2023) Expanding genome editing scopes with artificial intelligence. Sci Bull 23:2881–2883

Yang Y, Saand MA, Huang L, Abdelaal WB, Zhang J, Wu Y et al (2021) Applications of multiomics technologies for crop improvement. Front Plant Sci 12:563953

Yu G, Kim HK, Park J, Kwak H, Cheong Y, Kim D et al (2023) Prediction of efficiencies for diverse prime editing systems in multiple cell types. Cell 186(10):2256–2272

Zafar I, Rafique A, Fazal J, Manzoor M, Ain QU, Rayan RA (2021) Genome and gene editing by artificial intelligence programs. In: Advanced AI techniques and applications in bioinformatics. CRC Press, pp 165–188

Zhang F, Wen Y, Guo X (2014) CRISPR/Cas9 for genome editing: progress, implications and challenges. Hum Mol Genet 23(R1):R40–R46

Zhang C, Quan R, Wang J (2018) Development and application of CRISPR/Cas9 technologies in genomic editing. Hum Mol Genet 27(R2):R79–R88

Zhang W, Yin J, Zhang-Ding Z, Xin C, Liu M, Wang Y et al (2021) In-depth assessment of the PAM compatibility and editing activities of Cas9 variants. Nucleic Acids Res 49(15):8785–8795

Zhang C, Yang Y, Qi T, Zhang Y, Hou L, Wei J et al (2023) Prediction of base editor off-targets by deep learning. Nat Commun 14(1):5358

Chapter 3
Genome Editing: CRISPR-CAS System: A Promising Diagnostic Tool for Plant Pathogens

Kamel A. Abd-Elsalam, Aftab Ahmad, Hussien Abodalam, Mirza Abid Mehmood, Farah K. Ahmed, Hassan Almoammar, and Mythili Ravichandran

Abstract Plant diseases play a crucial role in agriculture since many of them affect a variety of food crops, including vegetables, fruits, and basic grains, lowering their quality and productivity. With enhanced yield, quality, and abiotic/biotic stress tolerance, the CRISPR/Cas systems hold promise as a valuable genome-editing technique for disease detection and the development of new cultivars of crops resistant to pathogens. This method can assist in the early diagnosis of various viral, fungal, and bacterial infections, enabling the prompt application of plant medicines to stop the spread of those diseases more effectively. A promising strategy for the quick and precise identification of pathogens is the creation of a diagnostic tool for plant diseases based on CRISPR-Cas systems. By enabling prompt interventions and lower-

K. A. Abd-Elsalam (✉)
Plant Pathology Research Institute, Agricultural Research Center (ARC), Giza, Egypt

A. Ahmad
Biochemistry/Center for Advanced Studies in Agriculture and Food Security (CASAFS), University of Agriculture, Faisalabad, Pakistan

H. Abodalam
Plant Pathology Department, Faculty of Agriculture, Cairo University, Giza, Egypt

M. A. Mehmood
Plant Pathology, Institute of Plant Protection, Muhammad Nawaz Shareef University of Agriculture, Multan, Pakistan

F. K. Ahmed
Biotechnology English Program, Faculty of Agriculture, Cairo University, Giza, Egypt

H. Almoammar
Wellness and Preventive Medicine Institute, Health Sector, King Abdulaziz City for Science and Technology (KACST), Riyadh, Saudi Arabia

M. Ravichandran
Department of Microbiology, Vivekanandha Arts and Science College for Women, Sankagiri, Tamil Nadu, India

K. A. Abd-Elsalam, A. Ahmad (eds.), *Gene Editing in Cereals*, Sustainability Sciences in Asia and Africa, https://doi.org/10.1007/978-981-95-6722-5_3

ing crop losses, this instrument has the potential to greatly enhance plant disease management. For this technology to be validated and improved for practical application in agricultural contexts, additional study and improvement are still needed. Together with CRISPR/Cas technology, next-generation colorimetric sensors can overcome these obstacles and develop into highly effective and user-friendly diagnostic instruments. This review will provide insights into a variety of CRISPR-based diagnostic tools, their applications in plant disease, challenges, and the future promise of the CRISPR-Cas system.

Keywords KCas effectors · Pathogen detection · Plant health · Diagnostic methods

3.1 Introduction

The three main threats to food safety are mycotoxins, plant pathogens, and food-borne diseases. At least 800 million people are underfed because of catastrophic plant disease in the existing food supply (Strange and Scott 2005). Plant infections can cause increasingly severe crop output losses globally, which can lead to food shortages and famine (Nicaise 2014). Plant infections are estimated to be the cause of 12.5% of losses and cause a threat to food security (Oerke 2006; Savary et al. 2019).

Diseases in our crops are brought on by many plant pathogens. Food crops cultivated over a large area are affected by pathogens, ranging from minor symptoms to complete devastation of cultivated areas. A global shift in the distribution of agricultural diseases, such as fungi and oomycetes, has probably been brought on by climate change. Some startling facts regarding the growing danger of diseases to crop security have been found by a series of meta-analyses that combed through vast data sets. The most numerous categories and those driving the global invasion of agriculture are fungi and oomycetes (Bebber and Gurr 2015). Commercial commodities and staple crops are affected by oomycetes and emerging fungal pathogens, as a result of which global food security is seriously threatened (Fones et al. 2020). Numerous bacteria in crops generate biotic stress, which results in illness and substantial yield loss. Plant diseases brought on by numerous pests and pathogens result in the loss of 20–40% of agricultural productivity annually (Savary et al. 2019; Khan et al. 2021). Climate change has resulted in direct or indirect output losses, including the spread and consequences of plant diseases (Chaloner et al. 2021).

Precise assessments of the disease's severity and its effects, as well as quick and accurate identification of the causative (Strange and Scott 2005). An important study area is the automatic detection of plant diseases for early detection, since it helps in monitoring vast fields covered with crops and identifying the disease symptoms soon after their emergence for targeted crop protection (Rumpf et al. 2010).

Therefore, it is quite important to search for quick, automatic, inexpensive, and accurate methods to identify plant disease situations (Al Hiary et al. 2011).

Early detection of pests or plant pathogens could help farmers take swift action to use fewer pesticides or herbicides and control epidemics globally. In addition to plants, monitoring cattle populations for changes in the microbiome, the appearance of infections, or antibiotic resistance could provide information on feeding schedules or provide surveillance of developing pathogens in herds (Mathew et al. 2007). In addition to studying host range dispersion, virulence, strain type, pathogenicity, diagnostics, and genetic diversity, several diseases strains were detected by the CRISPR system (Shami et al. 2021). Early knowledge of plant disease detection and the health of crops can help in disease control through appropriate management tactics, which in turn call for the creation of quick, perceptive, reasonably priced, and trustworthy disease detection tools (Kaur et al. 2023).

Nowadays, the CRISPR/Cas system is crucial for biosensing applications, especially since it was found that some Cas effectors have trans-cleavage activity (Fig. 3.1). We require a trustworthy point-of-care diagnostic kit that could be utilized at the infection site for the surveillance of new or re-emerging viruses. The CRISPR/Cas approach has been widely employed for this purpose. Based on the Cas9 protein, methods like FELUDA (Azhar et al. 2021), CRISPR/Cas9-triggered isothermal exponential amplification reaction (CAS-EXPAR), and CRISPR/Cas9-triggered nicking endonuclease-mediated strand displacement amplification (CRISDA) have been investigated for the SARS-CoV-2 and Zika virus. Diagnostics dependent on the CRISPR/Cas13 system, including the specific high-sensitivity enzymatic reporter unlocking (SHERLOCK) method, the Cas13-based effector Rugged, Equitable, Scalable Testing (CREST) (Rauch et al. 2021), and Combinatorial Arrayed Reactions for Multiplexed Evaluation of Nucleic acids (CREST) (Rauch et al. 2021). New CRISPR/Cas12-based detection systems that have been used to detect SARS-CoV-2 and other viruses are the one-hour DNA Endonuclease Targeted CRISPR Trans Reporter (DETECTR) and Low-cost Multi-purpose Highly Efficient System (HOLMES). Nevertheless, minimal work was done on the development of a CRISPR/Cas approach to detect plant viruses (Li et al. 2021; Broughton et al. 2020).

CRISPR/Cas system can target an extensive spectrum of targets, including bacteria, proteins, nucleic acids, and metal ions, among small molecules, which opens its potential applications and matures the system (Wu 2023). Field detection is very demanding and comprises early detection of pathogen, simple, specific, and precise mass screening, field-based alternative modules, and quick detection. Currently, different point-of-care (POC) nucleic acid diagnostics instruments have been developed with various CRISPR/Cas systems (Sánchez et al. 2022). It is now a common place investigation in the food science discipline because of the various advantages of CRISPR-based nucleic acid-based methods, such as rapid detection, convenience, low price, modularization, quick time consumption, high specificity, and high sensitivity (Taniguchi et al. 2021). These ideas also formed the basis of CRISPR subtyping/serotyping, CRISPR interference test (dCas9), and single guide RNA (sgRNA) diagnostic tests (Strich and Chertow 2019). Three Cas proteins (Cas13,

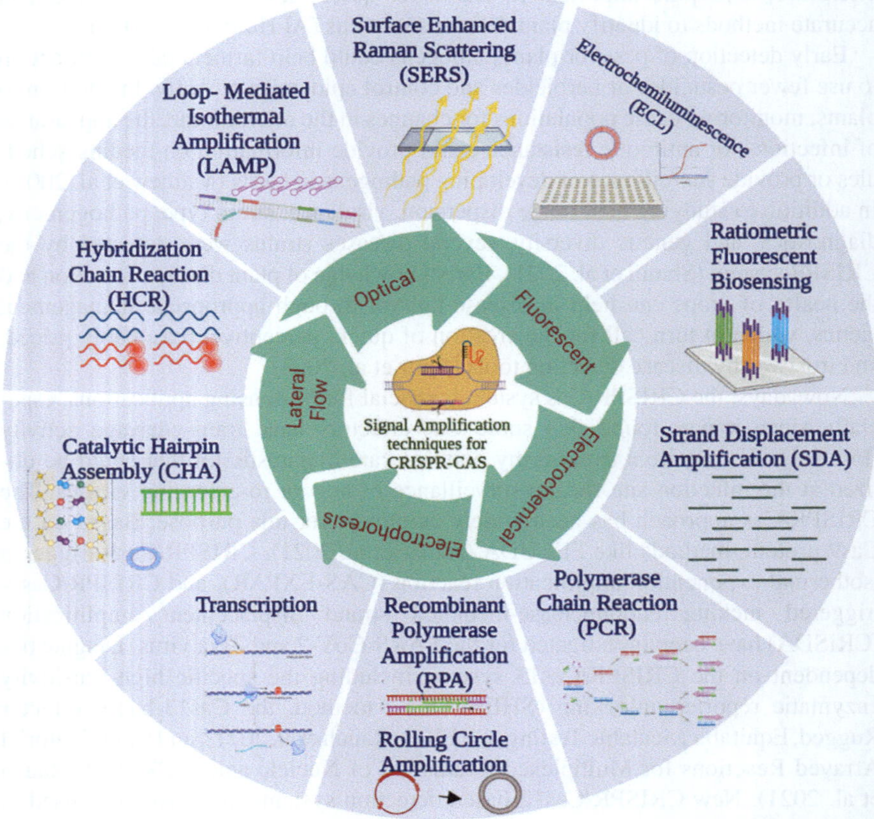

Fig. 3.1 Several methods exist to optimize signal amplification and output of CRISPR/Cas-based biosensing systems, such as enzymatic amplification, nucleic acid-based signal amplification, reporter molecules, signal transduction systems, and signal enhancement systems

Cas14, and Cas12) have been repurposed to detect nucleic acid by methods based on a nonspecific and collateral nuclease activity induced by both identifying the target (Gootenberg et al. 2018). CRISPR is a strong plant breeding tool that can support sustainable agriculture and food security for the increasing world population; therefore, scientific principles-based discussions should be carried out about the risks of genome editing instead of societal concerns (Varanda et al. 2021). To attain and address possible food safety, new detection technologies must be developed. The use of CRISPR-based biosensing analysis technologies to combat hazardous elements that target nucleic acids offers a significant advantage (Mao et al. 2022).

The study of plant–pathogen interactions is now made possible by CRISPR/Cas systems, which will aid future research on resilient, disease-resistant plants and agricultural production operations (Gosavi et al. 2020). In this case, we explained the development of a new CRISPR-Cas systems-based diagnostic tool for plant diseases to provide a faster, more accurate, and cost-effective method of detecting and identifying many plant diseases. The diagnostic device aims at enhancing the diagnosis of plant diseases and, eventually, assisting in directing the control of the disease and protection of crops based on the precision and performance of the CRISPR-Cas system.

3.2 CRISPR/Cas Biosensing Systems

The transcripts of the CRISPR array are processed to produce mature crRNA, which is a guide in a CRISPR/Cas system that directs Cas effectors. Cas effectors are enzyme subunits with target-dependent cleavage activity that may be an individual protein or a fusion of various proteins.

The existing CRISPR/Cas-based nucleic acid biosensing systems are implemented by multiple Cas effectors, notably Cas9, Cas12, and Cas13 (Li et al. 2019). The goal of the rapidly emerging CRISPR-based diagnostics field is to develop nucleic-acid-based POC diagnostic assays that rely on the simplicity, programmability, and specificity of CRISPR tools and are implemented in the field. The ribonucleoprotein effector complex differentiated CRISPR-Cas systems: Complex of effector proteins identified Class 1 systems; however, a single crRNA-binding protein comprises Class 2 systems (Kaminski et al. 2021). The combined use of nucleic acid and CRISPR-Cas RNA-programmable nucleases detection methods has significantly altered the field of disease diagnostics (Islam and Kasfy 2023).

3.2.1 Tools for Detecting Diseases Using CRISPR

Meganucleases: The precise and well-timed diagnosis of diseases and causative organisms is a critical component of implementing effective control measures. Therefore, precise disease diagnosis is an essential component of successful disease management. Because they are extremely specific in binding as well as cleaving nucleic acids and are programmable, Cas proteins are a great choice to be repurposed for nucleic acid detection. The advancement in ultrasensitive, portable, and affordable nucleic acid-based POC testing equipment has been made possible by recent discoveries and advances in our understanding of several Cas proteins. Despite the fact that dependable and robust platforms for nucleic acid detection have been developed by Cas9 proteins (Azhar et al. 2021; Jiao et al. 2021; Pardee et al. 2016; Quan et al. 2019), the recent finding that Cas13a and Cas12a have collateral cleavage activity has transformed nucleic acid detection (Chen et al. 2018; Gootenberg et al. 2018). Many detecting assays, which are CRISPR-based, also depend on Cas13a as well as Cas12a, which have already been developed. We will go over some of the selected CRISPR diagnostic platforms, their functions, and some cases of plant disease detection applications. We propose that readers first read an in-depth evaluation of CRISPR-based diagnostics (Kaminski et al. 2021), as well as the requirement for costly equipment and expert staff. The CRISPR diagnostic approaches presented below, on the other hand, combine affordability as well as usability with the sensitivity of qPCR.

3.2.1.1 Cas9

A noncoding tracrRNA (75–110 bp) oversees the formation of a pre-crRNA, which is paired with Cas9 to produce a complex of crRNA and tracrRNA in the Cas9 system. A usable tracrRNA:crRNA will be created by breaking down the tracrRNA: pre-crRNA using a non-Cas RNase III ribonuclease for the Cas9 interference complex (Bonini et al. 2021). Cas9 constructs the guide RNA-target duplex 10–12 nts over the protospacer adjacent motif (PAM) sequence (for example, 5′-NGG-3′) along the nontarget DNA strand to prevent the dsDNA duplex from reannealing (Sternberg et al. 2014). When metallic ions are present, the cleavage occurs, and resultantly a sharp break arises 3 bp upstream of the PAM (Bonini et al. 2021; Li et al. 2021). Currently, the CRISPR-Cas9 system's cleavage mechanism has been employed to discover a few viruses, including the African swine fever virus and Zika virus as well as Listeria monocytogenes (Wang et al. 2020: Pardee et al. 2016; Huang et al. 2018).

A protein-RNA complex called CRISPR/dCas9 can delicately and purposefully bind to DNA sequences. Moreover, you can employ DNA as a biorecognition receptor agent, and you can use this interaction to find certain DNA sequences (Uygun et al. 2020). Although CRISPR-Cas9 is widely recognized for its capacity to execute gene editing, the variant of CRISPR-dCas9 does not cut DNA but rather binds to a defined binding site securely, making it an ideal biosensing target (Yang et al. 2018). Specific nucleic acid sequences could be identified when CRISPR-Cas9 is combined with nucleic acid amplification, allowing genotyping infections while excluding SNPs (Li et al. 2021). Traditional CRISPR/Cas9 will help to address technological limitations in single universal primer recombinase polymerase amplification-based lateral flow biosensors, enabling speedy and accurate (Qiu et al. 2018; Wang et al. 2022).

3.2.1.2 dCas9

To determine the genotype of infections and SNP identification, Cas9 cleavage could be coupled with the amplification of nucleic acid to determine the exact sequence of the nucleic acid. As an example, Pardee and colleagues managed to use NASBA with a Cas9 effector to characterize and genotype the Zika virus, and it is an enormous achievement (Pardee et al. 2016). CRISPR/Cas9-based method utilizing the paper-based isothermal toehold switching biosensing device enhanced the specificity of a one-base resolution. After NASBA amplification had been performed, the amplification products were ligated with a trigger sequence to open the toehold sensor. This was followed by the production of the single-guide RNA (sgRNA) with the aid of a strain-specific protospacer-adjacent motif (PAM) sequence covering one SNP absent in the African Zika genome but present in the American Zika genome. The presence of sgRNA blocks the toehold activation of the American Zika because the trigger sequence omission was done after the cleavage of Cas9. Cas9 could only cleave the amplification products of the American strain of Zika virus. The discrimination of the diverse Zika genotypes was done with the help of the NASBACC technique (Pardee et al. 2016). This method has been successfully experimented in relation to its ability to detect *Listeria monocytogenes* hly mRNA and DNA methylation upon being refined to specificity, sensitivity, and amplification efficiency. The downstream EXPAR system that initiates the process of isothermal amplification uses the cleavage products as primers, and the outcome is many amplified double-stranded DNA (dsDNA) fragments that could be subsequently bound by SYBR Green I to read out fluorescent signals (Pardee et al. 2016). This was carried out through the development of one-of-a-kind antisense PAM-presenting the oligonucleotide.

The enzymatically deactivated Cas9 effector (dCas9), which may bind to sequences, has also been used to construct several innovative biosensing systems analogous to traditional in situ hybridization techniques via TALE or zinc finger proteins, such as CRISPR/Cas-based nucleic acid biosensing systems (Batista and Pacheco 2018). The recognition of DNA from *Mycobacterium tuberculosis* is an example of a paired dCas9 reporter system usage (Zhang et al. 2020).

A target DNA sequence with a pair of sgRNAs complementary to the downstream and upstream of that specific sequence directs the two dCas9 separately that are fused to the C- and N-terminal halves of the firefly luciferase. If the target DNA was there, PCR may have amplified it. A pair of neighboring dCas9 effectors bound the PCR products, resulting in a reconstituted luciferase with visible light signals (Fig. 3.2). Another technique for miRNA identification is the rolling circle amplification (RCA)-CRISPR-split-HRP (RCH) (Qiu et al. 2018). Two designed dCas9 effectors connected with half of the split horseradish peroxidase (HRP) protein are used to bind to the amplified target sequences after recruitment of RCA for isothermal miRNA target amplification. As a result of surrounding target sequences binding to dCas9 effectors, the addition of chromogenic substrate tetramethylbenzidine (TMB) in the final step can help reassemble split HRP and provide a colorimetric signal readout (Qiu et al. 2018).

3.2.1.3 Cas12

The CRISPR-Cas12a-based detection technology is now in the testing phase. CRISPR-ability Cas12a's to classify the existence of precise nucleic acid targets and their minute variation in sequence is critical for the detection and identification of pathogens (Mao et al. 2022). Cas12 is a type V Cas enzyme of Class 2 that has several subtypes, including Cas12f, Cas12c, Cas12b, and Cas12a (formerly known

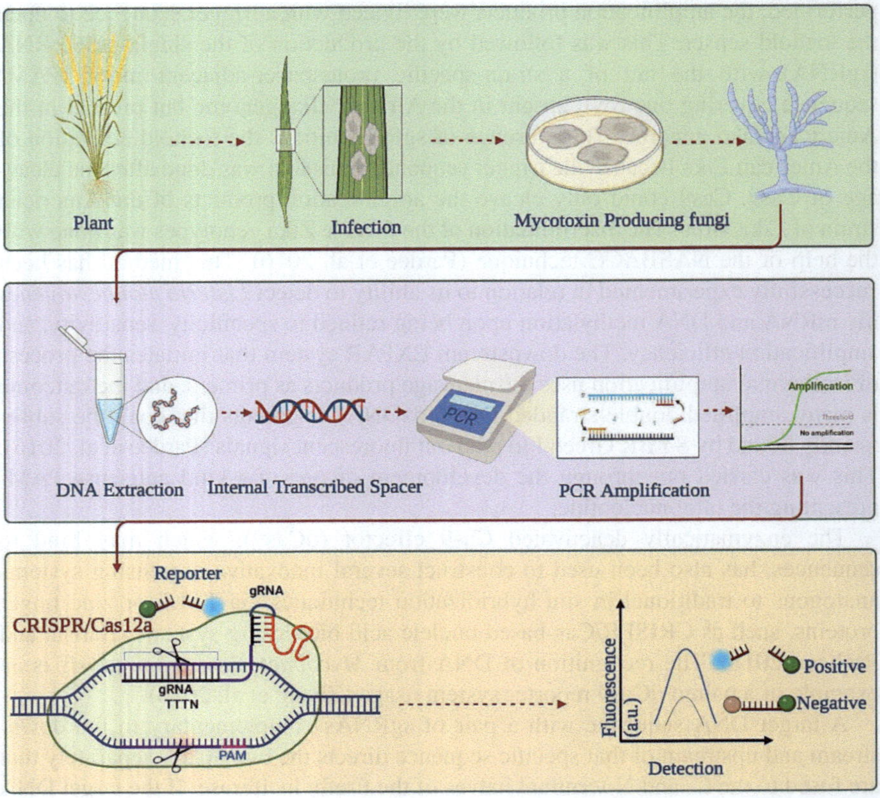

Fig. 3.2 CRISPR/Cas12a-driven nucleic acid test is another diagnostic tool that identifies certain nucleic acids in plant samples with the help of the CRISPR/Cas12a system. It works through the use of a guide RNA to target the sequence of interest and the process of activating Cas12a. In case of activating Cas12a, it cuts nearby DNA, which forms a fluorescent signal indicating the presence of the target sequence. This powerful assay provides a considerable possibility of fast and accurate diagnostic testing

as Cas14) (Makarova et al. 2020). The subtype Cas12a (Cpf1) has received the greatest attention. They may convert without the help of proteins or RNA, crRNA, into their nuclease domains.

Before the stem-loop, a pre-RNA can self-process immediately and provide a mature crRNA guide resultantly cut it just before the stem loop to construct an interference complex. Similar to the fit-induced mechanism in Cas9, a conformational change occurs upon the formation of the crRNA–Cas12a complex (Bonini et al. 2021). The nontarget DNA strand contains a T-rich PAM (5′-TTTN-3′) that has been recognized and cleaved by the crRNA: Cas12a at 8–19 nucleotides from the PAM's 3′ end. A 5′ sticky overhang is left behind when a target DNA is cut 23 bases away from PAM (Bonini et al. 2021). Furthermore, it was shown that the *Lactobacillus bacterium* Cas12a had nonspecific ssDNA collateral activity (trans-activity). The finding of trans-cleaving activity mediated by cis activity led to the development of biosensing devices. When Cas12 discovers its nucleic acid target, it generates trans-activity, which can be amplified and recorded using various readout techniques like fluorescence, colorimetric, and electrochemical technologies (Bonini et al. 2021). Cas12a is one of the effectors now being studied for disease detection (Marqués et al. 2022). Three RNA viruses with substantial economic and scientific significance in *Nicotiana benthamiana* plants were discovered: tobacco etch virus (TEV), potato virus X (PVX), and tobacco mosaic virus (TMV). Meanwhile, CRISPR-Cas12a has been found to precisely detect two varieties, tomato yellow leaf curl virus (TYLCV) and tomato leaf curl New Delhi virus (ToLCNDV) (Mahas et al. 2021).

3.2.1.4 Cas13

CRISPR-Cas13a protein biosensor can be utilized in developing new nucleic acid detection technologies, which are accurate, convenient, and sensitive. In terms of resolving public health problems and food safety, it is especially essential to mention that the *Salmonella* detection method using Cas13a protein can be further developed into a wide range of microbial detection applications (Gao et al. 2021). Molecular scientists have discovered that the cleavage activity of CRISPR-Cas13 nucleases can be utilized to identify pathogenic bacterial nucleic acids in vitro and for gene editing (Arslan et al. 2014; Zhou 2020). The targeted RNA activating Cas13, a type VI protein enzyme that is an RNA-guided Cas13 protein with RNAse activity, breaks nearby RNAs by trans-activity. Like the case with DNA targets using PAMs, the activity of the Cas13 enzyme is based on the presence of a protospacer flanking site (PFS) in which a PFS is required on the 3′-protospacer strand by Cas13a enzymes. By comparison, protospacer strands require a PFS to Cas13b enzymes. The LwaCas13a of *Leptotrichia wadei* is used to evade the PFS requirement. Cas12 and Cas13 have two categories of distinct trans-cleavage activities, so the three categories have brought significant progress in CRISPR-Dx that has simplified the signal generation (Fig. 3.3). The less popular Cas9 methodology is, however, not as popular in CRISPR diagnostics as it still requires complex guide RNA engineering and signaling cascades.

3.2.1.5 Cas14

CRISPR advances in genotyping of single-nucleotide polymorphisms (SNPs) in ssDNA viruses that infect eukaryotes have also been promising with Cas14 (Aquino-Jarquin 2019). Moreover, Cas14 is an excellent alternative to identify both ssDNA and dsDNA because it has the ability to identify ssDNA targets without the need of PAM recognition. The Cas14-DETECTR system has established a platform that possesses the high-fidelity of the DNA SNP using the nonspecific DNase active site of Cas14 (Harrington et al. 2018). It is precise and functional in comparison to its

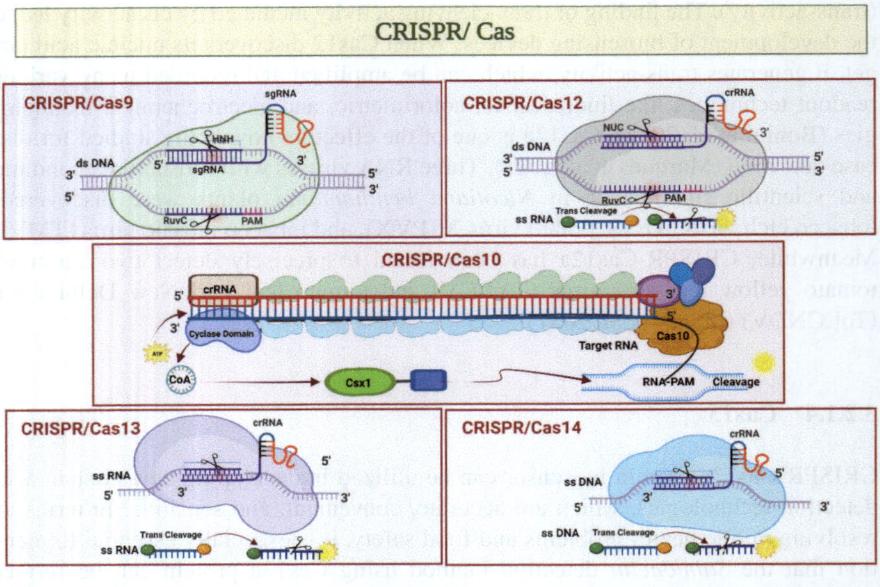

Fig. 3.3 Diagram of the CRISPR/Cas9, CRISPR/Cas10, CRISPR/Cas12, CRISPR/Cas13, and CRISPR/Cas14 genome editing protocol

analogue (Cas12a-DETECTR). Cas14-DETECTR can be used in SNP genotyping with high fidelity and ssDNA virus detection. The next-generation diagnostic tool FLASH, which is hybridization-based, makes use of Cas9's specificity to effectively enrich targeted sequences (Quan et al. 2019).

3.2.2 *Nucleic Acid Sequence-Based Amplification-CRISPR Cleavage (NASBACC)*

Genetic: The NASBACC instrument employs SpCas9 and CRISPR to diagnose illnesses. To distinguish between various genotypes, it uses the toehold switch principle and the selective Cas9 cleavage of the target DNA. This technique can distinguish between several genotypes by determining whether a PAM sequence is present or absent, which can differentiate between one base. The NASBACC technique generates a readout result by implementing the amplification, which is sequence-based, target detection, which is PAM-dependent, and cleavage of Cas9 with a toehold sensor. When the toehold sensor attaches to the trigger RNA, it produces a reporter protein that highlights the ribosome binding site (RBS). When the sgRNA-Cas9 identifies the PAM, it will form a shorter RNA, which is unable to activate the target site, toehold switch, and cleaves the DNA. Conversely, complete RNA is synthesized by a different genotype which does not contain the PAM sequence, which begins the toehold reaction resulting in a color change. This method involves the sgRNA, which has been thoroughly developed with stain-specific PAM locations to differentiate related virus types. When the sgRNA-Cas9 finds the PAM, it cleaves shorter RNA, which cannot activate the toehold switch at femtomolar concentrations of plasma of an infected monkey (Pardee et al. 2016).

3.2.3 LEOPARD

LEOPARD is a technology that makes use of modified DNAs that are the target and tracrRNAs to identify RNA dependence on CjCas9 (*Campylobacter jejuni* Cas9), which can reveal a variety of disease-related biomarkers simultaneously and is specific to a single nucleotide (Jiao et al. 2021). To investigate the dissimilarity between SARS-CoV-2 and its variant, LEOPARD was used by Jiao et al. (2021) to discover that in the case of CjCas9 in its natural host (ncrRNAs). The CjCas9 system tracrRNA of the cell interacts with cellular RNAs to make noncanonical crRNAs. They generated a reprogrammed version of the tracrRNA (Rptr) that could bind and cleave an RNA of interest after demonstrating that tracrRNA could be programmed to do so (Jiao et al. 2021). A part of the target RNA is cleaved, producing a functional guide RNA. A fluorescent DNA sensor is read out using a gel after being cleaved by ncrRNA and Rptr. Soon, plant disease diagnosis could also employ this highly effective platform.

3.2.4 FLASH

FLASH (Low Abundance Sequence Search by Hybridization) is a Cas9-based test that has been developed to find antimicrobial-resistant genes and genomic DNA in clinical samples (Quan et al. 2019). The FLASH readout, on the other hand, is dependent on next-generation sequencing. Cas9 gRNAs that have been specifically generated for the purpose that are used in this technique to cut the sequences of interest. In the first step of the process, genomic DNA is phosphorylated to inhibit its digestion in the following step, which is the designed gRNA complexes and Cas9 digesting the input DNA. The cleaved products are now eligible for ligation with universal sequence adapters as a result of this. The target sequences, followed by amplification, are brought to the forefront, elevated above the background, and prepared to bind to the sequencing flow cell. In a single test using multiplexing, we can detect 1000 targets (Quan et al. 2019).

3.2.5 FELUDA

An affinity-based method, FnCas9 Editor-Linked Uniform Detection Assay, or FELUDA, worked with a high sensitivity for detecting pathogenic RNA or DNA (Azhar et al. 2021). It involves an enzymatic readout of pathogenic biomarkers using FnCas9 (*Francisella novicida* Cas9) without producing any trans-cleavage of the reporter protein. The target sequence will be amplified in the first step of the method using biotin-labeled primers and recombinase polymerase amplification (RPA)/PCR. After that, items that have been biotinylated are fixed on magnetic streptavidin beads. The substrate is anchored to the fluorophore-tagged ribonucleoprotein complex, which is composed of dCas9 and a particular sgRNA. Observing fluctuations in the fluorescence intensity present in the supernatant caused by enzyme binding can provide an indicator of a favorable response. FELUDA was modified so that a lateral flow test kit could be developed (Azhar et al. 2021).

3.2.6 DETECTR

DETECTR is a Cas12a-based first diagnostic tool, which is a DNA endonuclease-targeted CRISPR trans reporter (Chen et al. 2018). Cas12a neglects its target and cleaves any surrounding single-stranded DNA. Cas12a disregards its goal and cleaves any single-stranded DNA in its way. DETECTR was developed using a function called collateral activity. A complementary guide for RNA directs Cas12a to a target dsDNA in this approach. A fluorophore is cleaved when Cas12a binds to the proper target, ssDNA reporter molecules attached to a quencher. The separation of the fluorophore and quencher generated the fluorescent signal that helps in the detection of undifferentiated cleavage. When paired with RPA preamplification, DETECTR displays greater sensitivity (Fig. 3.4). SARS-CoV-2 and human papillomavirus have been detected using the technique (Broughton et al. 2020). Similarly, the thermostable method HOLMESv2, based on Cas12b, was developed. HOLME Sv2 can be used to detect target DNA methylation (Li et al. 2019). DETECTR is a CRISPR-Cas12a-based system that was examined to detect BNYVV in sugarbeet roots (Chen et al. 2018). The CRISPR-Cas12a-based development diagnostic method involves amplification of viral fragment template DNA under isothermal conditions. A single-step, one-step isothermal RT method, which is low-cost, was designed to assist in determining the sequence of the BNYVV RT-RPA amplicon. Field soils were used to test the sensitivity of the CRISPR-based BNYVV by using the sugar beet root rhizomania.

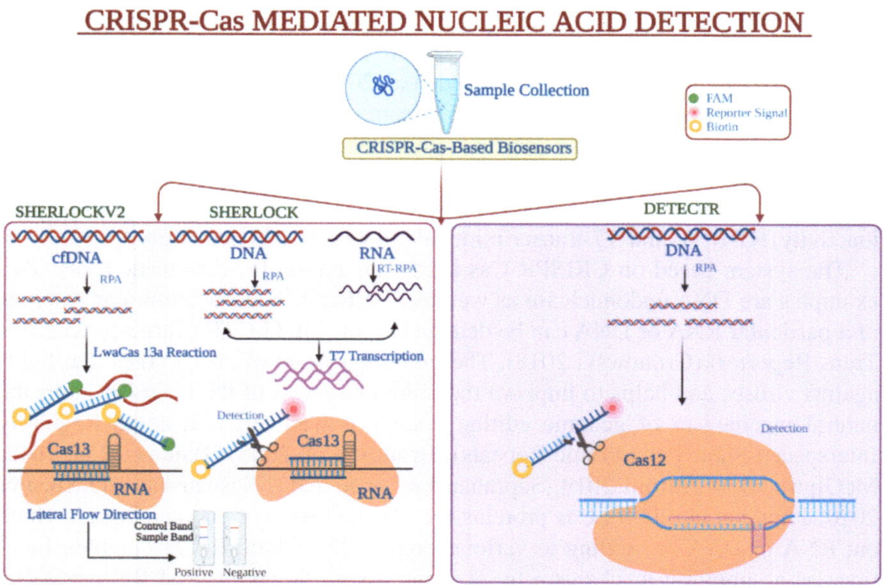

Fig. 3.4 Cas12 and Cas13 cleavage activity. The Cas12's collateral nuclease activity causes non-specific cleavage of the reporter molecule after binding the Cas12-crRNA complex to its target (dsDNA), enabling the detection of the fluorescent signal. In the SHERLOCK technology, Cas13a is guided by a single CRISPR RNA (crRNA) to cut mRNA or ssRNA, and the identical procedure takes place

3.2.7 SHERLOCK

With the help of the collateral cleavage activity of proteins with Cas13a, SHERLOCK (special high-sensitivity enzymatic reporter unlocking) was developed (Gootenberg et al. 2018). Cas13, also similar to Cas12a, recognizes the target RNA and cuts any of the adjacent single-stranded RNA randomly. SHERLOCK has successfully determined pathogens such as SARS-CoV-2 and bacteria such as *K. pneumoiae* and *M. tuberculosis* (Xiang et al. 2020). This process starts with isothermal amplification of the acquired samples. Subsequently, there is a guide RNA, an RNA reporter, and an engineered Cas13. Cas13 is designed based on RNA, and it is used to detect viral or bacterial RNA or DNA. A reporter probe is a separation of a fluorophore and has a quencher and a short oligonucleotide. Cas13 starts arbitrarily splicing surrounding RNA like the reporter once it splices with the appropriate target. Cas13 separates the reporter and quencher and generates individual signals across the sample. SHERLOCKv2, a new version, can be used to identify four different RNA targets using a single sample. The immunoprecipitation technique is a lateral flow readout of the immune chromatography-based test, where the cleaved and antibody-tagged reporter is recognized in a paper strip using an antibody-gold nanoparticle complex (Gootenberg et al. 2018). The HUDSON detection technology, which entails heat treatment of unextracted diagnostic products to inactivate nuclease, has been incorporated into the SHERLOCK system to make it even simpler and expand the functionality of the system (Myhrvold et al. 2018). SHERLOCK and HUDSON could detect ZIKV RNA in a variety of infectious samples, such as saliva, blood, plasma, and urine, even with a very low viral titer. In less than 2 hours, they enable instrument-free detection from samples. A single heat block can complete all the sequential processes required for HUDSON, such as nuclease inactivation, viral inactivity, RT-RPA, and T7-transcription, along with Cas13 detection.

The system based on CRISPR-Cas is the foundation for diagnostic tools. Two examples are DNA endonuclease as well as SHERLOCK. Low attomolar amounts of a particular RNA or DNA can be detected by the DETECTR (Targeted CRISPR Trans Reporter) (Gronowski 2018). The immune system of prokaryotes that fights against viruses and helps to improve the immune system of the organism, uses the natural mechanism of genome editing known as CRISPR (Clustered Regularly Interspaced Short Palindromic Repeats) (Brouns et al. 2008; Gasiunas et al. 2012; McGinn and Marraffini 2019; Sapranauskas et al. 2011; Sasnauskas and Siksnys 2020; Zhang et al. 2020). Cas proteins are ribonucleases or endonucleases, which cut RNA or DNA, according to various studies. The CRISPR-Cas complex functions as an antiviral mechanism by cleaving a specific target viral RNA or DNA (Gasiunas et al. 2012, 2014; Glemzaite et al. 2015; Makarova et al. 2020; Wilkinson et al. 2019). With the CRISPR-Cas system, researchers are able to cleave any RNA or DNA at the region of interest using the guide RNA and Cas protein (Gootenberg et al. 2017, 2018). The CRISPR-Cas nucleases can identify the nucleic acid sequences in a precise way. Thus, its use in fast diagnostic tests is under investigation worldwide. The formed pro-procedures can be performed during treatment, and they are sensitive and fast to react. Neither do they require costly equipment nor training. One of the methods of a portable nucleic acid detection and multiplexed platform (Chaijarasphong et al. 2019; Gootenberg et al. 2017; Khan et al. 2019). Both the CRISPR Cas13 cleavage mechanism and isothermal amplification are part of the SHERLOCK methodology, which is a detection tool. Even though the application of Cas12 or Cas13 enzymes can be used to detect nucleic acids (Kellner et al. 2019; Qin et al. 2019; Schindele et al. 2018; Yan et al. 2019), the Cas13 gene application is the most sensitive, and thus, the SHER-LOCK approach focuses on it. The technique involves the preamplification throughout the introduction of the T7 RNA polymerase promoter and using T7 RNA polymerase to produce RNA and initiate Cas13 trans-cleavage in the detection step. The Cas13a protein attaches itself to target RNA and cleaves it. Because of the nonspecific cleavage, SHERLOCK has evolved into an excellent technology to detect RNA viruses (Joung et al. 2020; Kim et al. 2020). Figure 3.4 shows SHERLOCK-Cas13, SHERLOCK, and DETECTR technologies, which involve ultrasensitive tests that do not need a huge amount of processing, could be employed to identify and detect infectious diseases in a short time (Mustafa and Makhawi 2021). Even though we demonstrated how to implement SHERLOCK swiftly and easily detect any characteristic in crops when selecting a breed, the concepts could be applied to a broad spectrum of other farming situations (Shami et al. 2021).

3.2.8 CARMEN

CARMEN (Combinatorial Arrayed Reactions for Multiplexed Evaluation of Nucleic Acids) is an approach that is used for multiplexed pathogen detection based on Cas13, which has been utilized to identify human viruses (169) and SARS-CoV-2 (Ackerman et al. 2020). CARMEN is a scalable, high-throughput detection approach. The first stage in this method is to amplify nucleotide targets using RPA and PCR. Following amplification, the Cas13 and the sample are colored and emulsified individually. The sample and detection sample are combined in a pairwise combination by combining droplets from each sample with the detection mix (cleavage reporter + Cas13 + a sequence-specific crRNA) and then placing the mixture on a microfabricated device. One pair is contained in each microwell. The droplets are mixed by an external electric field, allowing the reaction to occur. Fluorescence is produced when Cas13a cuts the reporter when identifying the target, indicating a positive sample reaction.

3.2.9 PlantDx

In addition to earlier toehold translational switches in human viral diagnostics, PLANT-Dx uses STARs to detect genomic sequences from CMV and PVY. Nucleic acid isolation is not necessary because these reactions can be carried out without the use of additional heating or visualization tools. The technique focuses on the potential of RNA synthetic biology approaches in practical applications by detecting only lower-replicating and genetically varied virus types. To increase sensitivity and incorporate PoUDs into other synthetic biology-based diagnostic systems, further sample preparation adjustments are required (Verosloff et al. 2019).

3.2.10 Imaging Techniques

Researchers are still grappling with the problem of how to use additional Cas proteins in intracellular imaging analyses (Wang et al. 2021). Tools for CRISPR can also be utilized for cutting-edge live imaging methods (Kalinina et al. 2020). The importance of live imaging is increasing because several molecular processes in plants are influenced by the movement of macromolecules like nucleic acids and their distribution (Turnbull et al. 2021). Agricultural biotechnology products are rapidly emerging as one of the largest trade commodities in agriculture in the world, covering our garments, our livestock, and even our eco-friendly automobiles. This is in spite of the asynchronous regulation schemes across the globe, whereby some countries have either moratoriums and bans on genetically modified (GM) organisms, or regulations that have the same regulations on both conventional and biotech novel plant products. Since the area that is under cultivation is vast, the issue of acceptance and the necessity of biotech crop varieties are no longer in question. The fact that the researchers have recently been awarded the Nobel Prize by the committee in recognition of the creation of a genome editing method, CRISPR/Cas9 is another milestone toward creating and breeding novel agricultural crops. The use of genome editing methods, which are precise, efficient, and cost-effective, has seen the introduction of new genome-edited crops in country regulatory schemes to be commercialized. The current leading countries in the production and export of GM crops are fast becoming aware of other varieties of gene-edited products through portraying the products in comparison with traditional breeds. This subtle legislative state of action, which was introduced in Argentina, and shortly thereafter in the group, demonstrates significant changes in agricultural biotechnology products landscape. The development of the law on gene-edited crops shows that the law is not fixed and that it needs to adapt to the mores of society based on the experience of 25 years of the cultivation and regulation of GM crops. The heart of this review is a synthesis of the world's legal position on GM crops, as it currently appears, based on previous literature, which deals specifically with the issue of how GM-edited crops will fit into the pre-existing systems. It is the first to compile a list of the relevant regulatory documents in all locations worldwide, including GM crop production specifically, and offers connections to original legislation on GM and gene edited crops (Turnbull et al. 2021; Kalinina et al. 2020) created an imaging technique that converts the location of targets using dormant catalytic Cas9 (dCas9) and a fluorescent marker and a sgRNA. This program allows you to view plant genomic regions in real-time. This has been observed to be used in RNA transcripts of living cells, where catalytically inactive Cas13 (dCas13) is used. This could be of special use in tracking the dispersion of viral RNA and the intra- and intercellular movement of viruses in plants prone and resistant to viruses in the study of plant virology.

3.2.11 Bio-SCAN Detection of Phytopathogens

Bio-SCAN test, as the abbreviation of the term biotin-coupled specific CRISPR-based assay and detection of nucleic acids, does not need technical expertise or any equipment and is developed as a dependable platform for pathogen detection. In this experiment, RPA (RT-RPA), bio-dCas9, and LFA were used. Bio-SCAN can identify the nucleic acid sequence, which is targeted, in just 1 hour, from the collection of samples to the outcome (Ali et al. 2021).

Bio-SCAN effectively identified the plant pathogenic fungi *Puccinia striiformis* f. sp. *tritici* in wheat, *Magnaporthe oryzae*, and *Triticum* (MoT), the viruses *Nicotiana benthamiana*, tomato yellow leaf curl virus (TYLCV), tobacco mosaic virus (TMV), and potato virus Y (PVY) (Sánchez et al. 2022). Bio-SCAN can precisely identify alleles that are found in nature (Ali et al. 2021). To find herbicide-resistant alleles that had "synthetically evolved," we wanted to improve

Bio-SCAN. OsmALS (bispyribac-resistant), SGR5 (SF3B1-GEX1A-Resistant 5), and SGR3 (SF3B1-GEX1A-Resistant 3) rice lines developed by our team were chosen for this purpose (Butt et al. 2019, 2020). There is a K1049 deletion and an H1048Q substitution in line SGR5 that differ from the wild type, while a deletion of a Lys residue at position 1050 (K1050-) in line SRG3 resulted from removal of the corresponding "AAG" codon in the WT (Fig. 3.5). The target sequences have been amplified using RPA and then sent to Bio-SCAN detection along with their associated sgRNAs while WT genomic DNA served as a negative control.

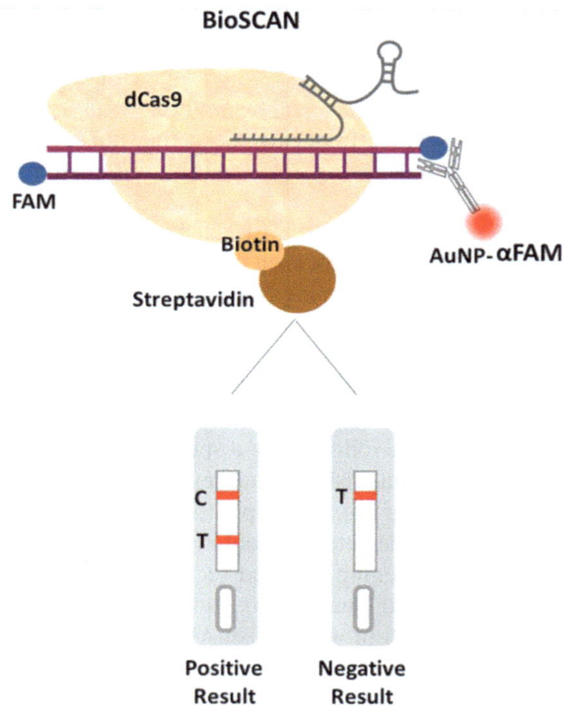

Fig. 3.5 Nucleic acid detection Bio-SCAN component assembly. The target DNA will be specifically recognized by the dCas9 (biotin-labeled, bio-dCas9)-sgRNA complex (FAM-labeled, FAM-DNA). The bio-dCas9-sgRNA-DNA combination has been captured by the streptavidin-coated line (T). After applying the FAM-DNA-bio-dCas9-sgRNA to the LFA strip, the concentration of gold (Au) nanoparticle-labeled anti-FAM antibody (AuNP-FAM antibody) throughout the lateral flow will allow the target nucleic acid to be visually recognized at the test line. The introduction of anti-anti-FAM antibodies in the control line (C) serves as a positive control by collecting the unbound residual AuNP FAM antibody

3.2.11.1 GMO Detection with BIO-SCAN

Both governmental regulations and the public's perception of GMOs for human consumption are widespread. In several countries, food, seeds, as well as plant-based products are routinely tested for the presence of GMOs and crops with altered genomes. Turnbull and coworkers (2021) revealed that these tests mostly rely on laboratory detection of transgenic DNA or immunological techniques like ELISA for protein identification encoded by transgenes (Salisu et al. 2017). We also provide Bio-SCAN, a rapid method for detecting the widely used maize Ubiquitin promoter (Ubipro) that is mostly used in monocots, as well as the CaMV 35S promoter, which is utilized for expressing transgenes in dicot plants for the same purpose. Whole genomic DNA, including aT-DNA with both CaMV 35S and Ubipro, was collected from transgenic *Nicotiana benthamiana* plants to assess the Bio-ability SCANs' propensity to recognize frequently occurring transgenic sequences. The target sequences were amplified using RPA, and the amplified products were then subjected to a Bio-SCAN assay.

The appearance of a test band on LFS revealed that the transgenic genes were targeted by designing sgRNAs complementary to CaMV 35S and Ubi promoters. Non-transgenic *N. benthamiana* amplified genomic DNA and sgRNAs were subjected to separate Bio-SCAN assays and did not produce a positive test band. Genomic DNA from non-transgenic and transgenic rice plants having both CaMV 35S and Ubi promoters was extracted and subjected to Bio-SCAN assay, where non-transgenic plants' DNA served as a control. Different sets of primers were used to amplify the target sequence, followed by a special sgRNA-coupled Bio-SCAN assay. Bio-SCAN can help distinguish the transgenic and non-transgenic rice plants. According to these findings, the Bio-SCAN platform might be deployed in the field as a low-cost, simply deployable detection tool that can meet worldwide regulatory standards for rapid GMO screening for food security and public safety (Bak and Emerson 2020; Christensen and Quail 1996). Bio-SCAN was used to successfully identify the Lr67 and Lr34 loci in wheat, OsmALS SGR3 and SGR5 in plants of rice (Sánchez et al. 2022).

3.2.11.2 Detection of Fungal Pathogens with BIOSCAN

Magnaporthe oryzae Triticum (MoT) and *Puccinia striformis* f. sp. *tritici* are the two most dangerous fungal infections that attack wheat. *P. striiformis* f. sp. *tritici* detection sequences help in avoiding the variation in the wheat genome. For the Bio-SCAN experiment to examine sgRNAs' specificity and activity, two sgRNAs targeting the ITS region, the fungal-tubulin gene, and two sets of FAM-labeled RPA primers were made using synthetic DNA. It is indicated that the ITS DNA segment was completely selective when the ITS sgRNA was used against tubulin and ITS-FAM-labeled amplicons. Synthetic DNA was spiked in the genomic DNA of wheat, and the Bio-SCAN experiment was performed to confirm that its DNA could be detected in total genomic DNA isolated from potentially infected plants of wheat.

While the addition of synthetic *P. striiformis* f. sp. *tritici* DNA to wheat genomic DNA did not result in a positive result, *P. striiformis tritici* DNA was uniquely recognized by Bio-SCAN.

Using Bio-SCAN on diseased wheat plants, *M. oryzae Triticum* (MoT) was recognized as another economically important fungal pathogen (Islam et al. 2016). In the recent past, sequences termed MoT-6098 and MoT-6099 were suggested as MoT pathogen-specific indicators (Kang et al. 2021). Two sgRNAs were developed as a result, one of which was directed at the conserved MoT-6098 sequence and the other at the MoT-6099 sequence. Wheat plants infected with MoT were subjected to a complete DNA analysis using an RPA-based Bio-SCAN technique. Wheat plant DNA that had not been infected was used as a negative control. MoT infection in wheat tissues was detected using *M. oryzae Triticum*-specific sgRNAs.

3.2.11.3 Detection of Pathogenic Bacteria with BIOSCAN

Using Bio-SCAN, bacteria such as *A. tumefaciens* and *Pseudomonas syringae* may be precisely and sensitively identified in *N. benthamiana* plants. For this purpose, RPA-based isothermal amplification of *A. tumefaciens* and *P. syringae* synthetic DNA fragments was carried out using specific FAM-labeled primers. The species-specific FAM-labeled amplicons were identified with the sgRNA-based Bio-SCAN test.

This procedure can identify bacterial pathogens in samples of infected plants, which is verified by performing an RPA-based Bio-SCAN test on the complete genomic DNA of *N. benthamiana* plants infected with *A. tumefaciens*. The Bio-SCAN assay was useful in the identification of the *Agrobacterium*-specific sequence on LFA in the *N. benthamiana* plant DNA samples. Such data indicate that Bio-SCAN is a great and effective methodology in detecting pathogens, which could be bacteria, viruses, or fungi. The numerous uses of bio-SCAN make it a precious low-cost resource in the simplification of screening processes in crop breeding and agricultural genetic engineering. It further shows that it can aid GMO food security regulations and enhance diagnostic methods of field crops.

Aman and his coworkers found potato virus X, tobacco mosaic virus, and potato virus Y by combining the RT-RPA technique with a one-step detection assay relying on CRISPR/Cas12a and created the iSCAN one-pot (iSCAN-OP) detection assay (Aman et al. 2018). It uses RT-RPA to transpose viral RNA into dsDNA, which is further used as a substrate of Cas12a cis action. Cas12a's collateral action is activated, which cuts the ssDNA reporter molecules and releases a signal, allowing for low-cost and rapid identification in the field when commercially available P51 fluorescence viewer equipment is combined with iSCAN-OP (Aman et al. 2018; Harrington et al. 2018). This test reliably detects plant viruses and can be completed in 30 minutes at the same temperature.

3.3 Detection of Foodborne Bacteria

The primary causes of events involving food safety are foodborne microorganisms. The preservation of human life and health greatly depends on the accurate and appropriate diagnosis of foodborne pathogens. The detection and quantification of phytopathogenic and spoilage bacteria in food, as well as the modification of bacterial communities in many agricultural operations, can be carried out by CRISPR-related applications. The main goal of the CRISPR detection approach is to achieve the POC testing of foodborne pathogens with accuracy like *L. monocytogenes*, *Vibrio parahaemolyticus*, and *S. aureus* (Ge et al. 2021). The low cost of recombinase polymerase amplification (RPA) with the precision of PCR, accompanied by visual strategies and simplicity of application, makes it suitable for detection purposes.

Beer spoilage bacteria were identified using a very portable CRISPR/Cas12a system in conjunction with RPA primer pairs. Beer spoilage bacteria-derived DNA may be quickly, precisely, and visually identified using fluorescent signals stimulated by blue light. The genomic DNAs resulting from spoilers might also be detected specifically by utilizing a fast experiment and a few basic tools. Future field applications can use assays with these conditions and capabilities (Meng et al. 2021). Early detection of foodborne contamination by pathogenic bacteria like *Salmonella* spp. can both stop the spread of foodborne disease outbreaks and help identify the source of the contamination. A CRISPR-based method called the allosteric probe-initiated catalysis and CRISPR-Cas13a (APC-Cas) system was created to identify *Salmonella enteritidis*, the most common cause of human foodborne salmonellosis (Gao et al. 2021).

This strategy was based on an allosteric ssDNA probe, a highly specific aptamer to *S. enteritidis*, a primer binding site, and a T7 promoter, which was used as the central element of the strategy. The probe was opened by following *S. enteritidis* and the primers bound, and the polymerase duplicated the probe. The probe was recognized and cleaved by Cas13a-crRNA in the presence of the target after replication as RNA by T7 RNA polymerase, followed by collateral cleavage and creation of a fluorescence signal. This method could detect even one bacterium because the produced bacteria were capable of growing repeated rounds of signal production (Gao et al. 2021; Shah et al. 2017). The detection of *Salmonella* spp. can be easy, fast, and specific by employing the one-tube RPA-Cas13a method, which was applied in the present research. The method was a sensitive and useful molecular detection aid for the early diagnosis of diarrheal diseases caused by *Salmonella* infection (An et al. 2021).

It was constructed, combined with (RPA), and assayed with two common foodborne bacteria, *Streptococcus aureus* and *E. coli* O157:H7, respectively, with the help of specific crRNA binding to the *rfbE* and *nuc* genes. OCTOPUS is a one-pot toolbox that is accurate and ultrasensitive, giving a special platform that detects foodborne pathogens in actual food matrices. OCTUPUS technology integrates Cas12a/crRNA cleavage test, preamplification using RPA, and bacterial genome extraction into a single tube [38]. The accuracy and specificity of CRISPR-Cas13a-based bacterial detection (CCB-Detection) to detect *S. aureus* were studied. The newly developed method was successfully applied in detecting *S. aureus* in real food samples, which was better than the best culture-based counting method in the industry (Zhou 2020). The CRISPR-Cas12 (PPCas12) test using a new proximal DNA probe was configured to detect 619 colony-forming units of *Salmonella enterica* without any nucleic acid amplification. The gene recognition of PPCas12 allowed distinguishing *S. enterica* among other foodborne pathogens (Zhang et al. 2021). DNA test is extensively utilized in the diagnosis of plant pathogens and in the administration of genetically modified organisms (GMOs). Even today, there is still a possibility of an easy-to-handle, field-ready, and cheap DNA detection method with high accuracy and sensitivity. RNA programmable nuclease of CRISPR-Cas is also being developed as an alternative nucleic acid detection system, which offers a promising and new DNA test strategy for in-field crop disease diagnosis and identification of GMOs. Here, we report the process of an all-paper-based DNA test with the help of CRISPR-Cas12a. It is a combination of filter-paper-based DNA purification followed by amplification with recombinase polymerase for detecting target genes with Cas12a and a lateral flow assay. This technique is simple to use, efficient, robust, and low-cost, which allows its easy implementation in the field to diagnose crop diseases and GMOs (Zhang et al. 2021).

3.4 Detection of Phytopathogenic Bacteria

It has been considered that CRISPR can have a certain impact on both pathogenicity and the virulence of this organism since it has been used as a clever genotyping technique to display the population structure of *Erwinia amylovora* (Louwen et al. 2014). CRISPR spacer duplication PCR assay in *E. amylovora* identified two major European populations of the pathogen. The analysis followed the Fire blight throughout Europe over decades, where two events of introduction were found about 20 years apart. The Eastern Mediterranean Basin and Northeastern Europe were the pathogen dispersal sites, and they invaded the Eurasian continent (Doolotkeldieva 2021).

The technique employing RPA and CRISPR/Cas12a helped to identify the *Xanthomonas arboricola pv. pruni* (Xap), the cause of peach bacterial spot. The same integrated strategy facilitated the detection of Xap from healthy and diseased peach samples. RPA/Cas12a test aided in separating the *X. arboricola* closely related pathovar Xaj from Xap and other peach pathogens if a high sensitivity of >10–17 M of Xap gDNA is maintained. Due to the thorough evaluation technique's speed, simplicity, efficiency, and adaptability, peach farmers and other users with a basic understanding can perform the test and evaluate the findings in the field (Luo et al. 2021). Three virulence indicators were found using CRISPR genotyping in conjunction with dot blot hybridization; the severity of symptoms and population of the fire blight pathogen (*E. amylovora*) were assessed by immature pear slice bioassays that played a havoc in Portugal. These findings highlight the need for continuous phytosanitary monitoring, particularly of the most dangerous varieties (Mendes et al. 2021a).

The development of the DETECTR test for the sensitive and precise identification of *Candidatus liberibacter asiaticus* (CLas) nucleic acids in disease samples was highlighted (Jaiswal et al. 2021). For future field-deployable diagnosis, the DETECTR test is capable of identifying the CLas from periwinkle and citrus psyllid samples and has been proven to be compatible via lateral flow assay technology (Wheatley et al. 2021). RPA and LAMP were utilized to identify *Erwinia amylovora*, the major cause of fire blight with quarantine concerns, followed by amplicon identification utilizing CRISPR/Cas and LFT with fluorescence detection techniques. They discovered *E. amylovora* in both infected and healthy plants. Primer sets were chosen for each of the three genes that were selected as targets. RPA-CRISPR/Cas, LAMP-LFT, and RPA-LFT were more sensitive than LAMP-CRISPR/Cas and were able to distinguish between individual cells in the reaction. When compared to previously published test systems, the developed RPA-LFT and RPA-CRISPR/Cas test systems have the lowest detection limit for *E. amylovora* (Ivanov et al. 2022). A novel Cas-PfLAMP assay was used to identify three rice pathogens: rice stripe virus (RSV), *Xanthomonas oryzae* pv. *oryzae*, and rice black-streaked dwarf virus (RBSD). By pairing quick DNA/RNA extraction and utilizing a lateral flow strip (LFS) device with the Cas-PfLAMP test to accomplish on-site detection, we expanded the usage of Cas-PfLAMP to point-of-care self-diagnosis. Other nucleic acid field testing, including the detection of new plant diseases, is also a possible use for the CRISPR-CAS assay (Zhu et al. 2022).

3.5 Detection of Fungal Diseases

Pathogenic fungi have been known as a significant threat to both the security of public health and agricultural output. CRISPR-Cas-based nucleic acid detection of fungal infection has been created. A new colorimetric assay was carried out using the CRISPR/Cas9 technique's signal enhancement. In addition, a 456-bp *Phytophthora infestans* dsDNA template sequence was selected, and Cas9/sgRNA was used to recognize and cut the desired sequence. The AuNP aggregation can be generated by interacting with the oligonucleotide-conjugated AuNPs (AuNP probes) with the RCA product (Chang et al. 2019). The ensuing color change can be either observed or recorded with a UV spectrophotometer.

The GMO testing and plant disease diagnosis were carried out by the Cas12a-based DNA detection technology. Two *Magnaporthe oryzae* genes were among the targets of 14 crRNAs, as was a synthetic Cry1C gene that was obtained from *Bacillus thuringiensis* and produces endotoxin and has been exploited to produce Bt-rice (transgenic) in China. The results showed that a small disc of rice leaf could detect transgenic rice and recognize *M. oryzae* (Zhang et al. 2021).

Some CRISPR/Cas12a-based diagnostic assays have been developed to identify plant diseases on a rapid basis. RPA-CRISPR/Cas12a-LFA system was designed to be used in diagnosing citrus scab caused by *Elsino fawcettii* through the incorporation of a lateral flow assay (LFA) readout device. As the template in the RPA reaction, crude DNA extract from diseased citrus fruits and leaves could be utilized. The future of this recently developed technique as a point-of-care diagnostic instrument is quite promising. On-site diagnosing citrus scab may help farmers come up with an accurate and effective control measure (Shin et al. 2021).

Magnaporthe oryzae Triticum might be detected with a precise, sensitive, and affordable approach. The Cas12a protein and guide RNAs particularly target two DNA sequences known as MoT-6098 and MoT-6099, which are absent in the *M. oryzae*-infected rice genome, while present in the *M. oryzae Triticum* genome. The results demonstrate that Cas12a activation displayed widespread ssDNase activity. The target-based Cas12a ssDNase activation with nucleic acid and RPA lateral flow immunoassay combination is then used to swiftly, precisely, and efficiently diagnose the disease in wheat. Mu et al. (2022) revealed a CRISPR-Cas12a-based nucleic acid assay, which is sufficient to differentiate other pathogenic fungi and *Fusarium graminearum* after 4 days of infection to diagnose *Fusarium* head blight (FHB) of wheat, early and quickly. Targets for the screening of guide RNA (gRNA) included the PCR amplicon of the ITS region and the transcription elongation factor 1 (EF1) of *F. graminearum*. Because the Cas12a transactivation protein cleaves ssDNA probes using the terminal fluorophore and quencher groups, by this we can confirm the presence of ITS and EF1 in *F. graminearum.*

In addition to significantly decreasing wheat production during the wheat-growing season, wheat scab is also known to produce several mycotoxins that can contaminate food (Zhou 2020). *F. graminearum* is mostly responsible for wheat scab. Early identification of *F. graminearum* infection was possible due to the combination of PCR and Cas12a. This assay enabled the identification of targets from apple, citrus, and tomato, as well as colorimetric nucleic acid diagnostics of *Alternaria* genes, which encode the citrus brown spot. However, the requirement for nucleic acid extraction, which is challenging to do in the field, hinders the on-site use of these CRISPR assays. A fast nucleic acid extraction method including microneedles was described by the Wei group (Paul et al. 2019), enabling them to gather pathogenic DNAs in 15 minutes. CRISPR-based nucleic acid assays for in-field crop health diagnostics would be possible with the simple sample-pretreatment method.

3.6 Detection of Viral Diseases

Plant viruses are very difficult to eradicate because they spread mechanically through insect vectors and act as necessary parasites. Recent plant virus pandemics and epidemics resulted in colossal losses of more than $30 billion annually, which poses a huge financial impact on the world economy (Jones 2021). Among these are begomoviruses, which are spread by whiteflies and are extremely contagious and frequently have more detrimental effects. Emerging plant viruses bring about disease outbreaks that lower crop quality and yield (Jones 2021). Plant viruses have drastically decreased crop productivity in South America, Asia, Africa, and Europe (Sastry and Zitter 2014). In contrast to other regularly used diagnostic platforms, the detection of targeted nucleic acids utilizing CRISPR-Cas-based platforms has considerably increased in the modern era for viral disease diagnoses. The double-stranded DNA (dsDNA)-detection CRISPR-Cas9 endoribonucleases were used to develop the very first CRISPR-Cas-based diagnosis (Zhang et al. 2017).

The turnip mosaic virus (TuMV), which expresses the green fluorescent protein (GFP), is inhibited by the CRISPR/Cas13a system in stable and transient overexpression lines of *Nicotiana benthamiana*. Pre-crRNAs can be converted by Cas13a into useful crRNAs. CRISPR/Cas13a may be utilized to design interference against RNA viruses, according to estimates (Aman et al. 2018). In just one and a half hours, CRISPR/Cas 13 engineering approaches can identify viroids and prevent their reproduction in plant tissue. As a result, these discoveries might be useful and effective methods for managing economically significant viroid infections. The investigation and diagnosis of viroids have the potential to use NGS CRISPR/Cas 13 approaches (Alon et al. 2021).

Compared to human health, there have been relatively few reports of plant viral disease identification using CRISPR-based diagnostics. In one study, the LAMP-Cas12a technique was used to quickly identify the ssDNA viruses: TYLCV and ToLCNDV. A low concentration of up to 100 aM (60 copies/mL) can be detected using in-tube fluorescence detection (Alon et al. 2021). CRISPR/Cas12a-based detection CRISPR/Cas12a-based detection is a new method for the efficient, sequence-specific detection of viruses. In this case, we are embracing the CRISPR/Cas12a technique in the identification of the tomato brown rugose fruit virus (ToBRFV), a newly identified and emerging tobamovirus that is causing significant losses to the world tomato industry. The detection of ToBRFV or the closely related tomato mosaic virus (ToMV) was designed to be detected by specific CRISPR RNAs (crRNAs). This technology has made it possible to differentiate ToBRFV and ToMV. Sensitivity experiments demonstrated that it is possible to detect viruses at 15–30 ng of RT-PCR product and that it is possible to specifically detect a mix of ToMV and To-BRFV. Moreover, we demonstrate that this approach can facilitate the detection of ToBRFV in the samples obtained in the commercial greenhouses. This evidence indicates that there is a new way to detect the presence of tobamovirus species specifically. A combination of this method in the future with isothermal amplification may offer a platform through which effective and easy methods of differentiating closely related strains and resistance-breaking pathogens may be achieved. CRISPR-Cas12a differentiates the tobamoviruses tomato brown rugose fruit virus (ToBRFV) and tomato mosaic virus (ToMV) (Alon et al. 2021; Mahas et al. 2021). In a similar fashion, LbCas12a was able to be used to introduce differentiation of ToMV and ToBRFV (Alon et al. 2021). The iSCAN-OP also identified two PVY and PVX, which are commercially relevant RNA viruses using Lbcas12a. Simply, the amplification and detection stages were combined in one tube. In less than 30 minutes, the test was able to identify tobacco mosaic virus (TMV) (Aman et al. 2020). Beet necrotic yellow vein virus (BNYVV) causes sugar beet rhizomania and is believed to be a significant issue for sugar beet growers across the globe. The results of a recent study allowed scientists to detect BNYVV at up to 0.1 pM (60200 copies/mL) of a standardized RT-RPA-Cas12a-based fluorescence assay (Ramachandran et al. 2021).

The recent introduction of fruit crops, however, puts the CRISPR-based diagnostics within a considerable range of plant domains. A strong LbCas12a-based assay was detected to identify the apple necrotic mosaic virus (ApNMV), apple chlorotic leaf spot virus (ACLSV), apple stem grooving virus (ASGV), apple scar skin viroid (ASSVd), and the apple stem pitting virus (ASPV).

After then, viral-specific crRNA was used to differentiate between each virus. Using oligonucleotide-conjugated gold nanoparticles, viruses can be recognized directly in crude leaf extract and detected by the human eye an hour later (Jiao et al. 2021). For the pervasive virus/viriose infection of an apple plantation, this assay offers a viable POC treatment. The Grapevine Red-Blotch viral infection was detected on-site by colorimetric method, while the aggregation behavior of AuNP-DNA probes was prolonged by different approaches. The two AuNP-DNA probes' ssDNA linkers were broken in the presence of the virus (as low as 10 aM; 6 copies/mL), which caused the virus to cleave. Due to their continued interaction, the two AuNP-DNA probes eventually got dissociated, giving blue and red colors in the negative and positive samples, respectively (Li et al. 2019). CRISPR-based diagnostics have become a popular alternative for molecular detection assays in plants due to their quick, sensitive, inexpensive, on-site detection; yet, their full promise in this field must be realized.

Recent news has utilized Cas14, Cas13a, and Cas12a, which are CRISPR-linked to recognize the nucleic acids (Chen et al. 2018; Gootenberg et al. 2018; Harrington et al. 2018). The identified method could identify one base mismatch and attomolar concentrations of RNA or DNA. One-Hour Low-Cost Multipurpose Extremely Efficient System (HOLMES) involves the Cas-12a effector system and enables fast and sensitive detection of a target DNA and RNA (Li et al. 2018). When the target is RNA, amplification of the sample and enrichment of the target DNA is performed by use of recombinase polymerase amplification (RPA) process. The RPA product is then changed to RNA by using T7 RNA polymerase. HOLMES, DETECTR, and SHERLOCK offer attomolar sensitivity of transgenes, viruses, and bacteria detection (Chaijarasphong et al. 2019; Myhrvold et al. 2018; Zhang et al. 2020).

The use of CRISPR-based virus detection has colossal potential, yet it is underused. Studies concerning the creation and the use of CRISPR-based diagnostics in the accurate identification of plant viruses were published in recent years. Gootenberg et al. (2018) invented a low-cost, quick, and sensitive lateral flow "paper strip" test method of detecting plant viruses on-site. A one-step RT-RPA-Cas12a test was utilized to determine plant viruses [93]. It involves fluorescent signal release to analyze, cleavage of ssDNA reporter molecules, and collateral action catalyzed by Cas12a endoribonuclease. The iSCAN-OP detection system has been integrated with commercially available fluorescence readers to form a fast and affordable in-field diagnostic system to detect plant RNA viruses. CRISPR-Cas12a-RT-RPA identified 250 copies of ASPV and ASGV per reaction, whereas ACLSV, ASSVd, and ApNMV found 2500 copies of each reaction compared to the previous method of detection, such as RT-qPCR. The ToLCNDV and TYLCV were also identified by the CRISPR-Cas12a method (Mahas et al. 2021). The test resulted in a low-cost on-site assay to diagnose ToLCNDV and TYLCV in 1 hour through the combination of LAMP and CRISPR-Cas12a technology.

CRISPR/Cas12a-based tests are under development to be used in a laboratory and potentially in the field to detect tomato brown rugose fruit virus (ToBRFV). Under specificity testing by other tobamoviruses, our method provided a clear positive outcome with samples that had ToBRFV, and no sign of cross-reactivity with distantly related viruses. CRISPR/Cas12a to identify the presence of plant RNA viruses uses a fluorescence visualizer. It can be done in the field within less than 30 minutes at one temperature. This technology can efficiently and quickly detect banana RNA viruses, so researchers can monitor their control tactics (Tripathi et al. 2021). The advantages of Cas12-based diagnostics to detect both DNA and RNA viruses are that they are required in direct diagnosis of DNA virus without the need of transcription of RNA (Table 3.1).

To visualize the maize chlorotic mottle virus (MCMV), a fast, accurate, and sensitive technique has been used. This RT-RAA-CRISPR/Cas12a detection system of MCMV has been effective according to specificity and sensitivity tests of detecting MCMV and differentiating it against many other non-related plant viruses within 45 minutes. This detection technique of the MCMV is highly effective and has a significantly less sensitive detection threshold than the classic RT-PCR technique (Duan et al. 2022). DETECTR assay using Cas12a is used in accurate and sensitive detection of potato witches' broom and potato purple top provoked by strains of *Phytoplasma trifolii*. The DETECTR test was able to detect 16S–23S ribosomal DNA (rDNA) ITS sequences in two diseased samples at the attomolar sensitivity threshold. The DETECTR system also allows obtaining test results with the help of a lateral flow assay or a fluorescent microplate reader, which can also be used in the diagnosis of the disease in the sphere (Wheatley et al. 2022). Plant health and success are reliant on virus surveillance and detection. Recent advances in CRISPR/Cas technology have led to the development of three key pillars of pathogen detection, namely, SHERLOCK, DETECTR, and SATORI. These pillars are highly versatile to detect numerous types of viral infections without affecting the sensitivity, in addition to addressing the nucleic acid diversity. They can be stored at room temperature over an extended period when compared to other PCR-based reagents. Then, they may be used in rural or distant locations that can help to counter the losses in agricultural yields (Prasad et al. 2022).

Table 3.1 CRISPR-systems employed for fast detection of plant viruses

CRISPR methods	Targeted genomic region	Plant	Plant pathogens	References
CRISPR-Cas12a	ORF1	Tomato	Tomato mosaic virus (ToMV)	Alon et al. (2021)
CRISPR-Cas12a	Coat protein (CP)	Tobacco	Tomato yellow leaf curl virus (TYLCV) and tomato leaf curl New Delhi virus (ToLCNDV)	Mahas et al. (2021)
CRISPR-Cas12a	RNA-1	Sugar beet	Beet necrotic yellow vein virus (BNYVV)	Ramachandran et al. (2021)
CRISPR-Cas12a	Coat protein (CP)	Tobacco	Potato virus X (PVX), potato virus Y (PVY), and tobacco mosaic virus (TMV)	Aman et al. (2020)
CRISPR•Cas9	Rep, p.C1	Tobacco	Cotton leaf curl virus (CLCuV)	Khan et al. (2020)
CRISPR-Cas12a	Cry1 C	Rice	Rice blast disease (*Magnaporthe oryzae*)	Zhang et al. (2020)
CRISPR-Cas9	ORF1. ORF2, and ORF3 of BSV	Banana	Banana streak virus (BSV)	Kimunye et al. (2021)
CRISPR (CR1-CR2-CR3)	T3SS, T3E	Apple, Pear	Fire blight (*Erwinia amylovora*)	Mendes et al. (2021b)
CRISPR-Cas12a	Coat protein (CP)	Apple	Apple chlorotic leaf spot virus (ACLSV), apple necrotic mosaic virus (ApNMV), apple stem grooving virus (ASGV), apple scar skin viroid (ASSVd), and apple stem pitting virus (ASPV)	Jiao et al. (2021)

Cited from (Sharma et al. 2021 Under a Creative Commons license)

3.7 Mycotoxins Detection

Mycotoxins are toxic substances that are generated by various species of fungi and may be measured using colorimetric biosensors, which are built on CRISPR/Cas12a and AuNP. Aflatoxin M1 (AFM1), which is carcinogenic in milk, can be detected by disrupting the AFM1 aptamer that identifies CRISPR/Cas12a (Malhotra et al. 2014). CRISPR-based and AuNP-based new colorimetric sensors have been developed to identify fungi, viruses, bacteria, cancer biomarkers, and poisons such as mycotoxins (Aldewachi et al. 2018). CRISPR/Cas has also shown the capacity to discriminate targets not only of nucleic acids but also of hormones, analytes, and proteins (Bhardwaj et al. 2022). The results demonstrated that the biosensor had a very high affinity to AFM1, and the limit of detection of this method was approximately 0.05 ng/L. It could also recognize AFM1 in polluted milk samples (Abnous et al. 2021). CRISPR-Cas12a-based form of analysis and double isothermal amplification can be used to sensitively and effectively detect toxins in flour samples using a

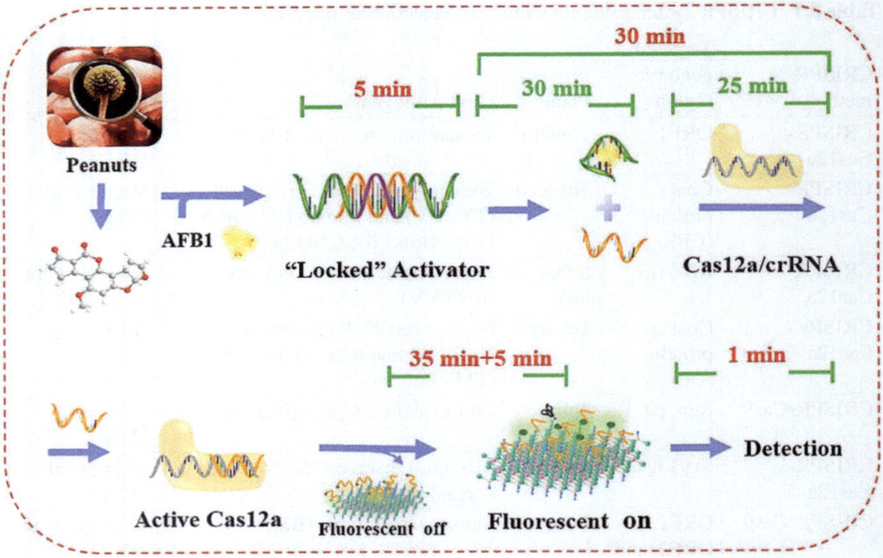

Fig. 3.6 A novel MXene-based fluorescence biosensor having integrated CRISPR/Cas12a for aflatoxin B1 detection. The activator may be released as a result of Apt-1 and Apt-2 preferentially binding to AFB1 when AFB1 is present. The released activator enhanced Cas12a's capacity to cleave ssDNA-FAM by utilizing base complementary pairing for binding to crRNA. The fragmentation of the adsorbed ssDNA-FAM on MXenes allowed for the recovery of fluorescence and the identification of AFB1 at a very low level. (Cited from Wu 2023 Under a Creative Commons license)

biobarcode. The extremely small limit of detection, as well as the simple fluorescence output, are very helpful for identifying toxins in foods (Zhang et al. 2021).

A focused and sensitive CRISPR-Cas12a-modulated LRET-based aptasensor that used MXene-Au as a better quencher and UCNPs as luminescence labels was successful in detecting the vomitoxin deoxynivalenol (DON). DON via aptamer-based target identification can regulate CRISPR-Cas12a's trans-cleavage activity, hence converting it into the DNA signals. The efficacy of quenching is enhanced by the decorating of AuNPs on MXene nanosheets (Lin et al. 2022).

Using nucleic acid aptamers, an upconversion-magnetic probe-DNA-Fe3O4 probe was created to take the place of conventional fluorescent probes in a CRISPR/Cas12a-powered biosensing technique. Ochratoxin A can be detected with great sensitivity, speed, and stability using the CRISPR/Cas12a technique (Mao et al. 2022). To specifically monitor the detection of zearalenone, an aptasensor for signal amplification using a dual enzyme was developed. The Nt.AlwI enzyme and the Cas12a enzyme's two signals are amplified by magnetic beads, which serve as a bridge. The ZEN addition concentration was calculated at 517 nm fluorescence intensity, which can monitor ZEN ranging from 1 pg/mL to 1 ng/mL with a 0.213 pg/mL detection limit. Additionally, in the samples of corn oil, the aptasensor demonstrated a good rate of recovery (Yao et al. 2022). A new fluorescence biosensor was created for the detection of AFB1 based on MXenes and CRISPR/Cas12a. The dual-AFB1 aptamers specifically locked the well-designed activator, Cas12a, which was directly connected to crRNA to produce inactivated complexes (Fig. 3.6), and

MXenes effectively adsorbed FAM fluorophore-modified single-stranded DNA (ssDNA-FAM), dimming its fluorescence (Wu 2023). Additionally, inactive complexes are produced by Cas12a as it is linked with crRNA. A CRISPR/Cas12a-assisted chemiluminescence sensor that can identify target uses in-pipet rolling circle amplification and functional nucleic acid-mediated target identification that can identify aflatoxin B1. Horseradish peroxidase is released as the sensor breaks down pipet-poly-HRP, and AFB1 is then quantified by chemiluminescence. With a detection limit of 5.2 pg/mL, this sensor provides a quick screening method for monitoring food security, i.e., ranging from 50 pg/mL to 100 ng/mL (Wang et al. 2015).

3.8 Challenges and Future Trends

CRISPR technology provides a new opportunity to explore crucial defense-related genes in plants that can enhance crop resistance to insects, pests, and diseases, both of which are necessary to safeguard food safety and long-term agricultural development. The application of Cas13, Cas12, and Cas9 in detection has advanced significantly, although these systems are still difficult, involve extra stages, and use delicate and unstable reagents. Most of these technologies are therefore inappropriate for their numerous field applications in agriculture due to these problems (Sánchez et al. 2022). The combination of CRISPR-Cas RNA-programmable nucleases and nucleic acid detection methods has significantly altered the field of disease diagnosis (Islam and Kasfy 2023).

First, the POCT is hampered by the inevitable usage of preamplification in femtomolar targets for CRISPR-based nucleic acid diagnostics. The ease of the test, the time frame, and aerosol contamination are all hampered by preamplification. To replace nucleic acid amplification, strategies utilizing tandem CRISPR nucleases and CRISPR in combination with other catalytic agents have been published. These techniques hold the potential to provide assays without the need for preamplification (Butt et al. 2019; Yang et al. 2022).

3.9 Conclusion

The CRISPR/Cas system is an imperative tool that helps to understand the interactions between hosts and diseases, develop gain-and-loss-of-function of mutants of genes, and reduce the damage caused by infections that disrupt agricultural ecosystems. Due to its rapid, sensitive, economical, on-site detection, plant molecular detection assays are usually carried out by CRISPR-based diagnostics; yet, their role in this area has not been realized. For POC diagnostics, the use of CRISPR-based technology may play a pivotal role in enabling early disease diagnosis in a variety of crops, the development of a successful control strategy, and eventually an increase in productivity and output. The ability to apply this method for mass diagnosis makes it potentially useful for machine learning, breeding research, and quarantine for artificial intelligence diagnostics. The CRISPR/Cas system and its variations enable innovative research into the difficult topic of plant–pathogen interaction. We anticipate that CRISPR technologies, along with ongoing shifts in farming operations and plant disease systems, will have a significant impact on understanding the relationships between plants and pathogens and developing resilient and multifaceted plants that are disease-resistant in the future.

References

Abnous K, Danesh NM, Ramezani M, Alibolandi M, Nameghi MA, Zavvar TS, Taghdisi SM (2021) A novel colorimetric aptasensor for ultrasensitive detection of aflatoxin M1 based on the combination of CRISPR-Cas12a, rolling circle amplification and catalytic activity of gold nanoparticles. Anal Chim Acta 1165:338549. https://www.sciencedirect.com/science/article/pii/S0003267021003755

Ackerman CM, Myhrvold C, Thakku SG, Freije CA, Metsky HC, Yang DK, Ye SH, Boehm CK, Kosoko-Thoroddsen T-SF, Kehe J, Nguyen TG, Carter A, Kulesa A, Barnes JR, Dugan VG, Hung DT, Blainey PC, Sabeti PC (2020) Massively multiplexed nucleic acid detection with Cas13. Nature 582(7811):277–282. https://doi.org/10.1038/s41586-020-2279-8

Al Hiary H, Bani Ahmad S, Reyalat M, Braik M, ALRahamneh Z (2011) Fast and accurate detection and classification of plant diseases. Int J Comput Appl 17(1):31–38. https://doi.org/10.5120/2183-2754

Aldewachi H, Chalati T, Woodroofe MN, Bricklebank N, Sharrack B, Gardiner P (2018) Gold nanoparticle-based colorimetric biosensors. Nanoscale 10(1):18–33. https://doi.org/10.1039/C7NR06367A

Alon DM, Hak H, Bornstein M, Pines G, Spiegelman Z (2021) Differential detection of the Tobamoviruses tomato mosaic virus (ToMV) and tomato Brown rugose fruit virus (ToBRFV) using CRISPR-Cas12a. Plants 10(6):1256. https://doi.org/10.3390/plants10061256

Aman R, Ali Z, Butt H, Mahas A, Aljedaani F, Khan MZ, Ding S, Mahfouz M (2018) RNA virus interference via CRISPR/Cas13a system in plants. Genome Biol 19(1):1. https://doi.org/10.1186/s13059-017-1381-1

Aman R, Mahas A, Marsic T, Hassan N, Mahfouz MM (2020) Efficient, rapid, and sensitive detection of plant RNA viruses with one-pot RT-RPA–CRISPR/Cas12a assay. Front Microbiol 11:610872. https://doi.org/10.3389/fmicb.2020.610872

An B, Zhang H, Su X, Guo Y, Wu T, Ge Y, Zhu F, Cui L (2021) Rapid and sensitive detection of salmonella spp. using CRISPR-Cas13a combined with recombinase polymerase amplification. Front Microbiol 12:732426. https://doi.org/10.3389/fmicb.2021.732426

Aquino-Jarquin G (2019) CRISPR-Cas14 is now part of the artillery for gene editing and molecular diagnostic. Nanomed Nanotechnol Biol Med 18:428–431. https://doi.org/10.1016/j.nano.2019.03.006

Arslan Z, Hermanns V, Wurm R, Wagner R, Pul Ü (2014) Detection and characterization of spacer integration intermediates in type I-E CRISPR–Cas system. Nucleic Acids Res 42(12):7884–7893. https://doi.org/10.1093/nar/gku510

Ali Z, Sánchez E, Tehseen M, Mahas A, Marsic T, Aman R, Sivakrishna Rao G, Alhamlan FS, Alsanea MS, Al-Qahtani AA, Hamdan S (2021) Bio-SCAN: A CRISPR/dCas9-based lateral flow assay for rapid, specific, and sensitive detection of SARS-CoV-2. ACS Synth Biol 11(1):406–419. https://doi.org/10.1021/acssynbio.1c00499

Azhar M, Phutela R, Kumar M, Ansari AH, Rauthan R, Gulati S, Sharma N, Sinha D, Sharma S, Singh S, Acharya S, Sarkar S, Paul D, Kathpalia P, Aich M, Sehgal P, Ranjan G, Bhoyar RC, Indian CoV2 Genomics & Genetic Epidemiology (IndiCovGEN) Consortium, Singhal K, Lad H, Patra PK, Makharia G, Chandak GR, Pesala B, Chakraborty D, Maiti S (2021) Rapid and accurate nucleobase detection using FnCas9 and its application in COVID-19 diagnosis. Biosens Bioelectron 183:113207. https://doi.org/10.1016/j.bios.2021.113207

Bak A, Emerson JB (2020) Cauliflower mosaic virus (CaMV) biology, management, and relevance to GM Plant detection for sustainable organic agriculture. Front Sustain Food Syst 4:21. https://doi.org/10.3389/fsufs.2020.00021

Batista AC, Pacheco LG (2018) Detecting pathogens with zinc-finger, TALE and CRISPR-based programmable nucleic acid binding proteins. J Microbiol Methods 152:98–104. https://www.sciencedirect.com/science/article/pii/S0167701218303932

Bebber DP, Gurr SJ (2015) Crop-destroying fungal and oomycete pathogens challenge food security. Fungal Genet Biol 74:62–64. https://doi.org/10.1016/j.fgb.2014.10.012

Bhardwaj P, Kant R, Behera SP, Dwivedi GR, Singh R (2022) Next-generation diagnostic with CRISPR/Cas: beyond nucleic acid detection. Int J Mol Sci 23(11):6052. https://doi.org/10.3390/ijms23116052

Bonini A, Poma N, Vivaldi F, Kirchhain A, Salvo P, Bottai D, Tavanti A, Di Francesco F (2021) Advances in biosensing: the CRISPR/Cas system as a new powerful tool for the detection of nucleic acids. J Pharm Biomed Anal 192:113645. https://doi.org/10.1016/j.jpba.2020.113645

Broughton JP, Deng X, Yu G, Fasching CL, Servellita V, Singh J, Miao X, Streithorst JA, Granados A, Sotomayor-Gonzalez A, Zorn K, Gopez A, Hsu E, Gu W, Miller S, Pan C-Y, Guevara H, Wadford DA, Chen JS, Chiu CY (2020) CRISPR–Cas12-based detection of SARS-CoV-2. Nat Biotechnol 38(7):870–874. https://doi.org/10.1038/s41587-020-0513-4

Brouns SJJ, Jore MM, Lundgren M, Westra ER, Slijkhuis RJ, Snijders APL, Dickman MJ, Makarova KS, Koonin EV, van der Oost J (2008) Small CRISPR RNAs guide antiviral defense in prokaryotes. Science 321(5891):960–964. https://doi.org/10.1126/science.1159689

Butt H, Eid A, Momin AA, Bazin J, Crespi M, Arold ST, Mahfouz MM (2019) CRISPR directed evolution of the spliceosome for resistance to splicing inhibitors. Genome Biol 20(1):73. https://doi.org/10.1186/s13059-019-1680-9

Butt H, Rao GS, Sedeek K, Aman R, Kamel R, Mahfouz M (2020) Engineering herbicide resistance via prime editing in rice. Plant Biotechnol J 18(12):2370–2372. https://doi.org/10.1111/pbi.13399

Chaijarasphong T, Thammachai T, Itsathitphaisarn O, Sritunyalucksana K, Suebsing R (2019) Potential application of CRISPR-Cas12a fluorescence assay coupled with rapid nucleic acid amplification for detection of white spot syndrome virus in shrimp. Aquaculture 512:734340. https://doi.org/10.1016/j.aquaculture.2019.734340

Chang W, Liu W, Liu Y, Zhan F, Chen H, Lei H, Liu Y (2019) Colorimetric detection of nucleic acid sequences in plant pathogens based on CRISPR/Cas9 triggered signal amplification. Microchim Acta 186(4):243. https://doi.org/10.1007/s00604-019-3348-2

Chen JS, Ma E, Harrington LB, Da Costa M, Tian X, Palefsky JM, Doudna JA (2018) CRISPR-Cas12a target binding unleashes indiscriminate single-stranded DNase activity. Science 360(6387):436–439. https://doi.org/10.1126/science.aar6245

Christensen AH, Quail PH (1996) Ubiquitin promoter-based vectors for high-level expression of selectable and/or screenable marker genes in monocotyledonous plants. Transgenic Res 5(3):213–218. https://doi.org/10.1007/BF01969712

Chaloner TM, Gurr SJ, Bebber DP (2021) Plant pathogen infection risk tracks global crop yields under climate change. Nat Clim Chang 11:710–715. https://doi.org/10.1038/s41558-021-01104-8

Doolotkeldieva T (2021) Tracking the dissemination of Erwinia amylovora in the Eurasian continent using a PCR targeted on the duplication of a single CRISPR spacer. https://publications.manas.edu.kg/yayin/1736952&dil=3

Duan X, Ma W, Jiao Z, Tian Y, Ismail RG, Zhou T, Fan Z (2022) Reverse transcription-recombinase-aided amplification and CRISPR/Cas12a-based visual detection of maize chlorotic mottle virus. Phytopathol Res 4(1):23. https://doi.org/10.1186/s42483-022-00128-y

Fones HN, Bebber DP, Chaloner TM, Kay WT, Steinberg G, Gurr SJ (2020) Threats to global food security from emerging fungal and oomycete crop pathogens. Nat Food 1(6):332–342. https://doi.org/10.1038/s43016-020-0075-0

Gao S, Liu J, Li Z, Ma Y, Wang J (2021) Sensitive detection of foodborne pathogens based on CRISPR-Cas13a. J Food Sci 86(6):2615–2625. https://doi.org/10.1111/1750-3841.15745

Gasiunas G, Barrangou R, Horvath P, Siksnys V (2012) Cas9–crRNA ribonucleoprotein complex mediates specific DNA cleavage for adaptive immunity in bacteria. Proc Natl Acad Sci 109(39). https://doi.org/10.1073/pnas.1208507109

Gasiunas G, Sinkunas T, Siksnys V (2014) Molecular mechanisms of CRISPR-mediated microbial immunity. Cell Mol Life Sci 71(3):449–465. https://doi.org/10.1007/s00018-013-1438-6

Ge X, Meng T, Tan X, Wei Y, Tao Z, Yang Z, Song F, Wang P, Wan Y (2021) Cas14a1-mediated nucleic acid detectifon platform for pathogens. Biosens Bioelectron 189:113350. https://doi.org/10.1016/j.bios.2021.113350

Gronowski AM (2018) Who or what is SHERLOCK? EJIFCC 29(3):201–204

Glemzaite M, Balciunaite E, Karvelis T, Gasiunas G, Grusyte MM, Alzbutas G, Jurcyte A, Anderson EM, Maksimova E, Smith AJ, Lubys A, Zaliauskiene L, Siksnys V (2015) Targeted gene editing by transfection of *in vitro* reconstituted *Streptococcus thermophilus* Cas9 nuclease complex. RNA Biol 12(1):1–4. https://doi.org/10.1080/15476286.2015.1017209

Gootenberg JS, Abudayyeh OO, Lee JW, Essletzbichler P, Dy AJ, Joung J, Verdine V, Donghia N, Daringer NM, Freije CA, Myhrvold C, Bhattacharyya RP, Livny J, Regev A, Koonin EV, Hung DT, Sabeti PC, Collins JJ, Zhang F (2017) Nucleic acid detection with CRISPR-Cas13a/C2c2. Science 356(6336):438–442. https://doi.org/10.1126/science.aam9321

Gootenberg JS, Abudayyeh OO, Kellner MJ, Joung J, Collins JJ, Zhang F (2018) Multiplexed and portable nucleic acid detection platform with Cas13, Cas12a, and Csm6. Science 360(6387):439–444. https://doi.org/10.1126/science.aaq0179

Gosavi G, Yan F, Ren B, Kuang Y, Yan D, Zhou X, Zhou H (2020) Applications of CRISPR technology in studying plant-pathogen interactions: overview and perspective. Phytopathol Res 2(1):21. https://doi.org/10.1186/s42483-020-00060-z

Huang M, Zhou X, Wang H, Xing D (2018) Clustered regularly interspaced short palindromic repeats/Cas9-triggered isothermal amplification for site-specific nucleic acid detection. Anal Chem 90(3):2193–2200. https://doi.org/10.1021/acs.analchem.7b04542

Harrington LB, Burstein D, Chen JS, Paez-Espino D, Ma E, Witte IP, Cofsky JC, Kyrpides NC, Banfield JF, Doudna JA (2018) Programmed DNA destruction by miniature CRISPR-Cas14 enzymes. Science 362(6416):839–842. https://doi.org/10.1126/science.aav4294

Islam T, Kasfy SH (2023) CRISPR-based point-of-care plant disease diagnostics. Trends Biotechnol 41(2):144–146. https://www.cell.com/trends/biotechnology/abstract/S0167-7799(22)00273-6

Islam MT, Croll D, Gladieux P, Soanes DM, Persoons A, Bhattacharjee P, Hossain MS, Gupta DR, Rahman MM, Mahboob MG, Cook N, Salam MU, Surovy MZ, Sancho VB, Maciel JLN, NhaniJúnior A, Castroagudín VL, Reges JTDA, Ceresini PC et al (2016) Emergence of wheat blast in Bangladesh was caused by a south American lineage of Magnaporthe oryzae. BMC Biol 14(1):84. https://doi.org/10.1186/s12915-016-0309-7

Ivanov AV, Safenkova IV, Drenova NV, Zherdev AV, Dzantiev BB (2022) Comparison of biosensing methods based on different isothermal amplification strategies: a case study with Erwinia amylovora. Biosensors 12(12):1174. https://doi.org/10.3390/bios12121174

Jaiswal D, Sidharthan VK, Sharma SK, Rai R, Choudhary N, Ghosh A, Baranwal VK (2021) Candidatus Liberibacter asiaticus manipulates the expression of vitellogenin, cytoskeleton, and endocytotic pathway-related genes to become circulative in its vector, Diaphorina citri (Hemiptera: Psyllidae). 3 Biotech 11(2):88. https://doi.org/10.1007/s13205-021-02641-x

Jiao J, Kong K, Han J, Song S, Bai T, Song C, Wang M, Yan Z, Zhang H, Zhang R, Feng J, Zheng X (2021) Field detection of multiple RNA viruses/viroids in apple using a CRISPR/Cas12a-based visual assay. Plant Biotechnol J 19(2):394–405. https://doi.org/10.1111/pbi.13474

Jones RAC (2021) Global plant virus disease pandemics and epidemics. Plants 10(2):233. https://doi.org/10.3390/plants10020233

Joung J, Ladha A, Saito M, Segel M, Bruneau R, Huang MW, Kim N-G, Yu X, Li J, Walker BD, Greninger AL, Jerome KR, Gootenberg JS, Abudayyeh OO, Zhang F (2020) Point-of-care testing for COVID-19 using SHERLOCK diagnostics. Infectious diseases (except HIV/AIDS). https://doi.org/10.1101/2020.05.04.20091231

Kalinina NO, Khromov A, Love AJ, Taliansky ME (2020) CRISPR applications in plant virology: virus resistance and beyond. Phytopathology 110(1):18–28. https://doi.org/10.1094/PHYTO-07-19-0267-IA

Kaminski MM, Abudayyeh OO, Gootenberg JS, Zhang F, Collins JJ (2021) CRISPR-based diagnostics. Nat Biomed Eng 5(7):643–656. https://www.nature.com/articles/s41551-021-00760-7

Kellner MJ, Koob JG, Gootenberg JS, Abudayyeh OO, Zhang F (2019) SHERLOCK: nucleic acid detection with CRISPR nucleases. Nat Protoc 14(10):2986–3012. https://doi.org/10.1038/s41596-019-0210-2

Khan M, Khan AU, Hasan MA, Yadav KK, Pinto MMC, Malik N, Yadav VK, Khan AH, Islam S, Sharma GK (2021) Agro-nanotechnology as an emerging field: A novel sustainable approach for improving plant growth by reducing biotic stress. Appl Sci 11(5):2282. https://doi.org/10.3390/app11052282

Khan H, Khan A, Liu Y, Wang S, Bibi S, Xu H, Liu Y, Durrani S, Jin L, He N, Xiong T (2019) CRISPR-Cas13a mediated nanosystem for attomolar detection of canine parvovirus type 2. Chin Chem Lett 30(12):2201–2204. https://doi.org/10.1016/j.cclet.2019.10.032

Khan S, Mahmood MS, Ur Rahman S, Rizvi F, Ahmad A (2020) Evaluation of the CRISPR/Cas9 system for the development of resistance against cotton leaf curl virus in model plants. Plant Prot Sci 56(3):154–162. https://doi.org/10.17221/105/2019-PPS

Kaur P, Harnal S, Gautam V, et al. (2023) A novel transfer deep learning method for detection and classification of plant leaf disease. J Ambient Intell Human Comput 14:12407–12424. https://doi.org/10.1007/s12652-022-04331-9

Kang H, Peng Y, Hua K, Deng Y, Bellizzi M, Gupta DR, et al. (2021) Rapid detection of wheat blast pathogen Magnaporthe oryzae triticum pathotype using genome-specific primers and Cas12a-mediated technology. Engineering 7:1326–1335. https://doi.org/10.1016/j.eng.2020.07.016

Kim D, Lee J-Y, Yang J-S, Kim JW, Kim VN, Chang H (2020) The architecture of SARS-CoV-2 transcriptome. Cell 181(4):914–921.e10. https://doi.org/10.1016/j.cell.2020.04.011

Kimunye J, Were E, Swennen R, Viljoen A, Mahuku G (2021) Sources of resistance to Pseudocercospora fijiensis, the cause of black Sigatoka in banana. Plant Pathol 70:1651–1664. https://doi.org/10.1111/ppa.13408

Li C, Zong Y, Wang Y, Jin S, Zhang D, Song Q, Zhang R, Gao C (2018) Expanded base editing in rice and wheat using a Cas9-adenosine deaminase fusion. Genome Biol 19(1):59. https://doi.org/10.1186/s13059-018-1443-z

Li Y, Li S, Wang J, Liu G (2019) CRISPR/Cas systems towards next-generation biosensing. Trends Biotechnol 37(7):730–743. https://doi.org/10.1016/j.tibtech.2018.12.005

Li P, Wang L, Yang J, Di L-J, Li J (2021) Applications of the CRISPR-Cas system for infectious disease diagnostics. Expert Rev Mol Diagn 21(7):723–732. https://doi.org/10.1080/14737159.2021.1922080

Lin X, Li C, Meng X, Yu W, Duan N, Wang Z, Wu S (2022) CRISPR-Cas12a-mediated luminescence resonance energy transfer aptasensing platform for deoxynivalenol using gold nanoparticle-decorated Ti3C2Tx MXene as the enhanced quencher. J Hazard Mater 433:128750. https://www.sciencedirect.com/science/article/pii/S0304389422005398

Louwen R, Staals RHJ, Endtz HP, Van Baarlen P, Van Der Oost J (2014) The role of CRISPR-Cas Systems in Virulence of pathogenic bacteria. Microbiol Mol Biol Rev 78(1):74–88. https://doi.org/10.1128/MMBR.00039-13

Luo M, Meng F-Z, Tan Q, Yin W-X, Luo C-X (2021) Recombinase polymerase amplification/Cas12a-based identification of Xanthomonas arboricola pv. Pruni on peach. Front Plant Sci 12:740177. https://doi.org/10.3389/fpls.2021.740177

Mahas A, Hassan N, Aman R, Marsic T, Wang Q, Ali Z, Mahfouz MM (2021) Lamp-coupled CRISPR–Cas12a module for rapid and sensitive detection of plant DNA viruses. Viruses 13(3):466. https://doi.org/10.3390/v13030466

Makarova KS, Wolf YI, Iranzo J, Shmakov SA, Alkhnbashi OS, Brouns SJJ, Charpentier E, Cheng D, Haft DH, Horvath P, Moineau S, Mojica FJM, Scott D, Shah SA, Siksnys V, Terns MP, Venclovas Č, White MF, Yakunin AF et al (2020) Evolutionary classification of CRISPR–Cas systems: a burst of class 2 and derived variants. Nat Rev Microbiol 18(2):67–83. https://doi.org/10.1038/s41579-019-0299-x

Malhotra S, Pandey AK, Rajput YS, Sharma R (2014) Selection of aptamers for aflatoxin M1 and their characterization. J Mol Recognit 27(8):493–500. https://doi.org/10.1002/jmr.2370

Mao Z, Chen R, Wang X, Zhou Z, Peng Y, Li S, Han D, Li S, Wang Y, Han T, Liang J, Ren S, Gao Z (2022) CRISPR/Cas12a-based technology: a powerful tool for biosensing in food safety. Trends Food Sci Technol 122:211–222. https://doi.org/10.1016/j.tifs.2022.02.030

Marqués MC, Sánchez-Vicente J, Ruiz R, Montagud-Martínez R, Márquez-Costa R, Gómez G, Carbonell A, Daròs Arnau JA, Rodrigo G (2022) Supporting information: diagnostics of infections produced by the plant viruses TMV, TEV, and PVX with CRISPR-Cas12 and CRISPR-Cas13. https://digital.csic.es/handle/10261/330872

Mathew AG, Cissell R, Liamthong S (2007) Antibiotic resistance in bacteria associated with food animals: a United States perspective of livestock production. Foodborne Pathog Dis 4(2):115–133. https://doi.org/10.1089/fpd.2006.0066

McGinn J, Marraffini LA (2019) Molecular mechanisms of CRISPR–Cas spacer acquisition. Nat Rev Microbiol 17:7–12. https://doi.org/10.1038/s41579-018-0071-7

Mendes RJ, Luz JP, Santos C, Tavares F (2021a) CRISPR genotyping as complementary tool for epidemiological surveillance of Erwinia amylovora outbreaks. PLoS One 16(4):e0250280. https://doi.org/10.1371/journal.pone.0250280

Mendes RJ, Luz JP, Santos C, Tavares F (2021b) CRISPR genotyping as complementary tool for epidemiological surveillance of Erwinia amylovora outbreaks. PLoS One 16(4 April):1–18. https://doi.org/10.1371/journal.pone.0250280

Meng Q, Yang H, Zhang G, Sun W, Ma P, Liu X, Dang L, Li G, Huang X, Wang X (2021) CRISPR/Cas12a-assisted rapid identification of key beer spoilage bacteria. Innov Food Sci Emerg Technol 74:102854. https://www.sciencedirect.com/science/article/pii/S1466856421002551

Mu K, Ren X, Yang H, Zhang T, Yan W, Yuan F, Wu J, Kang Z, Han D, Deng R, Zeng Q (2022) CRISPR-Cas12a-based diagnostics of wheat fungal diseases. J Agric Food Chem 70(23):7240–7247. https://doi.org/10.1021/acs.jafc.1c08391

Mustafa MI, Makhawi AM (2021) SHERLOCK and DETECTR: CRISPR-Cas systems as potential rapid diagnostic tools for emerging infectious diseases. J Clin Microbiol 59(3):e00745–e00720. https://doi.org/10.1128/JCM.00745-20

Myhrvold C, Freije CA, Gootenberg JS, Abudayyeh OO, Metsky HC, Durbin AF, Kellner MJ, Tan AL, Paul LM, Parham LA, Garcia KF, Barnes KG, Chak B, Mondini A, Nogueira ML, Isern S, Michael SF, Lorenzana I, Yozwiak NL et al (2018) Field-deployable viral diagnostics using CRISPR-Cas13. Science 360(6387):444–448. https://doi.org/10.1126/science.aas8836

Nicaise V (2014) Crop immunity against viruses: outcomes and future challenges. Front Plant Sci 5. https://doi.org/10.3389/fpls.2014.00660

Oerke E-C (2006) Crop losses to pests. J Agric Sci 144(1):31–43. https://doi.org/10.1017/S0021859605005708

Pardee K, Green AA, Takahashi MK, Braff D, Lambert G, Lee JW, Ferrante T, Ma D, Donghia N, Fan M, Daringer NM, Bosch I, Dudley DM, O'Connor DH, Gehrke L, Collins JJ (2016) Rapid, low-cost detection of Zika virus using programmable biomolecular components. Cell 165(5):1255–1266. https://doi.org/10.1016/j.cell.2016.04.059

Paul R, Saville AC, Hansel JC, Ye Y, Ball C, Williams A, Chang X, Chen G, Gu Z, Ristaino JB, Wei Q (2019) Extraction of plant DNA by microneedle patch for rapid detection of plant diseases. ACS Nano 13(6):6540–6549. https://doi.org/10.1021/acsnano.9b00193

Prasad D, Mani NK, Pandey DM (2022) CRISPR/Cas technology: opportunities for phytopathogenic viruses detection. J Biotechnol 360:211–217. https://www.sciencedirect.com/science/article/pii/S016816562200270X
Qin P, Park M, Alfson KJ, Tamhankar M, Patterson JL, Griffiths A, He Q, Yildiz A, Du K (2019) Rapid and fully microfluidic Ebola virus detection with CRISPR-Cas13a. ACS Sens 4(4):1048–1054
Qiu X-Y, Zhu L-Y, Zhu C-S, Ma J-X, Hou T, Wu X-M, Xie S-S, Min L, Tan D-A, Zhang D-Y, Zhu L (2018) Highly effective and low-cost MicroRNA detection with CRISPR-Cas9. ACS Synth Biol 7(3):807–813. https://doi.org/10.1021/acssynbio.7b00446
Quan J, Langelier C, Kuchta A, Batson J, Teyssier N, Lyden A, Caldera S, McGeever A, Dimitrov B, King R, Wilheim J, Murphy M, Ares LP, Travisano KA, Sit R, Amato R, Mumbengegwi DR, Smith JL, Bennett A et al (2019) FLASH: a next-generation CRISPR diagnostic for multiplexed detection of antimicrobial resistance sequences. Nucleic Acids Res 47(14):e83. https://doi.org/10.1093/nar/gkz418
Ramachandran V, Weiland JJ, Bolton MD (2021) CRISPR-based isothermal next-generation diagnostic method for virus detection in Sugarbeet. Front Microbiol 12:679994. https://doi.org/10.3389/fmicb.2021.679994
Rauch JN, Valois E, Solley SC, Braig F, Lach RS, Audouard M, Ponce-Rojas JC, Costello MS, Baxter NJ, Kosik KS, Arias C, Acosta-Alvear D, Wilson MZ (2021) A scalable, easy-to-deploy protocol for Cas13-based detection of SARS-CoV-2 genetic material. J Clin Microbiol 59(4):e02402–e02420. https://doi.org/10.1128/JCM.02402-20
Rumpf T, Mahlein A-K, Steiner U, Oerke E-C, Dehne H-W, Plümer L (2010) Early detection and classification of plant diseases with support vector machines based on hyperspectral reflectance. Comput Electron Agric 74(1):91–99. https://www.sciencedirect.com/science/article/pii/S0168169910001262
Sapranauskas R, Gasiunas G, Fremaux C, Barrangou R, Horvath P, Siksnys V (2011) The Streptococcus thermophilus CRISPR/Cas system provides immunity in Escherichia coli. Nucleic Acids Res 39(21):9275–9282. https://doi.org/10.1093/nar/gkr606
Sasnauskas G, Siksnys V (2020) CRISPR adaptation from a structural perspective. Curr Opin Struct Biol 65:17–25. https://doi.org/10.1016/j.sbi.2020.05.015
Salisu IB, Shahid AA, Yaqoob A, Ali Q, Bajwa KS, Rao AQ, Husnain T (2017) Molecular approaches for high throughput detection and quantification of genetically modified crops: a review. Front Plant Sci 8:1670. https://doi.org/10.3389/fpls.2017.01670
Sánchez E, Ali Z, Islam T, Mahfouz M (2022) A CRISPR -based lateral flow assay for plant genotyping and pathogen diagnostics. Plant Biotechnol J 20(12):2418–2429. https://doi.org/10.1111/pbi.13924
Sastry KS, Zitter TA (2014) Plant virus and viroid diseases in the tropics: volume 2: epidemiology and management. Springer Netherlands. https://doi.org/10.1007/978-94-007-7820-7
Savary S, Willocquet L, Pethybridge SJ, Esker P, McRoberts N, Nelson A (2019) The global burden of pathogens and pests on major food crops. Nat Ecol Evol 3(3):430–439. https://doi.org/10.1038/s41559-018-0793-y
Schindele P, Wolter F, Puchta H (2018) Transforming plant biology and breeding with CRISPR/Cas9, Cas12 and Cas13. FEBS Lett 592(12):1954–1967. https://doi.org/10.1002/1873-3468.13073
Shah DH, Elder JR, Chiok KL, Paul NC (2017) Genetic basis of salmonella Enteritidis pathogenesis in chickens. In: Producing safe eggs. Elsevier, pp 187–208. https://doi.org/10.1016/B978-0-12-802582-6.00010-0
Shami A, Mostafa M, Abd-Elsalam KA (2021) CRISPR applications in plant bacteriology: today and future perspectives. In: CRISPR and RNAi Systems, pp 551–577. https://www.sciencedirect.com/science/article/pii/B9780128219102000114
Sharma SK, Gupta OP, Pathaw N, Sharma D, Maibam A, Sharma P, Sanasam J, Karkute SG, Kumar S, Bhattacharjee B (2021) CRISPR-Cas-led revolution in diagnosis and management of emerging plant viruses: new avenues toward food and nutritional security. Front Nutr 8:751512. https://doi.org/10.3389/fnut.2021.751512

Shin K, Kwon S-H, Lee S-C, Moon Y-E (2021) Sensitive and rapid detection of citrus scab using an RPA-CRISPR/Cas12a system combined with a lateral flow assay. Plants 10(10):2132. https://doi.org/10.3390/plants10102132

Sternberg SH, Redding S, Jinek M, Greene EC, Doudna JA (2014) DNA interrogation by the CRISPR RNA-guided endonuclease Cas9. Biophys J 106(2):695a. https://doi.org/10.1016/j.bpj.2013.11.3848

Strange RN, Scott PR (2005) Plant disease: a threat to global food security. Annu Rev Phytopathol 43(1):83–116. https://doi.org/10.1146/annurev.phyto.43.113004.133839

Strich JR, Chertow DS (2019) CRISPR–Cas biology and its application to infectious diseases. J Clin Microbiol 57(4):e01307–18. https://doi.org/10.1128/JCM.01307-18

Taniguchi T, Ohki M, Urata A, Ohshiro S, Tarigan E, Kiatsomphob S, Vetchapitak T, Sato H, Misawa N (2021) Detection and identification of adhesins involved in adhesion of campylobacter jejuni to chicken skin. Int J Food Microbiol 337:108929. https://doi.org/10.1016/j.ijfoodmicro.2020.108929

Tripathi L, Ntui VO, Tripathi JN, Kumar PL (2021) Application of CRISPR/Cas for diagnosis and management of viral diseases of banana. Front Microbiol 11:609784. https://doi.org/10.3389/fmicb.2020.609784

Turnbull C, Lillemo M, Hvoslef-Eide TAK (2021) Global regulation of genetically modified crops amid the gene edited crop boom—a review. Front Plant Sci 12:630396. https://doi.org/10.3389/fpls.2021.630396

Uygun ZO, Yeniay L, Sağın FG (2020) CRISPR-dCas9 powered impedimetric biosensor for label-free detection of circulating tumor DNAs. Anal Chim Acta 1121:35–41. https://www.sciencedirect.com/science/article/pii/S0003267020303986

Varanda CMR, Félix MDR, Campos MD, Patanita M, Materatski P (2021) Plant viruses: from targets to tools for CRISPR. Viruses 13(1):141. https://doi.org/10.3390/v13010141

Verosloff M, Chappell J, Perry KL, Thompson JR, Lucks JB (2019) PLANT-dx: a molecular diagnostic for point-of-use detection of plant pathogens. ACS Synth Biol 8(4):902–905. https://doi.org/10.1021/acssynbio.8b00526

Wang J, Zhang Y, Li Y, Wang X, Nan W, Hu Y, Zhang H, Zhao C, Wang F, Li P, Shi H, Bi Y (2015) Endophytic microbes bacillus sp. LZR216-regulated root development is dependent on polar auxin transport in Arabidopsis seedlings. Plant Cell Rep 34(6):1075–1087. https://doi.org/10.1007/s00299-015-1766-0

Wang S-Y, Du Y-C, Wang D-X, Ma J-Y, Tang A-N, Kong D-M (2021) Signal amplification and output of CRISPR/Cas-based biosensing systems: a review. Anal Chim Acta 1185:338882. https://www.sciencedirect.com/science/article/pii/S000326702100708X

Wang J, Yang X, Wang X, Wang W (2022) Recent advances in CRISPR/Cas-based biosensors for protein detection. Bioengineering (Basel) 9(10):512. https://doi.org/10.3390/bioengineering9100512

Wang X, Ji P, Fan H, et al. (2020) CRISPR/Cas12a technology combined with immunochromatographic strips for portable detection of African swine fever virus. Commun Biol 3:62. https://doi.org/10.1038/s42003-020-0796-5

Wheatley MS, Duan Y-P, Yang Y (2021) Highly sensitive and rapid detection of citrus Huanglongbing pathogen ('*Candidatus* Liberibacter asiaticus') using Cas12a-based methods. Phytopathology 111(12):2375–2382. https://doi.org/10.1094/PHYTO-09-20-0443-R

Wheatley MS, Wang Q, Wei W, Bottner-Parker KD, Zhao Y, Yang Y (2022) Cas12a-based diagnostics for potato purple top disease complex associated with infection by '*Candidatus* Phytoplasma trifolii'-related strains. Plant Dis 106(8):2039–2045. https://doi.org/10.1094/PDIS-09-21-2119-RE

Wilkinson M, Drabavicius G, Silanskas A, Gasiunas G, Siksnys V, Wigley DB (2019) Structure of the DNA-bound spacer capture complex of a type II CRISPR-Cas system. Mol Cell 75(1):90–101.e5. https://doi.org/10.1016/j.molcel.2019.04.020

Wu Z (2023) A novel fluorescence biosensor based on CRISPR/Cas12a integrated MXenes for detecting aflatoxin B1. Talanta 252

Xiang J, Yan M, Li H, Liu T, Lin C, Huang S, Shen C (2020) Evaluation of enzyme-linked immunoassay and colloidal gold-Immunochromatographic assay kit for detection of novel coronavirus (SARS-Cov-2) causing an outbreak of pneumonia (COVID-19). Epidemiology. https://doi.org/10.1101/2020.02.27.20028787

Yan F, Wang W, Zhang J (2019) CRISPR-Cas12 and Cas13: the lesser known siblings of CRISPR-Cas9. Cell Biol Toxicol 35(6):489–492. https://doi.org/10.1007/s10565-019-09489-1

Yang W, Restrepo-Pérez L, Bengtson M, Heerema SJ, Birnie A, Van Der Torre J, Dekker C (2018) Detection of CRISPR-dCas9 on DNA with solid-state Nanopores. Nano Lett 18(10):6469–6474. https://doi.org/10.1021/acs.nanolett.8b02968

Yang X, Li J, Zhang S, Li C, Ma J (2022) Amplification-free, single-microbead-based Cas12a assay for one-step DNA detection at the single-molecule level. Analytical Chemistry 94(38):13076–13083. https://pubs.acs.org/doi/abs/10.1021/acs.analchem.2c02283

Yao X, Yang Q, Wang Y, Bi C, Du H, Wu W (2022) Dual-enzyme-based signal-amplified Aptasensor for Zearalenone detection by using CRISPR-Cas12a and Nt.AlwI. Foods 11(3):487. https://doi.org/10.3390/foods11030487

Zhang Y, Qian L, Wei W, Wang Y, Wang B, Lin P, Liu W, Xu L, Li X, Liu D, Cheng S, Li J, Ye Y, Li H, Zhang X, Dong Y, Zhao X, Liu C, Zhang HM et al (2017) Paired design of dCas9 as a systematic platform for the detection of featured nucleic acid sequences in pathogenic strains. ACS Synth Biol 6(2):211–216. https://doi.org/10.1021/acssynbio.6b00215

Zhang Y, Zhang Y, Xie K (2020) Evaluation of CRISPR/Cas12a-based DNA detection for fast pathogen diagnosis and GMO test in rice. Mol Breed 40(1):11. https://doi.org/10.1007/s11032-019-1092-2

Zhang Y-M, Yang Y, Xie K (2021) CRISPR-Cas12a-based DNA detection for fast pathogen diagnosis and GMO test in plants. In: Islam MT, Molla KA (eds) CRISPR-Cas methods. Springer US, pp 221–233. https://doi.org/10.1007/978-1-0716-1657-4_15

Zhou J (2020) CRISPR-Cas13a based bacterial detection platform: sensing pathogen Staphylococcus aureus in food samples. Anal Chim Acta 1127:225

Zhu Z, Li R, Zhang H, Wang J, Lu Y, Zhang D, Yang L (2022) PAM-free loop-mediated isothermal amplification coupled with CRISPR/Cas12a cleavage (Cas-PfLAMP) for rapid detection of rice pathogens. Biosens Bioelectron 204:114076. https://www.sciencedirect.com/science/article/pii/S0956566322001166

Chapter 4
CRISPR/Cas12a-Based Biosensors and Applications in Cereal Safety

Fatma Nur Basturk, Seran Kırkıncı, Nazlıcan Yıldırım, Yiğit Sabri Unlu, Hande Morgil, Emirhan Çağla, Berke Çevik, Kadir Boztas, Gül Cevahir Öz, and Yusuf Can Gerçek

Abstract The book chapter demonstrates how CRISPR/Cas12a genome editing technology enables agricultural breakthroughs that enhance food security. The worldwide food supply depends heavily on grains, but these crops encounter two major threats from climate change and pathogenic attacks. CRISPR-/Cas12a-based biosensors provide an efficient, portable, and affordable solution to fight these threats through their operation beyond conventional methods. The CRISPR/Cas12a system performs trans-cleavage through its guide RNA (crRNA), which guides the system to recognize particular DNA sequences. The system generates fluorescent or colored signals through probe cleavage, which occurs when the activity is activated. The system functions as a biosensor to detect nucleic acid contaminants and non-nucleic acid substances, including mycotoxins, pesticides, heavy metals, and bacteria. The detection capabilities of CRISPR/Cas12a biosensors surpass those of conventional methods because they provide better specificity and sensitivity. The technology will merge with the Internet of Things (IoT) and artificial intelligence systems to create better crops that fight diseases and stresses while ensuring safer food distribution throughout the supply chain. The combined systems enable real-time monitoring and transparency to stop food fraud and defend public health through integrated capabilities.

Fatma Nur Basturk and Seran Kırkıncı contributed equally with all other contributors.

F. N. Basturk · S. Kırkıncı · N. Yıldırım · Y. S. Unlu · E. Çağla · B. Çevik
Institute of Graduate Studies in Sciences, Istanbul University, Istanbul, Turkey

H. Morgil · K. Boztas · G. Cevahir Öz · Y. C. Gerçek (✉)
Department of Biology, Faculty of Science, Istanbul University, Istanbul, Turkey
e-mail: yusuf.gercek@istanbul.edu.tr

K. A. Abd-Elsalam, A. Ahmad (eds.), *Gene Editing in Cereals*, Sustainability Sciences in Asia and Africa, https://doi.org/10.1007/978-981-95-6722-5_4

4.1 Introduction

Agriculture is one of the most critical sectors of our planet in terms of sustaining human nutrition. Today, crop production is facing serious challenges due to climate change, various abiotic stresses (especially drought), and pathogenic damage. To overcome these challenges, plant scientists have developed and implemented various new molecular tools to increase crop yield and quality. In recent years, the CRISPR/Cas12a genome-editing system has been successfully applied to many field crops to increase stress tolerance and improve yield capacity. Cereal crops have been the main food sources providing energy and essential nutrients to humans for thousands of years, and they play a vital role in human life. Since they meet approximately 90% of the global population's food needs, cereals are grown in much larger areas than other plant species. Rice, wheat, and maize, in particular, constitute the staple cereal crops for the vast majority of the world's population. However, various environmental threats such as high temperature, drought, salinity, frost, bacteria, weeds, and viruses seriously threaten cereal production. To overcome these global challenges, various new and innovative molecular biotechnological approaches are being developed for the genetic improvement of cereal crops. In this review, the recently discovered CRISPR/Cas12a genome-editing system offers great potential in making cereals more resistant to adverse environmental conditions. In this review, the application of CRISPR/Cas12a technology to cereal crops is comprehensively discussed, and the effects of the system on agricultural productivity and climate resilience are discussed.

4.2 Cereal Food Safety: Challenges and Current Approaches

Cereal foods serve as primary dietary sources of macro- and micronutrients, which maintain essential roles in human metabolic wellness (Maqbool et al. 2024). The nutritional value of cereals stems from their composition of carbohydrates, proteins, fats, water-soluble fibers, vitamins, and minerals (Senarathna et al. 2024). Cereals serve as vital dietary components that contain essential bioactive compounds, including dietary fibers, lignans, sterols, phenolic acids, phytoestrogens, and antioxidants, which help prevent diseases (Varzakas 2016). Food safety involves safeguarding food products against microbial and chemical threats. The main focus of food safety consists of identifying potential hazards followed by evaluating their occurrence rates and impact levels and establishing risk reduction strategies (Alldrick 2017). All food categories, including cereal products, remain vulnerable to chemical as well as physical and microbiological hazards. The food industry has encountered major difficulties in food safety management because of disease outbreaks that result from contaminated food products (Djekic et al. 2022). The rising global population drives up food consumption, which makes it essential to maintain rigorous food safety and quality control measures to stop pathogen-related health risks in food production (Rahman et al. 2022). Serious outbreaks occur when people eat contaminated foods, according to previous research, which also forecasts more foodborne disease outbreaks because of rising environmental pollution and population density (Song et al. 2017).

Cereal plants are grown together, and their preservation environment determines food safety standards. Wang and Liu (2015) established that consumer preferences for cereal products depend on several factors, including brand reputation, safety attributes, nutritional value, and sensory qualities. Cereals encounter microbial threats in the field environment and during storage. The toxic compounds known as mycotoxins, which arise from various fungal species, pose dangerous health hazards to consumers, according to Bennett and Klich (2003) and Milani and Maleki (2014). Even proper storage practices will not eliminate mycotoxin contamination from grains that are vulnerable to fungal producers. According to the Food and Agriculture Organization (FAO), mycotoxins contaminate one-quarter of the world's cereal crops (Pankaj et al. 2018; Mahato et al. 2019). Human mycotoxicosis cases have become more prevalent during recent decades because people eat contaminated cereals, which endangers both human health and agricultural productivity (Kamle et al. 2019, 2022a, b). The toxic nature and carcinogenic properties of mycotoxins in cereal crops create a major food safety problem that leads to major economic losses (Campagnollo et al. 2016). The growing population demands urgent improvement of mechanical, biological, thermal, and non-thermal and biotechnological methods to achieve safe and quality cereals for human consumption (Senarathna et al. 2024). The safety of cereal-based foods is threatened by gluten proteins because they cause adverse reactions in people who have celiac disease or wheat allergy or gluten sensitivity (Wieser et al. 2021). The protection of these individuals from severe health risks depends on accurate labeling, along with preventing cross-contamination. The presence of pesticide residues stands as a significant safety issue that affects the quality of cereal-based foods. The presence of residual pesticides in agricultural products creates major health risks because they contain carcinogenic and allergenic and estrogenic and mutagenic properties (Wołejko et al. 2014; Valcke et al. 2017; Nardelli et al. 2021; Shalaby et al. 2021; Jankowska et al. 2022). The regulatory reports demonstrate that pesticide residue monitoring in cereals plays an essential role in maintaining food safety (EFSA, 2020; Jankowska and Łozowicka 2021). The development of environmentally friendly and cost-effective methods to decrease mycotoxin contamination in cereals through conventional breeding techniques remains essential because these methods do not compromise nutritional or sensory qualities (Mir et al. 2021). The development of cereal varieties with enhanced pathogen resistance and lower contamination risk becomes possible through modern breeding techniques, including CRISPR genome editing technology (Chen et al. 2019; Ghosh 2024).

4.3 Principles of CRISPR-/Cas12a-Based Biosensors

Biosensors are analytical devices that recognize specific target molecules and convert this recognition into measurable signals (Turner et al. 1987). Among various biosensing strategies, CRISPR/Cas systems have emerged as revolutionary tools for nucleic acid detection due to their programmability and high specificity.

CRISPR–Cas12a, derived from the immune systems of certain bacteria (e.g., *Acidaminococcus*), consists of a 1300-amino acid protein with key RuvC and Nuc domains responsible for DNA cleavage (Zetsche et al. 2015). Recognition starts with a crRNA (Fig. 4.1) containing a spacer region (19–24 nucleotides) complementary to the target DNA sequence (Zetsche et al. 2015). Cas12a protein binds to the direct repeat region at the 5′ end of the crRNAs, forming the Cas12a–crRNA complex. This complex searches for and recognizes the PAM motif on DNA in the

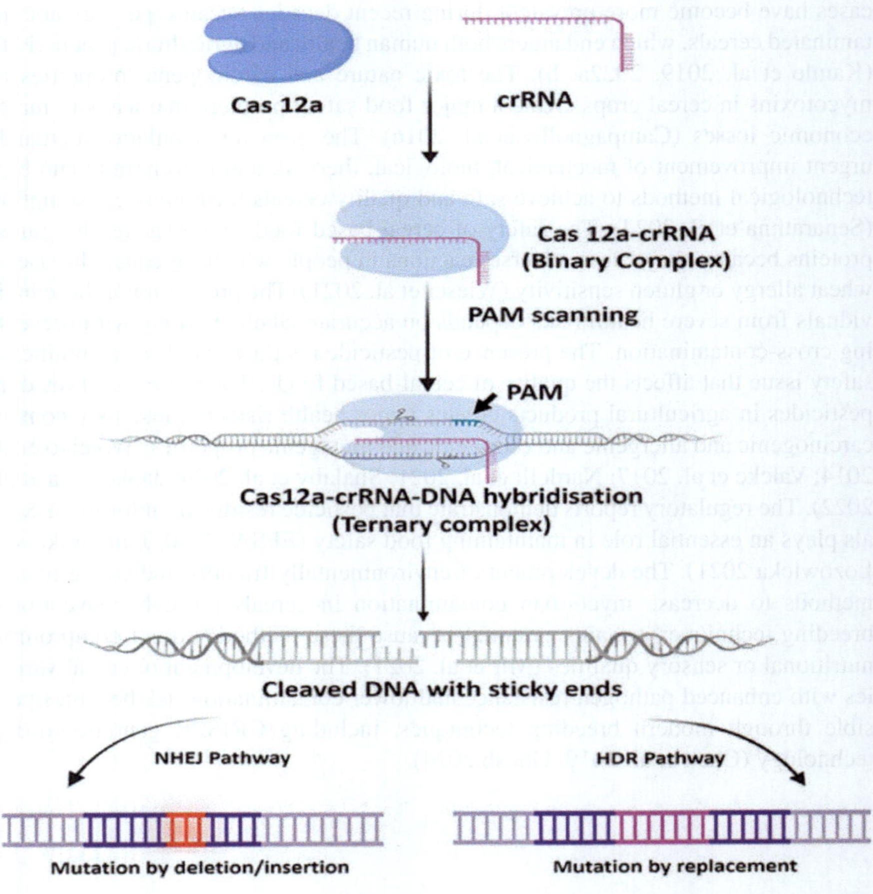

Fig. 4.1 General recognition and cleavage mechanism of CRISPR–Cas12a (Gupta et al. 2022)

form of TTTV (T: thymine; V: A/G/C). The PAM motif is located at the 3′ end of the spacer region where the crRNA matches the target sequence (Chen et al. 2018). Upon recognition of the PAM motif, Cas12a passively unwinds the DNA locally, without consuming energy, and the crRNA spacer hybridizes with the antisense strand of the DNA to form an RNA–DNA duplex. This hybridization triggers a conformational change in Cas12a, activating the RuvC domain. The catalytic DEDD motifs in the RuvC domain align with the DNA, performing a cleavage at the 3′ end of the PAM motif, resulting in either blunt ends or 5′ overhangs. After cis cleavage is completed, the Nuc domain becomes activated and begins to randomly cleave single-stranded DNAs (ssDNAs) present in the reaction mixture. The DEDD motifs of the Nuc domain, similar to those in the RuvC domain, catalyze the cleavage by coordinating with Mg^{2+} ions. These ions interact with the DNA's phosphate backbone, facilitating the nucleophilic attack. This entire process occurs without any external energy requirement.

Cas12a activity is demonstrated by the cleavage of fluorescently or otherwise labeled single-stranded DNA probes that are added to the reaction before it starts (Zhang et al. 2022). The reaction rate can be determined by measuring the time-dependent change in fluorescence intensity ($\Delta F/\Delta t$). The early phase of the reaction (linear region) reflects the initial enzyme activity, and the initial rate (V_0) is calculated during this phase. As the target DNA concentration increases, Cas12a activation accelerates, resulting in a higher $\Delta F/\Delta t$ value. For example, in the detection of *Sitophilus oryzae* DNA, $\Delta F/\Delta t = 1200$ RFU/min has been observed, confirming the high activity of Cas12a. As the reaction progresses, all the ssDNA probes are cleaved, leading to signal saturation at maximum fluorescence intensity.

In lateral flow systems, biotin-FAM/gold nanoparticle-labeled probes are used. Upon Cas12a activation, these probes are cleaved and generate a visible band at the test line (Wang et al. 2023a). For instance, a detection sensitivity of 10^{-18} M has been achieved for *Tribolium castaneum*. Accurate result interpretation critically depends on the specificity of the crRNA (to distinguish intraspecific variations) and PAM motif dependence (to prevent false positives) (Li et al. 2021a). The linear increase in fluorescence signal allows quantitative analysis through kinetic measurements, while the intensity of the lateral flow test band provides qualitative assessment. These systems offer revolutionary diagnostic tools for the agricultural sector with results available within 15–45 min, a field-compatible design, and low-cost protocols.

4.4 Detecting Cereal Contaminants

Cereal crops including wheat, rice, maize, and rye serve as global food staples that provide fundamental proteins, fats, and carbohydrates (Awika 2011; Deligeorgakis et al. 2023). Cereals face risks during cultivation from abiotic and biotic stresses that result in mycotoxin contamination and other harmful substances, which endanger both human and animal health (Milićević et al. 2010; Deligeorgakis et al. 2023). More than 500 different mycotoxins exist which show extreme toxicity toward humans and animals and plants but aflatoxins, ochratoxins, fumonisins, zearalenone, trichothecenes, patulin, and citrin are among the most well-known ones (Haque et al. 2020). The highly toxic, carcinogenic, and immunotoxic compounds known as mycotoxins emerge from fungal species such as *Fusarium*, *Aspergillus*, *Penicillium*, and *Alternaria* Mycotoxins can occur at any point, from plant growth in fields to storage facilities, because they can enter human food systems either through direct contamination or by contaminating animal-derived products such as milk, meat, and eggs (Deligeorgakis et al. 2023; Szelenberger et al. 2024). The detection of mycotoxins in food and feed has led scientists to develop multiple methods for identification purposes. The sensitivity and versatility of chromatographic techniques make thin-layer chromatography (TLC), high-performance liquid chromatography (HPLC), gas chromatography–mass spectrometry (GC–MS/MS), and liquid chromatography–mass spectrometry (LC–MS/MS) widely popular for mycotoxin detection (Turner et al. 2009). Immunoassays using antibodies serve as essential tools for detecting mycotoxins. TLC became the leading detection method between 1980 and 2000 because it offered both affordability and specific and sensitive outcomes. The stationary phase in TLC uses silica gel-coated glass plates as its base, while the mobile phase functions as the liquid phase, which relies on solid–liquid adsorption (Tiwari and Talreja 2022). Aiko and Mehta (2016) demonstrated the successful identification of mycotoxins in plant materials through TLC. LC–MS/MS provides superior specificity and detailed information compared to TLC, but its price point remains significantly higher. HPLC stands as the primary method for detecting food contaminants through its combination with UV-visible or fluorescence detectors and mass spectrometry, which enables precise and reliable measurements (Turner et al. 2009). The extraction of mycotoxins from samples uses three primary methods, including membrane filtration and solid-phase extraction (SPE) and immunoaffinity chromatography (IAC), which take advantage of antigen–antibody binding properties (Deligeorgakis et al. 2023). The ELISA test is commonly employed for mycotoxin detection because it enables quick, simultaneous detection of multiple targets through simple and reliable analysis. The ELISA procedure involves a primary antibody that binds to a specific molecule, which then produces a detectable signal through a substrate interaction (Szelenberger et al. 2024). Researchers have applied ELISA in their studies to detect aflatoxins in wheat, corn, and tea samples (Omar et al. 2020; Santos et al. 2009; Xing et al. 2020). The Lateral Flow Assays (LFAs) represent an inexpensive and portable solution for quick mycotoxin detection (Sajid et al. 2015; Bhalla et al. 2016). Paper-based biosensors function as fast detection tools for mycotoxins in foods, which makes them suitable for on-site testing (Sadeghi et al. 2024). Biosensors serve as crucial detection tools for contaminants that exist in cereal products. The basic components of a biosensor consist of a sensing element and a physicochemical detector and a signal processing system (Malhotra et al. 2017). The sensing section employs biological components such as tissues, microorganisms, cell receptors, enzymes, antibodies, and nucleic acids (Fig. 4.2). The signal detection by the sensor produces an electrical output that passes through amplification before reaching the signal processing system, which displays a colorimetric result (Phumlani et al. 2021). The food industry employs biosensors for quality control and safety monitoring because these devices enable fast and precise detection of contaminants alongside biological hazards. Allergenic reactions caused by gluten allergy represent another concern in cereal products. The analysis of allergenic proteins and peptides in cereals becomes difficult because cereal species together with their growing conditions and climate influence protein composition (Lu et al. 2023). Traditional immunological methods are widely used, but mass spectrometry-based approaches particularly LC–MS/MS deliver superior results for identifying and quantifying allergenic peptides. Two-dimensional electrophoresis separates allergenic proteins based on their isoelectric points, before LC–MS/MS detection performs rapid and precise analysis (Chňapek et al. 2023). Cereals are the most consumed food, but are also a major route for entry of genetically modified organisms (GMOs) into the human body and the environment. Detection of GMOs is most often by PCR after the extraction of DNA. However, near-infrared spectroscopy (NIRS) is an important technique for the qualitative and quantitative analysis of foods and feeds (Sohn et al. 2021). NIRS is faster and more reliable than PCR and ELISA for GMO detection.

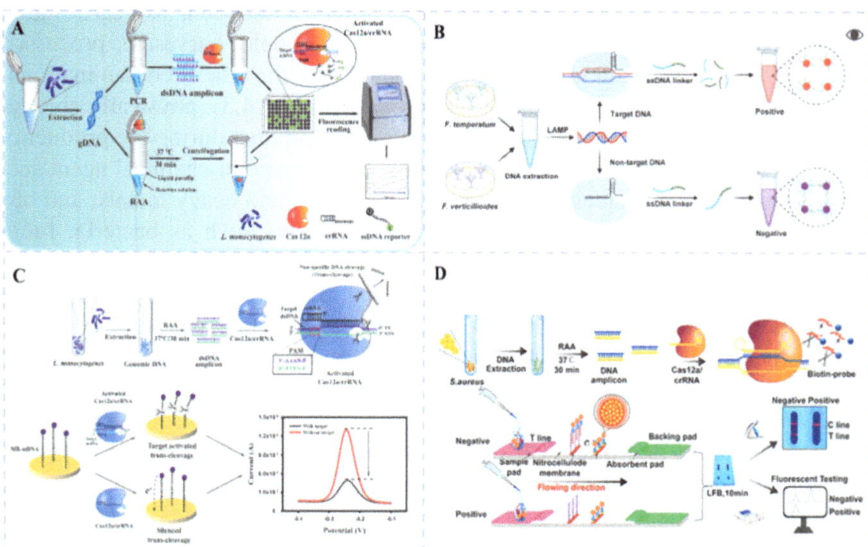

Fig. 4.2 Biosensing strategies of biological contaminants with the CRISPR-Cas12a system. (**a**) Schematic diagram of the CRISPR–Cas12a fluorescence biosensor based on PCR or RAA methods. (**b**) Schematic representation of CRISPR/Cas12a and AuNP-based biosensor constructed using LAMP. (**c**) Schematic illustration of a CRISPR-/Cas12a-based electrochemical biosensor combined with RAA. (**d**) LFB-based CRISPR-Cas12a fluorescence biosensor scheme combined with RAA (Shen et al. 2024)

The plant disease pathogens *Fusarium*, *Sclerotinia*, *Rhizopus*, *Pythium*, *Rhizoctonia,* and bacterial pathogens *Pectobacterium carotovorum* and *Xanthomonas campestris* can also infect cereals and cause yield losses (Khan et al. 2023; Khakimov et al. 2022). The detection methods for these pathogens include ELISA, PCR, aptamer-based assays, chromatographic techniques, fluorescence, and flow cytometry, which are faster and more reliable than the traditional methods (Gudkov et al. 2023). Cereals are among the most important staple foods in the human diet. The contaminants, such as mycotoxins, allergens, GMOs, and pathogens, pose a significant risk to human health and food safety. Accordingly, precise identification and quantification of these contaminants are critical for maintaining food quality and safety, as well as safeguarding public health.

4.5 Development of CRISPR-/Cas12a-Based Assays

From past to present, access to food and hunger have been one of the biggest problems for humanity and all living things. In addition to these problems, food safety is also a prominent issue in human nutrition. Cereals are the main food source in human nutrition and cover the largest area as a group (Suttie 2000). However, this food group, which has a very important place in human nutrition, also carries great health risks. These products, which are grown in the field, harvested, processed, packaged, and transported to tables in our homes, face the risk of contamination at any stage. Foods can be contaminated in three ways (Fig. 4.3): physically, through the introduction of a foreign substance into the food product during the production process; chemically, through the introduction of some type of chemical agent such as agrochemicals, pesticides, environmental pollutants and heavy metals; and biologically, through the introduction of biological agents such as bacteria, fungi, viruses and parasites, and substances derived from these agents such as toxins (Abdolshahi and Shokrollahi Yancheshmeh 2020; Sadiku et al. 2020). These contaminants have been detected by many different analytical methods (Fig. 4.3) developed to detect food contaminants, such as chromatographic methods (Elmastaş et al. 2023; Keskin and Eyupoğlu 2023), enzyme-linked immunosorbent assay (ELISA) (Zhang et al. 2023), polymerase chain reaction (PCR) (Pava-Ripoll et al. 2023), quantitative real-time PCR (qPCR) (Sohlberg et al. 2022), fluorescence and colorimetric techniques (Lin et al. 2018; Chen et al. 2022). Although the selectivity and sensitivity of these methods in detecting contaminants have improved, they are burdensome to apply due to disadvantages such as challenging pre-treatment of samples, the high cost of antibody preparation, the effect of the complex food matrix on the detection of very low amounts of substances, the need for specialized equipment and experts in the field, and the difficulty of field application (Liu et al. 2024a). However, methods developed in recent years with clustered regularly interspaced short palindromic repeats (CRISPR) and CRISPR-associated (Cas) proteins (CRISPR-Cas system) offer promising applications for the detection of cereal contaminants due to their improved specificity, selectivity, sensitivity, and on-site applicability (Table 4.1).

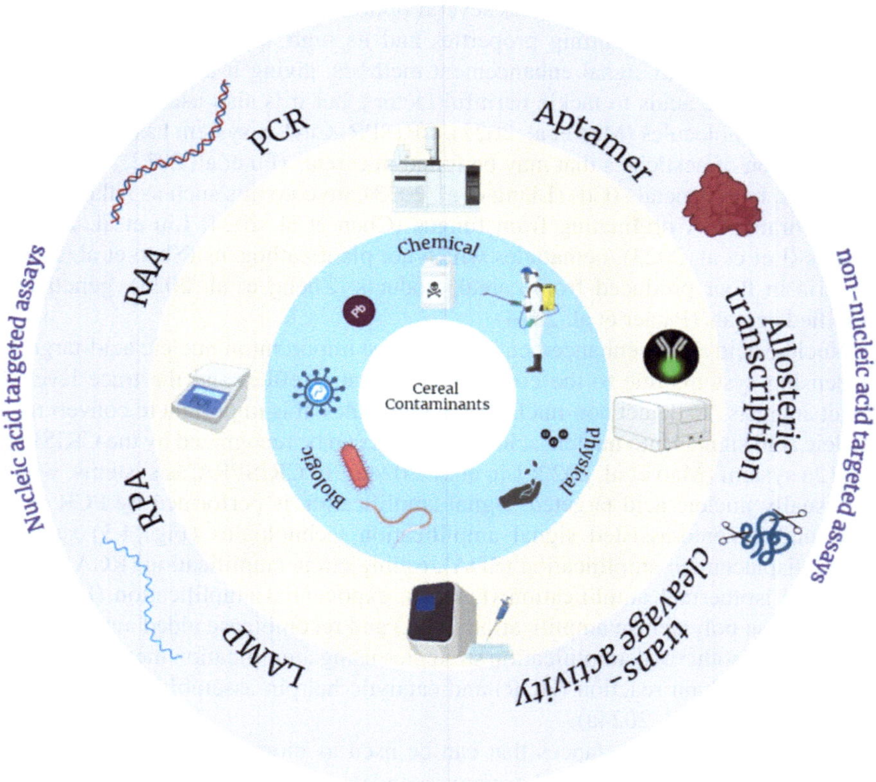

Fig. 4.3 Schematic representation of cereal contamination agents and methods for their detection

Table 4.1 Comparison of analytical methods and CRISPR–Cas12a for the detection of cereal contaminants

Feature	Chromatography (LC–MS/MS)	ELISA	CRISPR/Cas12a Biosensors
Specificity	High	Moderate	Very High
Sensitivity	High	Moderate	Very High (fM–aM levels)
Cost	High	Low	Medium–Low
Time to Result	Hours	1–2 h	<1 h
Field Applicability	No	Limited	Yes (portable devices)

CRISPR–Cas is a system encoded at genomic loci in the prokaryotic organisms bacteria and archaea, consisting of a series of direct repetitive sequences, Clustered Regularly Interspaced Short Palindromic Repeats (CRISPR), and a series of CRISPR-associated (Cas) genes adjacent to this sequence, which serve as a defense mechanism against invading nucleic acids such as phages and plasmids (Burmistrz et al. 2020; Koonin et al. 2023). There are multiple CRISPR–Cas systems and among the class 2 Cas proteins, Cas12a (type V) and Cas13 (type VI) are frequently used in areas such as gene editing, genome imaging and disease treatment.

CRISPR–Cas12a not only offers several options for biosensor design due to its homeopathic and trans-cutting properties and its high specificity and sensitivity when combined with signal enhancement methods, giving it a great advantage in targeting nucleic acids to tackle harmful factors, but it is also used to detect non-nucleic acid molecules (Mao et al. 2022). CRISPR–Cas12a system has been used in the detection of pesticides that may be found in cereals (Fu et al. 2022), vegetables and fruits, heavy metals (Cd) (Liang et al. 2023), mycotoxins such as Aflatoxin B1 and Ochratoxin A originating from fungus (Chen et al. 2024; Liu et al. 2024b), viruses (Lei et al. 2023), nematodes which are plant pathogens (Shao et al. 2023), bacteria in flour produced from cereal products (Zheng et al. 2023), genetically modified cereals (Pataer et al. 2024).

Nucleic acid signal enhancement methods are important in nucleic acid-targeted biosensing systems due to the complex food matrix effect and the trace level of target analysts. To detect non-nucleic acid molecules, it is important to convert non-nucleic acid signals into nucleic acid signals that can be recognized by the CRISPR–Cas12a system (Mao et al. 2022; Liu et al. 2024a). In CRISPR/Cas systems, which are usually nucleic acid targeted, signal amplification is performed by PCR reactions and enzyme-assisted signal amplification technologies (Fig. 4.3) such as strand displacement amplification (SDA), rolling circle amplification (RCA), loop-mediated isothermal amplification (LAMP), exponential amplification (EXPAR), recombinase polymerase amplification (RPA) and recombinase aided amplification (RAA), and isothermal amplification strategies using amplification methods such as hybridization chain reaction (HCR) and catalytic hairpin assembly (CHA) (Liang et al. 2018; Liu et al. 2024a).

Non-nucleic acid substances that can be used as biomarkers in environmental pollution, medical diagnosis and prognosis, and detection of food contaminants can be detected in CRISPR–Cas systems by DNAzymes, transcription factors, enzymatic reactions, aptamer-dependent or allosteric transcription factors (Fig. 4.3) (Cheng et al. 2022; Mao et al. 2022). CRISPR–Cas biosensor systems can be analyzed by converting the concentrations of target substances into detectable signal molecules, such as colorimetric, fluorescence, electrochemical, and SERS, which are frequently used in this field, and methods such as nanoparticle coating can be used to increase the signal (Yuan et al. 2021; Zhuang et al. 2022; Qian et al. 2024; Yang et al. 2024). The studies using the CRISPR–Cas12a system use techniques such as PCR, RAA, RPA, LAMP technologies and trans-clivage activity of CRISPR/Cas12a to amplify the signal (Mu et al. 2022; Lei et al. 2023, 2024a; Pataer et al. 2024; Zhu and Zhao 2024).

The CRISPR/Cas12a system is frequently used in biosensor designs as it targets substances with both nucleic acid and non-nucleic acid properties, with signal enhancement technologies, and provides results with high selectivity. In order to develop a portable and field-usable test kit, horizontal fluidic tests, microfluidic chips, smartphone applications, or mini test devices, and experimental designs that can be performed in a single chamber are being created (Li and Cheng 2023). Pataer et al. (2024) developed a system for the detection of genetically modified maize using the CRISPR-Cas12a system, which makes it possible to test even on-site in the presence of a simple incubator. Liang et al. (2024) introduced the RPA-Cas12a-LFD assay for *Fusarium verticillioides*, a maize pest fungus, providing rapid results with potential for application in the field. A paper-based microfluidic biosensor (Zhang et al. 2024c) was developed for the field detection of the heavy metal Pb, which can lead to grain contamination when present in high concentrations in soil. As a result, it was determined that this biosensor can detect Pb at low concentrations and also developed a smartphone application to support for on-site applications of analyzes. The test created by Li et al. (2024b) for the detection of enterogenic *Bacillus cereus* bacteria, which can be found in starchy foods such as rice and pasta and cause food poisoning, is reported to work with the RPA-assisted CRISPR/Cas12a primer, giving results in less than 60 min, and results can be obtained with a mini-smartphone application. The CRISPR/Cas12a system has been a popular and emerging system in recent years and is still open to many innovations and applications. In the field of food safety, it is important to develop these systems that are applicable in the field and provide fast results and to find a suitable test for each contaminant.

4.6 Advantages and Limitations in Cereal Food Safety

CRISPR/Cas12a biosensors offer a powerful approach for detecting foodborne contaminants and pathogens, leveraging the unique properties of the Cas12a enzyme. This enzyme can be programmed to recognize specific DNA or RNA sequences, triggering a detectable signal through its trans-cleavage activity (Zhang et al. 2024a; Li et al. 2018; Choi et al. 2020; Xie et al. 2024). While this technology offers high specificity, sensitivity, and speed, making it suitable for various applications, it also presents limitations, particularly concerning complex food matrices.

4.6.1 Sensitivity and Specificity

CRISPR/Cas12a biosensors exhibit high sensitivity, enabling the detection of even trace amounts of target analytes. The system's high specificity minimizes off-target effects, reducing the risk of false positives. Table 4.1 demonstrates the efficacy of CRISPR/Cas12a-based methods in detecting foodborne pathogens at low concentrations in cereals (Mu et al. 2022; Qiao et al. 2023; Chen et al. 2024; Liu et al. 2024b; Wu et al. 2023a, b, c; Yang et al. 2024; Wang et al. 2023b; Zheng et al. 2023; Meng et al. 2024; Zhang et al. 2024b), highlighting the technology's potential for pathogen detection (Shaizadinova et al. 2025). This sensitivity and specificity are crucial for ensuring food safety, allowing for the early detection of contaminants before they reach harmful levels.

Matrix Interference and Sample Preparation

CRISPR-/Cas-based biosensors hold great promise for rapid food safety screening, but matrix effects from complex food samples pose a significant challenge. These matrix components interfere with both target recognition (especially for non-nucleic acid targets, where functional nucleic acids are employed) and signal generation, hindering accurate trace detection (Sun et al. 2024). Therefore, integrating CRISPR/Cas biosensors with advanced sample pretreatment techniques, including simplified extraction and purification methods, is crucial for developing robust, cost-effective, and rapid on-site analysis platforms that minimize matrix interference and enhance detection efficiency (Xie et al. 2024).

4.6.2 Cost and Scalability for Commercial Applications

The cost-effectiveness of CRISPR/Cas12a compared to traditional methods depends on the specific application and scale of implementation. While initial setup costs, including crRNA synthesis and Cas12a enzyme acquisition, might be higher, these costs have decreased substantially due to readily available crRNA design platforms (Chen et al. 2018; Broughton et al. 2020). In contrast, traditional methods like PCR and ELISA, while having lower upfront costs, can accrue higher expenses over time due to recurring reagent and labor needs (Broughton et al. 2020).

Operationally, CRISPR/Cas12a offers significant cost advantages. Direct target DNA detection without amplification eliminates costs associated with PCR (Gootenberg et al. 2017). Platforms like SHERLOCK and DETECTR demonstrate the potential for low per-test costs (WHO 2021). In agriculture, CRISPR-based gene editing is more cost-effective than traditional hybridization (Bortesi and Fischer 2015). Furthermore, CRISPR/Cas12a's amenability to automation enhances its time and labor efficiency (Broughton et al. 2020). Scalability further increases its economic appeal, as reagent costs decrease with large-scale production (Chen et al. 2018).

CRISPR/Cas12a biosensors hold great promise for revolutionizing food safety diagnostics. While matrix effects and the need for optimized sample preparation remain significant challenges, ongoing research is addressing these limitations. The high sensitivity, specificity, and potential for cost-effectiveness, coupled with continuous improvements in sample preparation and assay development, position CRISPR/Cas12a biosensors as a powerful tool for rapid and accurate detection of foodborne contaminants and pathogens, ultimately contributing to improved public health (Table 4.2).

Table 4.2 CRISPR-/Cas12a-based methods in detecting foodborne pathogens

Classification	Target classification	Detection	Amplification method	Signal Reporter	Dynamic Range	Detection limit	Time	Natural samples	Reference
Fluorescence	Non-nucleic acid	Ochratoxin	CLASA	FQ-ssDNA	0.001–1 ng/mL	1,71E-04	80 min	Rice and oat	Chen et al. (2024)
Colorimetric	Non-nucleic acid	Aflatoxin B1	CRISPR/ Cas12a system's trans-cleavage activity	Fe-N-C SAzymes	10^{-6}–1 ng/μL	1.5×10^{-7} ng/ μL		Peanut and wheat	Liu et al. (2024b)
Colorimetric	Non-nucleic acid	Aflatoxin B1	CRISPR/ Cas12a cleaved G-quadruplex (G4)	G4-DNAzyme	0.001–0.1 ng·mL^{-1}	0.85 pg/mL (colourimetric signal)		Peanut, maize, and badam	Wu et al. (2023a)
Colorimetric	Non-nucleic acid	Aflatoxin B1	CRISPR/ Cas12a cleaved G-quadruplex (G4)	G4-DNAzyme	0.001–0.1 ng·mL^{-1}	0.79 pg/mL (SERS signal)		Peanut, maize, and badam	Wu et al. (2023a)
Colorimetric	Non-nucleic acid	Aflatoxin B1	CRISPR/ Cas12a cleaved G-quadruplex (G4)	G4-DNAzyme	0.001–0.1 ng·mL^{-1}	1.65 pg/mL (Fluorescent signal)		Peanut, maize, and badam	Wu et al. (2023a)

(continued)

Table 4.2 (continued)

Classification	Target classification	Detection	Amplification method	Signal Reporter	Dynamic Range	Detection limit	Time	Natural samples	Reference
Electrochemical	Non-nucleic acid	Zearalenone	CRISPR/Cas12a trans-cleavage activity	Mg2 + -DNAzyme	1×10^{-5}–10 ng/mL	6.27×10^{-6} ng/mL		Corn flour	Yan et al. (2024)
Chemiluminescence	Non-nucleic acid	Aflatoxin B1	RCA	HRP	50 pg/mL–100 ng/mL	5.2 pg/mL		Wheat, rice, and maize	Wang et al. (2023b)
Fluorescence	Nucleic acid	*Burkholderia gladioli*	RPA	FQ-ssDNA	1 –10^6 CFU/mL	10 CFU/mL	~1 h	Rice noodles, fresh noodles, and flutinous rice flour	Zheng et al. (2023)
Chromatography	Non-nucleic acid	Aflatoxin B1	Sp-ICPMS	AuNPs	1–1000 pM	1.1 pM		Peanut oil and beer	Chen et al. (2018)
Electrochemical	Non-nucleic acid	Ochratoxin A	Exo III-assisted recycling reaction	FC-ssDNA	1×10^{-5}, 0.001, and 0.1 ng/mL	0.74 fg/mL		Rice and peanuts	Wu et al. (2023c)
Fluorescence	Nucleic acid	*Bacillus cereus*	RPA	FQ-ssDNA	1.6×10^8–1.6 CFU/mL	10^2 CFU/mL	2–5 h	Rice, milk, and cooked meat	Meng et al. (2024)

Fluorescence	Non-nucleic acid	Aflatoxin B1	PCR	FAM	0.001–80 ng/mL	0.92 pg/mL	80 min	Peanut	Wu et al. (2023b)
Fluorescence	Nucleic acid	*Fusarium graminearum*	PCR	FQ-ssDNA	100 ng/μL–1 fg/μL	1 fg/μL	4 day	Wheat	Mu et al. (2022)
Fluorescence	Non-nucleic acid	Fumonisin B1	CRISPR/Cas12a trans-cleavage activity	FAM	25–500 nM	16.84 nM	40 min	Maize	Qiao et al. (2023)

4.7 Case Studies and Real-World Applications

Cereal safety should be assessed in two perspectives: one is the microorganisms (*Fusarium, Aspergillus,* etc.) that affect human and animal health, and the other one is the microorganisms (*Magnaporthe* and so on) and viruses (*rice black streaked dwarf virus, rice gall dwarf virus, Wheat dwarf virus, Barley yellow dwarf virus.* etc.) that infects the plant and causes yield loss. *Fusarium* and *Aspergillus* species produce extremely harmful mycotoxins such as deoxynivalenol (DON), trichothecenes, and various aflatoxins (Lal et al. 2024; Wang et al. 2023a, b). Cereals and their products are the main food and feed source in the world. According to FAO, 3.5×10^9 tons of cereals were produced worldwide in 2023 (FAOSTAT 2023). It is estimated that over %60 of the world's food supply, including cereals, is contaminated by mycotoxins (Yang et al. 2020). The maximum contamination levels were strictly regulated by various authorities. For instance, the maximum limits defined by the European Union for deoxynivalenol (DON) and aflatoxin are 200–1750 and 0.1–15 μg/kg, respectively (Commission Regulation (EC) No 1881/2006).

Today, chromatographic (Rauová 2010; Cunha and Fernandes 2010; Lin et al. 1998), microscopic (Hafizi et al. 2013), serologic (Meirelles et al. 2006), and PCR-based methods (González-Salgado et al. 2008; Ferrara et al. 2020; Ortega et al. 2020) have been developed to detect *Fusarium* and *Aspergillus* or their products. For the detection of other groups of bacteria, fungi, and viruses which are associated with plant disease, serological (Lister and Rochow 1979; Yang et al. 2013, 2014; Chen et al. 2012; Zhang et al. 2018) and PCR-based (Le et al. 2010; Balaji et al. 2003; French and Robertson 1994; Canning et al. 1996) techniques have been developed. Also, more recent methods such as digital PCR (dPCR) (Ballandras et al. 2024; Lee et al. 2024; Bühring et al. 2024) and MALDI-TOF (Dong et al. 2009; Eggert and Pawelzik 2011) have been implemented for cereal safety. These techniques are far advantageous in their own field (to PCR-based systems or chromatographic methods), yet their cost and requiring specialized expertise are the major obstacles to widespread application.

The CRISPR/Cas12a system is a promising tool as a biosensor for the detection of not only nucleic acids but also toxins (Wang et al. 2023a, b), pesticides (Li et al. 2024b), heavy metals (Liang et al. 2023), and antibiotics (Li et al. 2022). The tools such as PCR, LAMP, and RPA (recombinase polymerase amplification) coupled with CRISPR/Cas12a systems are highly sensitive and rapid. Especially, LFA coupled with CRISPR/Cas12a systems are serving as POCT (point-of-care test).

Xanthomonas oryzae pv. *oryzae*, rice stripe virus and black-streaked dwarf virus are the common pathogens of rice. Zhu et al. (2022) developed a LAMP-coupled CRISPR/Cas12a rapid assay. The test showed 100% specificity and high sensitivity (3–9 copies per reaction) as well as it can be used for on-site detection. Maize Chlorotic Mottle Virus (MCMV) was detected in a similar manner, coupled with reverse transcription recombinase aided amplification (RT-RAA) instead of LAMP. The sensitivity of this biosensor can reach up to 2.5 copies, whereas real-time PCR generates high CT values at low concentrations, which is not reliable (Lei et al. 2023). *Magnaporthe oryzae* is another pathogen of rice and has different pathotypes that also cause disease in other cereals such as wheat, oat, and barley. Recently, two RPA coupled with CRISPR/Cas12a tests have been developed for the detection of *Magnaporthe oryzae*. All the assays were highly specific, sensitive, rapid (under 1 h), and cost-effective ($3.5–6 per sample). Zhang et al. (2020) reported that *M. oryzae* can be detected in 0.00001 μg of genomic DNA and can also allow GMO detection in rice. On the other hand, Kang et al. (2021) detected *M. oryzae* at a concentration of 0.001 μg/μl and reported that it is 100 times more sensitive than other PCR methods. *Fusarium* species are the most threatening pathogens and have a wide range of hosts. Moreover, RPA coupled with the CRISPR/Cas12a test was implemented for the detection of nematodes by Shao et al. (2023).

Mu et al. (2022) achieved in detecting *F. graminearum* from 1 fg/μl pure *F. graminearum* DNA with the CRISPR/Cas12a test they developed, targeting the EF1α region. Furthermore, the assay was also able to detect *F. asiaticum* and *F. pseudograminearum,* and Mu et al. (2022) suggested that the test could be a promising early detection tool for *Fusarium*. Likewise, a similar methodology was applied for the detection of *F. fujikuroi*. This biosensor can detect even at the attomolar level (Zhang et al. 2024c). As mentioned above, CRISPR/Cas12a combined tools can also detect toxins, and the development of CRISPR-based tools for the detection of mycotoxins has accelerated. Lately, aptamer-based CRISPR/Cas12a biosensors have been implemented for the detection of Ochratoxin A (OTA) (Chen et al. 2024) and Aflatoxin B1 (AFB1) (Wang et al. 2023a, b). The lowest mycotoxin levels for detected AFB1 and OTA were 5.2×10^{-3} and 1.71×10^{-4} ng/mL, respectively, which is approximately 50-fold more sensitive than conventional ELISA (Wang et al. 2023a, b; Chen et al. 2024). Finally, heavy metals such as cadmium can be detected at the picomolar level by CRISPR/Cas12a-based biosensors without the need for complicated devices such as ICP–MS (Liang et al. 2023).

In summary, CRISPR/Cas12a coupled biosensors are highly specific, sensitive, rapid, cost-effective, and do not require any specialized equipment or expertise.

4.8 Future Directions and Emerging Trends

The CRISPR technology presents significant advancements in food safety and agricultural practices. This technology has facilitated the development of more resilient, nutritious, and sustainable crops through gene editing (Sampath et al. 2023). Enhancing disease resistance is one of the most critical aspects of such strategies. The use of CRISPR to render plants more resistant to harmful organisms provides several significant advantages. CRISPR technology has been utilized to develop resistance to Panama disease, which affects banana plants in tropical regions (Tripathi and Tripathi 2024). Furthermore, gene editing investigations are being conducted to develop wheat cultivars that exhibit resistance to rust diseases (Choudhary et al. 2022). In recent years, CRISPR technology has been utilized to enhance the nutritional composition of food crops. For example, Liang et al. (2014)

reported targeting genes encoding enzymes involved in the phytic acid (PA) synthesis pathway in maize (*Zea mays*) to mitigate issues associated with the inhibition of PA absorption, due to the formation of an insoluble complex in the digestive tract with essential elements such as Mg, Zn, Ca, and Fe. CRISPR technologies possess the potential to confer advantages for the elimination of allergens and toxins. CRISPR makes it possible to suppress the production of allergy-causing proteins or to eliminate harmful components. Biswas et al. (2022) endeavored to reduce allergenic proteins in peanuts utilizing CRISPR technology. CRISPR technology can additionally be employed to extend the shelf life of products. Gaining properties that delay the spoilage of products is promising in contributing to this situation. By editing genes that can cause rotting in tomatoes, varieties that remain fresh for longer periods have been developed (Adaskaveg and Blanco-Ulate 2023). Similarly, gene editing has been performed on lettuce and strawberries to increase their freshness (Kumar and Choudhary 2024). CRISPR technologies can also be used to increase plant yields against abiotic stresses, such as drought and salinity, indirectly increasing the yields of existing nutrient-rich crops. Consequently, this investigation contributes to the development of agricultural products that exhibit resilience to climate change. Cultivars resistant to drought and elevated salt concentrations have been developed in *Zea mays*, *Oryza sativa*, and *Triticum aestivum* (Abdallah et al. 2022; Nazir et al. 2022). CRISPR technology can be utilized to mitigate foodborne pathogens and enhance the safety of the food supply chain. Diagnostic assays and resistant products that are sensitive to *Listeria* and *Salmonella* are currently under development (Li et al. 2021b). Genetic modifications are being made to reduce mycotoxin levels in animal feeds and products (Thipe et al. 2021).

Unlike genetically modified organisms (GMOs), CRISPR provides a more targeted and natural editing process that facilitates consumer perception and regulatory processes. In 2023, CRISPR-edited food products (e.g., faster-growing tomatoes) were introduced in Japan and the United States (Naeem et al. 2024). Regarding food safety, novel CRISPR-based rapid test kits have been developed to detect pathogens and toxins (Li et al. 2022). CRISPR is anticipated to continue providing innovations in food safety and sustainability in the forthcoming years. Technological advances in data analysis and artificial intelligence have led to the development of the Internet of Things (IoT), Internet of Nanothings (IoNT), Industrial Internet of Things (IIoT), and Internet of Medical Things (IoMT) (Ibrahim et al. 2022). The integration of CRISPR technology with the IoT and smart devices presents significant potential in the field of food safety. This integration has the potential to enhance the safety and sustainability of food systems by facilitating real-time monitoring, analysis, and intervention in both agricultural processes and supply chains. CRISPR-based biosensors can be incorporated into IoT devices for the detection of pathogens in food products. Furthermore, CRISPR-programmed biosensors can be utilized for real-time pathogen detection. It can detect bacteria such as *Salmonella* and *Listeria*. Li et al. (2021a, b) developed an electrochemical, sensitive, and highly specific recombinase-assisted amplification (RAA)-based E-CRISPR biosensor for the detection of *Listeria monocytogenes* (Cossettini et al. 2022). With devices connected to IoT networks, these sensors can analyze the results in real time and send alerts in case of risk. Early detection can be highly advantageous in preventing large-scale outbreaks of foodborne diseases. IoT-based storage and logistics systems can improve product quality by processing the data obtained using CRISPR technology (Nayak and Dutta 2023). The spoilage risks of products whose shelf lives have been extended by CRISPR are monitored using IoT sensors. For example, parameters such as temperature, humidity, and gas release can be monitored, and appropriate conditions can be provided, thus reducing food waste, and products can reach the consumer in a fresh way. IoT-enabled agriculture will develop further in the coming years because the performance of CRISPR-genetically engineered plants can be optimized using smart farming systems. IoT devices can enable the real-time monitoring of the response of CRISPR-edited plants to soil nutrients, water use, and environmental stress (Wang et al. 2023a, b). Higher yields can be achieved with less water and fertilizer, and the environmental impact can be reduced. The journey of the product, whose shelf life is extended with CRISPR technology or whose agricultural durability is increased, from the field to the table, can be traced using IoT systems (Zheng et al. 2021). CRISPR-labelled food products were tracked using IoT devices. When integrated with blockchain technology, the origin, production process, and transportation conditions of a product can be transparently documented, thereby mitigating food fraud and mislabeling. CRISPR-based biosensors can be incorporated into smart packaging to indicate product freshness (Nayak et al. 2023). When packaging contains a CRISPR-based sensor that analyzes the microbiological status of food and the sensor is connected to IoT devices, the consumer can be informed about the safety of the food, reduce food waste, and provide access to safe products.

Portable genetic analysis systems can be developed by integrating CRISPR with IoT devices. In addition, IoT sensors can precisely apply the nutrients or protective chemicals required by CRISPR-edited plants, thus preventing excessive chemical use and minimizing environmental damage (Nayak et al. 2023). IoT devices can analyze genetic modifications or contamination in food products using CRISPR. These devices can be used on farms and in retail outlets. CRISPR-enhanced crops can be processed by artificial intelligence (AI) using data collected by IoT to support better decision-making. Data from CRISPR-edited plants can be transferred via IoT devices to a cloud platform, where AI can analyze the data and make recommendations to improve crop health and productivity. This can help optimize agricultural processes and increase productivity.

Although the integration of CRISPR technology with IoT and smart devices offers significant benefits in terms of food safety, it also requires regulatory and ethical considerations (Zheng et al. 2021). To implement such innovations effectively, fairly, and reliably, several issues must be addressed. Setting standards and legal frameworks is the main requirement. CRISPR-edited foods can be classified differently from genetically modified organisms (GMOs). However, regulatory frameworks should clearly define these differences. The legal protection of IoT data is another important issue. International data protection standards should be established for the security, privacy, and use of the data collected from IoT devices. A certification process should be developed for foods produced by the integration of

CRISPR and the IoT. This can increase transparency and consumer confidence. These results verified that CRISPR-based biosensors can be used in food safety audits. If biosensors do not function correctly in the food chain, inspection mechanisms should be strengthened to prevent the spread of pathogens. Data obtained through the integration of the IoT and CRISPR is critical in cross-border trade. Therefore, international food safety protocols should be harmonized. The World Trade Organization (WTO) and the Food and Agriculture Organization (FAO) can provide regulatory guidance for these technologies. Dissemination of information to consumers and transparency are essential for ensuring food safety. Foods traced with CRISPR and IoT must be clearly labelled by consumers. This provides consumers with the right to make informed decisions. Data collected through IoT devices should be transparently shared with stakeholders. IoT and CRISPR technologies are expensive for smallholder farmers and developing countries. Therefore, fair infrastructure for access to technology should be created. The risk of large agribusinesses owning and monopolizing these technologies could lead to the exclusion of small-scale producers from the market. The environmental impact of CRISPR-edited crops and IoT should be carefully assessed (Sahel et al. 2024). For example, the effects of genetic regulation on biodiversity should be examined. The carbon footprints of IoT devices and electronic waste management should also be considered. The advantages of technology should not be limited to wealthy countries. The same standards should be applied to developing regions. Social policies contributing to food justice should support systems that increase food security (Nayak et al. 2023). Gene editing with CRISPR may raise ethical concerns as it alters the genetic makeup of natural species. As the long-term health effects of gene-edited foods are not fully known, consumer perceptions should be given importance.

When considering potential solutions, comprehensive regulatory frameworks should be established for systems in which the IoT and CRISPR technologies work together. Educational programs should be organized for farmers, producers, and consumers regarding the benefits and risks of these technologies. Establish independent bodies to monitor and oversee IoT and CRISPR applications. Multidisciplinary committees could be established to assess the ethical implications of CRISPR and IoT. Subsidy and technology transfer programs should be implemented to facilitate smallholder farmers' access to these technologies. While the integration of CRISPR and IoT has the potential to revolutionize food security, the application of these technologies necessitates careful consideration of both ethical and regulatory dimensions. This process requires the active participation and collaboration of all stakeholders.

4.9 Conclusion

CRISPR/Cas12a-based genome editing system offers scientists many ways to change the sequence of interest in the plant genome. The CRISPR/Cas12a genome-editing system has recently become widespread in the field of plant science and has been used in many studies to improve cereal crops. Although off-target effects are taken into account, modifying agriculturally important cereal crops will bring a promising green revolution in the near future by solving problems such as nitrogen fixation, nutrient uptake improvement, biofuel production, and photosynthetic ability. Overall, the CRISPR-/Cas12a-based genome editing system is ready to offer various possibilities to improve cereal crops to overcome the adverse effects of climate change and can help strengthen food security in developing countries. CRISPR has become one of the most flexible genetic engineering tools in recent years and is being used for various genome editing applications.

References

Abdallah NA, Elsharawy H, Abulela HA, Thilmony R, Abdelhadi AA, Elarabi NI (2022) Multiplex CRISPR/Cas9-mediated genome editing to address drought tolerance in wheat. GM Crops Food 13(1):1–17. https://doi.org/10.1080/21645698.2022.2033977

Abdolshahi A, Shokrollahi Yancheshmeh B (2020) Food contamination. In: Mycotoxins and food safety. IntechOpen, pp 5–16. https://doi.org/10.5772/intechopen.90287

Adaskaveg JA, Blanco-Ulate B (2023) Targeting ripening regulators to develop fruit with high quality and extended shelf life. Curr Opin Biotechnol 79:102872. https://doi.org/10.1016/j.copbio.2023.102872

Aiko V, Mehta A (2016) Prevalence of toxigenic fungi in common medicinal herbs and spices in India. 3 Biotech 6(2):159. https://doi.org/10.1007/s13205-016-0485-8

Alldrick AJ (2017) Food safety aspects of grain and cereal product quality. In: Cereal grains. Woodhead Publishing, pp 393–424

Awika JM (2011) Advances in cereal science: implications to food processing and health promotion, vol 1089. American Chemical Society. https://doi.org/10.1021/bk-2011-1089

Balaji B, Bucholtz DB, Anderson JM (2003) Barley yellow dwarf virus and cereal yellow dwarf virus quantification by real-time polymerase chain reaction in resistant and susceptible plants. Phytopathology 93(11):1386–1392. https://doi.org/10.1094/PHYTO.2003.93.11.1386

Ballandras V, McNamara L, Carolan JC, Byrne S (2024) Development of a droplet digital PCR assay to detect and quantify BYDV-MAV and BYDV-PAS in their barley host and aphid vectors. Irish J Agric Food Res 63(1):54–65. https://doi.org/10.15212/ijafr-2024-0075

Bennett JW, Klich M (2003) Mycotoxins. Clin Microbiol Rev 16(3):497–516. https://doi.org/10.1128/CMR.16.3.497-516.2003

Bhalla N, Jolly P, Formisano N, Estrela P (2016) Biosensor technologies for the detection of biomolecules. Portland Press

Biswas S, Wahl NJ, Thomson MJ, Cason JM, McCutchen BF, Septiningsih EM (2022) Optimization of protoplast isolation and transformation for a pilot study of genome editing in peanut by targeting the allergen gene *Ara h 2*. Int J Mol Sci 23(2):837. https://doi.org/10.3390/ijms23020837

Bortesi L, Fischer R (2015) The CRISPR/Cas9 system for plant genome editing and beyond. Biotechnol Adv 33(1):41–52. https://doi.org/10.1016/j.biotechadv.2014.12.006

Broughton JP, Deng X, Yu G, Fasching CL, Servellita V, Singh J, Miao X, Streithorst JA, Granados A, Sotomayor-Gonzalez A, Zorn K (2020) CRISPR–Cas12-based detection of SARS-CoV-2. Nat Biotechnol 38(7):870–874. https://doi.org/10.1038/s41587-020-0513-4

Bühring S, Brunner A, Heeb K, Mergard MP, Schmauck G, Jacob S (2024) An array of signal-specific MoYpd1 isoforms determines full virulence in the pathogenic fungus *Magnaporthe oryzae*. Commun Biol 7(1):265. https://doi.org/10.1038/s42003-024-05889-3

Burmistrz M, Krakowski K, Krawczyk-Balska A (2020) RNA-targeting CRISPR–Cas systems and their applications. Int J Mol Sci 21(3):1122. https://doi.org/10.3390/ijms21031122

Campagnollo FB, Ganev KC, Khaneghah AM, Portela JB, Cruz AG, Granato D, Corassin CH, Oliveira CAF, Sant'Ana AS (2016) The occurrence and effect of unit operations for dairy products processing on the fate of aflatoxin M1: a review. Food Control 68:310–329. https://doi.org/10.1016/j.foodcont.2016.04.008

Canning ESG, Penrose MJ, Barker I, Coates D (1996) Improved detection of barley yellow dwarf virus in single aphids using RT-PCR. J Virol Methods 56(2):191–197. https://doi.org/10.1016/0166-0934(95)02000-6

Chen Z, Yin C, Liu J, Zeng M, Wang Z, Yu D et al (2012) Methodology for antibody preparation and detection of southern rice black-streaked dwarf virus. Arch Virol 157:2327–2333. https://doi.org/10.1007/s00705-012-1436-3

Chen JS, Ma E, Harrington LB, Da Costa M, Tian X, Palefsky JM, Doudna JA (2018) CRISPR-Cas12a target binding unleashes indiscriminate single-stranded DNase activity. Science 360(6387):436–439. https://doi.org/10.1126/science.aar6245

Chen K, Wang Y, Zhang R, Zhang H, Gao C (2019) CRISPR/Cas genome editing and precision plant breeding in agriculture. Annu Rev Plant Biol 70:667–697. https://doi.org/10.1146/annurev-arplant-050718-100049

Chen W, Zhang X, Zhang Q, Zhang G, Wu S, Yang H, Zhou Y (2022) Cerium ions triggered dual-readout immunoassay for fluorescent and colorimetric detection of ochratoxin A. Anal Chim Acta 1231:340445. https://doi.org/10.1016/j.aca.2022.340445

Chen S, Fang R, Li Y, Deng F, Liu X, Yang D (2024) CRISPR/Cas12a-powered CLASA towards OTA ultrasensitive detection in cereal samples. Microchem J 196:109691. https://doi.org/10.1016/j.microc.2024.109691

Cheng X, Li Y, Kou J, Liao D, Zhang W, Yin L et al (2022) Novel non-nucleic acid targets detection strategies based on CRISPR/Cas toolboxes: a review. Biosens Bioelectron 215:114559. https://doi.org/10.1016/j.bios.2022.114559

Chňapek M, Gálová Z, Balážová Ž, Hromadová Z, Mikolášová L, Vivodík M et al (2023) Application of two-dimensional electrophoresis and mass spectrometry for the detection of allergens in selected varieties of wheat, oats and buckwheat. J Microbiol Biotechnol Food Sci 13(1):e9934. https://doi.org/10.15414/jmbfs.9934

Choi JH, Lim J, Shin M, Paek SH, Choi JW (2020) CRISPR-Cas12a-based nucleic acid amplification-free DNA biosensor via Au nanoparticle-assisted metal-enhanced fluorescence and colorimetric analysis. Nano Lett 21(1):693–699. https://doi.org/10.1021/acs.nanolett.0c04273

Choudhary A, Kumar A, Kaur H, Pandey V, Singh B, Mehta S (2022) Breeding strategies for developing disease-resistant wheat: present, past, and future. In: Cereal diseases: nanobiotechnological approaches for diagnosis and management. Springer, pp 137–161. https://doi.org/10.1007/978-981-16-7716-2_7

Cossettini A, Vidic J, Maifreni M, Marino M, Pinamonti D, Manzano M (2022) Rapid detection of *Listeria monocytogenes*, *Salmonella*, *Campylobacter spp.*, and *Escherichia coli* in food using biosensors. Food Control 137:108962. https://doi.org/10.1016/j.foodcont.2022.108962

Cunha SC, Fernandes JO (2010) Development and validation of a method based on a QuEChERS procedure and heart-cutting GC-MS for determination of five mycotoxins in cereal products. J Sep Sci 33(4–5):600–609. https://doi.org/10.1002/jssc.200900694

Deligeorgakis C, Magro C, Skendi A, Gebrehiwot HH, Valdramidis V, Papageorgiou M (2023) Fungal and toxin contaminants in cereal grains and flours: systematic review and meta-analysis. Foods 12(23):4328. https://doi.org/10.3390/foods12234328

Djekic I, Nikolic A, Mujcinovic A, Blazic M, Herljevic D, Goel G et al (2022) How do consumers perceive food safety risks? Results from a multi-country survey. Food Control 142:109216. https://doi.org/10.1016/j.foodcont.2022.109216

Dong H, Kemptner J, Marchetti-Deschmann M, Kubicek CP, Allmaier G (2009) Development of a MALDI two-layer volume sample preparation technique for analysis of colored conidia spores of *Fusarium* by MALDI linear TOF mass spectrometry. Anal Bioanal Chem 395:1373–1383. https://doi.org/10.1007/s00216-009-3108-8

Eggert K, Pawelzik E (2011) Proteome analysis of Fusarium head blight in grains of naked barley (*Hordeum vulgare* subsp. *nudum*). Proteomics 11(5):972–985. https://doi.org/10.1002/pmic.201000521

Elmastaş A, Umaz A, Pirinç V, Aydın F (2023) Quantitative determination and removal of pesticide residues in fresh vegetables and fruits by LC–MS/MS and GC–MS/MS. Environ Monit Assess 195(2):277. https://doi.org/10.1007/s10661-023-11574-5

European Commission (2023) Regulation (EU) 2023/915 on maximum levels for certain contaminants in food. EUR-Lex. https://eur-lex.europa.eu/legal-content/EN/TXT/?uri=celex%3A32023R0915

European Food Safety Authority (EFSA) (2020) The 2018 European Union report on pesticide residues in food. EFSA J 18(4):e06057. https://doi.org/10.2903/j.efsa.2020.6057

Ferrara M, Logrieco AF, Moretti A, Susca A (2020) A loop-mediated isothermal amplification (LAMP) assay for rapid detection of fumonisin-producing *Aspergillus* species. Food Microbiol 90:103469. https://doi.org/10.1016/j.fm.2020.103469

French R, Robertson NL (1994) Simplified sample preparation for detection of wheat streak mosaic virus and barley yellow dwarf virus by PCR. J Virol Methods 49(1):93–99. https://doi.org/10.1016/0166-0934(94)90018-3

Fu R, Wang Y, Liu Y, Liu H, Zhao Q, Zhang Y et al (2022) CRISPR-Cas12a based fluorescence assay for organophosphorus pesticides in agricultural products. Food Chem 387:132919. https://doi.org/10.1016/j.foodchem.2022.132919

Ghosh S (2024) Current approaches and future potential for delivering CRISPR/Cas components in oilseeds and millets. Nucleus:1–16. https://doi.org/10.1007/s13237-024-00400-y

González-Salgado A, González-Jaén T, Vázquez C, Patiño B (2008) Highly sensitive PCR-based detection method specific for *Aspergillus flavus* in wheat flour. Food Addit Contam 25(6):758–764. https://doi.org/10.1080/02652030701773018

Gootenberg JS, Abudayyeh OO, Lee JW, Essletzbichler P, Dy AJ, Joung J, Verdine V, Donghia N, Daringer NM, Freije CA, Myhrvold C (2017) Nucleic acid detection with CRISPR-Cas13a/C2c2. Science 356(6336):438–442. https://doi.org/10.1126/science.aam9321

Gudkov SV, Matveeva TA, Sarimov RM, Simakin AV, Stepanova EV, Moskovskiy MN et al (2023) Optical methods for the detection of plant pathogens and diseases. AgriEngineering 5(4):1789–1812. https://doi.org/10.3390/agriengineering5040104

Gupta R, Ghosh A, Chakravarti R, Singh R, Ravichandiran V, Swarnakar S, Ghosh D (2022) Cas13d: a new molecular scissor for transcriptome engineering. Front Cell Dev Biol 10:866800

Hafizi R, Salleh B, Latiffah Z (2013) Morphological and molecular characterization of *Fusarium solani* and *F. oxysporum* associated with crown disease of oil palm. Braz J Microbiol 44:959–968. https://doi.org/10.1590/S1517-83822013000300005

Haque MA, Wang Y, Shen Z, Li X, Saleemi MK, He C (2020) Mycotoxin contamination and control strategy in human, domestic animal and poultry: a review. Microb Pathog 142:104095. https://doi.org/10.1016/j.micpath.2020.104095

Ibrahim AU, Al-Turjman F, Sa'id Z, Ozsoz M (2022) Futuristic CRISPR-based biosensing in the cloud and internet of things era: an overview. Multimed Tools Appl 81(24):35143–35171. https://doi.org/10.1007/s11042-022-12152-x

Jankowska M, Łozowicka B (2021) Natural and synthetic toxic substances occurring in agricultural plants and their products. Prog Plant Prot 61:24–30. https://doi.org/10.14199/ppp-2021-003

Jankowska M, Hrynko I, Łozowicka B (2022) Human health risk assessment of pesticide residues in fruit, vegetable and cereal samples from Poland—A 5-year survey. J Plant Prot Res 62(4):385–392. https://doi.org/10.24425/jppr.2022.143211

Kamle M, Mahato DK, Devi S, Lee KE, Kang SG, Kumar P (2019) Fumonisins: impact on agriculture, food, and human health and their management strategies. Toxins 11(6):328. https://doi.org/10.3390/toxins11060328

Kamle M, Mahato DK, Gupta A, Pandhi S, Sharma B, Dhawan K, Mishra S, Kumar M, Tripathi AD, Rasane P (2022a) Deoxynivalenol: an overview on occurrence, chemistry, biosynthesis, health effects and its detection, management, and control strategies in food and feed. Microbiol Res 13(1):292–314. https://doi.org/10.3390/microbiolres13010020

Kamle M, Mahato DK, Gupta A, Pandhi S, Sharma N, Sharma B, Mishra S, Arora S, Selvakumar R, Saurabh V (2022b) Citrinin mycotoxin contamination in food and feed: impact on agriculture, human health, and detection and management strategies. Toxins 14(2):85. https://doi.org/10.3390/toxins14020085

Kang H, Peng Y, Hua K, Deng Y, Bellizzi M, Gupta DR et al (2021) Rapid detection of wheat blast pathogen *Magnaporthe oryzae* Triticum pathotype using genome-specific primers and Cas12a-mediated technology. Engineering 7(9):1326–1335. https://doi.org/10.1016/j.eng.2021.06.011

Keskin E, Eyupoğlu OE (2023) Determination of mycotoxins by HPLC, LC-MS/MS and health risk assessment of mycotoxins in bee products from Turkey. Food Chem 400:134086. https://doi.org/10.1016/j.foodchem.2022.134086

Khakimov A, Salakhutdinov I, Omolikov A, Utaganov S (2022) Traditional and current-prospective methods of agricultural plant diseases detection: a review. IOP Conf Ser: Earth Environ Sci 951(1):012002). IOP Publishing. https://doi.org/10.1088/1755-1315/951/1/012002

Khan AM, Khan M, Salman HM, Ghazali HMZU, Ali RI, Hussain M et al (2023) Detection of seed-borne fungal pathogens associated with wheat (*Triticum aestivum* L.) seeds collected from farmer fields and grain market. J King Saud Univ - Science 35(4):102590. https://doi.org/10.1016/j.jksus.2023.102590

Koonin EV, Gootenberg JS, Abudayyeh OO (2023) Discovery of diverse CRISPR-Cas systems and expansion of the genome engineering toolbox. Biochemistry 62(24):3465–3487. https://doi.org/10.1021/acs.biochem.3c00345

Kumar J, Choudhary K (2024) Enhancing postharvest stability and shelf life of perishable fruits and vegetables using molecular approaches. In: Bioengineered fruit and vegetables. Apple Academic Press, pp 75–92

Lal D, Dev D, Kumari S, Pandey S, Aparna, Sharma N et al (2024) Fusarium wilt pandemic: current understanding and molecular perspectives. Funct Integr Genomics 24(2):41. https://doi.org/10.1007/s10142-023-00999-9

Le DT, Netsu O, Uehara-Ichiki T, Shimizu T, Choi IR, Omura T, Sasaya T (2010) Molecular detection of nine rice viruses by a reverse-transcription loop-mediated isothermal amplification assay. J Virol Methods 170(1–2):90–93. https://doi.org/10.1016/j.jviromet.2010.08.006

Lee HJ, Kim HJ, Kim SM, Jeong RD (2024) Development and application of reverse transcription nanoplate-based digital PCR assay for sensitive and accurate detection of rice black-streaked dwarf virus in cereal crops. Plant Pathol J 40(4):408–417. https://doi.org/10.5423/PPJ.FT.02.2024.0044

Lei R, Kuang R, Peng X, Jiao Z, Zhao Z, Cong H et al (2023) Portable rapid detection of maize chlorotic mottle virus using RT-RAA/CRISPR-Cas12a based lateral flow assay. Front Plant Sci 14:1088544. https://doi.org/10.3389/fpls.2023.1088544

Li SY, Cheng QX, Liu JK, Nie XQ, Zhao GP, Wang J (2018) CRISPR-Cas12a has both cis- and trans-cleavage activities on single-stranded DNA. Cell Res 28(4):491–493. https://doi.org/10.1038/s41422-018-0022-x

Li F, Ye Q, Chen M, Zhou B, Zhang J, Pang R et al (2021a) An ultrasensitive CRISPR/Cas12a-based electrochemical biosensor for Listeria monocytogenes detection. Biosens Bioelectron 179:113073. https://doi.org/10.1016/j.bios.2021.113073

Li T, Cheng N (2023) Sensitive and portable signal readout strategies boost point-of-care CRISPR/Cas12a biosensors. Acs Sensors 8(11):3988–4007

Li Y, Li S, Wang J, Liu G (2021b) CRISPR-Cas12a-assisted detection of stored-product pests: a case study on Rhyzopertha dominica. Pest Manag Sci 77(2):654–663. https://doi.org/10.1002/ps.6543

Li D, Ling S, Wu H, Yang Z, Lv B (2022) CRISPR/Cas12a-based biosensors for ultrasensitive tobramycin detection with single- and double-stranded DNA activators. Sensors Actuators B Chem 355:131329. https://doi.org/10.1016/j.snb.2021.131329

Li PR, Wang ZX, Xu ZK, Wang J, Li B, Shen X, Xu ZL (2024a) An RPA-assisted CRISPR/Cas12a assay combining fluorescence and lateral flow strips for rapid detection of Enterotoxigenic Bacillus cereus. J Agric Food Chem. https://doi.org/10.1021/acs.jafc.4c00829

Li W, Li Y, Zhao LD, Yuan R, Zhuo Y, Zhong X (2024b) Triple-helix as a target converter for trace pesticide detection based on CRISPR/Cas12a-based ECL biosensor. Sens Actuators B Chem 409:135599. https://doi.org/10.1016/j.snb.2023.135599

Liang Z, Zhang K, Chen K, Gao C (2014) Targeted mutagenesis in *Zea mays* using TALENs and the CRISPR/Cas system. J Genet Genomics 41(2):63–68. https://doi.org/10.1016/j.jgg.2013.11.010

Liang Z, Chen K, Yan Y, Zhang Y, Gao C (2018) Genotyping genome-edited mutations in plants using CRISPR ribonucleoprotein complexes. Plant Biotechnol J 16(12):2053–2062. https://doi.org/10.1111/pbi.12932

Liang P, Lv B, Chen K, Qiao W, Li D (2023) An ultrasensitive Cd^{2+} detection biosensor based on DNAzyme and CRISPR/Cas12a coupled with hybridization chain reaction. Anal Chim Acta 1283:341950. https://doi.org/10.1016/j.aca.2023.341950

Liang X, Zhang X, Xi K, Liu Y, Jijakli MH, Guo W (2024) Development of an RPA-based CRISPR/Cas12a assay in combination with a lateral flow strip for rapid detection of toxigenic *Fusarium verticillioides* in maize. Food Control 157:110172. https://doi.org/10.1016/j.foodcont.2023.110172

Lin L, Zhang J, Wang P, Wang Y, Chen J (1998) Thin-layer chromatography of mycotoxins and comparison with other chromatographic methods. J Chromatogr A 815(1):3–20. https://doi.org/10.1016/S0021-9673(98)00483-3

Lin C, Zheng H, Sun M, Guo Y, Luo F, Guo L et al (2018) Highly sensitive colorimetric aptasensor for ochratoxin A detection based on enzyme-encapsulated liposome. Anal Chim Acta 1002:90–96. https://doi.org/10.1016/j.aca.2017.12.009

Lister RM, Rochow WF (1979) Detection of barley yellow dwarf virus by enzyme-linked immunosorbent assay. Phytopathology 69(12):1238–1241. https://doi.org/10.1094/Phyto-69-1238

Liu J, Wu D, Chen J, Jia S, Chen J, Wu Y, Li G (2024a) CRISPR-Cas systems mediated biosensing and applications in food safety detection. Crit Rev Food Sci Nutr 64(10):2960–2985. https://doi.org/10.1080/10408398.2022.2156818

Liu M, Li X, Zhou S, Men D, Duan Y, Liu H et al (2024b) Ultrasensitive detection of mycotoxins using a novel single-atom, CRISPR/Cas12a-based nanozymatic colorimetric biosensor. Chem Eng J 497:154418. https://doi.org/10.1016/j.cej.2024.154418

Lu Y, Ji H, Chen Y, Li Z, Timira V (2023) A systematic review on the recent advances of wheat allergen detection by mass spectrometry: future prospects. Crit Rev Food Sci Nutr 63(33):12324–12340. https://doi.org/10.1080/10408398.2022.2149918

Mahato DK, Lee KE, Kamle M, Devi S, Dewangan KN, Kumar P, Kang SG (2019) Aflatoxins in food and feed: an overview on prevalence, detection and control strategies. Front Microbiol 10:2266. https://doi.org/10.3389/fmicb.2019.02266

Malhotra S, Verma A, Tyagi N, Kumar V (2017) Biosensors: principle, types and applications. Int J Adv Res Innov Ideas Educ 3(2):3639–3644

Mao Z, Chen R, Wang X, Zhou Z, Peng Y, Li S et al (2022) CRISPR/Cas12a-based technology: a powerful tool for biosensing in food safety. Trends Food Sci Technol 122:211–222. https://doi.org/10.1016/j.tifs.2022.02.036

Maqbool Z, Khalid W, Mahum, Khan A, Azmat M, Sehrish A et al (2024) Cereal sprout-based food products: industrial application, novel extraction, consumer acceptance, antioxidant potential, sensory evaluation, and health perspective. Food Sci Nutr 12(2):707–721. https://doi.org/10.1002/fsn3.3874

Meirelles PG, Ono MA, Ohe MCT, Maroneze DM, Itano EN, Garcia GT et al (2006) Detection of *Fusarium* sp. contamination in corn by enzyme-linked immunosorbent assay. Food Agric Immunol 17(2):79–89. https://doi.org/10.1080/09540100600772116

Meng JN, Xu ZK, Li PR, Zeng X, Liu Y, Xu ZL, Wang J, Ding Y, Shen X (2024) Universal and naked-eye diagnostic platform for emetic *Bacillus cereus* based on RPA-assisted CRISPR/Cas12a. J Agric Food Chem 72(15):8823–8830. https://doi.org/10.1021/acs.jafc.4c01380

Milani J, Maleki G (2014) Effects of processing on mycotoxin stability in cereals. J Sci Food Agric 94(12):2372–2375. https://doi.org/10.1002/jsfa.6600

Milićević DR, Škrinjar M, Baltić T (2010) Real and perceived risks for mycotoxin contamination in foods and feeds: challenges for food safety control. Toxins 2(4):572–592. https://doi.org/10.3390/toxins2040572

Mir SA, Dar BN, Shah MA, Sofi SA, Hamdani AM, Oliveira CAF et al (2021) Application of new technologies in decontamination of mycotoxins in cereal grains: challenges and perspectives. Food Chem Toxicol 148:111976. https://doi.org/10.1016/j.fct.2021.111976

Mu K, Ren X, Yang H, Zhang T, Yan W, Yuan F et al (2022) CRISPR-Cas12a-based diagnostics of wheat fungal diseases. J Agric Food Chem 70(23):7240–7247. https://doi.org/10.1021/acs.jafc.2c01909

Naeem M, Zaman W, Saqib S, Shahzad A, ur Rahman S, Ahmad N (2024) CRISPR/Cas-mediated genome editing for efficient tomato breeding: past achievements and future directions. S Afr J Bot 172:277–288. https://doi.org/10.1016/j.sajb.2024.01.018

Nardelli V, D'Amico V, Ingegno M, Della Rovere I, Iammarino M, Casamassima F, Calitri A, Nardiello D, Li D, Quinto M (2021) Pesticides contamination of cereals and legumes: monitoring of samples marketed in Italy as a contribution to risk assessment. Appl Sci 11(16):7283. https://doi.org/10.3390/app11167283

Nayak A, Dutta D (2023) A comprehensive review on CRISPR and artificial intelligence-based emerging food packaging technology to ensure "safe food.". Sustain Food Technol. https://doi.org/10.1016/j.susfood.2023.100097

Omar SS, Haddad MA, Parisi S (2020) Validation of HPLC and enzyme-linked immunosorbent assay (ELISA) techniques for detection and quantification of aflatoxins in different food samples. Foods 9(5):661. https://doi.org/10.3390/foods9050661

Ortega SF, Siciliano I, Prencipe S, Gullino ML, Spadaro D (2020) Development of PCR, LAMP and qPCR assays for the detection of aflatoxigenic strains of *Aspergillus flavus* and *A. parasiticus* in hazelnut. Toxins 12(12):757. https://doi.org/10.3390/toxins12120757

Pankaj SK, Shi H, Keener KM (2018) A review of novel physical and chemical decontamination technologies for aflatoxin in food. Trends Food Sci Technol 71:73–83. https://doi.org/10.1016/j.tifs.2017.10.008

Pataer P, Gao K, Zhang P, Wang X, Li Z (2024) A simple, portable and multiplex LAMP-based CRISPR/Cas12a assay for visually screening genetically modified crops. Sensors Actuators B Chem 403:Article 135124. https://doi.org/10.1016/j.snb.2023.135124

Pava-Ripoll M, Miller AK, Ziobro GC (2023) Development of a multiplex polymerase chain reaction (PCR) assay for the potential detection of insect contaminants in food. J Food Prot 86(8):100120. https://doi.org/10.4315/JFP-23-002

Phumlani T, Morgan SP, Zikhona NT (2021) Biosensors: design, development and applications. Nanopores 11:1–19

Qian H, Guo X, Yang H, Bao T, Wu Z, Wen W et al (2024) Enhancing CRISPR/Cas-mediated electrochemical detection of nucleic acid using nanoparticle-labeled covalent organic frameworks reporters. Biosens Bioelectron 261:116522. https://doi.org/10.1016/j.bios.2024.116522

Qiao Y, Wang X, Song Y, Zhang J, Han Q (2023) CRISPR-Cas12a-based aptasensor for sensitive and selective FB1 detection. J Food Compos Anal 123:105615. https://doi.org/10.1016/j.jfca.2023.105615

Rahman M, Hasan MS, Islam R, Rana R, Sayem ASM, Sad MAA et al (2022) Plasma-activated water for food safety and quality: a review of recent developments. Int J Environ Res Public Health 19(11):6630. https://doi.org/10.3390/ijerph19116630

Rauová D (2010) Single laboratory-validated HPLC methods for determination of ochratoxin A, fumonisin B1 and B2, zearalenone and deoxynivalenol in cereals and cereal-based foods. J Food Nutr Res 49:57–68

Sadeghi P, Sohrabi H, Majidi MR, Eftekhari A, Zargari F, de la Guardia M, Mokhtarzadeh AA (2024) Mycotoxins detection in food samples through lateral flow assays (LFAs): an update for status and prospect. TrAC Trends Anal Chem:117722. https://doi.org/10.1016/j.trac.2024.117722

Sadiku MN, Ashaolu TJ, Musa SM (2020) Food contamination: a primer. Int J Adv Sci Res Eng 6(3):1–7

Sahel DK, Giriprasad G, Jatyan R, Guha S, Korde A, Mittal A et al (2024) Next-generation CRISPR/Cas-based ultrasensitive diagnostic tools: current progress and prospects. RSC Adv 14(44):32411–32435. https://doi.org/10.1039/D4RA03754D

Sajid M, Kawde AN, Daud M (2015) Designs, formats and applications of lateral flow assay: a literature review. J Saudi Chem Soc 19(6):689–705. https://doi.org/10.1016/j.jscs.2014.09.001

Sampath V, Rangarajan N, Ch S, Deori M, Veeraragavan M, Ghodake BD, Kaushal K (2023) Advancing crop improvement through CRISPR technology in precision agriculture trends—A review. Int J Environ Clim Change 13(11):4683–4694. https://doi.org/10.9734/ijecc/2023/v13i113503

Santos L, Marín S, Sanchis V, Ramos AJ (2009) Screening of mycotoxin multicontamination in medicinal and aromatic herbs sampled in Spain. J Sci Food Agric 89(10):1802–1807. https://doi.org/10.1002/jsfa.3648

Senarathna S, Mel R, Malalgoda M (2024) Utilization of cereal-based protein ingredients in food applications. J Cereal Sci:103867. https://doi.org/10.1016/j.jcs.2024.103867

Shaizadinova A, Amanzholova M, Rukavitsina I, Abeldenov S, Zhumakayev AR (2025) CRISPR/Cas12a-based method coupled with isothermal amplification to identify *Alternaria* spp. isolated from wheat grain samples. Front Microbiol 15:1468336. https://doi.org/10.3389/fmicb.2024.1468336

Shalaby SE, Abdou GY, El-Metwally IM, Abou-elella GM (2021) Health risk assessment of pesticide residues in vegetables collected from Dakahlia, Egypt. J Plant Prot Res 61(3):254–264. https://doi.org/10.24425/jppr.2021.137951

Shao H, Jian J, Peng D, Yao K, Abdulsalam S, Huang W et al (2023) Recombinase polymerase amplification coupled with CRISPR-Cas12a technology for rapid and highly sensitive detection of *Heterodera avenae* and *Heterodera filipjevi*. Plant Dis 107(5):1365–1376. https://doi.org/10.1094/PDIS-09-22-2215-RE

Shen J, Zhang D, He Y, Shen Y, Duan M, Zhao Y et al (2024) Evolving CRISPR/Cas system for food safety monitoring across the food supply chain. TrAC Trends Anal Chem 181:118050

Sohlberg E, Virkajärvi V, Parikka P, Rämö S, Laitila A, Sarlin T (2022) TaqMan qPCR quantification and Fusarium community analysis to evaluate toxigenic fungi in cereals. Toxins 14(1):45. https://doi.org/10.3390/toxins14010045

Sohn SI, Pandian S, Oh YJ, Zaukuu JLZ, Kang HJ, Ryu TH et al (2021) An overview of near infrared spectroscopy and its applications in the detection of genetically modified organisms. Int J Mol Sci 22(18):9940. https://doi.org/10.3390/ijms22189940

Song Q, Zheng YJ, Xue Y, Sheng WG, Zhao MR (2017) An evolutionary deep neural network for predicting morbidity of gastrointestinal infections by food contamination. Nutrition 42–46:16–22. https://doi.org/10.1016/j.nut.2016.12.004

Sun Y, Wen T, Zhang P, Wang M, Xu Y (2024) Recent advances in the CRISPR/Cas-based nucleic acid biosensor for food analysis: a review. Foods 13(20):3222. https://doi.org/10.3390/foods13203222

Suttie JM (2000) Hay and straw conservation: for small-scale farming and pastoral conditions (No. 29). Food & Agriculture Organization

Szelenberger R, Cichoń N, Zajaczkowski W, Bijak M (2024) Application of biosensors for the detection of mycotoxins for the improvement of food safety. Toxins 16(6):249. https://doi.org/10.3390/toxins16060249

Thipe VC, Maloney V, Klein A, Gokul A, Keyster M, Katti KV (2021) RNA interference and CRISPR/Cas9 techniques for controlling mycotoxins. In: CRISPR and RNAi systems. Springer, pp 463–490. https://doi.org/10.1007/978-3-030-63147-2_19

Tiwari S, Talreja S (2022) Thin layer chromatography (TLC) vs. paper chromatography: a review. Acta Sci Pharm Sci 6(9)

Tripathi JN, Tripathi L (2024) Using CRISPR-Cas9 genome editing to enhance disease resistance in banana. CABI Rev 19(1):e001. https://doi.org/10.1079/PAVSNNR202419001

Turner APF, Karube I, Wilson GS (1987) Biosensors: fundamentals and applications. Oxford University Press

Turner NW, Subrahmanyam S, Piletsky SA (2009) Analytical methods for determination of mycotoxins: a review. Anal Chim Acta 632(2):168–180. https://doi.org/10.1016/j.aca.2008.11.010

Valcke M, Bourgault MH, Rochette L, Normandin L, Samuel O, Belleville D, Blanchet C, Phaneuf D (2017) Human health risk assessment on the consumption of fruits and vegetables containing residual pesticides: a cancer and non-cancer risk/benefit perspective. Environ Int 108:63–74. https://doi.org/10.1016/j.envint.2017.07.023

Varzakas T (2016) Quality and safety aspects of cereals (wheat) and their products. Crit Rev Food Sci Nutr 56(15):2495–2510. https://doi.org/10.1080/10408398.2013.835758

Wang H, Liu Y (2015) Empirical study on the factors influencing consumer satisfaction of online purchasing of fresh agricultural products. Consumer Economy 31(6):81–86

Wang X, Zhang F, Zhou J (2023a) Field-deployable CRISPR-Cas12a diagnostics for invasive grain beetles. Nat Commun 14(1):122. https://doi.org/10.1038/s41467-023-37822-0

Wang Z, Wei L, Ruan S, Chen Y (2023b) CRISPR/Cas12a-assisted chemiluminescence sensor for aflatoxin B1 detection in cereal based on functional nucleic acid and in-pipet rolling circle amplification. J Agric Food Chem 71(10):4417–4425. https://doi.org/10.1021/acs.jafc.2c08188

Wieser H, Segura V, Ruiz-Carnicer Á, Sousa C, Comino I (2021) Food safety and cross-contamination of gluten-free products: a narrative review. Nutrients 13(7):2244. https://doi.org/10.3390/nu13072244

Wołejko E, Łozowicka B, Kaczyński P (2014) Pesticide residues in berries and juices and potential risk for consumers. Desalin Water Treat 52(19–21):3804–3818. https://doi.org/10.1080/19443994.2013.862486

World Health Organization (WHO) (2021) Molecular diagnostics in low-resource settings. https://www.who.int/publications/i/item/9789240017502

Wu Z, Sun DW, Pu H (2023a) CRISPR/Cas12a and G-quadruplex DNAzyme-driven multimodal biosensor for visual detection of aflatoxin B1. Spectrochim Acta A Mol Biomol Spectrosc 302:123121. https://doi.org/10.1016/j.saa.2023.123121

Wu Z, Sun DW, Pu H, Wei Q (2023b) A novel fluorescence biosensor based on CRISPR/Cas12a integrated MXenes for detecting aflatoxin B1. Talanta 252:123773. https://doi.org/10.1016/j.talanta.2022.123773

Wu C, Wang X, Guo L, Huang X, Wu L, Huang H (2023c) An electrochemical aptasensor based on exonuclease III-assisted signal amplification coupled with CRISPR-Cas12a for ochratoxin A detection. Food Control 148:109631. https://doi.org/10.1016/j.foodcont.2023.109631

Xie S, Yue Y, Yang F (2024) Recent advances in CRISPR/Cas system-based biosensors for the detection of foodborne pathogenic microorganisms. Micromachines 15(11):1329. https://doi.org/10.3390/mi15111329

Xing KY, Peng J, Shan S, Liu DF, Huang YN, Lai WH (2020) Green enzyme-linked immunosorbent assay based on the single-stranded binding protein-assisted aptamer for the detection of mycotoxin. Anal Chem 92(12):8422–8426. https://doi.org/10.1021/acs.analchem.0c01538

Yang W, Chen J, Chen G, Wang S, Fu F (2013) Early diagnosis and fast detection of blast fungus, *Magnaporthe grisea*, in rice plant by using its chitinase as biochemical marker and a rice cDNA encoding mannose-binding lectin as recognition probe. Biosens Bioelectron 41:820–826. https://doi.org/10.1016/j.bios.2012.09.055

Yang W, Zhang H, Li M, Wang Z, Zhou J, Wang S et al (2014) Early diagnosis of blast fungus, *Magnaporthe oryzae*, in rice plant by using an ultra-sensitive electrically magnetic-

controllable electrochemical biosensor. Anal Chim Acta 850:85–91. https://doi.org/10.1016/j.aca.2014.08.019
Yang Y, Li G, Wu D, Liu J, Li X, Luo P et al (2020) Recent advances on toxicity and determination methods of mycotoxins in foodstuffs. Trends Food Sci Technol 96:233–252. https://doi.org/10.1016/j.tifs.2019.12.007
Yang Y, Li X, Wang X, Wang Z, Gong S (2024) CRISPR-Cas-based colorimetric strategies for nucleic acids detection. TrAC Trends Anal Chem:118058. https://doi.org/10.1016/j.trac.2024.118058
Yuan M, Ding R, Chen S, Duan G (2021) Advances in field detection based on CRISPR/Cas system. ACS Synth Biol 10(11):2824–2832. https://doi.org/10.1021/acssynbio.1c00294
Zetsche B, Gootenberg JS, Abudayyeh OO, Slaymaker IM, Makarova KS, Essletzbichler P et al (2015) Cpf1 is a single RNA-guided endonuclease of a class 2 CRISPR-Cas system. Cell 163(3):759–771. https://doi.org/10.1016/j.cell.2015.09.038
Zhang M, Chen R, Zhou X, Wu J (2018) Monoclonal antibody-based serological detection methods for wheat dwarf virus. Virol Sin 33:173–180. https://doi.org/10.1007/s12250-018-0012-1
Zhang YM, Zhang Y, Xie K (2020) Evaluation of CRISPR/Cas12a-based DNA detection for fast pathogen diagnosis and GMO test in rice. Mol Breed 40(1):11. https://doi.org/10.1007/s11032-020-01140-5
Zhang Y, Ren G, Liu Q (2022) CRISPR-Cas12a-mediated rapid detection of Sitophilus oryzae in stored grains. Biosens Bioelectron 204:114123. https://doi.org/10.1016/j.bios.2022.114123
Zhang L, Yin X, Yang H, Wen H, Han S, Pan X et al (2023) A sensitive and specific monoclonal antibody-based enzyme-linked immunosorbent assay for the rapid detection of pretilachlor in grains and the environment. Foods 13(1):12. https://doi.org/10.3390/foods13010012
Zhang C, Zhao T, Li J, Liu X, Wang W, Huang X et al (2024a) Ultrasensitive and on-site diagnosis of rice bakanae disease based on CRISPR-LbCas12a coupled with LAMP. Pest Manag Sci 80(12):6527–6534. https://doi.org/10.1002/ps.8078
Zhang C, Zhao X, Huang Z, Li Z, Hu J, Liu R, Lv Y (2024b) Highly sensitive detection of aflatoxin B1 by CRISPR/Cas12a-assisted single nanoparticle counting. Food Chem 443:138557. https://doi.org/10.1016/j.foodchem.2023.138557
Zhang Y, Yu Y, Yang X, Yuan X, Zhang J (2024c) Pb(II) inhibits CRISPR/Cas12a activation and application for paper-based microfluidic biosensor assisted by smartphone. Sens Actuators B Chem 398:134732. https://doi.org/10.1016/j.snb.2023.134732
Zheng M, Zhang S, Zhang Y, Hu B (2021) Construct food safety traceability system for people's health under the internet of things and big data. IEEE Access 9:70571–70583. https://doi.org/10.1109/ACCESS.2021.3077976
Zheng J, Liu L, Li X, Xu Z, Gai Z, Zhang X, Lei H, Shen X (2023) Rapid and simple detection of *Burkholderia gladioli* in food matrices using RPA-CRISPR/Cas12a method. Foods 12(9):1760. https://doi.org/10.3390/foods12091760
Zhu F, Zhao Q (2024) Competitive aptamer assay for aflatoxin B1 detection using CRISPR/Cas12a as a signal amplifier. Microchem J 207:112054. https://doi.org/10.1016/j.microc.2024.112054
Zhu Z, Li R, Zhang H, Wang J, Lu Y, Zhang D, Yang L (2022) PAM-free loop-mediated isothermal amplification coupled with CRISPR/Cas12a cleavage (Cas-PfLAMP) for rapid detection of rice pathogens. Biosens Bioelectron 204:114076. https://doi.org/10.1016/j.bios.2021.114076
Zhuang J, Zhao Z, Lian K, Yin L, Wang J, Man S et al (2022) SERS-based CRISPR/Cas assay on microfluidic paper analytical devices for supersensitive detection of pathogenic bacteria in foods. Biosens Bioelectron 207:114167. https://doi.org/10.1016/j.bios.2022.114167

Chapter 5
Bioinformatics Tools in Cereal Genome Editing

Nayla Munawar, Rana Ali Haider, Paulo Zaini, Baohong Zhang, Ahmed Al-Harrasi, and Aftab Ahmad

Abstract CRISPR has enabled researchers to introduce precise modifications into the genome, achieving desired traits in plants. It offers unprecedented control over DNA and RNA editing in plants with precision, flexibility, and versatility. Although the CRISPR system contains three components, including gRNA, Cas nuclease, and a PAM sequence to introduce the (double-strand break) DSB in the genome, gRNA holds a key importance in driving Cas nuclease to the target site. Over the past decade, several bioinformatics and machine learning tools have been developed to design efficient gRNAs for various applications of CRISPR–Cas technology. In this chapter, we highlight the importance of gRNA in CRISPR–Cas. Similarly, we also discuss different deep learning tools to design an efficient gRNA for different CRISPR applications. In addition, we also summarize the integration of machine learning and AI tools to improve CRISPR–Cas applications for gene editing. We also show how the integration of AI with CRISPR–Cas is influencing the gRNA design in different CRISPR applications. Finally, we summarize the changing dynamics of gRNA design with applications of AI in CRISPR–Cas studies.

These authors contributed equally to this chapter

N. Munawar
Department of Chemistry, College of Science, United Arab Emirates University, Al-Ain, United Arab Emirates

R. A. Haider
Department of Biochemistry, University of Agriculture, Faisalabad, Pakistan

P. Zaini
Department of Plant Sciences, University of California, Davis, CA, USA

B. Zhang
Department of Biology, East Carolina University, Greenville, NC, USA

A. Al-Harrasi · A. Ahmad (✉)
Natural and Medical Sciences Research Center (NMSRC), University of Nizwa, Nizwa, Oman
e-mail: aftab.ahmad@unizwa.edu.om

K. A. Abd-Elsalam, A. Ahmad (eds.), *Gene Editing in Cereals*, Sustainability Sciences in Asia and Africa, https://doi.org/10.1007/978-981-95-6722-5_5

5.1 Introduction

Today, much of the biological research is focused on the applications of of CRISPR–Cas systems. The first discovery of CRISPR came nearly three decades ago, while studying the gene for the isozyme conversion of the alkaline phosphatase in Escherichia coli (Ishino et al. 2018). At the time of this discovery, limited of DNA sequence data, particularly regarding mobile genetic elements, made it extremely difficult to determine the biological role of these odd repetitive elements. Consequently the biological function of this enigmatic repetitive motif remain unresolved for many years (Mojica and Montoliu 2016). CRISPR immune system functions were not elucidated untill the early 2000s, when spacer regions of CRISPR were found to share sequence homology with bacteriophages, archaeal viruses, and plasmids. Untill the mid-2000s, the exact functions of this sequence remained largely unknown (Makarova et al. 2018). Around the same time, a number of genes previously thought to encode DNA repair proteins specific to hyperthermophilic archaea were identified as being consistently linked with CRISPR and were subsequently designated as Cas (CRISPR-associated) genes. Comparative genomics studies ultimately revealed that, similar to the eukaryotic RNAi system, CRISPR and Cas proteins (products of the cas genes) function cooperatively, forming an acquired immune system that protect prokaryotic cells from invading viruses and plasmids (Xu and Li 2020).

The remarkable diversity of CRISPR–Cas system was revealed in 2012, when these systems were introduced as genome-editing tools defined by their unprecedented ability to enable targeted genetic modifications across virtually all living organisms (Moreno-Mateos et al. 2015). Shortly thereafter, in early 2013, definitive proof of their ability to make precise modifications in mammalian genomes was reported, marking a major turning point in molecular biology. The goal of this chapter is to summarize CRISPR components, design tools, improving specificity, reducing off-targets, and how AI tools are contributing to improving CRISPR–Cas technology for precise gene editing in plants and animals (Ishino et al. 2018).

5.1.1 Overview of CRISPR–Cas Technology

CRISPR–Cas technology represent a transformative breakthrough in genetic engineering, owing to its unprecedented precision, programming, and efficiency in gene editing. Originally discovered as a natural defense mechanism in bacteria that confer protection against viral infections, CRISPR–Cas system has been repurposed for a wide range of biotechnological and medical applications (Tyagi et al. 2020). The discovery and development of the CRISPR based gene editing is largly attributed to poineering work of Jennifer Doudna and Emmanuelle Charpentier, whose work has fundamentally revolutionized our ability to manipulate DNA with high precision and simplicity. Their contributions have had a profound impact across multiple scientific disciplines and continue to shape advances in medicine, agriculture, and biotechnology (Mojica and Montoliu 2016).

Essentially, the CRISPR–Cas system operates through a relatively simple but efficient mechanism. Target specificity is achieved through short guide RNA (gRNA) that directs Cas protein to complementary DNA sequences, to make precise cleavage at the target site (Zhu 2015). Unlike earlier gene-editing methods such as ZFNs and TALENs, CRISPR–Cas is distinguished by its precision in targeting and editing efficiency, underscoring its potential as a useful and highly versatile tool for biological research and biotechnology (Li et al. 2020). The applications of CRISPR–Cas are extensive, ranging from transformative medical approaches for the treatments of genetic disorder to agricultural innovations, such as the development of pest- and disease resistant crops (Makarova et al. 2018).

Although CRISPR is described as precise but the rapid development and widespread deployment of CRISPR–Cas technology also raise significant ethical and social concerns, especially in health, medical, and therapeutic applications. Current debates focuses on the off-target effects, potential long-term effects of gene editing, and the ethical concerns of manipulating the human germline (Li et al. 2023). These concerns have prompted calls for proactive and inclusive governance frameworks. In particular, cooperation among scientists, ethicists, and policymakers has been emphasized to ensure that the applications of CRISPR technology aligns with principles of safety, ethical responsibility, and social equity (Jiang and Doudna 2015).

The future of CRISPR–Cas technology is rich with transformative potential. Continuous efforts to improve the efficiency, specificity and modularity of CRISPR-Cas has lef to development of new tools like base editing, prime editing, and epigenome editing, which enable targeted modifications without introducing DSBs (Uniyal et al. 2019). These developments have significantly broaden the scope of CRISPR–Cas, especially in synthetic biology, computational biology and gene therapy. As research continues to deepen and real world applications expand, the potential of CRISPR–Cas to revolutionize genetic research and therapeutic strategies expected to become increasingly evident, ushering in a new era of innovation and scientific discovery (Yang and Patel 2017).

5.1.2 Importance of Precision in Genome Editing

Genome editing referes to the targeted modification of genomic DNA at specific loci across different cell types and organisms. It enables the precise addition, deletion, and substitution of DNA sequences, resulting in the inactivation of a particular gene, the introduction of new genetic traits, and the correction of deleterious mutations associated with genetic disorders (Javed et al. 2021). Driven by the accelerated advancement in life sciences, genome editing technology has become the most promising tool to investigate the pathophysiological mechanisms of genetic diseases, study gene functions, identifying new therapeutic targets, and develop new crop varieties, among many other applications (Naeem and Alkhnbashi 2023).

At present, the three most widely used genome editing technologies are zinc finger nucleases (ZFNs), transcription activator-like effector nucleases (TALENs), and the RNA-guided CRISPR (clustered regularly interspaced short palindromic repeats)–Cas (CRISPR-associated) nuclease systems (Yang and Patel 2017). Among these platforms, CRISPR–Cas systems are the most dominant genome editing tools in molecular biology laboratories worldwide, due to their cost-effectiveness, efficiency, scalability, and rapid implementation. This chapter focuses on CRISPR–Cas systems in general, including how they are programmed and used to manipulate genes in plants, animals, and health sciences (Xu and Li 2020).

The effectiveness of genome editing largely depends the precision with which CRISPR–Cas systems are programed to the target site in the genome. Accurate targeting of the intended target site is essential to avoid off-target effects, which cause alterations and could pose risks especially in health and therapeutic applications (Ryan et al. 2018). This requirement is very significant in therapeutic applications, where precise modifications are necessary to correct pathogenic mutations without introducing additional harms to patients (Hille and Charpentier 2016). In agriculture, precise genome editing facilitates crop improvement by introducing desired traits, such as resistance to pests or improved yield, without compromising nutritional value and ecological balance (Waddington et al. 2016). Precise genome editing is also important in developing robust disease models and studying genetic diseases, thus contributing to fundamental and translational research. Moreover, public acceptance of genome editing tools depends on demonstrated reliability, safety, and ethical responsibility, underscoring the central importance of precision in genome editing (Chakrabarti et al. 2019). Greater precision will not only reduce the off targets, but also minimize the regulatory and validation costs. As CRISPR–Cas systems continue to evolve, precision will remaine the key factor to unlock their full potential in agriculture, medicine, and biotechnology (Barrangou and Marraffini 2014).

5.2 CRISPR–Cas Technology

CRISPR–Cas technology represents a revolutionary development in genetic engineering that has fundamentally transformed how researchers manipulate DNA. The term CRISPR is an acronym for Clustered Regularly Interspaced Short Palindromic Repeats, and the associated Cas (CRISPR-associated) proteins, originally identified as part of the bacterial adaptive immune defense mechanism against viral infection (Leenay and Beisel 2017). In genome editing applications, the Cas9 enzyme is directed to a target site in the genome using a guide RNA (gRNA), where it creates a double-strand break in the DNA. The cell's endogenous repair processes subsequently repair this DSB, enabling precise gene editing including gene knock out, correction, or precise insertion (Chakrabarti et al. 2019).

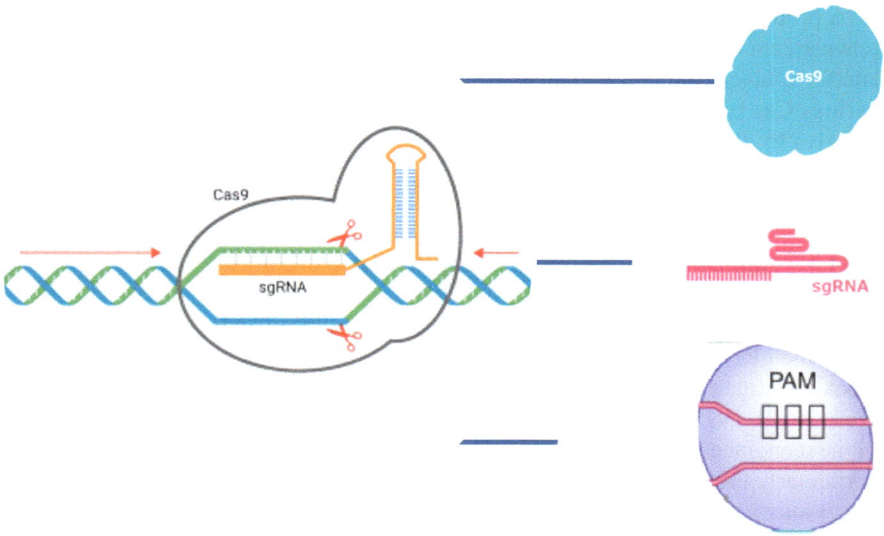

Fig. 5.1 Component of CRISPR–Cas system

5.2.1 Components and Mechanism of CRISPR–Cas

The guide RNA (gRNA) binds to the complementary target site within the genome and drives the Cas protein to the specific locus, thus activating CRISPR–Cas system. The Cas protein then makes a break in the double strand at the target site, depending on the presence of an adjacent PAM sequence. The cell's intrinsic repair systems, either homology-directed repair (HDR) or non-homologous end joining (NHEJ), repair the double-strand break. HDR can make precise insertions with the use of a donor template, but NHEJ can make minor indels that may disrupt gene function and let to gene knockout (Zhu et al. 2020).

The key functional components of the CRISPR-Cas system are the Cas enzymes which act as programmable nucleases. Among these, Cas9 is the most extensively characterized endonuclease and can introduce the precise DSB in DNA, recruited by gRNA to the complementary target site (Fig. 5.1). In addition to Cas9, several other Cas nucleases have been identified including Cas12, Cas14, and Cas13. These Cas nucleases are also demonstrated play an important role in DNA and RNA editing thus expanding the scope of CRISPR-based gene editing technologies (Gori et al. 2015).

A gRNA can be engineered to precisely direct the Cas protein to the specific DNA sequence intended for editing. The gRNA for CRISPR-Cas9 system is essentially composed of two important components: a structural scaffold that binds to the Cas enzyme and a variable spacer, complementary to the target DNA (Akbari et al. 2019).

A protospacer adjacent motif (PAM) is a small DNA motif located immediately downstream of the target site in the genome. Recognition of the PAM is a prerequisite for efficient binding and cleavage by the Cas protein. The target specificity of various Cas proteins is not only determined by the gRNA-DNA complementarity but also by their requirement for particular PAM sequences, which varies among Cas protein variants (Kant 2021) (Fig. 5.1).

5.2.1.1 Importance of gRNA in CRISPR–Cas

Beyond intrinsic nuclease activity, target site accessibility and the binding specificity of the gRNA to the target sequence, are two of the most critical parameters for DNA cleavage efficiency, influencing both on-target and off-target effects (Dimauro et al. 2022). Despite substantial progress in CRISPR-Cas, the molecular dynamics of the reengineered Cas9 nuclease in living cells are still not completely understood, particularly the key aspects such as interaction kinetics with DNA, resident time at target loci, and cell cycle-dependent cleavage activity (Waddington et al. 2016).

5.2.1.2 gRNA and Its Components (the Driver of Cas)

The guide RNA (gRNA) is at the core component of the CRISPR–Cas system, enabling accurate and precise genome editing by directing the the Cas protein, most notably the Cas9 protein, to specific loci in the genome (Chen et al. 2017b). The high specificity of the gRNA plays a crucial role in minimizing the potential of off-target effects. Researchers can easily program gRNA to target a complex genome, enabling wide range of applications such as base editing, activation, repression, and deletion of genes (Ai et al. 2022). The gRNA–Cas complex is crucial for efficient recognition, binding, and cleavage of the target site thereby enhancing both speed and accuracy of genome editing. In a therapeutic context, high gRNA specificity is essential to ensure safe and effective gene therapy for genetic disorders. Adding chemical changes to the gRNA is a recent advancement in gRNA engineering, which improves the effectiveness, stability and and functional performance (Luo et al. 2022). The programmability and versatility of the gRNA represent the defining features of the CRISPR–Cas systems facilitating transformative advances across medicine, biotechnology, and basic research (Konstantakos et al. 2022).

5.2.1.3 PAM

The Protospacer Adjacent Motif (PAM) was described by Horvath and his team in 2008 while working on CRISPR/Cas systems in *Streptococcus thermophilus*. Using bioinformatic tools they confirmed the existence of PAM and demonstrated their essential role in the CRISPR–Cas mediated immunity (Lee 2023). The early findings indicated that PAMs are a conserved elements and also play a crucial role in

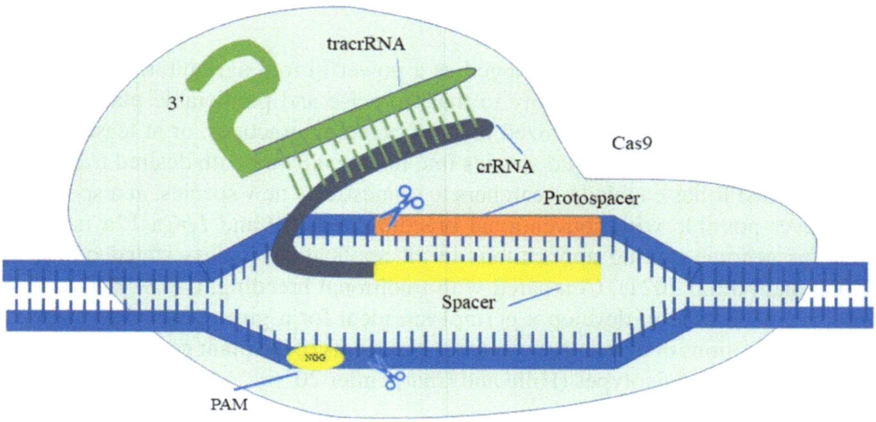

Fig. 5.2 Structure of PAM sequence

CRISPR–Cas systems. Subsequent studies provided further insights into the role of PAMs by showing that, in *Staphylococcus epidermidis*, PAM enable the differentiation between foreign DNA and self-DNA, thus allowing the CRISPR system to target foreign invaders while exempting cleavage of the host genome (Gaudelli et al. 2017). Structural and biochemical studies have elucidated the molecular basis of PAM dependent target recognition. The studies show that Cas effector proteins initially bind with the PAM sequence and then recognize the adjacent DNA for complementarity with the CRISPR RNA (Fig. 5.2). The essential role of PAM's in self–nonself distinction is further supported by the fact that the target will not be recognized and cleaved in its absence (Anzalone et al. 2019) (Fig. 5.2).

For this very reason, it is a prerequisite for successful spacer acquisition, ensuring that newly acquired spacers can effectively recognize and target invading DNA. Regardless of changes in their composition, the importance of PAMs in specifying the target sequence plays a vital role in CRISPR–Cas systems (Ryan et al. 2018).

5.2.2 Impact of CRISPR–Cas in Various Fields

CRISPR–Cas technology has emerged as an unparalleled tool in genetic engineering because of its extreme accuracy, versatility, and efficiency in modifying the living organisms' DNA. The development of CRISPR–Cas has revolutionized modern biology and multiple applied desciplines including health and therapeutic by allowing accurate genome editing (Zhu 2022). Derived from bacterial adaptive immune system, CRISPR/Cas9 has made precision genome editing widely accessible with accelerated developments in biotechnology, agriculture, biomedicine, and beyond (Li et al. 2020).

5.2.2.1 Crop Improvement

CRISPR–Cas technology has emerged as a powerful tool agricultural biotechnology due to its unparalleled ability to make precise and predictable plant genome modifications. It has revolutionized modern breeding practices, or at least allowed the breeding of previously wild species into novel cultivars with desired traits (Zhu 2022). CRISPR has enabled researchers to domesticate new species, at a speed that was never possible with conventional breeding. *Sp*Cas9 and *Lb*Cas12a has been used for genome editing applications in all applications unless stated otherwise (Goell and Hilton 2021). Compared with traditional breeding, CRISPR–Cas technology can be used to develop a germplasm ideal for a given context by gain-of-function mutations or deleting unwanted components of the plant genome associated with undesirable phenotypes (Hille and Charpentier 2016).

5.2.2.2 Diagnostics

CRISPR–Cas9 has been widely used to develop innovative diagnostic platforms for infectious diseases. In one notable study, using a macaque model and in vitro experiments, CRISPR–Cas9 was skilfully combined with nucleic acid sequence-based amplification (NASBA), an isothermal amplification method that allowed scientists to accurately discriminate between closely related Zika virus strains. Following NASBA amplification of the potential viral RNA, a synthetic trigger sequence was introduced and sgRNA–Cas9 combination was used to cleave the resulting dsDNA (double-stranded DNA). Cas9 mediated cleavage produced either full-length or shortened DNA fragments, depending upon the presence of a strain-specific PAM. Full-length DNA fragments activated the trigger switch while truncated fragments did not, enabling accurate strain-specific discrimination. This molecular distinction was ultimately visualized as a color change on a paper based disc (Strich and Chertow 2019).

In a different study, optical DNA mapping was combined with CRISPR–Cas9 to identify the genes encoding drug resistance in bacteria. In this study, a fluorescent dye (YOYO-1) and the minor groove binder netropsin bind independently to DNA specifically in AT-rich regions. This produces a characteristic fluorescence intensity relative to each DNA segment generating a bar code like optical map. Meanwhile a guide RNA (gRNA)–Cas9 complex binds and cleaves nucleic acid sequences on plasmids containing the resistance genes. By using this approach, researchers were able to distinguish the plasmids encoding different carbapenemases, such as Klebsiella pneumoniae carbapenemase (KPC) and New Delhi metallo-β-lactamase 1 (NDM-1), and extended-spectrum β-lactamases (ESBLs), such as cefotaxime 15 (CTX-M-15) and CTX-M-14. In addition, by introducing multiple crRNAs, the system enabled detection of multiple resistance genes in the same reaction (Leenay and Beisel 2017).

5.3 Deep Learning Tools for gRNA Design in Plants and Animal Systems

CRISPR–Cas9 is a powerful genome-editing technology that has revolutionized the field of molecular biology by enabling site specific modifications in the genome (Zhu 2015). The CRISPR–Cas9 system relies on gRNA, as it directs the Cas9 nuclease to the targeted sequence, where it creates the DSB in the DNA (Grosso et al. 2013). Accurate design of gRNA is therefore critical for precise outcomes. In recent years, deep-learning-based approaches have been developed for gRNA design, offering better on-target efficiency and specificity compared with traditional methods (Strich and Chertow 2019).

Deep learning methods outperform traditional bioinformatics tools across multiple dimensions, such as accuracy, efficiency, and specificity, and have emerged as valuable tools for gRNA design in both plant and animal model system (LaFountaine et al. 2015). By leveraging large datasets deep learning technologies and complex neural networks can predict accurate gRNA activity, off-target effects, and specificity, allowing rational design of optimal gRNA sequences for CRISPR applications (Zhu 2015). Looking ahead, continuous developments in deep learning algorithms, together with rapid expansion in genomic data, are expected to play a significant role in enhancing the utility and reliability of gRNA design tools, thereby accelerating progress in the fields of genome editing and molecular biology (Luo et al. 2022).

Table 5.1 On-target selection tools

On-target software	URL
CRISPick	https://portals.broadinstitute.org/gppx/crispick/public
WU-CRISPR	https://crisprdb.org/wu-crispr-website/index.html
TUSCAN	https://gt-scan.csiro.au/apps/tuscan
TSAM	http://www.aai-bioinfo.com/CRISPR
SSC	http://crispr.dfci.harvard.edu/SSC/
SgRNAScorer	https://crispr.med.harvard.edu/sgRNAScorerV2/
ge-CRISPR	http://bioinfo.imtech.res.in/manojk/gecrispr/
DeepCRISPR	https://bio.tools/t?page=1&q=DeepCRISPR&sort=score
DeepCpf1	http://deepcrispr.info
DeepCas9	https://github.com/lje00006/DeepCas9
CRISPRScan	http://www.crisprscan.org/?page=welcome
CRISPRpred	https://bio.tools/crisprpred
CRISPRater	https://crispr.cos.uni-heidelberg.de/

5.3.1 *On-Target Selection Tools*

Among various sequence features, deep learning methods have been widely applied to predict gRNA performance. Programs such as DeepCRISPR and sgRNA Designer utilize convolutional neural networks (CNNs) to detect sequence patterns associated with successful Cas9 cleavage (Javed et al. 2021). These methods predict on-target activity of gRNAs by integrating features related to nucleotide context, secondary structure, and sequence composition. Trained on large-scale experimental datasets of gRNA-Cas9 cleavage outcomes, deep-learning-based models consistely outperformed the traditional rule-based methods in selecting highly active and specific gRNAs (Wilson et al. 2018b) (Table 5.1).

5.3.2 *Off-Target Prediction Tools*

One of the most significant limitation of CRISPR based genome editing is the potential to introduce unintended mutations at genomic sites other than the target site, commonly termed as off targets, which can compromise genomic integrity (Zhu 2022). To address these challenges, CRISPRoff and DeepHF deep learning algorithms have been developed to predict off-targets. These algorithms utilize

Table 5.2 Off-target prediction tools

Tool	Description	Website (accessed Jul 10, 2025)
CHOP CHOP	A pioneering online instrument developed by Harvard University for the creation of gRNA sequences based on both matches and mismatches	https://chopchop.cbu.uib.no/
WU-CRISPR	Machine-learning-algorithm-based tool that can predict off-target effects by providing sequences between 20 and 30,000 bp	https://crisprdb.org/wu-crispr-website/index.html
sgRNACas9	A web-based tool to predict off-target effects	http://www.biootools.com/software.html
Png Designer	A newly developed tool for creating guide RNA for both base editing and prime editing	https://www.crisprindelphi.design/
GuideScan	Predicts off-target effects based on sequence and structural features	https://guidescan.com/
EuPaGDT	Machine-learning-based tool to predict on- and off-target effects simultaneously	http://grna.ctegd.uga.edu/
CRISPRz	Trained on large datasets from zebrafish, humans, and mice to generate a gRNA dataset	https://research.nhgri.nih.gov/CRISPRz/
CRISPRPOR	Design gRNA dataset based on match/match in seed regions	http://crispor.tefor.net/
CRISPR-P	Web-based off-target and on-target prediction tools for a wide range of plant species	http://crispr.hzau.edu.cn/CRISPR2/
Cas-OFFinder	Forecasts a range of off-target and on-target effects using the Cas-OFFinder model	http://www.rgenome.net/cas-offinder/

networks with long-short-term memory (LSTM) and Recurrent Neural Networks (RNNs) to compare the candidate sequence with the intended target site and identify sequence similarities associated with off-target (Naeem and Alkhnbashi 2023). These models accurately predict the likelihood of off-targets by analyzing the factors like sequence and contextual motifs. Overall deep learning based off target prediction tools offer better prediction of off-target events compared with traditional bioinformatic tools (Wilson et al. 2018a) (Table 5.2).

5.3.3 *Large-Scale Design Tools*

Effective gRNA design methods require the ability to process extensive genomic data for large-scale applications (Li et al. 2019). DeepCRISPR and sgRNAcas9 are the deep learning based tools to predict gRNA performance across the whole genome or at specific genomic regions. They employ tree-based or ensemble learning approaches to evaluate gRNA activity (Konstantakos et al. 2022). By rapidly scanning a large number of potential target sites, they facilitate high-throughput gRNA design. Large scale gRNA design tools typically rank gRNAs, depending on their specificity, as well as the likelyhood of off target cleavage at unintended genomic sites (Tycko et al. 2016).

5.3.4 *Specificity*

Genome editing using CRISPR is effective only if it is highly precise, meaning that cleavage occur only at the target site with minimal affect on other genomic sequence (Gori et al. 2015). Deep learning tools like CRISPR–SKIP and DeepSpCas9 use advanced architectures, like transformer models or recurrent neural networks (RNNs), trained on sequence samples provided by the users. These approaches can analyze for possible off-target areas and make a good prediction regarding the gRNA's specificity (Ahmadzadeh et al. 2019). These models enhance the specificity of gRNA design by integrating features like sequences, epigenetic changes, and chromatin accessibility. Consequently, these models reduce the risk of unintended effects and increase the confidence in CRISPR-based genome editing Applications (Ryan et al. 2018).

5.4 The Convergence of Artificial Intelligence and CRISPR

Integrating CRISPR–Cas9 with artificial intelligence (AI) will improve gene editing applications. AI applications include discovering information, epitope forecasting, and modeling-based methods (Maserat 2022). This approach can easily be programed to improve gene editing applications in health and agriculture. In the area of genome engineering, CRISPR/Cas9 technology is making quick progress (Akbari et al. 2019). The convergence of AI and CRISPR technologies will revolutionize the genome engineering particularly in gene therapy and translational applications. AI will help improving CRISPR–Cas systems to improve accuracy and repair outcomes through data driven approaches (Zhang et al. 2022).

AI algorithms help designing an optimal gRNA for CRISPR–Cas systems. These developments enhance the specificity of CRISPR-Cas systems and reduce the likelihood of off-target effects (Gaudelli et al. 2017). More effective and safer gene-editing therapies can be developed with the help of AI tools that analyze genomic sequences and identify the appropriate target sites. This is very significant in therapeutic applications, where promoting accuracy at the intended target sites, with minimal off-target effects, are critical for safety and efficacy (Long et al. 2023). AI has also accelerated identification and characterization of new and improved CRISPR-Cas variants. The new variants expand the scope of CRISPR technology by providing unique advantages such as improved efficiency, increased specificity, and the capibility to target RNA (Du-Harpur et al. 2020).

5.4.1 What Is Artificial Intelligence?

Artificial intelligence referes to the development of computational systems and software that can perform tasks, typically requiring human intelligence. It can accomplish tasks that usually require human intelligence, like speech recognition, visual perception, language translation, problem-solving, and decision-making. These systems use algorithms to analyze data and make plausible conclusions based on learned patterns from training data (Bock et al. 2020). AI aims to replicate human cognitive abilities across a range of applications. Key AI techniques include natural-language processing, which enable human-machine communication, deep learning, which is based on neural networks that can identify complex patterns in data, and machine learning, which refers to data-learned algorithms (Hamet and Tremblay 2017). Innovations and efficient applications of AI offer transformative opportunities across multiple fields. To ensure the positive impact of AI applications, potential ethical and societal challenges need careful consideration (Wang et al. 2020).

5.4.2 AI Can Help in Optimizing Factors Influencing gRNA

The choice of gRNA design greatly influence the efficiency of genome editing. While some gRNAs may exhibit low or no activity, others can potentially disrupt several unintended target sequences in a genome. To address this challenge, numerous gRNA design tools have been developed using many deep learning and machine learning (ML) approaches (Dixit et al. 2024a).

A number of gRNA efficacy prediction models have been developed with increasing methodological sophistication. Advance classifiers that utilize feature-to-feature interactions, such as Support Vector Machines (SVM) and Gradient Boosted Trees (GBT) models, have been shown superior performance compared with earlier models, such as Linear Regression (LR) (Mohammadpour et al. 2019). For example, a CNN based TIGER model, outperforms other AI models in identifying the differences between essential and control genes by predicting gRNA activity based on both sequence and context (Lin et al. 2024). Similarly, CRISPRon is an excellent model, trained on a database containing >23,000 gRNAs to accurately predict on-target activity (Khoshandam et al. 2024). Prediction of both on-target and off-target activity was also improved using deep learning models including DeepHF and attention-based CNNs CRISPR-ONT and CRISPR-OFFT. Advances have also been made for algorithms that predict off-targets, which are typically based in vitro cleavage experiments with immortalized cell lines. These off-target prediction algorithms usually utilize a combination of data from the experiments, and will likely use, but are not limited to, sequence-based information (Carlson-Stevermer et al. 2017). There is a fine correlation between the sequence-based information and target site activity, so these algorithms largely depend on sequence-based information. The physical genomic characteristics, such as methylation patterns and chromatin accessibility, can utilized to enhance the ability to predict genome editing, as demonstrated by piCRISPR (Haghir Ebrahim Abadi et al. 2024).

5.4.3 AI-Driven Predictive Modeling for On-Targeting Specificity

Predictive modeling using AI has emerged as a crucial tool to improve the specificity of CRISPR–Cas systems. Researchers need to predict which gRNAs would produce better and more accurate editing in a CRISPR experiment. These forecasts are achievable because of machine learning and deep learning algorithms (Tyagi et al. 2020). ML and DL models analyze large datasets to find patterns that influence gRNAs performance on their target sites. These models incorporate multiple features, like the DNA sequence of the target and the nucleic acid sequence of the gRNA itself (Kant 2021). Convolutional neural networks (CNNs), gradient increased trees (GBT), and support vector machines (SVMs) are some of the more advanced technologies that have demonstrated strong performance in forecasting the efficiency and specificity of gRNAs (Wilson et al. 2018b). For instance, the CRISPRon model, trained on large datasets, can accurately predict how effectively gRNA an perform at the target site. These model-based predictions not only provide accurate predictions of on-target activity, but they can also help reducing off-target impacts (Wang et al. 2023b).

Also, these prediction models continue to improve in accuracy and productive with developments in AI. Several methods, such as DeepHF and attention-based CNNs, that have been developed to make gene editing more precise and less likely to produce off targets (Lee et al. 2021).

5.5 AI and Minimizing Off-Targets

Data-driven machine learning methods have emerged over the past few years as a powerful tool for modeling CRISPR-Cas studies, including precise prediction of target site, gRNA, off-targets, repair outcomes, and potential risks associated with these mutations. They models perform better than traditional score prediction methods like MIT CRISPR Design Tool2, CCTop algorithm, CRISPR Design, E-CRISP, and CHOPCHOP. A major limitation of these earlier approaches is inability to improve prediction accuracy as the sample size increases (Moreno-Mateos et al. 2015). Conversely, one of the main advantages of modern data-driven models particularly those trained on deep learning is their ability to improve predictive performance with an increasing number of samples. Current state-of-the-art research in this field focuses on developing robust CRISPR/Cas9 applications suitable for clinical and translational use by: (i) enhancing on-target efficiency, (ii) improving off-target specificity, and (iii) jointly maximizing on-target activity with minimum off-target effects simultaneously (Dimauro et al. 2022). There are a number of machine learning and deep learning-based tools available to accurately predict cleavage sites in both animals and plants. These tools integrate various dimensions, encompassing target site's features, the thermodynamic properties of the sgRNA, and the sequence similarity between the sgRNA and its target. Thus, machine learning methodologies are very sophisticated and efficient approaches to precisely identify cleavage positions in genes during CRISPR-Cas studies (Luo et al. 2022).

Although CRISPR is predicted as precise but one of the challenges associated with CRISPR is the occurrence of off-target effects. These off-target effects limit the safe use of CRISPR applications particularly in health, medical, therapeutics, and translational applications (Hiranniramol et al. 2020). In order to predict and minimize off-target effects, AI-based approaches are being increasingly used. Machine learning algorithms can analyze complex genomic data to identify features associated with off-target activity, leading to improved specificity of CRISPR and facilitating development and more reliable predicitve tools for the future (Goell and Hilton 2021).

5.5.1 Addressing Challenges of Off-Target Effects with AI

By using deep learning and machine learning techniques, large and complex genomic data can be effectively analyzed and interpreted. Advances in these approaches have improved the ability of AI models' to predict editing outcomes, including off-target editing, by enabling more informative features for the learning of AI models (Gaudelli et al. 2017). The sequence features of the gRNA are critical determinants of the genome editing efficiency. While some gRNA sequences can disrupt almost every target allele within a cell population, others exhibit little or no activity. To address this variability, several gRNA design tools have emerged, primarily based on ML and deep learning algorithms to optimize the gRNA efficiency and efficacy (Yang et al. 2024).

5.5.2 *Example of AI Algorithms Reducing Off-Targeting*

Several recent studies have demonstrated the capability of deep learning models to improve the prioritization of sgRNA structure and activity predictions for CRISPR-based genome editing (Anzalone et al. 2019). For example, in DeepGuide, use a neural network architecture trained on genome-wide CRISPR activity profiles to improve CRISPR-based mutagenesis efficacy in *Yarrowia lipolytica* (Frangoul et al. 2020). DeepHF surpasses earlier models, showing that combining biological feature engineering with RNNs can improve gRNA design for high-fidelity SpCas9 variants. To enhance the predictive accuracy of sgRNA design for bacterial CRISPR interference, convolutional neural network were used to evaluate the effects of guide-target mismatches on dCas9-sgRNA binding activity (Long et al. 2023). Related studies also provided additional proof of concept, by rigorously assessing PAM-sequence compatibilities as well as on-target/off-target activities for SpCas9 and its variants in human cells with and without new non-NGG PAM sequences. These analysis also include xCas9 and SpCas9-NG with expanded PAM requirements, and integrate functional predictions using deep-learning modelling. Collectively these studies demonstrate the potential of deep learning to enhance CRISPR system applications and improve sgRNA design (Chen et al. 2017a).

5.6 AI and Influence of Chromatin on CRISPR Efficiency

The efficacy of CRISPR editing is determined, in part, by the chromatin landscape. There are several factors that can affect the efficiency of gRNA, such as how open the chromatin is in the target area, how much gene expression is there, and how many nucleosomes are there (Yang et al. 2023). These features allow researchers to comprehend the impacts of CRISPR efficiency to target a particular sequence in the genome. This is really important for epigenome editing since it has to be very precise. While developing sgRNAs for CRISPR based epigenetic alterations, scientists can add chromatin knowledge to the prediction algorithms. This makes CRISPR based epigenetic editing more accurate and helpful (Ai et al. 2022).

CRISPR–Cas9 based editing of DNA is influenced by different factors. For example, Cas9 activity varies with nucleotide composition, chromatin accessibility, and thermodynamic stability of the target site in the genome (Ryan et al. 2018).

It’s intriguing to think about the way chromatin affects gRNA design instruments when it is under consideration that how AI can improve CRISPR. Researchers will be able to apply hypothesis-driven, machine learning, as well as deep learning to improve gRNA selection for making precise and accurate alterations to genomes in a range of biological situations (Zheng et al. 2021).

5.6.1 *Impact of Chromatin on CRISPR-Cas Activity*

Different studies have shown that chromatin can affect how successfully CRISPR-based genome editing works for a particular target site. Scientists are trying to engineer CRISPR–Cas enzymes that only specifically act on the target site without producing any off targets. However, access to the chromatin and more complex chromatin can continue to influence how well CRISPR–Cas9 works. In future scientists will be able to engineer CRISPR-Cas systems that are more precise and applicable in fundamental research, industry, and medicinal therapies, by studying how chromatin and CRISPR function together in a highly complex manner (Chen et al. 2017b).

5.6.2 *AI Contributes in Understanding Chromatin Effects*

Machine learning is critical for understanding and reducing the effects of chromatin on CRISPR because it lets researchers utilize in-silico prediction methods to modulate transcription. These methods use machine learning to make predictions about how mutations would impact transcription. They are far better than ordinary motif-based investigations (Chen et al. 2017a).

5.6.2.1 Data Utilization

Machine learning algorithms employ public transcriptome datasets and human epigenetic data, like how accessible chromatin is and how histones alter the access to the target site. This indicates that researchers can use massive omics datasets like the ENCODE and FANTOM projects without having to have genotype data (Lee 2023).

5.6.2.2 Model Training

For gene editing studies in health, researchers use DNA sequences from the human genome, such as GRCh37 or GRCh38, to create and teach machine learning models. After that, these models can predict patterns in transcriptional events. Convolutional neural networks (CNNs) and deep learning models are for this purpose (Nayak and Dutta 2023).

5.6.2.3 In Silico Mutagenesis

In silico mutagenesis focuses on how modifications to the input DNA sequences alter the predictions and probability after the machine learning models have been trained. This approach make it possible to learn what alterations affect transcription without having to perform extensive experimental testing (Ibrahim and Al-Turjman 2022).

5.6.2.4 Advancements

Models using deep learning, like convolutional neural networks (CNNs), have made it possible to analyze longer DNA sequences (more than 1 kb) due to their improved predictive efficiency. This increased accuracy makes it easier to comprehend the complex rules that control chromatin effects (Long et al. 2023).

5.6.2.5 Integration with Rich Datasets

The field of AI and its convergence with CRISPR is emerging rapidly due to the development of more powerful techniques for machine learning and availibility of high quality datasets. By training machine learning algorithms on multiple datasets, researchers can learn deeper into how mutations impact transcriptional regulation in various kinds of cells and tissues (Wang et al. 2023a).

5.7 Changing Dynamics of CRISPR Design with AI

Genome editing techniques like CRISPR–Cas9 has revolutionized biotechnology with tremendous potential in crop genetic improvement. CRISPR enables the accurate and precise editing of the genetic code of an organism including bacteria, plants, and animals. CRISPR-based genome editing techniques provide exciting opportunities for gene discovery, gene therapy and advance health care (Tyagi et al. 2020).

CRISPR–Cas9 is the most common genome editing tool because of its efficiency, specificity, and versatility. Recent advances in CRISPR–Cas9 therapies have led to initiation of several clinical studies resulting in dynamic and rapidly evolving field of gene and cell therapy. CRISPR several potential applications for the treatment of human diseases (Li et al. 2023).

CRISPR-based genome editing technologies including Base-editing (BED), epigenome-editing (epi-GED), and prime-editing (PED) offer several benefits over conventional gene editing applications such as knockout applications. Each of these strategies has their own advantages and disadvantages and can be used for precise base editing and introducing precise modifications to the epigenetic signatures in the genome (Lee 2023).

With the convergence of CRISPR technology with artificial intelligence (AI), marks a major milestone in the field of gene editing. The convergence of AI with CRISPR not only enhance the capabilities of gene editing, but it also opens up new opportunities for exploring and altering the genetic code (Harms et al. 2014). In many cases, AI is essential for the analysis of medical data, particularly in multiomics technologies that potentially produce large amounts of data across various modalities, including genomics, transcriptomics, proteomics, metabolomics, and mutation studies. Several CRISPR–Cas tasks such as gRNA design activities, including gRNA selection, evaluation of off-target effects, and interpretation of experimental results, will be facilitated by AI models, particularly deep learning models (Hamet and Tremblay 2017).

5.7.1 Transformative Potential of AI inCRISPR/Cas Designing

The revolution of artificial intelligence (AI) has transformed the design and implementation of CRISPR–Cas systems, ushering in a new era of genome editing technologies that are more efficient, precise, and accessible. The AI driven innovations in CRISPR–Cas systems has allowed scientists to overcome the limitations of CRISPR and explore the previously uncharted possibilities in genetic engineering (Dixit et al. 2024b).

5.7.1.1 Automated Systems Powered by Deep Learning

Deep learning algorithms has enabled the development of automated systems that has significantly enhanced the precision and effectiveness of CRISPR design. AI-enabled tools such as CRISTA, DeepCRISPR, and DeepHF recommend either pre-designed gRNA or user defined gRNA that are tailored for specific target sequences, considering parameters such as Cas protein type, genomic context, on-target and off-target scores. The predictability and reliability of CRISPR–Cas9 based genome editing have been greatly enhanced by these AI-powered tools (Bhat et al. 2022).

5.7.1.2 Predictive Models and Enhanced Decision-Making

Automated CRISPR design systems depend on deep learning algorithms to process and analyze large amount of empirical and theoretical data. By integrating training and validation datasets, these learning models can predict on-target efficiency and off target effects with increasing efficiency. Recently, three major types of machine learning tools have been applied to predict CRISPR target prediciton: (a) regression-based methods that estimate continuous editing outcome (Moreno-Mateos et al. 2015), (b) classification-based techniques that discriminate between effective and ineffective gRNA (Wilson et al. 2018b), and (c) ensemble-based approaches that combine multiple algorithms to improve predictive performance (Hiranniramol et al. 2020). These models can use diverse sequence and context-dependent features to identify the target site. In particular, deep learning models such as artificial neural networks, or ANN, that are developed for the CRISPR–Cas9 system have demonstrated high accuracy in predicting gRNA performance, thereby enhancing the precision and advancing the overall reliability of machine learning based genome editing (Mohammadpour et al. 2019).

5.7.1.3 Integration of AI in Advanced Genome Editing Techniques

The design of gRNA and applying CRISPR in diverse applications is increasingly driven by AI-based deep learning and machine learning (MDL) models. These AI models facilitate the selection of best gRNA and delivery strategies by integrating the sequence-specific features with environment and contextual factors that influences the gRNA performance. In doing so, AI models accelerate the optimization of CRISPR-Cas system by enabling high throughput analysis of experimental datasets and improving the accuracy of off-target effect prediction, thereby enhancing the precision and safety of CRISPR applications (Udegbe et al. 2024).

5.7.2 Studies Showcasing AI-Driven CRISPR–CAS

Numerous case studies have shown that integrating artificial intelligence (AI) with CRISPR–Cas systems substantially enhances genome editing. These studies have proven that AI algorithms significantly improve editing efficiency, target specificity, and overall accuracy of CRISPR technology by optimising gRNA design and minimizing off targets (Maddalo et al. 2014).

5.7.2.1 Explainable Artificial Intelligence (XAI) for Enhanced Predictability

The first case study shows how explainable artificial intelligence (XAI) can enhance interpretability of CRISPR–Cas genome editing. By integrating XAI with CRISPR technology, genome editing has become more precise and transparent. The study shows that precise gene editing cab be achieved by using an RNA aptamer-based strategy to link CRISPR ribonucleoproteins with biotinylated oligonucleotides, enabling controlled and efficient gene modifications. These findings suggest that XAI informed approaches can improve CRISPR-based tools by optimizing gRNA design and enhancing editing performance, thereby addressing the inherent complexity of genetic traits and regulatory networks better and making gRNA. This is why genetic qualities are so hard to understand (Carlson-Stevermer et al. 2017).

5.7.2.2 AI CRISPR–Cas Systems in Cancer Immunotherapy and Vaccine Development

A study published in *Cancer Informatics* discussed potential synergy between CRISPR/Cas9 system and artificial intelligence (AI) in cancer therapy and vaccine development. The researchers examined how integration of AI and the CRISPR/Cas9 system could impact molecular modeling, gene mutations, and knowledge discovery approaches in biomedical research. The authors highlighted that AI-powered CRISPR–Cas systems could accelerate developing vaccines and improve precision in cancer treatment. Researchers AI techniques, including epitope prediction, agent-based modeling, and data driven knowledge discovery, were applied to enhance the accuracy and efficiency of gene editing, tumor genome modification, and molecular cloning processes (Maserat 2022).

5.7.2.3 AI-Enhanced CRISPR–Cas Systems in Drug Discovery

Researchers reporting in the Journal of Translational Medicine demonstrated that convergence of CRISPR/Cas9 technology and AI may accelerate drug development. In this study, AI algorithms were applied to CRISPR-based gene editing to identify drug targets, interrogate non-coding genomic regions, and predict mechanism underlying the emergence of drug-resistant. This case study revealed that the AI-powered CRISPR–Cas system could transform pharmaceutical discovery by enabling more precise, efficient and systematic target and validation (Bhat et al. 2022).

5.7.2.4 Precision Medicine Applications of AI and CRISPR Technology

Researchers have explored how the convergence of AI and CRISPR could advance precision medicine. Several studies have demonstrated that AI-mediated CRISPR systems have the potential to transform personalized medicine by enabling disease modeling, patient specific genetic modifications, and personalized therapeutic treatments (Platt et al. 2014). Researchers have also evaluated the merits and limitations of this AI-driven CRISPR/Cas9 based approaches highlighting their critical role in improving the accuracy, scalability and clinical relevance of genome editing for precision medicine applications. Combining CRISPR gene editing with AI could make it possible to make personalized and effective treatment strategies for a wide range of ailments (Harms et al. 2014). Collectively these studies demonstrate that convergence of AI and CRISPR technology represent a major step forward in precision medicine, facilitating the development of drugs tailored for each patient and enhancing treatment adherence through more effective and personalized therapeutic inventions (Khoshandam et al. 2024).

5.8 AI and Future of Design Tools

Artificial intelligence (AI) applies computational tools to predict an individual's risk of developing diseases like breast cancer by scanning the mammogram images and extracting information from the patient record (Wilson et al. 2018b).

In this context "Intelligent agents" refer agents are computational technologies that are capable of perceiving their environment, interpreting complex data and taking adaptive actions to achieve specific objectives. Unlike human intelligence these agents operate through machine learning mechanisms that mimics key aspects of human cognitive processes enabling them to solve complex problems (Huang et al. 2015).

Artificial intelligence is used in a wide range of disciplines with particular impact in medicine and agricultural biotechnology. For example, the medical industry AI driven tools are increasing used to detect cancer cells and disease diagnosis. AI also support the study of chronic diseases by integrating the laboratory data and other clinical data to guarantee early diagnosis. By combining large scale historical data sets with domain specific medical knowledge, AI has become a powerful tool to facilitate drug discovery and development (Sajja and Sajja 2021).

5.8.1 AI-Based Risk Prediction in CRISPR–Cas Applications

Genome editing has been revolutionized by the development of CRISPR–Cas9 technology, which provides an efficient and versatile means of modifying gene activity in mammalian cells. Unlike previous techniques like RNA interference (RNAi) which results in transient gene silencing, CRISPR–Cas9 enables permanent genome alterations, resulting in either loss-or-gain-of-function phentotypes (Gilbert et al. 2014). This RNA-guided system is comparatively inexpensive, quick, and efficient, allowing modifications in multiple genes simultaneously. In contrast, zinc-finger nucleases (ZFNs) and transcription activator-like effector nucleases (TALENs) were more expensive, time-consuming, and technically challenging to employ at a scale for high throughput screening (Rahman 2023).

CRISPR–Cas9 originates from an adaptive immune system in bacteria and archaea cleaves invasive nucleic acids. The most widely used type II CRISPR system is derived from *Streptococcus pyogenes*, and employs a single Cas9 endonuclease, guided by a 20-nucleotide sequence encoded within a single gRNA, formed by the fusion of both trans-acting RNA and CRISPR RNA. Web-based tools are used to identify potential CRISPR target sites within genes of interest (Zhao et al. 2014).

Upon target recognition, CRISPR-Cas9 induce double-strand DNA breaks (DSBs) which are subsequently repaired by the Non-Homologous End Joining (NHEJ) or Homology Directed Repair (HDR) pathways. NHEJ repair typically generates insertions/deletions (indels) that lead to gene knockout, whereas HDR enables the precise introduction of defined nucleotide changes using an exogenous repair template (Grosso et al. 2013).

When paired gRNAs are used in conjunction with a linear homologous repair donor, genomic deletions of up to 10 kb can be effectively achieved. Advances in large-scale oligonucleotide synthesis have enabled synthesis of genome wide libraries, facilitating systemic exploration of gene functions (Lee 2023). In combination with Cas9, these gRNA libraries generate diverse cell populations carrying knockout mutations, offering advantages over shRNA libraries. Cas9 and gRNAs can be delivered either transiently or stably into mammalian cells, using methods such as lentiviral vectors, nucleofection, electroporation, and Lipofectamine-mediated transfection. Depending on the cell types and experimental methods, one vector- or two-vector delivery systems are used for CRISPR cargos (Khoshandam et al. 2024). In the two-vector method, cells are first transfected with Cas9 and then with a sgRNA library. The one-vector approach uses a single vector to transfect both Cas9 and sgRNA into cells. Using these strategies, comprehensive sgRNA libraries targeting human or mouse genes with continual improvements aimed at reducing off-target effects, an increase in target gene coverage and enabling the disruption of non-coding RNAs including miRNAs (Freer 2015). Genome-wide lentiviral sgRNA libraries targeting the protein coding regions of the mouse genome have further expanded the scope of functional genomics. Collectively, these advancements in CRISPR–Cas9 technology, has enabled large scale knockout studies and comprehensive functional genomics screening (Zhu 2022).

5.8.2 *Prediction of Long Non-coding RNAs Based on Deep Learning*

Recently, novel prediction techniques for distinguish messenger RNAs (mRNAs) from long non-coding RNAs (lncRNAs), driven in part by sequencing errors and limitations in biosequencing technology. Machine learning, particularly support vector machines (SVMs), has been used to train identification models for mRNAs and lncRNAs, using features drived from the Coding Potential Calculator (CPC) and related algorithms (Wilson et al. 2018b).

Among commonly used approaches, Coding-Non-Coding Index (CNCI) and PLEK have demostrated strong predictive potential. PLEK employs SVM classification and improved k-mer frequency to achieve above 90% accuracy across diverse datasets, while CNCI uses a score matrix associated with species to discriminate coding from non-coding transcripts (Moreno-Mateos et al. 2015). Additional techniques, including the PhyloCSF algorithm and the codon substitution frequency

(CSF) algorithm, categorize sequences based on the evolutionary pattern of codon replacement. Despite their widespread use, traditional shallow learning approaches may not be adequate to capture complex lncRNA properties (Mohammadpour et al. 2019). However, deep learning methods have shown substantial promise for lncRNA discovery across multiple biological domains. Deep neural networks incorporating hybrid feature have significantly improved the prediction accuracy; with some models achieving 97.1% accuracy rates. The open reading frame (ORF) is a common characteristic that is still used to differentiate between mRNAs and lncRNAs (Li et al. 2020). The rapid and precise identification of lncRNAs from large RNA sequencing datasets, particularly without annotation information remains a major challenge and continues to drive the development of novel computational methods and machine learning technologies (Tyagi et al. 2020).

References

Ahmadzadeh V, Farajnia S, Baghban R, Rahbarnia L, Zarredar H (2019) CRISPR-Cas system: toward a more efficient technology for genome editing and beyond. J Cell Biochem 120:16379–16392

Ai Y, Liang D, Wilusz JE (2022) CRISPR/Cas13 effectors have differing extents of off-target effects that limit their utility in eukaryotic cells. Nucleic Acids Res 50:e65–e65

Akbari P, Gavidel P, Gardaneh M (2019) The revolutionizing impact of artificial intelligence on breast cancer management. Arch Breast Cancer:1–3

Anzalone AV, Randolph PB, Davis JR, Sousa AA, Koblan LW, Levy JM, Chen PJ, Wilson C, Newby GA, Raguram A (2019) Search-and-replace genome editing without double-strand breaks or donor DNA. Nature 576:149–157

Barrangou R, Marraffini LA (2014) CRISPR-Cas systems: prokaryotes upgrade to adaptive immunity. Mol Cell 54:234–244

Bhat AA, Nisar S, Mukherjee S, Saha N, Yarravarapu N, Lone SN, Masoodi T, Chauhan R, Maacha S, Bagga P (2022) Integration of CRISPR/Cas9 with artificial intelligence for improved cancer therapeutics. J Transl Med 20:534

Bock DE, Wolter JS, Ferrell OC (2020) Artificial intelligence: disrupting what we know about services. J Serv Mark 34:317–334

Carlson-Stevermer J, Abdeen AA, Kohlenberg L, Goedland M, Molugu K, Lou M, Saha K (2017) Assembly of CRISPR ribonucleoproteins with biotinylated oligonucleotides via an RNA aptamer for precise gene editing. Nat Commun 8:1711

Chakrabarti AM, Henser-Brownhill T, Monserrat J, Poetsch AR, Luscombe NM, Scaffidi P (2019) Target-specific precision of CRISPR-mediated genome editing. Mol Cell 73:699–713

Chen X, Liu J, Janssen JM, Gonçalves MAFV (2017a) The chromatin structure differentially impacts high-specificity CRISPR-Cas9 nuclease strategies. Mol Ther Nucleic Acids 8:558–563

Chen Y, Zeng S, Hu R, Wang X, Huang W, Liu J, Wang L, Liu G, Cao Y, Zhang Y (2017b) Using local chromatin structure to improve CRISPR/Cas9 efficiency in zebrafish. PLoS One 12:e0182528

Dimauro G, Barletta VS, Catacchio CR, Colizzi L, Maglietta R, Ventura M (2022) A systematic mapping study on machine learning techniques for the prediction of CRISPR/Cas9 sgRNA target cleavage. Comput Struct Biotechnol J 20:5813–5823

Dixit S, Kumar A, Srinivasan K, Vincent PMDR, Ramu Krishnan N (2024a) Advancing genome editing with artificial intelligence: opportunities, challenges, and future directions. Front Bioeng Biotechnol 11:1335901

Dixit S, Kumar A, Srinivasan K, Vincent PMDR, Krishnan NR (2024b) Advancing genome editing with artificial intelligence: opportunities, challenges, and future directions. Front Bioeng Biotechnol 11:1335901

Du-Harpur X, Watt FM, Luscombe NM, Lynch MD (2020) What is AI? Applications of artificial intelligence to dermatology. Br J Dermatol 183:423–430

Frangoul H, Bobruff Y, Cappellini MD, Corbacioglu S, Fernandez CM, De la Fuente J, Grupp SA, Handgretinger R, Ho TW, Imren S (2020) Safety and efficacy of CTX001 in patients with transfusion-dependent β-thalassemia and sickle cell disease: early results from the climb THAL-111 and climb SCD-121 studies of autologous CRISPR-CAS9-modified CD34+ hematopoietic stem and progenitor cells. Blood 136:3–4

Freer PE (2015) Mammographic breast density: impact on breast cancer risk and implications for screening. Radiographics 35:302–315

Gaudelli NM, Komor AC, Rees HA, Packer MS, Badran AH, Bryson DI, Liu DR (2017) Programmable base editing of A• T to G• C in genomic DNA without DNA cleavage. Nature 551:464–471

Gilbert LA, Horlbeck MA, Adamson B, Villalta JE, Chen Y, Whitehead EH, Guimaraes C, Panning B, Ploegh HL, Bassik MC (2014) Genome-scale CRISPR-mediated control of gene repression and activation. Cell 159:647–661

Goell JH, Hilton IB (2021) CRISPR/Cas-based epigenome editing: advances, applications, and clinical utility. Trends Biotechnol 39:678–691

Gori JL, Hsu PD, Maeder ML, Shen S, Welstead GG, Bumcrot D (2015) Delivery and specificity of CRISPR/Cas9 genome editing technologies for human gene therapy. Hum Gene Ther 26:443–451

Grosso C, Valentão P, Ferreres F, Andrade PB (2013) The use of flavonoids in central nervous system disorders. Curr Med Chem 20:4694–4719

Haghir Ebrahim Abadi MH, Ghasemlou A, Bayani F, Sefidbakht Y, Vosough M, Mozaffari-Jovin S, Uversky VN (2024) AI-driven covalent drug design strategies targeting main protease (mpro) against SARS-CoV-2: structural insights and molecular mechanisms. J Biomol Struct Dyn:1–29

Hamet P, Tremblay J (2017) Artificial intelligence in medicine. Metabolism 69(S36):–S40. Urban G, et al. Deep learning localizes and identifies polyps in real time with 96% accuracy in screening colonoscopy

Harms DW, Quadros RM, Seruggia D, Ohtsuka M, Takahashi G, Montoliu L, Gurumurthy CB (2014) Mouse genome editing using the CRISPR/Cas system. Curr Protoc Hum Genet 83:15–17

Hille F, Charpentier E (2016) CRISPR-Cas: biology, mechanisms and relevance. Philos Trans R Soc Lond B Biol Sci 371:20150496

Hiranniramol K, Chen Y, Liu W, Wang X (2020) Generalizable sgRNA design for improved CRISPR/Cas9 editing efficiency. Bioinformatics 36:2684–2689

Huang G, Huang G-B, Song S, You K (2015) Trends in extreme learning machines: a review. Neural Netw 61:32–48

Ibrahim AU, Al-Turjman F, Sa'id Z, Ozsoz M (2022) Futuristic CRISPR-based biosensing in the cloud and internet of things era: an overview. Multimed Tools Appl 81:35143–35171

Ishino Y, Krupovic M, Forterre P (2018) History of CRISPR-Cas from encounter with a mysterious repeated sequence to genome editing technology. J Bacteriol 200:10–1128

Javed MR, Farooq R, Hussain K, Rashid K, Bashir A, Saif H (2021) Tricks and trends in CRISPR/Cas9-based genome editing and use of bioinformatics tools for improving on-target efficiency. In: CRISPR and RNAi systems. Elsevier, pp 441–462

Jiang F, Doudna JA (2015) The structural biology of CRISPR-Cas systems. Curr Opin Struct Biol 30:100–111

Kant T (2021) Identity, advertising, and algorithmic targeting: Or how (not) to target your "ideal user"

Khoshandam M, Soltaninejad H, Hosseinkhani S, Saltanatpour Z, Goudarzi MTH, Hamidieh AA (2024) CRISPR/Cas and artificial intelligence to improve precision medicine: future perspectives and potential limitations. Hum Gene 42:201356

Konstantakos V, Nentidis A, Krithara A, Paliouras G (2022) CRISPR–Cas9 gRNA efficiency prediction: an overview of predictive tools and the role of deep learning. Nucleic Acids Res 50:3616–3637

LaFountaine JS, Fathe K, Smyth HDC (2015) Delivery and therapeutic applications of gene editing technologies ZFNs, TALENs, and CRISPR/Cas9. Int J Pharm 494:180–194

Lee C-K, Liao C-W, Meng S-W, Wu W-K, Chiang J-Y, Wu M-S (2021) Lipids and lipoproteins in health and disease: focus on targeting atherosclerosis. Biomedicine 9:985

Lee M (2023) Deep learning in CRISPR-Cas systems: a review of recent studies. Front Bioeng Biotechnol 11:1226182

Leenay RT, Beisel CL (2017) Deciphering, communicating, and engineering the CRISPR PAM. J Mol Biol 429:177–191

Li C, Chu W, Gill RA, Sang S, Shi Y, Hu X, Yang Y, Zaman QU, Zhang B (2023) Computational tools and resources for CRISPR/Cas genome editing. Genomics Proteomics Bioinformatics 21:108–126

Li H, Yang Y, Hong W, Huang M, Wu M, Zhao X (2020) Applications of genome editing technology in the targeted therapy of human diseases: mechanisms, advances and prospects. Signal Transduct Target Ther 5:1

Li X-F, Zhou Y-W, Cai P-F, Fu W-C, Wang J-H, Chen J-Y, Yang Q-N (2019) CRISPR/Cas9 facilitates genomic editing for large-scale functional studies in pluripotent stem cell cultures. Hum Genet 138:1217–1225

Lin L, Cui Z, Wang S, Chen Y, Zong Y (2024) AI-driven protein engineering for DNA sequence modification. J Theory Pract Eng Sci 4:183–190

Long E, Wan P, Chen Q, Lu Z, Choi J (2023) From function to translation: decoding genetic susceptibility to human diseases via artificial intelligence. Cell Genom 3

Luo M, Wang J, Dong Z, Wang C, Lu G (2022) CRISPR-Cas9 sgRNA design and outcome assessment: bioinformatics tools and aquaculture applications. Aquac Fish 7:121–130

Maddalo D, Manchado E, Concepcion CP, Bonetti C, Vidigal JA, Han Y-C, Ogrodowski P, Crippa A, Rekhtman N, de Stanchina E (2014) In vivo engineering of oncogenic chromosomal rearrangements with the CRISPR/Cas9 system. Nature 516:423–427

Makarova KS, Wolf YI, Koonin EV (2018) Classification and nomenclature of CRISPR-Cas systems: where from here? CRISPR J 1:325–336

Maserat E (2022) Integration of artificial intelligence and CRISPR/Cas9 system for vaccine design. Cancer Inform 21

Mohammadpour A, Karan E, Asadi S (2019) Artificial intelligence techniques to support design and construction. In: ISARC. Proceedings of the international symposium on automation and robotics in construction. IAARC Publications, pp 1282–1289

Mojica FJM, Montoliu L (2016) On the origin of CRISPR-Cas technology: from prokaryotes to mammals. Trends Microbiol 24:811–820

Moreno-Mateos MA, Vejnar CE, Beaudoin J-D, Fernandez JP, Mis EK, Khokha MK, Giraldez AJ (2015) CRISPRscan: designing highly efficient sgRNAs for CRISPR-Cas9 targeting in vivo. Nat Methods 12:982–988

Naeem M, Alkhnbashi OS (2023) Current bioinformatics tools to optimize CRISPR/Cas9 experiments to reduce off-target effects. Int J Mol Sci 24:6261

Nayak A, Dutta D (2023) A comprehensive review on CRISPR and artificial intelligence based emerging food packaging technology to ensure "safe food". Sustain Food Technol 1:641–657

Platt RJ, Chen S, Zhou Y, Yim MJ, Swiech L, Kempton HR, Dahlman JE, Parnas O, Eisenhaure TM, Jovanovic M (2014) CRISPR-Cas9 knockin mice for genome editing and cancer modeling. Cell 159:440–455

Rahman MM (2023) Stroke prediction using machine learning techniques

Ryan DE, Taussig D, Steinfeld I, Phadnis SM, Lunstad BD, Singh M, Vuong X, Okochi KD, McCaffrey R, Olesiak M (2018) Improving CRISPR–Cas specificity with chemical modifications in single-guide RNAs. Nucleic Acids Res 46:792–803

Sajja PS, Sajja PS (2021) Introduction to artificial intelligence. In: Illustrated computational intelligence: examples and applications, pp 1–25

Strich JR, Chertow DS (2019) CRISPR-Cas biology and its application to infectious diseases. J Clin Microbiol 57:10–1128

Tyagi S, Kumar R, Das A, Won SY, Shukla P (2020) CRISPR-Cas9 system: a genome-editing tool with endless possibilities. J Biotechnol 319:36–53

Tycko J, Myer VE, Hsu PD (2016) Methods for optimizing CRISPR-Cas9 genome editing specificity. Mol Cell 63:355–370

Udegbe FC, Ebulue OR, Ebulue CC, Ekesiobi CS (2024) Synthetic biology and its potential in US medical therapeutics: a comprehensive review: exploring the cutting-edge intersections of biology and engineering in drug development and treatments. Eng Sci Technol J 5:1395–1414

Uniyal AP, Mansotra K, Yadav SK, Kumar V (2019) An overview of designing and selection of sgRNAs for precise genome editing by the CRISPR-Cas9 system in plants. 3 Biotech 9:223

Waddington SN, Privolizzi R, Karda R, O'Neill HC (2016) A broad overview and review of CRISPR-Cas technology and stem cells. Curr Stem Cell Rep 2:9–20

Wang H, Fu T, Du Y, Gao W, Huang K, Liu Z, Chandak P, Liu S, Van Katwyk P, Deac A (2023a) Scientific discovery in the age of artificial intelligence. Nature 620:47–60

Wang J, Xiang X, Bolund L, Zhang X, Cheng L, Luo Y (2020) GNL-Scorer: a generalized model for predicting CRISPR on-target activity by machine learning and featurization. J Mol Cell Biol 12:909–911

Wang Z, Xiao C, Sun J (2023b) Spot: sequential predictive modeling of clinical trial outcome with meta-learning. In: Proceedings of the 14th ACM International Conference on Bioinformatics, Computational Biology, and Health Informatics, pp 1–11

Wilson LOW, O'Brien AR, Bauer DC (2018a) The current state and future of CRISPR-Cas9 gRNA design tools. Front Pharmacol 9:749

Wilson LOW, Reti D, O'Brien AR, Dunne RA, Bauer DC (2018b) High activity target-site identification using phenotypic independent CRISPR-Cas9 core functionality. CRISPR J 1:182–190

Xu Y, Li Z (2020) CRISPR-Cas systems: overview, innovations and applications in human disease research and gene therapy. Comput Struct Biotechnol J 18:2401–2415

Yang H, Patel DJ (2017) New CRISPR-Cas systems discovered. Cell Res 27:313–314

Yang Q, Wu L, Meng J, Ma L, Zuo E, Sun Y (2023) EpiCas-DL: predicting sgRNA activity for CRISPR-mediated epigenome editing by deep learning. Comput Struct Biotechnol J 21:202–211

Yang Z, Zhang Z, Li J, Chen W, Liu C (2024) CRISPRlnc: a machine learning method for lncRNA-specific single-guide RNA design of CRISPR/Cas9 system. Brief Bioinform 25:bbae066

Zhang L, Lei Z, Xiao C, Du Z, Jiang C, Yuan X, Hu Q, Zhai S, Xu L, Liu C (2022) Ai-boosted crispr-cas13a and total internal reflection fluorescence microscopy system for sars-cov-2 detection. Front Sens 3:1015223

Zhao Y, Dai Z, Liang Y, Yin M, Ma K, He M, Ouyang H, Teng C-B (2014) Sequence-specific inhibition of microRNA via CRISPR/CRISPRi system. Sci Rep 4:3943

Zheng X, Cui J, Wang Y, Zhang J, Wang C (2021) CRSIPR-AI: a webtool for the efficacy prediction of CRISPR activation and interference. bioRxiv. 2012–2021

Zhu H, Li C, Gao C (2020) Applications of CRISPR–Cas in agriculture and plant biotechnology. Nat Rev Mol Cell Biol 21:661–677

Zhu LJ (2015) Overview of guide RNA design tools for CRISPR-Cas9 genome editing technology. Front Biol (Beijing) 10:289–296

Zhu Y (2022) Advances in CRISPR/Cas9. Biomed Res Int 2022:9978571

Part II
Cereals Crop Breeding

Chapter 6
CRISPR/Cas Gene Editing and Improvement of Abiotic Stress Tolerance in Cereals

Theivanayagam Maharajan, T. P. Ajeesh Krishna, and Stanislaus Antony Ceasar

Abstract Cereals constitute a major source of energy, protein, and other necessary nutrients in both developed and developing countries. It makes cereals one of the most important food crops for human nutrition and global food security. However, abiotic stress has a profound impact on the growth and output of all major cereals. Molecular breeding and functional genetics have made significant contributions to understanding the complex mechanisms behind cereal responses to abiotic stresses. The development of genome editing tools in recent years has completely transformed precision plant breeding. The CRISPR/Cas system is a powerful and precise tool for editing specific genes related to abiotic stress tolerance. The CRISPR/Cas tools offer a chance to create non-transgenic cereals that are more resilient to harsh climatic factors. This is a crucial tactic to guarantee food security in the face of persistent climate change. New versions of next-generation CRISPR/Cas systems have increased the flexibility of genome engineering technologies. Examining the role of abiotic stress-responsive genes through the application of the CRISPR/Cas genome editing system holds great promise for enhancing the growth and yield of cereals under adverse environmental conditions. By enabling precise functional validation and targeted modification of key genes, CRISPR/Cas facilitates the development of new cultivars with enhanced tolerance to abiotic stresses. This approach not only accelerates the breeding process but also supports the creation of improved, climate-resilient cereal varieties. This chapter discusses the impact of abiotic stress on cereals, the mechanisms and variations of genome editing tools, and the genes targeted by genome editing tools to improve abiotic stress in cereals. Application of CRISPR/Cas technology to enhance stress tolerance in cereals can contribute significantly to global food security, particularly in the face of climate change and an increasing population.

Keywords Abiotic stress · Cereals · Genome editing

T. Maharajan · T. P. Ajeesh Krishna · S. A. Ceasar (✉)
Division of Plant Molecular Biology and Biotechnology, Department of Biosciences, Rajagiri College of Social Sciences, Cochin, Kerala, India

K. A. Abd-Elsalam, A. Ahmad (eds.), *Gene Editing in Cereals*, Sustainability Sciences in Asia and Africa, https://doi.org/10.1007/978-981-95-6722-5_6

6.1 Introduction

Cereal crops provide more than half of the world's food energy consumption and are the cornerstone of global agriculture and human nutrition (Shiferaw et al. 2013). Major cereals are widely grown in a variety of agro-ecological zones. These include rice (*Oryza sativa*), wheat (*Triticum aestivum*), maize (*Zea mays*), barley (*Hordeum vulgare*), and sorghum (*Sorghum bicolor*). These crops are crucial for food security and economic stability in developed and developing countries because they provide essential proteins, carbohydrates, vitamins, and minerals. Abiotic stresses such as drought, salinity, extreme and low temperatures, heavy metal toxicity, and nutrient deficiencies pose major challenges to cereal crop production by adversely affecting plant growth, development, and yield potential (Kumar 2020; Shabbir et al. 2022). Yield losses due to these stresses are estimated to range from 10% to 15%, with reports of losses reaching up to 42% in severely affected regions (Oyebamiji et al. 2024; Reddy et al. 2024). These abiotic environmental challenges, exacerbated by changing climate conditions can disrupt key developmental stages of the plant, including germination, vegetative growth, and grain filling (Sanchez-Bermudez et al. 2022). In addition to reducing yield, abiotic stresses also negatively affect grain quality and overall crop productivity, posing a serious threat to global food security (Wang and Frei 2011). Among various abiotic stresses, drought stress is particularly detrimental to cereals, leading to significant reductions in morphological traits such as leaf size and width, shoot length, and plant height (Zia et al. 2021; Mishra et al. 2025). It also disrupts key physiological and biochemical processes, including respiration, photosynthesis, carbohydrate accumulation, nutrient metabolism, and overall plant growth (Yang et al. 2021; Seleiman et al. 2021). Salinity is another major abiotic stress that poses a serious threat to global crop production (Atta et al. 2023). Approximately 20% of agricultural land worldwide is affected by saline soils, and this figure continues to rise due to poor irrigation practices and climate change (Negacz et al. 2022; Mukhopadhyay et al. 2021). Salinity stress has been reported to reduce crop yield by more than 45% (Atta et al. 2023). It adversely impacts key yield components such as the number of grains per spike, 1000-grain weight, and overall grain yield in rice (Li et al. 2023; Hussain et al. 2019). Exposure to high temperatures, especially during the reproductive and grain-filling stages, can lead to pollen sterility, grain-filling defects, and reduced milling quality (Yu et al. 2024; Kaushal et al. 2016). These effects significantly reduce both the yield and market value of cereal crops (Wang et al. 2018). Prolonged flooding limits oxygen availability to the root zone, disrupting root respiration and nutrient absorption (Manghwar et al. 2024). This stress can severely inhibit plant growth and development, often resulting in significant yield reductions (Qari et al. 2022; Feng et al. 2025). Nutrient deficiencies, especially deficiencies of essential elements such as nitrogen, phosphorus, potassium, and micronutrients, disrupt physiological and metabolic processes, ultimately limiting plant growth and reducing crop yields (Ahmed et al. 2024; White and Brown 2010). Heavy metal toxicity can cause cellular membrane damage, impairing the plant's ability to regulate water uptake and transpiration through the roots and stomata (Vasilachi et al. 2023; Guo et al. 2023). This disruption in water balance can lead to reduced water transport, wilting, and increased water stress, affecting plant growth and productivity (Yadav 2010). These findings collectively indicate that abiotic stresses significantly reduce cereal crop productivity. In response, researchers are actively working to develop new cultivars with enhanced tolerance to abiotic stresses. Various biotechnological tools such as genetic engineering, marker-assisted selection, and genome editing are being employed to improve stress resilience and ensure sustainable cereal production under changing environmental conditions.

Traditional and modern plant breeding techniques play a vital role in developing stress-tolerant cereal varieties (Breseghello and Coelho 2013). Conventional breeding methods, including hybridization and selection, have been widely used to introduce and enhance stress-tolerant traits by crossing high-yielding cultivars with stress-resistant donor lines (Lamichhane and Thapa 2022). Marker-assisted selection has further accelerated the breeding process by enabling the identification and selection of desirable traits linked to specific genetic markers (Collard and Mackill 2008). In addition, quantitative trait loci (QTL) mapping helps breeders understand the genetic basis of complex traits related to abiotic stress tolerance, such as root architecture, osmotic adjustment, and antioxidant activity (Collins et al. 2008; Raj and Nadarajah 2022). More recently, genomic selection and speed breeding have emerged as powerful tools for improving stress resilience in cereals by combining high-throughput phenotyping and genotyping (Sinha et al. 2023). These plant breeding strategies, when integrated with advanced molecular tools, offer promising potential to develop high-yielding, climate-resilient cereal varieties suitable for sustainable agriculture under changing environmental conditions. Functional genomics plays a crucial role in understanding and improving abiotic stress tolerance in cereal crops (Langridge et al. 2006). This approach focuses on identifying and characterizing the functions of genes and regulatory elements involved in plant responses to abiotic stresses (Shinozaki and Yamaguchi-Shinozaki 2022). High-throughput technologies, including transcriptomics, proteomics, metabolomics, and gene expression profiling, enable researchers to explore global changes in gene expression and metabolic pathways under stress conditions (Raza et al. 2024; Naik et al. 2023). Functional characterization of stress-responsive genes through reverse genetics tools helps elucidate their roles in stress tolerance mechanisms. Moreover, the identification of key transcription factors, signaling molecules, and stress-inducible genes provides valuable targets for crop improvement. Integration of functional genomics with phenomics and systems biology offers a comprehensive understanding of stress adaptation, facilitating the development of genetically enhanced cereal varieties with improved resilience to abiotic stresses (Mansoor et al. 2025). These advances contribute significantly to sustainable agriculture in the face of climate variability and environmental challenges. Genetic engineering offers a powerful and targeted approach to enhancing abiotic stress tolerance in cereal crops (Parmar et al. 2017). Unlike conventional breeding, genetic engineering allows for the direct manipulation of specific genes associated with stress responses, enabling the introduction of novel traits from diverse genetic sources (Abdulraheem et al. 2024).

Through the insertion of stress-responsive genes, transcription factors, and signaling components, cereals can be engineered to better withstand abiotic stresses. For instance, overexpression of genes involved in osmotic adjustment, antioxidant defense, or ion homeostasis has shown promising results in improving tolerance levels in various cereals (Reddy et al. 2017; Terfa et al. 2025). In addition, transgenic approaches allow for the development of crops with enhanced root architecture, water-use efficiency, and stress-induced metabolic reprogramming (Wan et al. 2017). Advances in tissue culture, transformation methods, and regulatory control of gene expression (e.g., stress-inducible promoters) have further increased the precision and efficiency of genetic engineering in cereals (Wang et al. 2025; Anjanappa and Gruissem 2021). When integrated with genomics and systems biology tools, genetic engineering presents a valuable strategy for developing climate-resilient cereal varieties, contributing to food security in stress-prone environments. Genome editing has emerged as a revolutionary tool for enhancing abiotic stress tolerance in cereal crops with unprecedented precision and efficiency. Among various genome-editing tools, CRISPR-/Cas system-based tools have gained significant attention due to their simplicity, accuracy, and versatility (Anjanappa and Gruissem 2021). This approach allows targeted modifications in stress-responsive genes, enabling either gene knockout, insertion, or replacement to improve plant performance under adverse environmental conditions. In cereals like rice, wheat, and maize, the genome editing approach has been successfully used to alter genes involved in abscisic acid signaling, ion transport, osmotic balance, and reactive oxygen species (ROS) detoxification, contributing to enhanced stress resilience (Yadav et al. 2023; Li et al. 2022). Unlike traditional genetic engineering, genome editing can create transgene-free plants, which may help in easing regulatory constraints and increasing public acceptance. Additionally, multiplex genome editing allows simultaneous targeting of multiple genes, which is particularly beneficial for improving complex traits controlled by several loci (Abdelrahman et al. 2021). As genome editing technologies continue to advance, their integration with high-throughput phenotyping, genomics, and systems biology offers great potential for the development of climate-resilient, high-yielding cereal varieties tailored for future agricultural challenges (Chavhan et al. 2025).Genome editing approaches, particularly those based on CRISPR/Cas technology, offer distinct advantages over traditional plant breeding, and genetic engineering, for improving abiotic stress tolerance in cereals (Chen et al. 2024). Unlike conventional breeding, which is time-consuming and limited by the availability of natural genetic variation, genome editing enables precise, targeted modifications in specific genes responsible for stress responses, greatly accelerating the development of improved varieties (Hamdan and Tan 2024). Compared to genetic engineering, genome editing can create transgene-free plants by directly editing endogenous genes without introducing foreign DNA, potentially reducing regulatory hurdles and improving public acceptance (Bhattacharjee et al. 2023). While functional genomics provides critical insights into gene function and stress response mechanisms, it primarily serves as a discovery tool rather than a direct crop improvement method (Zhang et al. 2021). Genome editing effectively bridges this gap by applying functional genomics knowledge to precisely modify key genes, thereby enabling rapid and efficient development of stress-resilient cereal crops (Cardi et al. 2023). Consequently, genome editing represents a powerful and versatile approach for addressing the challenges posed by abiotic stresses in modern agriculture. This chapter begins by outlining the effects of various abiotic stresses on cereal crops, highlighting their impact on growth, development, and yield. It then discusses the principles and mechanisms of genome editing tools, with a particular focus on CRISPR/Cas systems and related tools. Finally, the chapter explores the application of genome editing in improving abiotic stress tolerance in cereals, emphasizing recent advances, target genes, and future prospects for developing climate-resilient cereal varieties. It provides insights into how genome editing can be strategically used to develop stress-resilient cultivars, thereby contributing to sustainable cereal production in the face of environmental challenges.

6.2 Effect of Abiotic Stresses in Cereals

Cereal growth, development, and productivity are all significantly affected by abiotic stresses. These stresses interfere with key physiological and biochemical functions such as respiration, water balance, nutrient uptake, and photosynthesis (Fig. 6.1). For example, salinity causes osmotic stress and ion toxicity, which inhibit nutrient uptake, while drought stress reduces leaf size, shoot length, and plant height. According to Hassan et al. (2023), drought stress reduced rice production and affected key traits such as panicle length, number of grains, and spikelet's in each panicle, spikelet fertility, tillers producing ears, and overall biological yield per plant. Drought stress reduced barley yield by 56.8% due to fewer spikes, tillers, and grains per plant (Samarah 2005). Under drought stress, pollen sterility caused a >51% reduction in grain number in wheat (Dong et al. 2017). In addition, drought stress also affects plant size and shape, including leaf size and width, shoot length, and plant height of maize (Deribe 2025), sorghum (Demissie et al. 2023), and finger millet (Rakkammal et al. 2023). Several studies have reported that salinity stress negatively affects important grain yield components (Zheng et al. 2023; Atta et al. 2023). Salinity stress reduces the number of fertile tillers, spikelet number, grain weight, and grain yield in wheat (El Sabagh et al. 2021; Seleiman et al. 2022). Similarly, salt stress in rice reduces plant growth, biomass accumulation, and grain yield components by inhibiting seed germination and leaf area growth (Hussain et al. 2017). According to Abbas et al. (2013), salinity also reduces the concentration of key nutrients such as phosphorus, potassium, fat, protein, and fibre in wheat grains. A significant issue during rice grain development is panicle sterility under salt stress, which reduces stigma surface area, pollen viability, and grain structure (Khan and Abdullah 2003). Other research has reported that salt stress reduces 1000-grain weight, number of flowers, tiller number, and panicle length of rice (Zhang et al. 2022). All these results point to the negative effects of salinity on grain quality and grain productivity. For cereal to grow, develop, and produce better, macro- and micronutrients are essential. Crop production is ultimately affected by

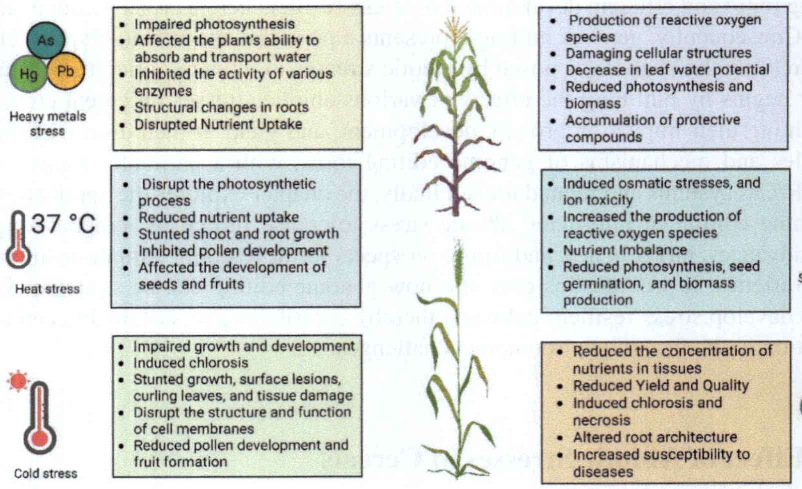

Fig. 6.1 Major effects of abiotic stress on cereals. Many abiotic stressors, such as drought, salinity, nutrient deficiencies, high temperatures, cold and heavy metal toxicity, have a substantial impact on cereal growth and yield. By lowering photosynthetic activity, these stresses significantly hinder plant growth and development, which ultimately results in significant drops in yield quality and quantity. Each stress operates through distinct physiological and biochemical mechanisms to affect cereal development, as illustrated in the rectangular boxes

deficiencies of essential nutrients, which severely disrupt several physiological and biochemical processes (Ahmed et al. 2024). Nutrient deficiencies alter the structure of roots, making it more difficult for the plant to absorb nutrients and water effectively. For all major cereal crops, this leads to reduced plant biomass, leaf area, chlorophyll content, and grain yield (Maharajan et al. 2018; Krishna et al. 2017). For example, phosphorus deficiency has a significant negative impact on the growth, biomass, and yield of millets (Ceasar et al. 2014; Maharajan et al. 2019). Nitrogen deficiency often causes chlorosis and stunted growth, but micronutrient deficiencies such as boron, iron, and zinc can interfere with the development and enzymatic activity of reproductive organs in various cereals (Francis et al. 2023; Zuluaga and Sonnante 2019). Therefore, in cereal production systems, controlling nutrient availability is essential to ensure both production and quality. High temperatures negatively affect rice growth and yield, impacting various growth stages including germination, plant growth, flowering, and grain filling. In particular, high temperatures reduce seedling vigor, reduce tillering, spikelet sterility, and grain quality, resulting in significant yield losses. High temperatures reduce germination rates, seedling growth, and overall yield, affecting the size, number, and quality of wheat.

6.3 Genome-Editing Tools

Genome editing is the process of altering the genetic code of an organism. Genome-editing tools have been used to induce specific mutations and regulate transcription without altering the underlying genetic code. Zinc finger nucleases (ZFNs), transcription activator-like effector nucleases (TALENs), and clustered regularly interspaced short palindromic repeats (CRISPR)/CRISPR-associated protein (Cas) are among the three kinds of tools majorly utilized in genome editing technologies. The double-strand break (DSB) was produced in traditional genome editing techniques by enzymes that were directed by proteins to cut DNA at precise locations (Mohanta et al. 2017). Non-Homologous End Joining (NHEJ) and homology-directed repair (HDR) are the two methods used to repair DSBs. While HDR uses extra DNA to create a desired sequence within the genome (gene knock-in), NHEJ creates random mutations (gene knockout). But in addition to basic breaks, CRISPR/Cas has given rise to a number of alternative genome-editing techniques. ZFNs are targetable, specially made DNA cleavage proteins that snip DNA sequences at particular locations (Carroll 2011). By causing the DSB in DNA to replace the gene by homologous recombination, they enable targeted gene editing. According to Gupta et al. (2012), ZFN has two domains: DNA binding and cleaving. A FokI nuclease is part of the DNA-cleaving domain, while the DNA-binding domain recognizes a specific 6-base pair in the DNA sequence. A zinc finger protein is created by joining these two domains. The combination of the two domains creates a highly specialized genomic scissor. Because the ZFNs are costly, time-consuming, and challenging to manage, their broad application is restricted, particularly in high-throughput investigations.

The genomes of bacteria (45%) and archaea (84%) include the repetitive DNA sequences known as CRISPR. It was initially identified in *Escherichia coli* downstream of the alkaline phosphatase isozyme gene (Ishino et al. 1987). Previously referred to as short, regularly spaced repeats (SRSRs), it aids in identifying and eliminating the virus's DNA. The CRISPR system uses small guide RNAs (gRNAs) to interfere with invasive nucleic acids in a sequence-specific manner. CRISPR is a collection of brief repeats (repeats) divided by distinct sequences (spacers). The nucleic acid and plasmids of the virus are the source of the spacers and repeats. CRISPR/Cas is a group of proteins that are part of the CRISPR mechanism. They can search for, cut, and ultimately change phage DNA in a certain way. The enzyme Cas significantly breaks DNA sequences and CRISPR arrays with nuclease activity (Mali et al. 2013; Liu et al. 2017). Two classifications (classifications I and II) have been established for the CRISPR/Cas system (Makarova et al. 2020). Although both classes of CRISPR systems have the potential to be utilized for genome editing, class II is preferred for this purpose because of its significantly simpler approach. Class II CRISPR/Cas systems are divided based on three key proteins: Cas1, Cas2, and Cas9, and class II has three types (II-A, II-B, and II-C). The most used CRISPR system for editing genomes in different organisms is Cas9, out of the three proteins. CRISPR/Cas12a and Cas12b, which were recently developed, belong to classes VA and VB, respectively.

6.4 Mechanism and CRISPR/Cas System Variations

With numerous benefits over alternative genome editing technologies, the CRISPR/Cas system is an easy-to-use genome editing technology. The gRNA and Cas protein are two of its most important constituents. The RNA-dependent DNA-cutting Cas protein combines with gRNA to produce a complex (Jinek et al. 2012). The joint work of Emmanuelle Charpentier and Jennifer A. Doudna (Jinek et al. 2012), for which they were also granted the 2020 Nobel Prize, established the concept of employing gRNA and making use of CRISPR as a genome editing tool. The two components that make up the gRNA are CRISPR RNA (crRNA) and trans-activating CRISPR RNA (tracrRNA). TracrRNA is a long part that helps the Cas protein attach, while crRNA is shorter, about 18–20 base pairs, and connects to the target DNA sequence. The target sequence in the gene of interest is recognized by the 5′crRNA complement base pair component of the designed gRNA, which also activates Cas9. The (protospacer adjacent motif) PAM sequence at 5′-NGG-3″ is usually recognized by the widely used Cas9 from *Streptococcus pyogenes* (SpCas9), where N can be any nucleotide base. The gRNA creates an RNA-DNA hybrid after it identifies a target site that has the appropriate PAM. The HNH and RuvC parts of Cas9 cut the matching and non-matching strands of the target DNA, respectively. To enhance plant growth under a range of biotic and abiotic stress, numerous researchers have employed the CRISPR/Cas system (Hillary and Ceasar 2019, 2024; Maharajan et al. 2022). As a result, several publications that evaluate and forecast this approach to crop development are currently accessible. The first version of CRISPR/Cas9 technology had several problems, mainly because the natural Cas9 from *S. pyogenes* would randomly cut DNA in unintended places, leading to unwanted effects. Many high-fidelity Cas9 with altered residues, including HF-Cas9, eSpCas9, HypaCas9, and Sniper Cas9, with less off-target effects, were introduced in an effort to address this issue (Lee et al. 2018, 2019).

Qi et al. (2013) created the dead Cas9 (dCas9) system to enhance the CRISPR/Cas9 system. To deactivate the nuclease function in dCas9, mutations H840A and D10A were added to the HNH and RuvC domains of Cas9, respectively (Qi et al. 2013). When directed by gRNA, the dCas9 system can target and bind DNA with the same specificity as it can break it. Numerous dCas9-based gene-editing tools have surfaced. The dCas9 technology is being employed in many molecular genetic applications, including chromatin topology, epigenome editing, base and primer editing, gene expression regulation, chromatin imaging, and minimizing off-target effects. The CRISPR-dCas9 components, dCas9, and effectors (repressor and activator) enhance the efficacy of transcriptional regulation (Fig. 6.2) (Xu and Qi 2019). The dCas9 protein helps turn on and boost the gene's activity when it connects with transcriptional activators like VP64; this process is called CRISPR activator (CRISPRa). The expression of the target gene is effectively inhibited when dCas9 attaches to the gene's target site. CRISPR interference (CRISPRi) is the term for this phenomenon. In general, gRNAs play a key role in CRISPRa and CRISPRi because they guide dCas9 activators and repressors, which can either boost or lower

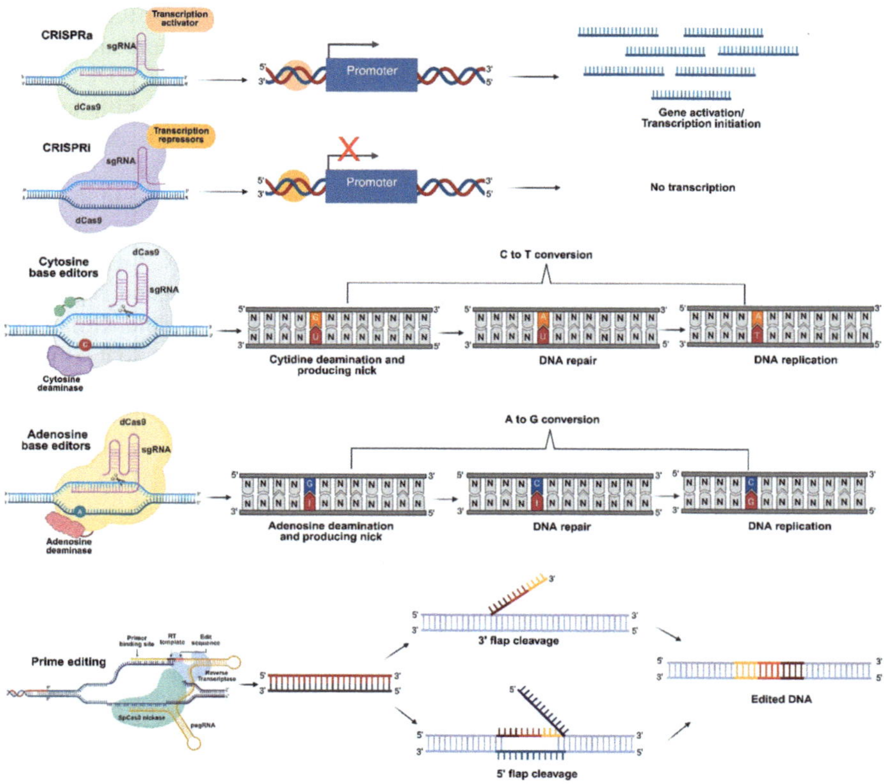

Fig. 6.2 Mechanisms of CRISPR/dCas9 system- based gene editing tools. Various genome editing tools have emerged based on dCas9 system. It is mainly used in various molecular genetic applications, such as regulation and suppression of gene expression, base editing, primer editing and so on. In dCas9, activator (CRISPRa) and repressor (CRIPSRi) are involved in activating and inhibiting the expression levels of genes, respectively. In CRIPSRa, when dCas9 binds to transcriptional activators (VP64/VPR) in the promoter region of gene, it can increase the expression level of target gene. On the other hand, when dCas9 binds to transcriptional repressors (KRAB/LSD1) in the promoter region of the gene, it can inhibit the expression of the target gene. The guide RNAs (gRNAs) play a significant role in both CRISPRa and CRISPRi because activators and repressors are guided by gRNA. The base editing system allows for the direct conversion of one target DNA base to another without double-stranded break. Two base editors (cytosine base editor (CBE) and adenosine base editor (ABE)) have been developed that are involved in all four types of transition mutations (C-T, G-A, A-G, and T-C) at target positions in DNA without making double-stranded break. CBE allows to change C-G to T-A when dCas9 linked to cytosine deaminases or uracil glycosylase inhibitor domains. Among these, uracil glycosylase inhibitor prevents base excision repair which changes the U back to a C mutation. When dCas9 fused with adenosine deaminase which allows A-T to G-C conversion. The deamination converts T-A to T-I, which is then recognized by DNA repair and replication machinery and converted to C-G. Prime editing consists of SpCas9 nickase (HB40A), reverse transcriptase (RT), prime editing gRNA (pegRNA) and RT primer site. SpCas9 is linked to RT via a linker. The RT then copies the pegRNA extension directly from the template into genomic DNA, inserting the edited sequence into the target locus

the activity of the target gene. The gRNAs need to be aimed at the specific area of the gene that controls its activity when dCas9 activators or repressors are used to turn the target gene on or off.

One target DNA base can be directly changed into another base without DSB using the CRISPR/Cas-mediated base editing technology. As a result, donor DNA templates, HDR, and DSB are not necessary for base editing technologies. Through protein engineering, two base editors, the adenine base editor (ABE) and the cytosine base editor (CBE) have been developed (Molla and Yang 2019). CBE and ABE can change or repair all four types of transition mutations (C-T, G-A, A-G, and T-C) at specific spots in DNA without causing DSB. A-T to G-C conversion is possible with dCas9 fused with ABE, but C-G to T-A transformation is possible with dCas9 fused with CBE (Bharat et al. 2020). This editing will affect proteins, so it will help introduce point mutations and change how proteins work in plants. A variety of base editor systems have also been introduced, including multiplex base editor, transversion base editor, PAMless base editor, dual base editor/saturated targeted endogenous mutagenesis editor (STEME), and others (Azameti and Dauda 2021). STEME enables the conversion of A-T to G-C and C-G to T-A. It includes four enzymes: uracil DNA glycosylase inhibitor (UGI), ecTadA (adenosine deaminase), D10A (nCas9), and APOBEC3A (cytidine deaminase). Cytidines and adenosines are changed into uridine and inosine, respectively, via the STEME. It is necessary to recognize PAM sequences in order to manipulate DNA using the CRISPR/Cas system, which includes ABE and CBE. The CRISPR/Cas system is hence severely limited. Walton et al. (2020) developed the PAMless base editor, often known as the SpRY-base editor, to get around this restriction. They have created SpG and SpRY, two distinct SpCas9 enzyme variations that may be able to modify any gene without the need for PAMs. Multiplex base editing is a highly precise method of modifying two or more particular loci in a genome (Abdelrahman et al. 2021). It makes it possible to produce exact direct mutations at several loci in one or more genes at the same time. Prime editing is a next-generation genome editing technology that allows for precise, versatile, and efficient gene modifications without the need for DSBs or donor DNA templates. Prime editing's ability to precisely delete, add, and substitute sequences gives it numerous advantages over other techniques (Hillary and Ceasar 2022). Prime editing uses SpCas9 nickase (HB40A), a reverse transcriptase fusion protein, and prime editing gRNA (pegRNA) instead of a deaminase. Since prime editing does not require a PAM sequence close to the target site, its development was revolutionary. Additionally, it is capable of insertions (up to 44 bp) and deletions (up to 80 bp) in addition to all 12 forms of point mutations. Crucially, compared to previous CRISPR/Cas systems, prime editing displayed decreased off-target activity. As a result, it might aid in enhancing cereals' resistance to abiotic stresses.In addition to causing stress, epigenetic modifications can alter gene expression, which can persist for many generations. To target epigenetic factors (histones), CRISPR-/Cas-mediated epigenetic engineering may be employed (Pulecio et al. 2017). Methylation or demethylation at the target site is activated by DNA methyltransferase (DNMT3a and b) coupled with dCas9. Histone modifiers and dCas9 can also be used together to modify histones. Important considerations for gene expression research include gene structure and chromatin folding in the nucleus. Targeting and modifying chromatin structure and DNA loop creation, which changes gene expression, is made easier by the CRISPR-/Cas-mediated chromatin topology modifier system. A potent genetic manipulation tool, the CRISPR/Cas protein and the catalytically inactive or dCas9 system can direct biological research according to the capacity to accomplish certain modifications, which is essential for phenotypic modification. By fusing dCas9 with transcriptional regulators, gene expression levels can be altered. The expression of the desired downstream genes is changed when these transcriptional modulators are recruited to the promoter region near the transcriptional start point. Similarly, dCas9-based DNA-chromatin-modifying enzymes for precision epigenome editing guided by sgRNA will be created by combining dCas9 with epigenetic modulators, such as methylation and deacetylation enzymes.

Additionally, the Cas12a (formerly known as Cpf1) and Cas12b (formerly known as C2c1) systems have been found. Because of its shorter gRNA nucleotide length and smaller Cas protein, CRISPR/Cas12a is superior to CRISPR/Cas9 (Matres et al. 2021). Because its gRNA only needs crRNA, CRISPR/Cas12a prefers T-rich PAMs and does not require tracrRNA or crRNA (Haroon et al. 2022). Cas12a's staggered cuts distinguish it from Cas9's double-strand breaks. The finding of the CRISPR/Cas12a system expands genome editing options by allowing changes in AT-rich regions, such as promoter and untranslated areas. Cas12a needs a specific sequence called PAM, which is "TTTV" or "TTV," that helps it target areas rich in AT sequences and promoters in genes. The inability to identify and modify other locations devoid of TTTN motifs limits the application of this technique in plants. However, recent research has focused on engineering Cas12a variants with altered or deleted PAM requirements. These advances aim to broaden the targeting scope of Cas12a and improve its efficiency in various applications. Like CRISPR/Cas12a, CRISPR/Cas12b also contains an RNA-guided endonuclease. However, compared to CRISPR/Cas9 and CRISPR/Cas12a, CRISPR/Cas12b has a reduced protein size. Similar to CRISPR/Cas9, CRISPR/Cas12b needs tracrRNA and crRNA to target any DNA or gene. Of all the CRISPR systems, CRISPR/Cas12b has the longest sticky ends, producing DSBs of six to eight nucleotides. Because of its capacity to induce temperature changes, it aids in the development of plants' resilience to extreme heat.

6.5 Improving Abiotic Stress Tolerance in Cereals by CRISPR/Cas System

One of the biggest threats to crop production worldwide is abiotic stress, which has a negative impact on growth, yield, and food security. As a result, many genes that respond to abiotic stress have been identified, allowing cereals to withstand environmental stress and continue to produce. Researchers have worked hard to use functional genomics techniques to harness these genes for crop development. However, several obstacles, including genetic complexity and limited transformation potential in some cereal crops, have made development relatively slow. The advent of genome-editing technologies, particularly the CRISPR/Cas system, has opened up new avenues for precise and efficient manipulation of target genes to improve abiotic stress tolerance in cereals (Table 6.1). Targeted mutagenesis and gene regulation have been made possible by CRISPR/Cas, hastening functional gene validation and improving traits. CRISPR/Cas technology has already been used to define and functionally validate a large number of abiotic stress-responsive genes in cereals (Table 6.1). Notably, the application of CRISPR/Cas has been most extensively explored in rice, owing to its relatively well-established transformation protocols and genomic resources. In contrast, there are comparatively fewer studies available in maize, wheat, and barley, indicating a need for further advancement and optimization of genome editing systems in these crops.

6.6 Drought Tolerance

The CRISPR/Cas system has revealed the function of several drought-responsive genes in cereals. Abscisic acid activates SNF1-related protein kinase 2 (SnRK2) members of the stress-activated protein kinase (SAPK) family. Under a variety of abiotic stress, such as drought, osmatic, and salinity, the members of the SAPK family aid in promoting plant growth, especially in the early phases. To describe the functional characteristics of SAPK2, Lou et al. (2017) used the CRISPR/Cas9 technology to create *sapk2* mutants in rice. They found that compared to the wild-type plants, the *sapk2* mutants were more vulnerable to ROS and drought stress. This finding showed that SAPK2 plays a crucial role in rice's ability to respond to drought. Additional research in the same study showed that *sapk* mutants had higher levels of expression for some genes that respond to abiotic stress. According to these results, SAPK2 may be a viable gene candidate for upcoming crop enhancement under drought stress in rice and other cereals. Rice drought stress may be mitigated by creating and growing new crops that use the SAPK2 locus. The architecture of plants is mostly determined by leaf rolling, which has a major impact on crop yields and the accumulation of dry matter that follows. Under drought stress, leaf rolling reduces water loss and stomatal conductance. Creating genotypes with semi-rolled leaves could be a way to increase agricultural yield in drought-stressed environments. By influencing epidermal growth and structure, the genes for semi-rolled leaf 1 (SRL1) and SRL2 play important roles in leaf rolling. By focusing on the *SRL1* and *SRL2* genes, the CRISPR/Cas9 system was utilized to create rice with semi-rolled leaves (Liao et al. 2019). Under drought stress, rice with the *SRL1* and *SRL2* genes knocked out had a greater survival rate than the wild type. When compared to wild-type plants under drought stress, proteomic profiling showed that 107 proteins were elevated in the mutant line (Liao et al. 2019). Additionally, they created a genotype of semi-rolled leaves from wild-type and mutant rolled leaves,

Table 6.1 Details on the CRISPR/Cas system-based studies used to study the role of abiotic stress-responsive genes in various cereals

Name of the stress	Plant name	Genotypes	Name of knockout gene	Result	References
Drought	Rice	Japonica	*SAPK2*	Improved growth of shoot and root and contents of proline and soluble sugars.	Lou et al. (2017)
		GXU16, GXU20, and GXU28	*SRL1* and *2*	Improved grain filling percentage, survival rate, contents of abscisic acid and catalase and superoxide dismutase activities	Liao et al. (2019)
		Japonica	*OsERA1*	Improved leaf length and primary root growth and abscisic acid content	Ogata et al. (2020)
		IR-96	*OsPYL9*	Improved yield and upregulated differentially expressed proteins	Usman et al. (2020)
		Japonica	*OsNAC006*	Improved plant growth and Upregulated the expression levels of pathways genes	Wang et al. (2020)
		Japonica	*OsmiR535*	Improved urvival rate	Yue et al. (2020)
	Maize	Hi-II	*ARGOS8*	Improved growth and grain yield	Shi et al. (2017)
Salinity	Rice	Japonica	*OsRR22*	Improved plant height and shoot fresh and dry weight	Zhang et al. (2019a)
		Kitaake	*OsOTS1*	Improved shoot and root biomass	Zhang et al. (2019b)
		Kitaake	*OsRR9 and 10*	Improved shoot and root biomass and upregulated potassium transporters	Wang et al. (2019)
		MTU1010	*OsDST*	Improved grain yield and contents of chlorophyll	Kumar et al. (2020)
	Rice	A91	*OsbHLH024*	Improved shoot weight, total chlorophyll content, and chlorophyll fluorescence	Alam et al. (2022)

(continued)

Table 6.1 (continued)

Name of the stress	Plant name	Genotypes	Name of knockout gene	Result	References
Heat	Rice	Nipponbare	*OsHSA1*	Delayed chloroplast activity	Qiu et al. (2018)
Cold	Rice	Taipei309	*OsAnn3*	Improved growth and yield	Shen et al. (2017)
	Rice	GUX2–44	*OsPRP1*	Reduced survival rate, fresh and dry weight of shoot and root, proline, abscisic, and ascorbic acid contents in leaf and root tissues, and activities of antioxidant enzymes	Nawaz et al. (2019)
	Rice	Nipponbare	*OsPIN5b,* Os*GS3* and *OsMYB30*	Improved panicle length and grain size	Zeng et al. (2020)
Heavy metal	Rice	Huazhan and Longke 638S	*OsNramp5*	Development of new lines with low Cd accumulation	Tang et al. (2017)
Nutrients deficiency	Tomato	Micro-Tom	*SlPHO1;1* (Solyc09g090360)	Decreased shoot fresh weight and phosphate content in shoots and identified SlPHO1;1 plays an important role in Pi transport in the tomato at seedling stage	Zhao et al. (2019)

Abbreviations: *Ann* annexin, *ARGOS* auxin-regulated gene involved in organ size content, *CAT* catalase, *DST* drought and salt tolerance, *ERA* enhanced response to abscisic acid, *HSA* heat sensitive albino1miR, micro RNA, *OTS* overly tolerant to salt, *PRP* proline-rich proteins, *PYL* pyrabactin resistance like, *RR* response regulator, *SAPK* stress-activated protein kinase, *SOD* superoxide dismutase, *SRL* semi rolled leaf

which enhanced the number of panicles, the number of grains per panicle, and the yield per plant when drought stress was present (Liao et al. 2019). This investigation leads that both the *SRL1* and *SRL2* genes play a role in the development of unrolled leaf genotypes during drought stress. By modifying H_2O_2 homeostasis, the zinc finger transcription factor encoded by the *drought and salt tolerance* (*DST*) gene increases leaf breadth and stomatal closure while decreasing stomatal density in rice. Under drought stress, rice with the *OsDST* gene knocked out using CRISPR/Cas9 showed decreased stomatal density, increased flag leaf growth, and leaf water retention (Kumar et al. 2020). As a result, the *OsDST* gene may be crucial in creating future Indica rice varieties that can withstand drought. The β-subunit of the farnesyltransferase gene, which controls abscisic acid signaling and the dehydration response, is encoded by *enhanced response to abscisic acid* (ERA). To determine the function of the *OsERA1* gene under drought stress, rice mutant lines were created using the CRISPR/Cas9 method (Ogata et al. 2020). Under drought stress, stomatal conductance was increased by *OsERA1* gene knockout. This finding showed that rice's resistance to drought was improved by OsERA1 gene deletion. Under abiotic stress, the *pyrabactin resistance-like* (*PYL*) gene's abscisic acid receptor controls plant growth and development. In order to clarify the function of *OsPYL9* gene under drought stress, Usman et al. (2020) created *pyl9* mutants in rice using the CRISPR/Cas9 technology. Plant height, panicle number, panicle length, flag leaf length and breadth, number of grains per panicle, grain weight, grain length and width, and yield per plant were all higher in the mutant plants under drought stress. The findings of this study could provide a useful basis for the future creation of high-yielding and drought-tolerant rice varieties. Plant growth and development under abiotic stresses are mediated by genes known as *no apical meristem* (*NAC*). Rice exhibited increased drought sensitivity when *OsNAC006* was knocked out using CRISPR/Cas9. Numerous biological processes linked to plant growth, development, and abiotic stress responses require microRNAs (miRNAs). In rice, the OsmiR535 regulates grain length, panicle architecture, and plant height in response to abiotic stressors. Using the CRISPR/Cas9 technology, the role of OsmiR535 in rice under drought stress was discovered. The survival rate of seedlings under drought stress was considerably raised by *OsmiR535* gene deletion. The developmental gene known as *epidermal patterning factor like 9* (*OsEPFL9*) aids in controlling the stomatal density of the leaf. Under drought stress, rice with the *OsEPFL9* gene knocked out using the CRISPR/Cas9 system had fewer stomata and more efficient water utilization (Yin et al. 2019).

In maize, the CRISPR/Cas9 system has been used to exploit the function of a single gene under drought stress. Members of the auxin-regulated gene implicated in organ size (ARGOS) regulate ethylene signal transduction, improve drought tolerance, and act as negative regulators of ethylene responses. In order to create two mutants (ARGOS8-v1 and ARGOS8-v2), Shi et al. (2017) employed the HDR route to introduce the maize native GOS2 promoter into the 5′ untranslated region of the *ARGOS8* gene. In multi-location fields, the growth and yield of hybrids created from these two mutants were assessed. When maize was subjected to drought stress, the hybrid plants outperformed the wild variety in terms of growth and yield (Shi et al. 2017). According to this study, the CRISPR/Cas system enhances maize growth in drought-stressed conditions and could result in the creation of novel cultivars. To create improved maize varieties that are resistant to abiotic stresses, more research is needed to analyze the functions of other important genes that contribute to drought tolerance. The function of drought-responsive genes in other cereals has not been determined using the CRISPR/Cas system. The majority of cereals under drought stress have several genes and transcription factors that have been discovered and functionally defined among under abiotic several stress. Therefore, using CRISPR/Cas variants to target additional genes and transcription factors in other cereals may help increase the production and growth of all cereals under drought stress.

6.7 Salinity Tolerance

The CRISPR/Cas system was used to determine the roles of a few salt-responsive genes in rice, much as it is under drought stress. The B-type response regulator transcription factor is encoded by the two-component *response regulator* (*RR*) gene, which also plays a role in the metabolism and transmission of cytokinin signals. The lack of function of the *OsRR22* gene has greatly improved salt tolerance. Rice's ability to withstand salt was improved by *OsRR22* gene knockout using the CRISPR/Cas9 technology (Zhang et al. 2019a). At the seedling stage, the transgenic plants demonstrated a notable resilience to salinity. Another team described the functions of *OsRR9* and *OsRR10* genes under salinity stress using the CRISPR/Cas9 method (Wang et al. 2019). Using a multiplexed CRISPR/Cas9 system, they produced double mutants of the *osrr9* and *osrr10* genes, improving their ability to withstand salinity. The two potassium transporter genes (*OsHKT1;1* and *2;1*) in rice leaves under salt stress were further validated in the same study. The transgenic plants' leaves showed high expression of both genes. This finding showed that, via controlling HKT family transporters, the *OsRR9* and *OsRR10* genes improved resistance to salinity stress. According to this study, we can use these genes to create a new rice cultivar that will thrive in soils with salinity stress. The rice gene known as *OsOTS1* (Small Ubiquitin-like Modifier (SUMO) protease family member) was knocked out by Zhang et al. (2019a) using the CRISPR/Cas9 technology. The transgenic lines increased shoot and root biomass and salinity tolerance, suggesting that the *OsOTS1* gene is crucial for rice's ability to withstand salt stress. These reports show how the CRISPR/Cas system could be used to efficiently determine the function of genes in other cereals that can react to salinity stress.

6.8 Heat and Cold Tolerance

Chloroplast activity is impacted by high temperatures, which lowers photosynthesis and, ultimately, plant yields. Using CRISPR/Cas9, Qiu et al. (2018) described the function of the *heat-sensitive albino1* (*HSA1*) gene in rice. When compared to wild-type rice, rice with the *HSA1* gene knocked out showed a considerable delay in chloroplast activity at high temperatures (Qiu et al. 2018). These findings suggested that chloroplast activity during heat stress is significantly influenced by the *HSA1* gene. To increase the development of rice and maize under higher temperature conditions, Malzahn et al. (2019) used the CRISPR/Cas12a system to target two genes: *dense and erect panicle 1* (*OsDEP1*) and *rice outermost cell-specific* (*OsROC5*). Because of delayed heading and pollen sterility, cold stress has an impact on seed germination, seedling growth, tillering, and yield. This issue might be resolved by using the CRISPR/Cas system to create cereals that can withstand cold stress. Under cold stress, rice grew more when the *annexin 3* (*OsAnn3*) gene was knocked out (Shen et al. 2017). Another study used CRISPR/Cas9 to identify the function of *proline-rich proteins 1* (*OsPRP1*) gene in rice under cold stress. In comparison to the wild type, mutant rice with *OsPRP1* knockout had lower survival rates, fresh and dry weight of shoots and roots, proline, abscisic, and ascorbic acid concentrations in leaf and root tissues, and antioxidant enzyme activities under cold stress (Nawaz et al. 2019). According to this study, rice grows more when exposed to cold stress when *OSPRP1* is present. The CRISPR/Cas9 system in rice increased panicle length and grain size under cold stress by targeting three genes, such as *OsPIN5b* (a gene for panicle length), *OsGS3* (a gene for grain size), and *OsMYB30* (a gene for cold tolerance) (Zeng et al. 2020). The CRISPR/Cas12b system is one of the CRISPR/Cas variations that have been shown to produce grains that can withstand high temperatures. So, the CRISPR/Cas12 system might help create heat-resistant cereals by focusing on genes that deal with heat stress, making it better than other CRISPR/Cas variations.

6.9 Nutrient Deficiency (Macro and Micro) Tolerance

The physiological and biochemical functions of plants depend on the availability of both macro- and micronutrients. Numerous macro- and micro-nutrient transporters have been functionally characterized in various cereals (Maharajan et al. 2022; Krishna et al. 2020, 2022; Roch et al. 2019). These include nitrate transporter (NRT), phosphate transporter (PHT), potassium transporters (K+ transporters), sulfate transporters (SULTR), ammonium transporters (AMT), iron transporters (IRT), copper transporters (COPT), zinc-regulated and iron-regulated transport-like proteins (ZIP), cation diffusion facilitator (CDF), copper transporter (COPT), iron-regulated transporter (IRT), heavy metal ATPase (HMA), molybdate transporter type 1 (MOT1), natural resistance-associated macrophage protein (NRAMP), and yellow stripe-like proteins (YSL). Base editing has been used to target the *NRT1* family gene (*OsNRT1;1b*) (Li et al. 2018; Lu and Zhu 2017). More recently, the CRISPR/Cas system can be used to improve plant nutrient transport, according to two review studies (Ceasar et al. 2022; Sathee et al. 2022). Using CRISPR/Cas9, the function of *antioxidant protein 1* (*ATX1*) in rice has been determined. Cu buildup in roots and older leaves was enhanced in rice when the *OsATX1* gene was knocked out (Zhang et al. 2018). According to these findings, OsATX1 is essential for promoting the movement of copper from roots to shoots and remobilizing copper from aged leaves to growing tissues. Certain micronutrients (cadmium (Cd), chromium (Cr), lead (Pb), aluminum (Al), and mercury (Hg)) are regarded as non-essential since they are not engaged in any physiological or biochemical processes in plants. Even at very low quantities, these non-essential micronutrients are extremely hazardous to plants. Common harmful effects of non-essential micronutrients on plants include decreased biomass, altered water balance, nutrient assimilation, induced chlorosis, and growth and photosynthetic processes, all of which lead to plant death. According to a study, utilizing the CRISPR/Cas9 technology to delete the *OsNramp5* gene in rice decreased Cd accumulation in the roots and shoots of mutant plants more than in wild-type plants, while having no effect on yield in soil contaminated with Cd (Tang et al. 2017). Under high (40 μM) As stress, rice plants with *arsenate (As)-responsive MYB1* (*OsARM1*) knockout had longer roots and greater plant height (Wang et al. 2017). In the same study, under low (2 μM) As circumstances, the knockout rice plants had higher As concentrations in their shoot and root tissues than the wild-type plants. These findings suggest that the *OsARM1* gene may have a role in both improved plant growth (in high As conditions) and As translocation from root to shoot (in low As conditions).

6.10 Conclusion and Future Direction

Genome editing has provided new opportunities for enhancing abiotic stress tolerance in cereals. This approach offers unprecedented precision in modifying genes associated with stress responses such as drought, heat, and more. By implementing targeted changes at specific locations, genetic editing accelerates the development of cereal varieties that are better equipped to withstand adverse environmental conditions, thus contributing to higher agricultural productivity and food security in the face of climate change. Although challenges remain to be addressed, such as off-target effects, regulatory issues, and the need for more efficient delivery systems, the successes achieved in model crops demonstrate the potential for widespread use in major cereals. As genome editing technologies continue to develop, they hold the promise of creating crops that are not only adaptable but also resilient to various stress conditions. Future research should aim to change several genes or regulatory elements simultaneously to create cereals that can resist various types of stress over a long time and help plants tolerate tough conditions. Creating better ways to deliver CRISPR tools specifically to certain tissues without using transgenic methods should help get regulatory approvals faster, allowing for quicker production of better cereal varieties. Combining CRISPR with transcriptomics, phenomics, and genomics allows us to investigate the underlying stress response networks and pinpoint new editing targets. We need to conduct more comprehensive field trials in various agro-climatic settings to evaluate the practical effectiveness and environmental safety of CRISPR-edited cereals. Harmonizing international legislation and raising public awareness are crucial for the acceptance and commercialization of gene-edited cereals. CRISPR/Cas gene editing will continue to transform cereal crop development, ensuring sustainable agriculture in the face of growing abiotic stresses caused by climate change.

Acknowledgments We thank Rajagiri College of Social Sciences for all the support and help for the research.

Funding This work was financially supported by Rajagiri College of Social Sciences (Autonomous), under Seed Money for Faculty Minor Research.

References

Abbas G, Saqib M, Rafique Q, Rahman AU, Akhtar J, Haq MAU, Nasim M (2013) Effect of salinity on grain yield and grain quality of wheat (Triticum aestivum L.). Pak J Bot 50:185–189

Abdelrahman M, Wei Z, Rohila JS, Zhao K (2021) Multiplex genome-editing technologies for revolutionizing plant biology and crop improvement. Front Plant Sci 12:721203

Abdulraheem MI, Xiong Y, Moshood AY, Cadenas-Pliego G, Zhang H, Hu J (2024) Mechanisms of plant epigenetic regulation in response to plant stress: recent discoveries and implications. Plants 13(2):163

Ahmed N, Zhang B, Chachar Z, Li J, Xiao G, Wang Q et al (2024) Micronutrients and their effects on horticultural crop quality, productivity and sustainability. Sci Hortic 323:112512

Alam MS, Kong J, Tao R, Ahmed T, Alamin M, Alotaibi SS et al (2022) CRISPR/Cas9 mediated knockout of the OsbHLH024 transcription factor improves salt stress resistance in rice (Oryza sativa L.). Plants 11(9):1184

Anjanappa RB, Gruissem W (2021) Current progress and challenges in crop genetic transformation. J Plant Physiol 261:153411

Atta K, Mondal S, Gorai S, Singh AP, Kumari A, Ghosh T et al (2023) Impacts of salinity stress on crop plants: improving salt tolerance through genetic and molecular dissection. Front Plant Sci 14:1241736

Azameti MK, Dauda WP (2021) Base editing in plants: applications, challenges, and future prospects. Front Plant Sci 12:1531

Bharat SS, Li S, Li J, Yan L, Xia L (2020) Base editing in plants: current status and challenges. Crop J 8:384–395

Bhattacharjee S, Bhowmick R, Kant L, Paul K (2023) Strategic transgene-free approaches of CRISPR-based genome editing in plants. Mol Gen Genomics 298(3):507–520

Breseghello F, Coelho ASG (2013) Traditional and modern plant breeding methods with examples in rice (Oryza sativa L.). J Agric Food Chem 61(35):8277–8286

Cardi T, Murovec J, Bakhsh A, Boniecka J, Bruegmann T, Bull SE et al (2023) CRISPR/Cas-mediated plant genome editing: outstanding challenges a decade after implementation. Trends Plant Sci 28(10):1144–1165

Carroll D (2011) Genome engineering with zinc-finger nucleases. Genetics 188:773–782

Ceasar SA, Hodge A, Baker A, Baldwin SA (2014) Phosphate concentration and arbuscular mycorrhizal colonisation influence the growth, yield and expression of twelve PHT1 family phosphate transporters in foxtail millet (Setaria italica). PLoS One 9(9):e108459

Ceasar SA, Maharajan T, Hillary E, Krishna TPA (2022) Insights to improve the plant nutrient transport by CRISPR/Cas system. Biotechnol Adv 59:107963

Chavhan R, Jaybhaye S, Hinge V, Deshmukh AS, Uwais S, Jadhav P et al (2025) Emerging applications of gene editing technologies for the development of climate-resilient crops. Front Genome Ed 7:1524767

Chen F, Chen L, Yan Z, Xu J, Feng L, He N et al (2024) Recent advances of CRISPR-based genome editing for enhancing staple crops. Front Plant Sci 15:1478398

Collard BC, Mackill DJ (2008) Marker-assisted selection: an approach for precision plant breeding in the twenty-first century. Philos Trans R Soc Lond B Biol Sci 363(1491):557–572

Collins NC, Tardieu F, Tuberosa R (2008) Quantitative trait loci and crop performance under abiotic stress: where do we stand? Plant Physiol 147(2):469–486

Demissie HS, Mindaye TT, Teklu DN, Kebede FG (2023) Root system architecture analysis of sorghum genotypes and its effect on drought adaptation. Rhizosphere 27:100772

Deribe H (2025) Review on effects of drought stress on maize growth, yield and its management strategies. Commun Soil Sci Plant Anal 56(1):123–143

Dong B, Zheng X, Liu H, Able JA, Yang H, Zhao H et al (2017) Effects of drought stress on pollen sterility, grain yield, abscisic acid and protective enzymes in two winter wheat cultivars. Front Plant Sci 8:1008

El Sabagh A, Islam MS, Skalicky M, Ali Raza M, Singh K, Anwar Hossain M et al (2021) Salinity stress in wheat (Triticum aestivum L.) in the changing climate: adaptation and management strategies. Front Agron 3:661932
Feng Y, Li Z, Kong X, Khan A, Ullah N, Zhang X (2025) Plant coping with cold stress: molecular and physiological adaptive mechanisms with future perspectives. Cells 14(2):110
Francis B, Aravindakumar CT, Brewer PB, Simon S (2023) Plant nutrient stress adaptation: a prospect for fertilizer limited agriculture. Environ Exp Bot 213:105431
Guo Z, Gao Y, Yuan X, Yuan M, Huang L, Wang S et al (2023) Effects of heavy metals on stomata in plants: a review. Int J Mol Sci 24(11):9302
Gupta A, Christensen RG, Rayla AL, Lakshmanan A, Stormo GD, Wolfe SA (2012) An optimized two-finger archive for ZFN-mediated gene targeting. Nat Met 9:588–590
Hamdan MF, Tan BC (2024) Genetic modification techniques in plant breeding: a comparative review of CRISPR/Cas and GM technologies. Hortic Plant J
Haroon M, Wang X, Afzal R, Zafar MM, Idrees F, Batool M, Imran M (2022) Novel plant breeding techniques shake hands with cereals to increase production. Plants 11:1052
Hassan MA, Dahu N, Hongning T, Qian Z, Yueming Y, Yiru L, Shimei W (2023) Drought stress in rice: morpho-physiological and molecular responses and marker-assisted breeding. Front Plant Sci 14:1215371
Hillary VE, Ceasar SA (2019) Application of CRISPR/Cas9 genome editing system in cereal crops. Open Biotechnol J 13:173–179
Hillary VE, Ceasar SA (2022) Prime editing in plants and mammalian cells: mechanism, achievements, limitations, and future prospects. BioEssays 44:2200032
Hillary VE, Ceasar SA (2024) CRISPR/Cas system-mediated base editing in crops: recent developments and future prospects. Plant Cell Rep 43(11):271
Hussain S, Shaukat M, Ashraf M, Zhu C, Jin Q, Zhang J (2019) Salinity stress in arid and semi-arid climates: effects and management in field crops. In: Climate change and agriculture. IntechOpen
Hussain S, Zhang JH, Zhong C, Zhu LF, Cao XC, Yu SM et al (2017) Effects of salt stress on rice growth, development characteristics, and the regulating ways: a review. J Integr Agric 16(11):2357–2374
Ishino Y, Shinagawa H, Makino K, Amemura M, Nakata A (1987) Nucleotide sequence of the iap gene, responsible for alkaline phosphatase isozyme conversion in Escherichia coli, and identification of the gene product. J Bacteriol 169:5429–5433
Jinek M, Chylinski K, Fonfara I, Hauer M, Doudna JA, Charpentier E (2012) A programmable dual-RNA–guided DNA endonuclease in adaptive bacterial immunity. Science 337:816–821
Kaushal N, Bhandari K, Siddique KH, Nayyar H (2016) Food crops face rising temperatures: an overview of responses, adaptive mechanisms, and approaches to improve heat tolerance. Cogent Food Agric 2(1):1134380
Khan MA, Abdullah Z (2003) Salinity–sodicity induced changes in reproductive physiology of rice (Oryza sativa) under dense soil conditions. Environ Exp Bot 49(2):145–157
Krishna TA, Ceasar SA, Maharajan T, Ramakrishnan M, Duraipandiyan V, Al-Dhabi NA, Ignacimuthu S (2017) Improving the zinc-use efficiency in plants: a review. SABRAO J Breed Genet 49(3):211–230
Kumar VS, Verma RK, Yadav SK, Yadav P, Watts A, Rao MV, Chinnusamy V (2020) CRISPR-Cas9 mediated genome editing of drought and salt tolerance (OsDST) gene in indica mega rice cultivar MTU1010. Phy Mol Biol Plants 26:1099–1110
Kumar S (2020) Abiotic stresses and their effects on plant growth, yield and nutritional quality of agricultural produce. Int J Food Sci Agric 4(4):367
Krishna TPA, Maharajan T, Roch GV, Ignacimuthu S, Ceasar SA (2020) Structure, function, regulation and phylogenetic relationship of ZIP family transporters of plants. Front Plant Sci 11:662
Krishna TPA, Maharajan T, Ceasar SA (2022) Application of CRISPR/Cas9 genome editing system to reduce the pre-and postharvest yield losses in cereals. Open Biotech J 16:1–9

Li C, Zong Y, Wang Y, Jin S, Zhang D, Song Q, Gao C (2018) Expanded base editing in rice and wheat using a Cas9-adenosine deaminase fusion. Genome Biol 19:1–9
Lu Y, Zhu JK (2017) Precise editing of a target base in the rice genome using a modified CRISPR/Cas9 system. Mol Plant 10:523–525
Lamichhane S, Thapa S (2022) Advances from conventional to modern plant breeding methodologies. Plant Breed Biotechnol 10(1):1–4
Langridge P, Paltridge N, Fincher G (2006) Functional genomics of abiotic stress tolerance in cereals. Brief Funct Genomics 4(4):343–354
Lee J, Jung MH, Jeong E, Lee JK (2019) Using Sniper-Cas9 to minimize off-target effects of CRISPR-Cas9 without the loss of on-target activity via directed evolution. J Vis Exp 144:e59202
Lee JK, Jeong E, Lee J, Jung M, Shin E, Kim YH, Kim JS (2018) Directed evolution of CRISPR-Cas9 to increase its specificity. Nat Commun 9:1–10
Li Y, Wu X, Zhang Y, Zhang Q (2022) CRISPR/Cas genome editing improves abiotic and biotic stress tolerance of crops. Front Genom Edit 4:987817
Li Z, Zhou T, Zhu K, Wang W, Zhang W, Zhang H et al (2023) Effects of salt stress on grain yield and quality parameters in rice cultivars with differing salt tolerance. Plants 12(18):3243
Liao S, Qin X, Luo L, Han Y, Wang X, Usman B, Li R (2019) CRISPR/Cas9-induced mutagenesis of semi-rolled Leaf1, 2 confers curled leaf phenotype and drought tolerance by influencing protein expression patterns and ROS scavenging in Rice (Oryza sativa L.). Agronomy 9:728
Liu C, Zhang L, Liu H, Cheng K (2017) Delivery strategies of the CRISPR-Cas9 gene-editing system for therapeutic applications. J Control Rel 266:17–26
Lou D, Wang H, Liang G, Yu D (2017) OsSAPK2 confers abscisic acid sensitivity and tolerance to drought stress in rice. Front Plant Sci 8:993
Maharajan T, Ceasar SA, Ajeeshkrishna TP, Ramakrishnan M, Duraipandiyan V, Naif Abdulla AD, Ignacimuthu S (2018) Utilization of molecular markers for improving the phosphorus efficiency in crop plants. Plant Breed 137(1):10–26
Maharajan T, Ceasar SA, Krishna TPA, Ignacimuthu S (2019) Phosphate supply influenced the growth, yield and expression of PHT1 family phosphate transporters in seven millets. Planta 250:1433–1448
Maharajan T, Krishna TA, Rakkammal K, Ceasar SA, Ramesh M (2022) Application of CRISPR/Cas system in cereal improvement for biotic and abiotic stress tolerance. Planta 256(6):106
Makarova KS, Wolf YI, Iranzo J, Shmakov SA, Alkhnbashi OS, Brouns SJ, Koonin EV (2020) Evolutionary classification of CRISPR–Cas systems: a burst of class 2 and derived variants. Nat Rev Microbiol 18:67–83
Mali P, Yang L, Esvelt KM, Aach J, Guell M, DiCarlo JE, Church GM (2013) RNA-guided human genome engineering via Cas9. Science 339:823–826
Malzahn AA, Tang X, Lee K, Ren Q, Sretenovic S, Zhang Y, Qi Y (2019) Application of CRISPR-Cas12a temperature sensitivity for improved genome editing in rice, maize, and Arabidopsis. BMC Biol 17:1–14
Manghwar H, Hussain A, Alam I, Khoso MA, Ali Q, Liu F (2024) Waterlogging stress in plants: unraveling the mechanisms and impacts on growth, development, and productivity. Environ Exp Bot 224:105824
Mansoor S, Karunathilake EM, Tuan TT, Chung YS (2025) Genomics, phenomics, and machine learning in transforming plant research: advancements and challenges. Hortic Plant J 11(2):486–503
Matres JM, Hilscher J, Datta A, Armario Najera V, Baysal C, He W, Slamet-Loedin IH (2021) Genome editing in cereal crops: an overview. Trans Res 30:461–498
Mishra UN, Chauhan J, Singhal RK, Anuragi H, Dey P, Lal D et al (2025) Abiotic stress responses in forage crops and grasses: the role of secondary metabolites and biotechnological interventions. Front Plant Sci 16:1542519
Mohanta TK, Bashir T, Hashem A, Abd-Allah EF, Bae H (2017) Genome editing tools in plants. Genes 8(12):399

Molla KA, Yang Y (2019) CRISPR/Cas-mediated base editing: technical considerations and practical applications. Trends Biotechnol 37:1121–1142

Mukhopadhyay R, Sarkar B, Jat HS, Sharma PC, Bolan NS (2021) Soil salinity under climate change: challenges for sustainable agriculture and food security. J Environ Manag 280:111736

Naik B, Kumar V, Rizwanuddin S, Chauhan M, Choudhary M, Gupta AK et al (2023) Genomics, proteomics, and metabolomics approaches to improve abiotic stress tolerance in tomato plant. Int J Mol Sci 24(3):3025

Nawaz G, Han Y, Usman B, Liu F, Qin B, Li R (2019) Knockout of OsPRP1, a gene encoding proline-rich protein, confers enhanced cold sensitivity in rice (Oryza sativa L) at the seedling stage. 3 Biotech 9:1–18

Negacz K, Malek Ž, de Vos A, Vellinga P (2022) Saline soils worldwide: identifying the most promising areas for saline agriculture. J Arid Environ 203:104775

Ogata T, Ishizaki T, Fujita M, Fujita Y (2020) CRISPR/Cas9-targeted mutagenesis of OsERA1 confers enhanced responses to abscisic acid and drought stress and increased primary root growth under nonstressed conditions in rice. PLoS One 15:e0243376

Oyebamiji YO, Adigun BA, Shamsudin NAA, Ikmal AM, Salisu MA, Malike FA, Lateef AA (2024) Recent advancements in mitigating abiotic stresses in crops. Horticulturae 10(2):156

Parmar N, Singh KH, Sharma D, Singh L, Kumar P, Nanjundan J et al (2017) Genetic engineering strategies for biotic and abiotic stress tolerance and quality enhancement in horticultural crops: a comprehensive review. 3 Biotech 7:1–35

Pulecio J, Verma N, Mejia Ramírez E, Huangfu D, Raya A (2017) CRISPR/Cas9-based engineering of the epigenome. Cell Stem Cell 21:431–447

Qari SH, Hassan MU, Chattha MU, Mahmood A, Naqve M, Nawaz M et al (2022) Melatonin induced cold tolerance in plants: physiological and molecular responses. Front Plant Sci 13:843071

Qi LS, Larson MH, Gilbert LA, Doudna JA, Weissman JS, Arkin AP, Lim WA (2013) Repurposing CRISPR as an RNA-guided platform for sequence-specific control of gene expression. Cell 152:1173–1183

Qiu Z, Kang S, He L, Zhao J, Zhang S, Hu J, Zhu L (2018) The newly identified heat-stress sensitive albino 1 gene affects chloroplast development in rice. Plant Sci 267:168–179

Raj SRG, Nadarajah K (2022) QTL and candidate genes: techniques and advancement in abiotic stress resistance breeding of major cereals. Int J Mol Sci 24(1):6

Rakkammal K, Maharajan T, Shriram RN, Ram PJ, Ceasar SA, Ramesh M (2023) Physiological, biochemical and molecular responses of finger millet (Eleusinecoracana) genotypes exposed to short-term drought stress induced by PEG-6000. S Afr J Bot 155:45–59

Raza A, Salehi H, Bashir S, Tabassum J, Jamla M, Charagh S et al (2024) Transcriptomics, proteomics, and metabolomics interventions prompt crop improvement against metal (loid) toxicity. Plant Cell Rep 43(3):80

Reddy INBL, Kim BK, Yoon IS, Kim KH, Kwon TR (2017) Salt tolerance in rice: focus on mechanisms and approaches. Rice Sci 24(3):123–144

Reddy KS, Khapte PS, Changan SS (2024) Abiotic stress management in agriculture. Indian J Fertilisers 20(4):336–349

Roch GV, Maharajan T, Ceasar SA, Ignacimuthu S (2019) The role of PHT1 family transporters in the acquisition and redistribution of phosphorus in plants. Critical Reviews in Plant Sciences 38(3):171–198

Samarah NH (2005) Effects of drought stress on growth and yield of barley. Agron Sustain Dev 25(1):145–149

Sanchez-Bermudez M, Del Pozo JC, Pernas M (2022) Effects of combined abiotic stresses related to climate change on root growth in crops. Front Plant Sci 13:918537

Sathee L, Barman D, Nagar S, Tripathi S, Jha SK, Chinnusamy V (2022). Genome editing targets for improving nutrient use efficiency and nutrient stress adaptation. Front Gen 13: 900897

Seleiman MF, Al-Suhaibani N, Ali N, Akmal M, Alotaibi M, Refay Y et al (2021) Drought stress impacts on plants and different approaches to alleviate its adverse effects. Plants 10(2):259

Seleiman MF, Aslam MT, Alhammad BA, Hassan MU, Maqbool R, Chattha MU et al (2022) Salinity stress in wheat: effects, mechanisms and management strategies. Phyton. (0031-9457) 91(4)
Shabbir R, Singhal RK, Mishra UN, Chauhan J, Javed T, Hussain S et al (2022) Combined abiotic stresses: challenges and potential for crop improvement. Agronomy 12(11):2795
Shen C, Que Z, Xia Y, Tang N, Li D, He R, Cao M (2017) Knock out of the annexin gene OsAnn3 via CRISPR/Cas9-mediated genome editing decreased cold tolerance in rice. J Plant Biol 60:539–547
Shi J, Gao H, Wang H, Lafitte HR, Archibald RL, Yang M, Habben JE (2017) ARGOS 8 variants generated by CRISPR-Cas9 improve maize grain yield under field drought stress conditions. Plant Biotechnol J 15:207–216
Shiferaw B, Smale M, Braun HJ, Duveiller E, Reynolds M, Muricho G (2013) Crops that feed the world 10. Past successes and future challenges to the role played by wheat in global food security. Food Secur 5:291–317
Shinozaki K, Yamaguchi-Shinozaki K (2022) Functional genomics in plant abiotic stress responses and tolerance: from gene discovery to complex regulatory networks and their application in breeding. Proc Jpn Acad Ser B 98(8):470–492
Sinha D, Maurya AK, Abdi G, Majeed M, Agarwal R, Mukherjee R et al (2023) Integrated genomic selection for accelerating breeding programs of climate-smart cereals. Genes 14(7):1484
Tang L, Mao B, Li Y, Lv Q, Zhang L, Chen C et al (2017) Knockout of OsNramp5 using the CRISPR/Cas9 system produces low Cd-accumulating indica rice without compromising yield. Sci Rep 7(1):14438
Terfa GN, Pan W, Hu L, Hao J, Zhao Q, Jia Y, Nie X (2025) Mechanisms of salt and drought stress responses in foxtail millet. Plants 14(8):1215
Usman B, Nawaz G, Zhao N, Liao S, Liu Y, Li R (2020) Precise editing of the ospyl9 gene by rna-guided cas9 nuclease confers enhanced drought tolerance and grain yield in rice (Oryza sativa l.) by regulating circadian rhythm and abiotic stress responsive proteins. Int J Mol Sci 21:7854
Vasilachi IC, Stoleru V, Gavrilescu M (2023) Analysis of heavy metal impacts on cereal crop growth and development in contaminated soils. Agriculture 13(10):1983
Walton RT, Christie KA, Whittaker MN, Kleinstiver BP (2020) Unconstrained genome targeting with near-PAMless engineered CRISPR-Cas9 variants. Science 368:290–296
Wan TENG, Xue HE, Yi-ping TONG (2017) Transgenic approaches for improving use efficiency of nitrogen, phosphorus and potassium in crops. J Integr Agric 16(12):2657–2673
Wang WC, Lin TC, Kieber J, Tsai YC (2019) Response regulators 9 and 10 negatively regulate salinity tolerance in rice. Plant Cell Physiol 60:2549–2563
Wang J, Vanga SK, Saxena R, Orsat V, Raghavan V (2018) Effect of climate change on the yield of cereal crops: a review. Climate 6(2):41
Wang P, Si H, Li C, Xu Z, Guo H, Jin S, Cheng H (2025) Plant genetic transformation: achievements, current status and future prospects. Plant Biotechnol J 23(6):2034–2058
Wang Y, Frei M (2011) Stressed food–the impact of abiotic environmental stresses on crop quality. Agric Ecosyst Environ 141(3–4):271–286
White PJ, Brown P (2010) Plant nutrition for sustainable development and global health. Ann Bot 105(7):1073–1080
Wang FZ, Chen MX, Yu LJ, Xie LJ, Yuan LB, Qi H, Chen QF (2017) OsARM1, an R2R3 MYB transcription factor, is involved in regulation of the response to arsenic stress in rice. Front Plant Sci 8:1868
Wang B, Zhong Z, Wang X, Han X, Yu D, Wang C, Zhang Y (2020) Knockout of the OsNAC006 transcription factor causes drought and heat sensitivity in rice. Int J Mol Sci 21:2288
Xu X, Qi LS (2019) A CRISPR–dCas toolbox for genetic engineering and synthetic biology. J Mol Biol 431:34–47
Yadav RK, Tripathi MK, Tiwari S, Tripathi N, Asati R, Chauhan S et al (2023) Genome editing and improvement of abiotic stress tolerance in crop plants. Life 13(7):1456
Yadav SK (2010) Heavy metals toxicity in plants: an overview on the role of glutathione and phytochelatins in heavy metal stress tolerance of plants. S Afr J Bot 76(2):167–179

Yang X, Lu M, Wang Y, Wang Y, Liu Z, Chen S (2021) Response mechanism of plants to drought stress. Horticulturae 7(3):50
Yin X, Anand A, Quick P, Bandyopadhyay A (2019) Editing a stomatal developmental gene in rice with CRISPR/Cpf1. In: Qi Y (ed) Plant genome editing with CRISPR systems. Humana Press, New York, NY, pp 257–268
Yue E, Cao H, Liu B (2020) OsmiR535, a potential genetic editing target for drought and salinity stress tolerance in Oryza sativa. Plants 9:1337
Yu J, Du T, Zhang P, Ma Z, Chen X, Cao J et al (2024) Impacts of high temperatures on the growth and development of rice and measures for heat tolerance regulation: a review. Agronomy 14(12):2811
Zeng Y, Wen J, Zhao W, Wang Q, Huang W (2020) Rational improvement of rice yield and cold tolerance by editing the three genes OsPIN5b, GS3, and OsMYB30 with the CRISPR–Cas9 system. Front Plant Sci 10:1663
Zhang A, Liu Y, Wang F, Li T, Chen Z, Kong D, Luo L (2019a) Enhanced rice salinity tolerance via CRISPR/Cas9-targeted mutagenesis of the OsRR22 gene. Mol Breed 39:1–10
Zhang C, Srivastava AK, Sadanandom A (2019b) Targeted mutagenesis of the SUMO protease, Overly Tolerant to Salt1 in rice through CRISPR/Cas9-mediated genome editing reveals a major role of this SUMO protease in salt tolerance. bioRxiv:555706
Zhang D, Zhang Z, Unver T, Zhang B (2021) CRISPR/Cas: a powerful tool for gene function study and crop improvement. J Adv Res 29:207–221
Zhang R, Wang Y, Hussain S, Yang S, Li R, Liu S et al (2022) Study on the effect of salt stress on yield and grain quality among different rice varieties. Front Plant Sci 13:918460
Zhang Y, Chen K, Zhao FJ, Sun C, Jin C, Shi Y, Lian X (2018) OsATX1 interacts with heavy metal P1B-type ATPases and affects copper transport and distribution. Plant Physiol 178:329–344
Zhao P, You Q, Lei M (2019) A CRISPR/Cas9 deletion into the phosphate transporter SlPHO1; 1 reveals its role in phosphate nutrition of tomato seedlings. Physiol Plant 167(4):556–563
Zheng C, Liu C, Liu L, Tan Y, Sheng X, Yu D et al (2023) Effect of salinity stress on rice yield and grain quality: a meta-analysis. Eur J Agron 144:126765
Zia R, Nawaz MS, Siddique MJ, Hakim S, Imran A (2021) Plant survival under drought stress: implications, adaptive responses, and integrated rhizosphere management strategy for stress mitigation. Microbiol Res 242:126626
Zuluaga DL, Sonnante G (2019) The use of nitrogen and its regulation in cereals: structural genes, transcription factors, and the role of miRNAs. Plants 8(8):294

Chapter 7
An Outlook of CRISPR-Mediated Gene Editing in Wheat

Veysel Süzerer, Selin Gül Ünsal, and Engin Tilkat

Abstract CRISPR technology has reshaped plant genetic modification by facilitating targeted and effective changes in defined DNA segments. Wheat, a globally vital staple crop, is expected to gain substantially from these advancements, especially through enhanced productivity, greater tolerance to environmental stresses, and improved nutritional value. This chapter offers a holistic summary of ongoing advancements and cutting-edge implementations of CRISPR-based genome editing in wheat. The chapter elaborates on recent technological progress, experimental strategies, and real-world implementations of CRISPR tools in the context of wheat genetic enhancement. Furthermore, the chapter addresses key obstacles and constraints linked to CRISPR applications and examines existing regulatory frameworks and ethical implications. Lastly, we highlight emerging directions and anticipated innovations in CRISPR research that may accelerate the genetic advancement of wheat.

Keywords CRISPR/Cas9 · Gene editing · Wheat · Molecular breeding · Regulatory landscape

V. Süzerer
Bingöl University, Vocational School of Health Services, Department of Pharmacy Services, Bingöl, Türkiye

S. G. Ünsal
Istanbul Technical University, Faculty of Science and Letters, Department of Molecular Biology and Genetics, Istanbul, Türkiye

E. Tilkat (✉)
Batman University, Faculty of Science and Literature, Department of Biology, Batman, Türkiye
e-mail: engin.tilkat@batman.edu.tr

K. A. Abd-Elsalam, A. Ahmad (eds.), *Gene Editing in Cereals*, Sustainability Sciences in Asia and Africa, https://doi.org/10.1007/978-981-95-6722-5_7

7.1 Introduction

Nutrition is a fundamental requirement for sustaining human life. In today's dynamic world, ensuring adequate and balanced nutrition remains one of the most pressing global challenges. For nations striving toward economic and social development, proper nutrition is a cornerstone for enhancing public health and quality of life (Nalinci 2013). Among staple foods, cereals occupy a central role worldwide, with wheat being especially significant due to its dual use as both a food source and animal feed. Rich in vitamins, minerals, fiber, and carbohydrates, wheat supports digestive health and fulfills a considerable portion of daily caloric intake. Globally, wheat and its derivatives contribute to over one-fifth of average daily energy consumption (Peng et al. 2011a, b; Anonymous 2017).

Domesticated over 10,000 years ago in the Fertile Crescent, wheat has undergone extensive artificial selection to meet diverse agricultural demands (Venske et al. 2019; FAO 2020; De Sousa et al. 2021). Its broad genetic diversity enables successful cultivation across a wide range of climates—from temperate to subtropical—and is supported by traits such as high yield potential and excellent storage capacity (William et al. 2011; Shewry 2018; De Sousa et al. 2021). Although around 300,000 plant species are considered potentially edible, fewer than 100 are widely cultivated, with wheat being among the most prominent. Remarkably, just three cereals—wheat, rice, and maize—collectively supply nearly 60% of global dietary energy, with wheat alone providing around 20% of the world's total calorie and protein intake. As a member of the Poaceae family, wheat shares its botanical classification with other major cereals like maize and rice, yet it continues to be a leading contributor to both energy and protein supply in the human diet (Patniak and Khurana 2001; FAO 2004; Braun et al. 2010).

Wheat is one of the most widely cultivated and consumed cereal grains globally, accounting for approximately 20% of dietary energy intake. Therefore, it plays a critical role in global food security. (Peng et al. 2011a; FAO 2020). According to 2024 FAO statistics, wheat production spans 220 million hectares worldwide, yielding nearly 799 million tons annually. Based on 30-year averages, the top 10 wheat-producing countries are China (115.9 Mt), India (83.1 Mt), the United States (56.8 Mt), Russia (55.1 Mt), France (36.1 Mt), Canada (27.2 Mt), Australia (22.8 Mt), Germany (22.3 Mt), Pakistan (22.2 Mt), and Ukraine (19.9 Mt) (Fig. 7.1; FAO 2024).

Over the last two decades, international trade in agricultural products has grown by 133%, outpacing production and driven largely by major crops such as wheat, rice, maize, and soybean. These four crops alone account for over 60% of global staple food consumption and 28% of total grain demand (FAO 2022; Ji et al. 2024). However, recent global crises have severely disrupted these trends. The COVID-19 pandemic, followed by the Russia–Ukraine conflict, caused the most significant food supply chain disruptions since World War II, impacting nearly two billion people worldwide (Lin et al. 2023). Notably, Russia and Ukraine together supply 24%

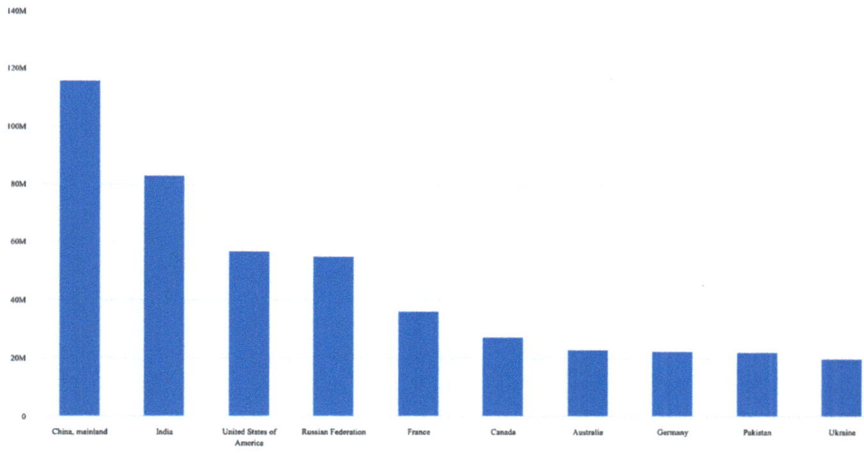

Fig. 7.1 The ten leading wheat-producing countries worldwide, ranked by their average production volume from 1994 to 2023 (FAO 2024)

of global wheat exports, particularly to low-income nations. The escalation of geopolitical tensions has further destabilized global wheat markets and triggered a surge in wheat prices (FAO 2023; Kahraman and Akova 2024).

Wheat's adaptability and high productivity are largely attributable to its rich genetic diversity. However, despite its global significance, wheat breeding continues to face persistent challenges. These include the complexity of its polyploid genome, vulnerability to a wide range of biotic and abiotic stresses, and stagnation in yield gains. To address these constraints, cutting-edge genome editing technologies—particularly CRISPR/Cas systems—have opened new avenues for precision-driven crop improvement.

CRISPR/Cas technology facilitates diverse applications in cereal crops, including yield improvement, stress resilience, and quality enhancement, as illustrated in Fig. 7.2.

This chapter explores the transformative impact of CRISPR-based genome editing on wheat improvement. It provides a comprehensive overview of recent technological advances, experimental strategies, and key applications in CRISPR research on wheat. The chapter also examines the biological, technical, and regulatory barriers associated with genome editing, and addresses the ethical dimensions of these emerging biotechnologies. Finally, it offers forward-looking perspectives and discusses innovative directions for integrating CRISPR/Cas systems into future wheat breeding and development programs.

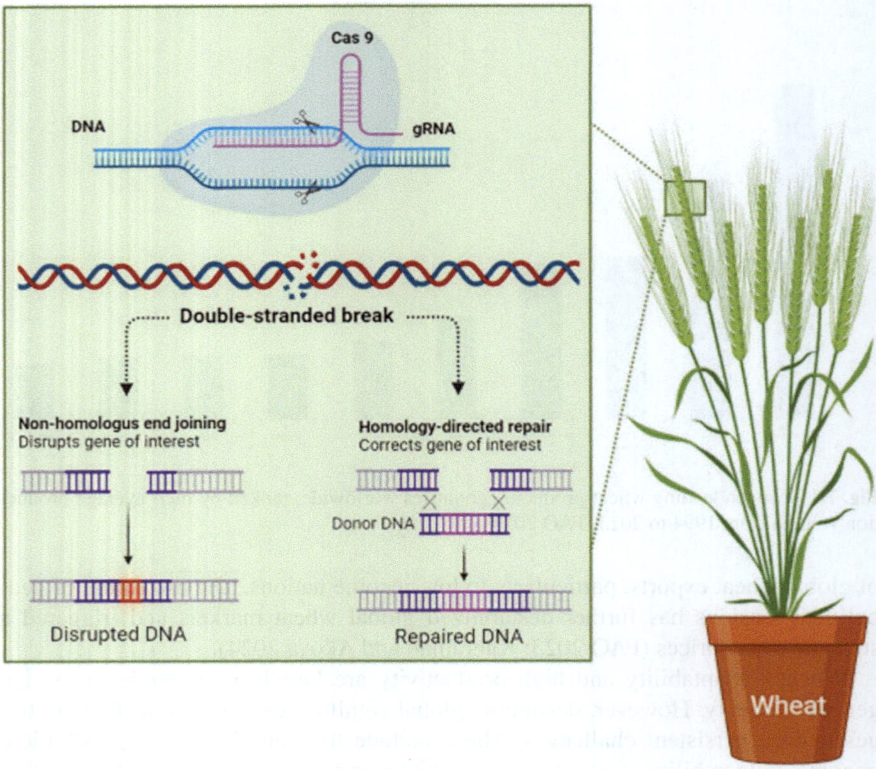

Fig. 7.2 Schematic representation of key applications of genome editing in cereal crops, including yield improvement, stress tolerance, grain quality enhancement, and resistance to diseases. This holistic view highlights the potential of CRISPR/Cas systems to address multiple breeding objectives in wheat. (Adapted from Elsharawy and Refat (2023))

7.1.1 Modern Genome Editing Strategies in Plants: Spotlight on CRISPR/Cas

Unlike classical mutagenesis methods that rely on random mutations, modern genome editing (GE) technologies enable highly specific modifications within plant genomes. These approaches employ programmable DNA-binding modules in combination with nucleases to generate double- or single-stranded DNA cuts at specific loci. The resulting DNA lesions are resolved through endogenous pathways, including non-homologous end joining (NHEJ), typically resulting in indels, or homology-directed repair (HDR), and in some cases, prime editing enables precise sequence substitutions (Wang and Akhunov 2022). Previous generations of gene modification tools, such as ZFNs and TALEN-based systems, laid the groundwork for site-specific genetic alterations; however, their use has been constrained by high development costs, time-intensive design, and limited flexibility (Hillary and Ceasar 2023, 2024). The advent of the CRISPR/Cas system, derived from prokaryotic immune defense, has substantially transformed genome editing through the introduction of a single-guide RNA (sgRNA) mechanism (Doudna and Charpentier 2014). Its ease of design, multiplexing ability, and high efficiency have established CRISPR–Cas systems as the preferred tool in plant functional genomics.

CRISPR/Cas9's discovery and applications build on earlier identification of CRISPR loci in *Escherichia coli* (Ishino et al. 1987), with key insights into their immune function emerging in the early 2000s (Mojica et al. 2005; Pourcel et al. 2005). Jinek et al. (2012) demonstrated that Cas9, guided by synthetic sgRNA, could induce precise site-specific DNA cleavage, sparking rapid adoption in crop science. As shown in Fig. 7.3, the technology's trajectory has included major milestones that have revolutionized gene editing in plants (Shah 2024).

Fig. 7.3 Discovery and development of CRISPR: A look through time. (Modified from Shah 2024)

7.2 Development and Functional Dynamics of CRISPR Mechanisms

7.2.1 Advances in the Emergence of CRISPR Systems: From Past to Present

Genome editing has evolved substantially, with early innovations including zinc finger nucleases (ZFNs) and transcription activator-like effector nucleases (TALENs) paving the way. While effective, these earlier tools were hampered by high costs and design complexities (Hillary and Ceasar 2023). The discovery and subsequent development of CRISPR–Cas systems offered a more accessible, precise, and flexible alternative.

The origins of CRISPR (Clustered Regularly Interspaced Short Palindromic Repeats) trace back to its first observation in *Escherichia coli* by Ishino et al. (1987), noted repetitive DNA sequences interspaced with unique regions. These sequences were later linked to fragments of viral DNA, suggesting a defense mechanism in bacteria and archaea. Over the following years, similar loci were identified in other prokaryotes, including *Shigella dysenteriae*, *Salmonella enterica* (Nakata et al. 1989), and *Mycobacterium tuberculosis* (Hermans et al. 1991). Although the functional role of these repeats was initially unclear, foundational insights began to emerge in the early 2000s.

Mojica et al. (2005), Pourcel et al. (2005), and van der Oost et al. (2009) demonstrated that these sequences and their linked CRISPR-associated (Cas) proteins comprised an adaptive defense mechanism that used short CRISPR-derived RNAs (crRNAs) to direct interference against invading genetic elements. This sequence-specific targeting capability positioned CRISPR as a naturally evolved gene-silencing mechanism.

A landmark breakthrough came in 2012, when Jinek et al. reconstituted the CRISPR–Cas9 complex in vitro using a synthetic single-guide RNA (sgRNA) to program Cas9 nuclease to initiate targeted DNA double-strand cleavage (DSBs). This pivotal advance marked a turning point from a biological curiosity to a transformative genome editing platform, later popularized by Doudna and Charpentier (2014).

Today, CRISPR–Cas architectures are classified into two major groups: Class 1, which employs multiple protein assemblies (e.g., Cas3, Cas10), and Class 2, which uses individual, multidomain proteins (e.g., Cas9, Cas12a, Cas13) that are more advantageous for molecular biology applications (Burstein et al. 2017). Among them, the Class 2 variants—especially Type II (Cas9), Type V (Cas12), and Type VI (Cas13)—have been widely adopted for editing DNA and RNA across a range of organisms, including crops like wheat.

Figure 7.2 summarizes the key scientific milestones that have driven the evolution of CRISPR from its early discovery to its application as a central tool in plant genome engineering (Shah 2024).

7.2.2 Key Components and Mechanisms of CRISPR–Cas Systems

CRISPR–Cas systems serve as an adaptive immune mechanism in prokaryotic organisms, enabling the detection and degradation of invasive genetic elements through RNA-guided endonucleases. Based on the composition and organization of their effector modules, these systems are broadly grouped into two classes: Class 1, characterized by multi-protein effector complexes, and Class 2, defined by the use of a single, multidomain effector protein (Burstein et al. 2017).

In plant biotechnology, Class 2 systems, particularly Type II (Cas9) and Type V (Cas12a, formerly Cpf1), have gained prominence due to their functional simplicity and broad applicability across species (Shan et al. 2013; Reynolds and Braun 2022). These systems enable precise manipulation of plant genomes by directing site-specific modifications, typically by induction double-stranded DNA breaks (DSBs) at user-defined loci.

The CRISPR–Cas9 system comprises three essential components:

1. CRISPR array harbors short repetitive DNA sequences (palindromes) separated by unique spacers derived from prior exposure to foreign DNA.
2. Cas9 protein, a dual nuclease enzyme that generates double-strand breaks at targeted genomic locations.
3. Guide RNA (gRNA) provides sequence specificity by binding to the target DNA and directing Cas9 activity.

In engineered CRISPR systems, a synthetic guide RNA—known as sgRNA—merges the roles of both crRNA and tracrRNA to direct the Cas9 protein to its target site. Cas9 identifies target sequences by binding to a specific 5′-NGG protospacer adjacent motif (PAM) and introduces double-stranded breaks (DSBs) through its dual nuclease domains: RuvC and HNH. These breaks are typically resolved by either the error-prone non-homologous end joining (NHEJ) pathway, which may generate insertions or deletions that impair gene function, or by the homology-directed repair (HDR) pathway when a suitable DNA repair template is present (Schiml et al. 2014; Belhaj et al. 2015; Kumar et al. 2019).

- *CRISPR–Cas12a System:*
- *Cas12a*, another widely used Class 2 effector, differs structurally and functionally from Cas9:
  - It recognizes a *T-rich PAM sequence (typically TTTV).*
  - It introduces *staggered DSBs with 5′ overhangs*, which favor certain repair pathways.
  - Unlike Cas9, Cas12a can process its own crRNAs independently, without requiring a tracrRNA, making it well-suited for multiplex genome editing applications.

These distinctions make Cas12a a valuable alternative to Cas9, particularly in targets where PAM constraints or cleavage patterns are critical. Both systems have demonstrated high efficiency in editing agriculturally important crops such as wheat (Wang and Akhunov 2022).

- *Classification Overview:*
  The classification of CRISPR–Cas systems is summarized in Table 7.1, which outlines the types, signature Cas proteins, and target molecules associated with each subtype.
  This classification provides a framework for selecting the appropriate CRISPR system based on the editing objective, such as whether the target is DNA or RNA, and whether a multi-component or single-effector system is preferred.

7.2.3 Advancements Specific to Wheat Genome Editing

Wheat, a fundamental crop for ensuring global food supply, contributes nearly one-fifth of global caloric and protein requirements (Reynolds and Braun 2022). The two main cultivated forms, tetraploid durum wheat (*Triticum turgidum* ssp. Durum L.) and hexaploid bread wheat (*Triticum aestivum* L.), possess extensive and complex genomes, shaped by historic hybridization events (Smedley et al. 2021).

Given the rising global food needs and the negative impacts of climatic variability, improving both harvest efficiency and quality traits in wheat has become a pressing goal. Modern breeding strategies focus not only on enhancing the nutritional value of wheat but also on minimizing harmful compounds such as toxins and anti-nutrients (Nigro et al. 2024). Despite its complex genome structure, wheat remains amenable to genome editing—offering promising opportunities for innovation beyond what has been achieved in other cereals (Guo et al. 2023).

Gene-editing platforms, particularly CRISPR/Cas9, have remarkably boosted the potential for agronomic enhancement and final product quality in wheat by facilitating highly targeted genomic modifications (Nigro et al. 2024). Since its first application in wheat in 2013, CRISPR/Cas9 has been widely adopted for generating elite, DNA-free cultivars (Elsharawy and Refat 2023). Owing to its precision and functional robustness, it enables the induction of favorable genetic changes without introducing transgenes, making it a valuable asset in modern wheat breeding programs (Haque et al. 2018). In complex genomes such as polyploid wheat, CRISPR/

Table 7.1 CRISPR–Cas system classification

Class	Type	Signature Cas protein	Target
1	I (A–F), III, IV	Cas3, Cas10, Csf1	DNA / RNA
2	II (A–C)	Cas9	DNA
	V (A–E)	Cas12, Cas14	DNA
	VI	Cas13	RNA

Adapted from Reynolds and Braun (2022), Burstein et al. (2017)

Cas9 supports coordinated manipulation of several gene loci, accelerating gene function elucidation and complex trait dissection—ultimately strengthening the efficiency of improvement strategies (Nigro et al. 2024).

Given the high degree of sequence conservation across wheat subgenomes, CRISPR/Cas9 allows the concurrent targeting of multiple homologous genes either separately or simultaneously by employing single or dual-guide RNAs (Smedley et al. 2021). These technological advancements have substantially refined our capacity to characterize and improve genetic traits, leading to the development of superior genotypes with enhanced yield and improved grain quality.

7.3 Methodological Approaches in Wheat Gene Editing

7.3.1 CRISPR Component Delivery of Wheat

Efficient delivery remains a major bottleneck in wheat transformation, primarily because of its inherently recalcitrant genetic background. Numerous approaches have been devised to reliably introduce CRISPR/Cas reagents into wheat cells, allowing for genotype-independent transformation. Zhang et al. (2019a) implemented an optimized Agrobacterium-mediated transfer protocol, successfully producing 68 edited lines with modifications in key grain-yield regulatory loci (TaCKX2-1, TaGLW7, TaGW2, and TaGW8) and reporting a 10% editing success rate without any detectable off-target mutations.

In a related study, Zhang et al. (2018a) employed the CRISPR/Cas9 system in wheat protoplasts and regenerated T0 lines, attaining a 54.17% gene-editing success rate. Building on this, Liu et al. (2020) further validated Agrobacterium delivery methods by assessing the relative performance of SpCas9, LbCpf1, and xCas9, and comparing promoter elements (OsU6a, TaU3, and TaU6) used to regulate sgRNA expression.

For wheat varieties resistant to Agrobacterium, silicon carbide (SiC) whiskers have provided a stable alternative for gene transfer (Kumagai et al. 2022). Plant virus-based vectors have also proven effective for transient protein expression and virus-induced gene silencing (VIGS) in wheat (Rössner et al. 2022; Wang et al. 2020). Gil-Humanes et al. (2017) introduced a replicon-based system using a segmented wheat dwarf virus (WDV), which enhanced reporter gene expression 110-fold and improved editing efficiencies approximately 12-fold compared to non-viral methods.

Virus-based vectors like BSMV and WDV for transient expression are also used for *CRISPR* component delivery in wheat. Barley Stripe Mosaic Virus (BSMV) has emerged as a potent vector for delivering gene-editing reagents into wheat leaves. Wang et al. (2022) demonstrated BSMV-mediated delivery of sgRNAs into Cas9-expressing wheat, enabling heritable promoter deletions and multi-gene editing.

Another study utilizing an engineered BSMV-based sgRNA vector achieved up to 100% gene regulation efficiency in the next generation.

The biolistic method, involving high-speed microprojectile delivery, has also been effective. Ghogare et al. (2021) and Hamada et al. (2018) reported in vivo genome editing via the shoot apical meristem pathway, identifying 11 transgenic lines from 210 wheat plants, with stable transmission of the targeted mutation to the next generation, free of detectable Cas9 and guide RNA.

Another innovative approach under investigation is carbon nanotube-mediated delivery. To overcome the limitations of conventional transformation techniques, researchers have explored nanomaterial-based strategies—particularly the use of carbon nanotube nanoparticles—for delivering CRISPR components into wheat leaves (Demirer et al. 2019a). While these methods offer potential advantages in efficiency and precision, their widespread adoption is still in its early stages. In parallel, alternative techniques such as in planta particle bombardment (iPB) (Liu et al. 2021) and the iPB-ribonucleoprotein (RNP) method (Brauer et al. 2020) have demonstrated considerable promise, offering practical pathways for broader application and potential commercialization.

7.3.2 Target Selection and sgRNA Design for Wheat Genome Editing

Effective CRISPR-mediated genome editing in wheat begins with the precise identification of target sequences and the rational design of single-guide RNAs (sgRNAs). This process involves a series of carefully coordinated steps:

(a) *Target Identification*

Candidate sequences are identified using wheat genome databases such as the URGI Wheat Genome Repository, Ensembl Plants, and Wheat GDB. These resources offer essential genomic annotations, facilitating accurate localization of start codons, exons, and introns—key for directing CRISPR/Cas activity (Tsakirpaloglou et al. 2023).

(b) *Sequence Acquisition*

Transcript and coding sequences of the gene of interest are retrieved to define the target region precisely, ensuring accurate sgRNA design (Tsakirpaloglou et al. 2023).

(c) *Nuclease Selection*

The most commonly used nuclease is SpCas9 from *Streptococcus pyogenes*, known for its broad application and NGG PAM requirement. However, Cas12a (Cpf1) is preferred in thymine-rich regions due to its distinct PAM sequence and cutting pattern (D'Orso et al. 2023).

(d) *sgRNA Design*

sgRNAs—typically 18–21 nucleotides in length—are designed with key parameters in mind: GC content, presence of restriction sites or nucleotide repeats, predicted secondary structure, and off-target potential. Tools such as WheatCRISPR, CRISPOR, CRISPR-P, and Cas-OFFinder streamline this process.

- Target sites are ideally located in conserved exonic regions near the 5′ end to induce frameshift mutations.
- For editing all homeologs, conserved sequences across sub-genomes are aligned to design a universal sgRNA (Kumar et al. 2019; Tsakirpaloglou et al. 2023).
- sgRNAs are then ranked based on predicted efficiency, target suitability, and mutagenic potential.
- Primers flanking the target region are designed to validate edits by PCR and sequencing (Tsakirpaloglou et al. 2023).

(e) *CRISPR–Cas Construct Assembly*

Expression cassettes are constructed using Golden Gate cloning, where synthesized oligonucleotides encoding the sgRNA sequence are flanked by Type IIS restriction sites. This method facilitates modular and efficient assembly of multiple sgRNAs for multiplex editing (Nigro et al. 2024).

(f) *Delivery into Plant Cells*

The delivery of CRISPR constructs into wheat cells is commonly achieved through Agrobacterium-mediated transformation or particle bombardment, both of which facilitate the stable incorporation of editing elements into the plant genome (Zhou et al. 2023; Akanmu et al. 2024).

The direct introduction of ribonucleoprotein (RNP) complexes is widely preferred for generating transgene-free edited plants. This transient delivery method ensures temporary Cas9 activity and significantly reduces the risk of off-target modifications (Liang et al. 2017; Zhou et al. 2023).

(g) *Regeneration and Validation*

Following transformation, callus cultures are induced and regenerated into T_0 plants. Genome edits are validated through PCR and sequencing, and potential off-target mutations are assessed. Confirmed T_0 lines are self-pollinated to generate homozygous T_1 progeny (Zhou et al. 2023; Akanmu et al. 2024).

7.3.3 Mitigating Unintended Off-Target Edits in CRISPR Applications

Although CRISPR/Cas9 is widely praised for its targeting accuracy, unintended edits still pose a significant concern in polyploid crops such as wheat. The Cas9 enzyme may occasionally cut genomic sequences exhibiting homology to the guide RNA, resulting in unplanned alterations. While the likelihood of these off-target changes is generally low-estimated to be under 1% (Graham et al. 2020)—even rare instances can interfere with gene function and influence experimental validity. Encouragingly, the phenotypic risks associated with off-targets can often be reduced via robust screening protocols and confirmation assays (Akanmu et al. 2024).

In polyploid crops such as wheat, which contains A, B, and D subgenomes, the complexity of the genome increases the likelihood of unintended cleavage events. Accurate off-target prediction thus begins with in silico alignment of sgRNA sequences against the wheat reference genome. This step is crucial for distinguishing between target and off-target loci across homeologs (Li et al. 2021a).

For multiplex genome editing, gRNA cassettes—carrying one or more promoters within a single vector—allow simultaneous targeting of multiple genes (Hyun 2020). However, this strategy can exacerbate off-target effects if not precisely optimized, especially in a hexaploid context like wheat.

To minimize such risks, sgRNA sequences should be carefully designed using computational tools, followed by experimental validation. Recommended general-purpose tools include:

- CRISPOR, CRISPR-P, RGEN Cas Designer, and CHOPCHOP—with CRISPOR being particularly versatile across species (Hassan et al. 2021).
  For wheat-specific analyses, the following tools offer high accuracy:
- CRISPR-Cereal, Cas-OFFinder, E-CRISP, CRISPR-Direct, and WheatCRISPR—among which CRISPR-Cereal and Cas-OFFinder are strongly recommended (He et al. 2021).
  Further improvements in targeting specificity can be achieved by:
- Modifying sgRNA architecture: Truncating the 3′ end, shortening the 5′ end by up to three nucleotides, or adding two guanine (G) nucleotides at the 5′ end have been shown to reduce off-target cleavage by up to 5000-fold (Zhang et al. 2015).
- Optimizing the Cas9–sgRNA complex concentration, thereby limiting excessive or prolonged activity.

Moreover, modified versions of Cas9 enzymes—such as eSpCas9, SpCas9-HF1, and LZ3—have been engineered to lower their affinity for non-target DNA sites, thereby increasing editing accuracy and potentially enhancing the efficiency of homology-directed repair (HDR) mechanisms (Matsumoto et al. 2024).

However, even with high-fidelity variants like SpCas9–HF1, the complexity of wheat's polyploid genome continues to pose significant challenges for eliminating off-target mutations entirely. As Guo et al. (2023) highlight, unintended edits in such repetitive genomes can still affect critical regulatory regions, leading to unpredictable phenotypic effects. This underscores the necessity of combining high-fidelity enzymes with rational gRNA design and validation protocols.

One of the most effective strategies for limiting off-target activity is the delivery of Cas9–sgRNA as ribonucleoprotein (RNP) complexes. This transient, DNA-free approach alter Cas9 expression and duration of exposure (Wienert and Cromer 2022).

In summary, combining rational sgRNA design, advanced Cas9 variants, computational prediction tools, and transient delivery methods enables precise and efficient genome editing in wheat while significantly minimizing off-target risks.

Approaches to Enhance Targeting Specificity in CRISPR Systems:

- Use of high-fidelity Cas9 variants (eSpCas9, SpCas9–HF1)
- RNP-based delivery of CRISPR components (Liang et al. 2017)
- In silico sgRNA design with tools like CRISPOR and Cas-OFFinder, followed by experimental validation

7.4 CRISPR Applications in Wheat Breeding

7.4.1 Yield and Biomass Enhancement

The rising global population continues to drive increasing food demand, necessitating improvements in wheat productivity. Enhancing morphological traits such as grain weight, length, and width has become a focal point for boosting yield. Genetic manipulation offers a promising avenue to achieve these improvements. Early research identified three TaGW2 homoeologs in hexaploid wheat (Qin et al. 2014), with subsequent studies confirming that TaGW2-A1 influences grain weight, length, and width, while TaGW2-B1 and -D1 primarily affect grain weight (Simmonds et al. 2016). Using CRISPR/Cas9, Zhang et al. (2018b) investigated TaGW2-B1 and -D1, generating knockout mutants that demonstrated significant effects on thousand-grain weight and grain protein content.

Similarly, editing TaGW7—homologous to rice OsGW7—via CRISPR/Cas9 revealed its dosage-dependent role in grain morphology. Mutations in the B and D genome homoeologs increased grain width and weight but reduced grain length. CRISPR/Cas9-mediated editing of TaGASR7 using ribonucleoproteins (RNPs) further confirmed its role in regulating grain length (Liang et al. 2017).

In another approach, SQUAMOSA promoter-binding protein-like (SPL) genes have been targeted to improve yield. Gupta et al. (2023) disrupted microRNA 156 recognition sites in the TaSPL13 gene using CRISPR/Cas9, enhancing grain size and number through increased TaSPL13 expression. Efforts to boost nutrient use efficiency have also shown promise; Zhang et al. (2021a) developed TaARE1 mutants with improved nitrogen use efficiency, leading to higher yields and delayed senescence. Furthermore, Song et al. (2023) advanced grain yield potential by developing semi-dwarf wheat varieties through simultaneous knockout of Rht-B1 and ZnF-B, encoding a RING-type E3 ligase.

In sum, CRISPR/Cas9 genome editing has proven to be a powerful tool to enhance yield and biomass in wheat, offering precise and efficient pathways to meet rising global food demands.

Moreover, CRISPR/Cas9 has been effectively employed to enhance wheat's resistance to yellow mosaic virus caused by *Polymyxa graminis* through the targeted modification of the TaPDIL5-1 gene in hexaploid wheat (Zhang et al. 2021b). Furthermore, suppression of the TaCIPK14 gene has increased wheat's resistance to stripe rust (He et al. 2023). To combat herbicide resistance, the TaALS gene has been edited to produce herbicide-resistant wheat varieties (Zhang et al. 2019b).

In addition, biolistic-mediated and CRISPR/Cas-based gene editing have generated mutants with suppressed S genes, improving resistance to biotic stress in bread wheat (Taj et al. 2022). The CRISPR/Cas9-guided knockout of TaeIF4E has also enhanced resistance to wheat yellow mosaic virus (Kan et al. 2023). Finally, CRISPR-edited TaPHT1;9 wheat mutants with increased arsenate tolerance have been developed to address arsenic contamination in groundwater and soil, improving both health and yield (Wang et al. 2023).

7.4.2 Resistance to Biotic Stresses

Biotic stresses, including pathogens (such as viruses, bacteria, fungi, and nematodes), along with parasitic weeds, significantly impair wheat health, growth, and yield (Glick and Gamalero 2021). CRISPR/Cas9 has played a transformative role in accelerating the breeding of disease-resilient wheat by facilitating precise gene-level modifications.

For example, powdery mildew resistance (*Blumeria graminis* f. sp. tritici) has been accomplished through targeted disruption of six MLO genes (Wang et al. 2014), while editing of TaEDR1 has further improved defense capacity (Zhang et al. 2017). Silencing of TaLpx and TaLox, two lipoxygenase genes, has led to increased resistance to *Fusarium graminearum* (Nalam et al. 2015), and inactivation of NFXL1 has similarly enhanced Fusarium head blight resistance (Brauer et al. 2020).

CRISPR-based strategies have also been applied to combat viral pathogens. For instance, editing the TaPDIL5-1 gene has improved resistance against wheat yellow mosaic virus transmitted by *Polymyxa graminis* (Zhang et al. 2021b), while editing of TaCIPK14 was shown to enhance resistance to stripe rust (He et al. 2023). Moreover, targeting the TaALS gene has supported the generation of wheat lines tolerant to herbicides (Zhang et al. 2019b).

Furthermore, both biolistic and CRISPR-mediated interventions have been explored to reduce pathogen-related susceptibility in wheat (Taj et al. 2022). These approaches have broadened resistance against multiple pathogens. As a case in point, genome editing of the *TaEIF4E* gene using the CRISPR/Cas9 system has been shown to enhance viral resistance in wheat, notably improving tolerance to yellow mosaic virus (Kan et al. 2023). In addition, CRISPR-driven improvements in genes such as TaPHT1;9 have yielded arsenate-tolerant wheat genotypes, providing agronomic and food safety advantages by limiting arsenic accumulation in plant tissues and soil (Wang et al. 2023).

7.4.3 Improving Wheat Resilience to Abiotic Stressors

Abiotic stress conditions—such as drought, salinity, extreme heat or cold, and toxic levels of heavy metals—present major challenges to wheat farming by leading to substantial reductions in crop productivity. Plants react to these environmental pressures through intricate signaling cascades and induction of defense-related genes. CRISPR/Cas9 has gained attention as a powerful approach for enhancing abiotic stress resistance by enabling targeted editing of key regulatory elements (Imran et al. 2021).

Experiments conducted in wheat protoplasts involving CRISPR-based editing of transcriptional regulators TaDREB2 and TaERF3 have shown notable improvements in drought tolerance and related abiotic stress responses (Kim et al. 2018). Abdallah et al. (2025) successfully edited five TaSal1 homologs in the drought-sensitive Giza 168 variety, achieving a 34% editing efficiency. The resulting mutants exhibited enhanced drought tolerance, with improved germination and growth compared to wild-type plants, and the mutations were stably inherited.

Another critical target is the TaMBF1c gene, known for its upregulation under multiple abiotic stress conditions. Overexpression of this gene, enabled via CRISPR/Cas9 editing, significantly enhanced drought and stress tolerance in wheat (Tian et al. 2022; Yadav et al. 2022).

Taken together, these studies highlight the potential of CRISPR/Cas9 to produce wheat varieties resilient to environmental stresses, contributing to sustainable agriculture under increasingly variable climatic conditions.

7.4.4 Improving Nutritional Quality and Food Safety

A globally consumed staple food, accounting for approximately 30% of total cereal consumption, wheat plays a central role in human nutrition by providing energy, starch, protein, vitamins, fiber, and phytochemicals. In addition, it is rich in a variety of macro and micronutrients compared to other cereal crops (Islam et al. 2024).

Improving its nutritional value and reducing allergenic or harmful compounds have become priority goals in wheat breeding.

One significant advance targets gluten-related disorders. The α-gliadin protein family is implicated in celiac disease. Using CRISPR/Cas9, Sánchez-León et al. (2018) edited conserved regions across α-gliadin genes, generating 21 mutant lines with drastically reduced α-gliadin content and up to 85% less immunoreactivity. To reduce free asparagine—a precursor of the potentially carcinogenic compound acrylamide formed during high-temperature food processing—Raffan et al. (2021) knocked out the TaASN2 gene, achieving a marked decline in free asparagine levels. Similarly, Camerlengo et al. (2020) used multiplex CRISPR/Cas9 editing to disrupt WTAI-CM3 and WTAI-CM16, two subunits of α-amylase/trypsin inhibitors, reducing wheat allergenicity.

Enhancements in dough and starch quality have also been achieved. Sun et al. (2022) edited TaXip homologs (TaXip-6A, -6B, -6D), resulting in improved dough performance. Li et al. (2021b) targeted TaSBEIIa to increase amylose and resistant starch content, improving the nutritional profile and glycemic properties of wheat.

7.5 Challenges and Limitations of CRISPR Applications in Wheat

7.5.1 Technical Barriers

The inception of genome modification technologies can be traced back to the advent of zinc finger nucleases (ZFNs) in the mid-1990s (Kim et al. 1996). While somewhat effective, ZFNs present notable design challenges, particularly in AT-rich genomic sequences, and exhibit susceptibility to unintended genomic alterations. TALENs (Transcription activator-like effector nucleases), developed from Xanthomonas-derived effectors, offered improvements over ZFNs by providing enhanced specificity and customizable targeting (Jankele and Svoboda 2014). Nonetheless, TALEN-based systems also encounter limitations, including reduced activity on methylated DNA, thymine dependency, and issues with suboptimal delivery efficiency (Ansari et al. 2020).

The advent of CRISPR/Cas9 marked a transformative step in plant biotechnology with a 30% increase since 2014 and the relevance of gene editing has increased significantly since 2022 (Aljabali et al. 2024). Despite its success, several technical hurdles still limit its efficiency in wheat. These include the difficulty of delivering CRISPR components into wheat cells, genotype-dependent tissue culture response, and low regeneration efficiency.

Current delivery methods, such as Agrobacterium-mediated transformation and biolistic bombardment, are widely used but inefficient in wheat. Agrobacterium transformation suffers from a limited host range and low infectivity in many wheat varieties. Biolistic methods, while more broadly applicable, often result in random DNA integration and may cause genomic damage. Additionally, wheat requires longer regeneration periods, making the entire transformation process more time-consuming and labor-intensive.

7.5.2 Biological Constraints

The complex biology of wheat presents several inherent challenges to the effective use of CRISPR–Cas systems. As an allohexaploid species (AABBDD), wheat contains three closely related subgenomes, each potentially harboring up to six homeologs of a single gene. This redundancy makes it difficult to achieve complete gene knockouts, as edits in one or two copies may be compensated by the remaining functional copies (Zhang et al. 2019a). Therefore, obtaining a true loss-of-function phenotype often requires simultaneous and precise editing across all homeologs.

Another major limitation involves off-target activity, where unintended DNA modifications may arise at genomic loci sharing sequence homology with the guide RNA. Although advances in engineered high-fidelity Cas9 enzymes (e.g., SpCas9–HF1) have enhanced targeting precision, achieving absolute specificity remains a challenge—particularly within the context of wheat's highly repetitive and polyploid genome (Guo et al. 2023). Off-target mutations may impact essential regulatory elements or unrelated genes, potentially causing unintended phenotypic outcomes.

Moreover, tissue culture and plant regeneration, which are integral to CRISPR workflows in wheat, are prone to introducing somaclonal variation—spontaneous genetic or epigenetic changes unrelated to genome editing. These background mutations can confound experimental results or reduce the uniformity of edited lines. The regeneration efficiency is also highly genotype-dependent, limiting the applicability of established protocols across diverse wheat cultivars (Neres and Wright 2024).

Beyond technical issues, edited genes may have pleiotropic effects. For example, a gene targeted for improving stress resistance might also influence yield-related traits, growth, or flowering time, leading to unexpected trade-offs. Additionally, environmental interactions may further alter the expression or phenotypic impact of edits (Neres and Wright 2024).

In sum, biological limitations such as genome complexity, off-target risks, regeneration bottlenecks, pleiotropic effects, and instability of edits present significant challenges. Addressing these requires continued advances in multiplex editing, genotype-independent transformation systems, and long-term field validation.

7.5.3 Socio-Economic and Regulatory Hurdles

While CRISPR–Cas9 presents transformative potential for wheat improvement, its widespread application faces substantial socio-economic and regulatory challenges.

7.5.3.1 Legal Barriers and Disparities in Access

The CRISPR–Cas9 technology is governed by a complex and fragmented intellectual property (IP) landscape. Patents are held primarily by a few academic institutions and large corporations, creating barriers for public-sector breeding programs and researchers in low- and middle-income countries. High licensing fees, legal uncertainties, and restricted access hinder innovation and limit the equitable use of this technology (Glenna 2023). Such IP concentration risks widening global disparities in seed development, potentially excluding smallholder farmers from the benefits of genome editing (Pixley et al. 2022; Beumer and de Roij 2023).

7.5.3.2 Public Perception and Market Acceptance

Public attitudes toward gene-edited crops often mirror those toward GMOs, despite scientific distinctions. Misunderstanding of the technology, coupled with concerns over food safety, ecological risks, and corporate control, fuels skepticism—especially in Europe and parts of Asia. Surveys indicate that although consumers support transparency, many feel that current labeling practices are insufficient (Ewa et al. 2022). Without effective science communication and public engagement, CRISPR applications in wheat may struggle to gain broad societal acceptance.

7.5.3.3 Regulatory Discrepancies Across Countries

Global inconsistencies in regulatory policies present another obstacle. While the United States excludes certain CRISPR-modified crops from GMO oversight, the European Union treats all such organisms as subject to strict GMO regulations (Court of Justice of the EU 2018). Nations like Japan and Argentina pursue more lenient frameworks, whereas countries such as India and China are still in the process of formulating their regulatory systems. This regulatory fragmentation complicates international collaboration, delays commercialization, and increases costs (Kalaitzandonakes et al. 2022).

7.5.3.4 Capacity Gaps in Developing Regions

The successful adoption of CRISPR also depends on institutional capacity—access to genomic tools, trained personnel, laboratory infrastructure, and integration with traditional breeding programs. Many regions lack these resources, limiting their participation in advanced genome editing initiatives. Building interdisciplinary collaborations that bridge molecular biology, plant breeding, policy, and social science is crucial to ensuring that CRISPR technologies address both agronomic needs and societal goals (Verdezoto-Prado et al. 2025).

7.5.4 Regulatory and Ethical Considerations

The widespread deployment of CRISPR–Cas technologies in agriculture has triggered significant discussions around regulation, ethics, and societal implications. While the potential benefits—such as enhanced yield, stress tolerance, and nutritional quality—are clear, their realization depends largely on coherent legal frameworks and public acceptance.

7.5.4.1 Diverse Global Regulatory Landscapes

Regulations for gene-edited crops vary considerably among countries, often depending on whether foreign DNA is introduced. Many countries distinguish between different types of genome editing based on site-directed nuclease (SDN) classifications:

- *SDN-1:* Small insertions/deletions without a repair template.
- *SDN-2:* Precise edits using a short repair template.
- *SDN-3:* Larger insertions involving foreign DNA.

A comparison of international regulations is shown below (Table 7.2):

Table 7.2 Regulatory status of SDN-1, SDN-2, and SDN-3 genome editing techniques across selected countries

Country	Regulatory status	SDN-1	SDN-2	SDN-3	References
United States	USDA, APHIS, FDA, EPA Certain CRISPR-edited plants are exempt from regulatory requirements.	DRG	DRG	CBC	Ahmad et al. (2021), Turnbull et al. (2021)
Canada	Health Canada and Canadian Food Inspection Agency (CFIA). Some CRISPR-edited crops are excluded from regulatory requirements.	CBC-N	CBC-N	CBC-N	Ahmad et al. (2021)
European Union	European Court of Justice. Regulation applies if crop could not be produced conventionally.	REG	REG	REG	Zimny et al. (2019), Menz et al. (2020)
China	MARA and NBC. Some CRISPR-edited crops are excluded.	UD	UD	UD	Wong and Chan (2016)
Japan	MAFF, MHLW, MOE. Some CRISPR-edited crops are excluded.	DRG	DRG	DRG	Kasai and Oshawa (2021), Menz et al. (2020)
Australia	FSANZ	DRG	DRG	DRG*	Menz et al. (2020)
Argentina	NRG	DRG	DRG	DRG	Whelan and Lema (2019)
Brazil	CTNBio	DRG	DRG	DRG*	Gatica-Arias (2020)
India	Ministry of Science and Technology, GEAC	UD	UD	UD	Turnbull et al. (2021)
New Zealand	EPA, FSANZ	REG	REG	REG	Turnbull et al. (2021)
Pakistan	National Biosafety Committee	UD	UD	UD	Babar et al. (2020)
Chile	Ministry of Agriculture's SAG	DRG	DRG	DRG*	Turnbull et al. (2021)

Abbreviations: DRG Deregulated, *REG* Regulated, *CBC* Case-by-case, *CBC-N* Case-by-case by novelty, *UD* Under Development, *DRG** Deregulated (if not transgenic), *NCR* Not Currently Regulated

These varying regulatory approaches create uncertainty for international trade and collaboration. Harmonizing science-based policies would help promote both innovation and biosafety.

7.5.4.2 Ethical Concerns and Governance

The application of CRISPR in agriculture prompts critical ethical considerations, such as:

- *Ecological impacts:* Potential unintended effects on biodiversity or non-target species.
- *Equity issues:* Concerns over the monopolization of gene-editing technologies by corporations.
- *Consent and transparency:* The right of consumers to know whether their food has been gene-edited.

Harfouche et al. (2021) emphasize the importance of transparent risk assessment, participatory policymaking, and multi-stakeholder engagement. Similarly, Ewa et al. (2022) highlight that ethical frameworks should reflect cultural values and food security needs across different regions.

7.5.4.3 Public Perception and Market Dynamics

Public understanding of CRISPR is limited, and many consumers confuse gene editing with older GMO technologies. In Turkey, for example, 80% of respondents support GMO labeling, but 65% consider current packaging information inadequate (Ewa et al. 2022). In South Korea, only 36% are willing to purchase gene-edited fruits, compared to 70% of respondents in Pakistan who emphasize the importance of proper labeling.

These findings suggest that improved communication, transparent labeling policies, and inclusive outreach are critical for market acceptance of CRISPR-edited crops.

7.5.4.4 Biosecurity and International Policy Harmonization

CRISPR technology also raises concerns related to dual-use risks and biosecurity. Baltimore et al. (2015) and the U.S. Senate Intelligence Committee have called for global oversight mechanisms to address potential misuse. De Lecuona et al. (2017) argue for embedding ethical safeguards and transparency into all stages of genome editing regulation and research.

7.6 Future Prospects and Potential Breakthroughs

The swift evolution of CRISPR–Cas technologies continues to redefine the frontiers of plant biotechnology. Several innovative directions are emerging that promise to enhance agricultural productivity, environmental sustainability, and resilience to climate change.

7.6.1 Precision Trait Engineering and Multiplex Editing

Recent advances in CRISPR technologies have made it possible to edit multiple genomic sites concurrently, opening new possibilities for the rapid improvement of complex traits such as drought tolerance, pathogen resistance, and nutrient uptake efficiency within a single breeding cycle. Additionally, next-generation editing tools—including base editors and prime editing systems—provide increased accuracy and help reduce unintended genetic alterations (Anzalone et al. 2019).

In polyploid crops like wheat, achieving efficient and stable edits across multiple homoeologous loci remains a major technical challenge. To address this, sequence-specific nucleases (SSNs) are being refined for improved targeting and reduced off-target activity. May et al. (2023) highlighted the progress and remaining barriers in deploying SSNs in polyploid species, emphasizing the importance of integrating advanced guide design, repair template optimization, and delivery innovations to achieve reliable trait engineering.

7.6.2 Synthetic Biology and De Novo Domestication

The use of wild species in traditional breeding remains a major challenge. The combination of CRISPR tools with synthetic biology has led to the emergence of de novo domestication strategies. These approaches open a new avenue for adapting wild species to modern agricultural needs, meeting specific breeding objectives, and rapidly transforming them into climate-resistant crops (Zhang et al. 2023).

7.6.3 Epigenome Editing and Gene Regulation

Emerging CRISPR-based systems, such as dead Cas9 (dCas9) fused with transcriptional regulators, allow modulation of gene activity without introducing permanent changes to the DNA sequence. This transient and reversible regulation of traits could offer a more acceptable path under certain regulatory regimes while providing fine-tuned control over complex plant phenotypes (Papikian et al. 2019).

7.6.4 Delivery Technologies

A key limitation in plant genome modification is achieving reliable and genotype-independent transport of CRISPR-based components. Innovative strategies such as nanoparticle-mediated delivery, viral vectors, and RNP complexes provide transgene-free editing, increasing the potential for regulatory acceptance and field-level implementation (Demirer et al. 2019b).

7.6.5 Artificial Intelligence and Predictive Design

Thanks to the development of new computational approaches with AI-powered platforms and deep learning, the design accuracy and rapid high-yield acquisition of crRNAs, which are critical for the CRISPR/Cas system, are being facilitated. In the studies conducted, the generated crRNA datasets have achieved superior performances in terms of specificity and excellent results in activity determination tests (Lee 2023; Sapoval et al. 2022; Greener et al. 2022).

7.7 Conclusion

Integrating CRISPR/Cas9 tools into wheat breeding has significantly transformed the field by allowing targeted, adaptable, and diverse genomic alterations. This has facilitated genotype-to-phenotype optimization and strengthened data-informed approaches in contemporary crop development (Li et al. 2021c). Through this strategy, major agronomic features—such as yield potential, disease resistance, abiotic stress tolerance, grain quality, and even allergen reduction—can be selectively enhanced. For instance, editing disease-susceptibility loci such as *MLO* and *EDR1* has enhanced tolerance against fungal pathogens (Zhang et al. 2021b), while modifications to *TaGW2*, *gliadin*, and *TaXip* genes have led to superior grain quality and reduced allergenic potential (Lyzenga and Kagale 2021; Elsharawy and Refat 2023).

In addition, CRISPR has enabled yield improvements through the manipulation of genes associated with grain size and nitrogen use efficiency. The technology has also facilitated the development of low-gluten wheat, opening possibilities for allergen-friendly food products. Beyond these trait-specific advances, CRISPR is reshaping the breeding paradigm itself—shifting it from phenotype-based selection to gene-targeted, data-driven improvement. Integration with omics data, non-transgenic editing techniques, and high-throughput phenotyping platforms is making wheat improvement more precise and efficient.

Nevertheless, unlocking the complete potential of CRISPR in wheat breeding demands overcoming technical limitations, particularly those related to genome complexity, transformation efficiency, and regulatory hurdles. Future progress will depend on the integration of AI-driven gene discovery, open-access innovation ecosystems, and inclusive regulatory frameworks (Rafiei et al. 2024; Farooq et al. 2024). With responsible application and equitable access, CRISPR technology promises to revolutionise wheat production—contributing significantly to global food security and sustainable agriculture.

References

Abdallah NA, Elsharawy H, Abulela HA, Thilmony R, Abdelhadi AA, Elarabi NI (2025) Multiplex CRISPR/Cas9-mediated genome editing to address drought tolerance in wheat. GM Crops Food 16(1):1–17

Ahmad A, Munawar N, Khan Z, Qusmani AT, Khan SH, Jamil A, Ashraf S, Ghouri MZ, Aslam S, Mubarik MS, Munir A, Sultan Q, Abd-Elsalam KA, Qari SH (2021) An outlook on global regulatory landscape for genome-edited crops. Int J Mol Sci 22(21):11753

Akanmu AO, Asemoloye MD, Marchisio MA, Babalola OO (2024) Adoption of CRISPR-Cas for crop production: present status and future prospects. PeerJ 12:e17402

Aljabali AA, El-Tanani M, Tambuwala MM (2024) Principles of CRISPR-Cas9 technology: advancements in genome editing and emerging trends in drug delivery. J Drug Deliv Sci Technol 92:105338

Anonymous (2017) Global wheat research. CIMMYT. http://www.cimmyt.org/global-wheat-research/ Accessed 11 August 2023

Ansari WA, Chandanshive SU, Bhatt V, Nadaf AB, Vats S, Katara JL, Sonah H, Deshmukh R (2020) Genome editing in cereals: approaches, applications and challenges. Int J Mol Sci 21(11):4040

Anzalone AV, Randolph PB, Davis JR, Sousa AA, Koblan LW, Levy JM, Chen PJ, Wilson C, Newby GA, Raguram A, Liu DR (2019) Search-and-replace genome editing without double-strand breaks or donor DNA. Nature 576(7785):149–157

Babar U, Nawaz MA, Arshad U, Azhar MT, Atif RM, Golokhvast KS, Tsatsakis AM, Shcerbakova K, Chung G, Rana IA (2020) Transgenic crops for the agricultural improvement in Pakistan: a perspective of environmental stresses and the current status of genetically modified crops. GM Crops & Food 11(1):1–29

Baltimore D, Berg P, Botchan M, Carroll D, Charo RA, Church G, Corn JE, Daley GQ, Doudna JA, Fenner M, Greely HT, Jinek M, Martin GS, Penhoet E, Puck J, Sternberg SH, Weissman JS, Yamamoto KR (2015) A prudent path forward for genomic engineering and germline gene modification. Science 348(6230):36–38

Belhaj K, Chaparro-Garcia A, Kamoun S, Patron NJ, Nekrasov V (2015) Editing plant genomes with CRISPR/Cas9. Curr Opin Biotechnol 32:76–84

Beumer K, de Roij S (2023) Inclusive innovation in crop gene editing for smallholder farmers: status and approaches. Elementa: Sci Anthrop 11(1)

Brauer EK, Balcerzak M, Rocheleau H, Leung W, Schernthaner J, Subramaniam R, Ouellet T (2020) Genome editing of a deoxynivalenol-induced transcription factor confers resistance to *Fusarium graminearum* in wheat. Mol Plant-Microbe Interact 33(3):553–560

Braun H, Atlin G, Payne T (2010) Multi-location testing as a tool to identify plant response to global climate change. In: Reynolds M (ed) Climate change and crop production. CABI, UK, pp 115–138

Burstein D, Harrington LB, Strutt SC, Probst AJ, Anantharaman K, Thomas BC, Doudna JA, Banfield JF (2017) New CRISPR–Cas systems from uncultivated microbes. Nature 542(7640):237–241

Camerlengo F, Frittelli A, Sparks C, Doherty A, Martignago D, Larré C, Lupi R, Sestili F, Masci S (2020) CRISPR-Cas9 multiplex editing of the α-amylase/trypsin inhibitor genes to reduce allergen proteins in durum wheat. Front Sustain Food Syst 4:104

CJEU (Court of Justice of the European Union) (2018) Judgment of the Court (Grand Chamber) 25 July 2018 in Case C-528/16. http://www.curia.europa.eu/juris/liste.jsf?language=en&td=ALL&num=C-528/16. Accessed 12 Sept 2020

D'Orso F, Forte V, Baima S, Possenti M, Palma D, Morelli G (2023) Methods and techniques to select efficient guides for CRISPR-mediated genome editing in plants. In: Ricroch A, Eriksson D, Miladinović D, Sweet J, Van Laere K, Woźniak-Gientka E (eds) A roadmap for plant genome editing. Springer Nature Switzerland, pp 89–117

De Lecuona I, Casado M, Marfany G, Baroni ML, Escarrabill M (2017) Gene editing in humans: towards a global and inclusive debate for responsible research. Yale J Biol Med 90(4):673-681

De Sousa T, Ribeiro M, Carolina S, Gilberto I (2021) The 10,000-year success story of wheat! Foods 10(9):2124

Demirer GS, Zhang H, Goh NS, González-Grandío E, Landry MP (2019a) Carbon nanotube–mediated DNA delivery without transgene integration in intact plants. Nat Protoc 14:2954–2971

Demirer GS, Zhang H, Matos JL, Goh NS, Cunningham FJ, Sung Y, Chang R, Aditham AJ, Chio L, Cho MJ, Staskawich B, Landry MP (2019b) High aspect ratio nanomaterials enable delivery of functional genetic material without DNA integration in mature plants. Nat Nanotechnol 14(5):456–464

Doudna JA, Charpentier E (2014) The new frontier of genome engineering with CRISPR-Cas9. Science 346:1258096

Elsharawy H, Refat M (2023) CRISPR/Cas9 genome editing in wheat: enhancing quality and productivity for global food security—a review. Funct Integr Genomics 23:265

Ewa WG, Agata T, Milica P, Anna B, Dennis E, Nick V, Godelieve CS, Naghmeh A, Tomasz T (2022) Public perception of plant gene technologies worldwide in the light of food security. GM Crops Food 13(1):218–231

FAO (2004) What is agrobiodiversity? Fact sheet. Building on gender, agrodiversity and local knowledge. Food and Agriculture Organization of the United Nations

FAO (2022) The state of agricultural commodity markets 2022. Food and Agriculture Organization of the United Nations. https://www.fao.org/publications/soco/2022/en/

FAOSTAT (2020) Food and Agriculture Organization. http://www.fao.org/faostat/en/#home

FAOSTAT (2023) Food and Agriculture Organization. http://www.fao.org/faostat/en/#home

FAOSTAT (2024) Food and Agriculture Organization. http://www.fao.org/faostat/en/#home

Farooq MA, Gao S, Hassan MA, Huang Z, Rasheed A, Hearne S, Prasanna B, Li X, Li H (2024) Artificial intelligence in plant breeding. Trends Genet 40(10):891–908

Gatica-Arias A (2020) The regulatory current status of plant breeding technologies in some Latin American and the Caribbean countries. Plant Cell Tiss Organ Cult 141:229–242

Ghogare R, Ludwig Y, Bueno GM, Slamet-Loedin IH, Dhingra A (2021) Genome editing reagent delivery in plants. Transgenic Res 30:321–335

Gil-Humanes J, Wang Y, Liang Z, Shan Q, Ozuna CV, Sánchez-León S, Baltes NJ, Starker C, Barro F, Gao C, Voytas DF (2017) High-efficiency gene targeting in hexaploid wheat using DNA replicons and CRISPR/Cas9. Plant J 89(6):1251–1262

Glenna LL (2023) Are intellectual property policies for gene-edited crops fit for purpose? The perspectives of German scientists. Plants, People, Planet 5(6):976–984

Glick BR, Gamalero E (2021) Recent developments in the study of plant microbiomes. Microorganisms 9(7):1533

Graham N, Patil GB, Bubeck DM, Dobert RC, Glenn KC, Gutsche AT, Kumar S, Lindbo JA, Maas L, May GD, Vega-Sanches ME, Stupar RM, Morrell PL (2020) Plant genome editing and the relevance of off-target changes. Plant Physiol 183(4):1453–1471

Greener JG, Kandathil SM, Moffat L, Jones DT (2022) A guide to machine learning for biologists. Nat Rev Mol Cell Biol 23:40–55

Guo C, Ma X, Gao F, Guo Y (2023) Off-target effects in CRISPR/Cas9 gene editing. Front Bioeng Biotechnol 11:1143157

Gupta A, Hua L, Zhang Z, Yang B, Li W (2023) CRISPR-induced miRNA156-recognition element mutations in *TaSPL13* improve multiple agronomic traits in wheat. Plant Biotechnol J 21:536–548

Hamada H, Liu Y, Nagira Y, Miki R, Taoka N, Imai R (2018) Biolistic-delivery-based transient CRISPR/Cas9 expression enables *in planta* genome editing in wheat. Sci Rep 8:14422

Haque E, Taniguchi H, Hassan MM, Bhowmik P, Karim MR, Śmiech M, Zhao K, Rahman M, Islam T (2018) Application of CRISPR/Cas9 genome editing technology for the improvement of crops cultivated in tropical climates: recent progress, prospects, and challenges. Front Plant Sci 9:617

Harfouche AL, Petousi V, Meilan R, Sweet J, Twardowski T, Altman A (2021) Promoting ethically responsible use of agricultural biotechnology. Trends Plant Sci 26(6):546–559

Hassan MM, Zhang Y, Yuan G, De K, Chen JG, Muchero W, Tuskan GA, Qi Y, Yang X (2021) Construct design for CRISPR/Cas-based genome editing in plants. Trends Plant Sci 26(11):1133–1152

He C, Liu H, Chen D, Xie WZ, Wang M, Li Y, Gong X, Yan W, Chen LL (2021) CRISPR-cereal: a guide RNA design tool integrating regulome and genomic variation for wheat, maize and rice. Plant Biotechnol J 19(11):2141

He F, Wang C, Sun H, Tian S, Zhao G, Liu C, Wan C, Guo J, Huang X, Zhan G, Yu X, Kang Z, Guo J (2023) Simultaneous editing of three homoeologues of *TaCIPK14* confers broad-spectrum resistance to stripe rust in wheat. Plant Biotechnol J 21:354–368

Hermans PW, Van Soolingen D, Bik EM, De Haas PE, Dale JW, Van Embden JD (1991) Insertion element IS987 from *Mycobacterium bovis* BCG is located in a hot-spot integration region for insertion elements in *Mycobacterium tuberculosis* complex strains. Infect Immun 59(8):2695–2705

Hillary VE, Ceasar SA (2023) A review on the mechanism and applications of CRISPR/Cas9/Cas12/Cas13/Cas14 proteins utilized for genome engineering. Mol Biotechnol 65:311–325

Hillary VE, Ceasar SA (2024) CRISPR/Cas system-mediated base editing in crops: recent developments and future prospects. Plant Cell Rep 43:271

Hyun TK (2020) CRISPR/Cas-based genome editing to improve abiotic stress tolerance in plants. Botanica Serbica 44(2):121–127

Imran QM, Falak N, Hussain A, Mun BG, Yun BW (2021) Abiotic stress in plants: stress perception to molecular response and role of biotechnological tools in stress resistance. Agronomy 11(8):1579

Islam MZ, Alam MN, Rahman MM, Islam MZ, Rahman A (2024) Nutritional values of wheat and the roles and functions of its compositions in health

Ishino Y, Shinagawa H, Makino K, Amemura M, Nakata A (1987) Nucleotide sequence of the iap gene, responsible for alkaline phosphatase isozyme conversion in Escherichia coli, and identification of the gene product. J Bacteriol 169(12):5429-5433.

Jankele R, Svoboda P (2014) TAL effectors: tools for DNA targeting. Brief Funct Genomics 13(5):409–419

Ji G, Zhong H, Feukam Nzudie HL, Wang P, Tian P (2024) The structure, dynamics, and vulnerability of the global food trade network. J Clean Prod 434:140439

Jinek M, Chylinski K, Fonfara I, Hauer M, Doudna JA, Charpentier E (2012) A programmable dual-RNA–guided DNA endonuclease in adaptive bacterial immunity. Science 337(6096):816–821

Kasai M, Ohsawa R (2021) Biotechnology and its regulatory system in Japan. In: Gujar GT, Trisyono YA, Chen M (eds) Genetically modified crops in Asia Pacific. CISRO Publishing Clayton, Australia, pp 215–231.

Kahraman M, Akova İ (2024) Ukraine-Russia war and global wheat crisis. Karadeniz Araştırmaları 21(83):681–700

Kalaitzandonakes N, Willig C, Zahringer K (2022) The economics and policy of genome editing in crop improvement. Plant Genome 16:e20248

Kan J, Cai Y, Cheng C, Chen S, Jiang C, He Z, Yang P (2023) CRISPR/Cas9-guided knockout of *eIF4E* improves wheat yellow mosaic virus resistance without yield penalty. Plant Biotechnol J 21(5):893

Kim YG, Cha J, Chandrasegaran S (1996) Hybrid restriction enzymes: zinc finger fusions to FokI cleavage domain. Proc Natl Acad Sci 93(3):1156–1160

Kim D, Alptekin B, Budak H (2018) CRISPR/Cas9 genome editing in wheat. Funct Integr Genomics 18:31–41

Kumagai Y, Liu Y, Hamada H, Luo W, Zhu J, Kuroki M, Nagira Y, Taoka N, Katoh E, Imai R (2022) Introduction of a second "Green Revolution" mutation into wheat via *in planta* CRISPR/Cas9 delivery. Plant Physiol 188(4):1838–1842

Kumar R, Kaur A, Pandey A, Mamrutha HM, Singh GP (2019) CRISPR-based genome editing in wheat: a comprehensive review and future prospects. Mol Biol Rep 46(3):3557–3569

Lee M (2023) Deep learning in CRISPR-CasSystems: a review of recent studies. Front Bioeng Biotechnol 11:1226182

Li J, Li Y, Ma L (2021a) Recent advances in CRISPR/Cas9 and applications for wheat functional genomics and breeding. Abiotech 2(4):375–385

Li J, Jiao G, Sun Y, Chen J, Zhong Y, Yan L, Jiang D, Ma Y, Xia L (2021b) Modification of starch composition, structure and properties through editing of *TaSBEIIa* in both winter and spring wheat varieties by CRISPR/Cas9. Plant Biotechnol J 19(5):937–951

Li S, Zhang C, Li J, Yan L, Wang N, Xia L (2021c) Present and future prospects for wheat improvement through genome editing and advanced technologies. Plant Commun 2(4):100211

Liang Z, Chen K, Li T, Zhang Y, Wang Y, Zhao Q, Liu J, Zhang H, Liu C, Ran Y, Gao C (2017) Efficient DNA-free genome editing of bread wheat using CRISPR/Cas9 ribonucleoprotein complexes. Nat Commun 8(1):14261

Lin F, Li X, Jia N, Feng F, Huang H, Huang J, Fan S, Ciais P, Song XP (2023) The impact of Russia–Ukraine conflict on global food security. Glob Food Secur 36:100661

Liu H, Wang K, Jia Z, Gong Q, Lin Z, Du L, Pei X, Ye X (2020) Efficient induction of haploid plants in wheat by editing of *TaMTL* using an optimized *Agrobacterium*-mediated CRISPR system. J Exp Bot 71(4):1337–1349

Liu Y, Luo W, Linghu Q, Abe F, Hisano H, Sato K, Kamiya Y, Kawaura K, Onishi K, Endo M, Toki S, Hamada H, Nagira Y, Taoka N, Imai R (2021) *In planta* genome editing in commercial wheat varieties. Front Plant Sci 12:648841

Lyzenga WJ, Kagale S (2021) Wheat improvement using genome editing technology. BioTechniques 71(6):577–579

Matsumoto D, Matsugi E, Kishi K, Inoue Y, Nigorikawa K, Nomura W (2024) SpCas9-HF1 enhances accuracy of cell cycle-dependent genome editing by increasing HDR efficiency, and by reducing off-target effects and indel rates. Mol Ther Nucleic Acids 35:1

May D, Paldi K, Altpeter F (2023) Targeted mutagenesis with sequence-specific nucleases for accelerated improvement of polyploid crops: progress, challenges, and prospects. Plant Genome 16:e20298

Menz J, Modrzejewski D, Hartung F, Wilhelm R, Sprink T (2020) Genome edited crops touch the market: a view on the global development and regulatory environment. Front Plant Sci 11:586027

Mojica FJ, Díez-Villaseñor CS, García-Martínez J, Soria E (2005) Intervening sequences of regularly spaced prokaryotic repeats derive from foreign genetic elements. J Mol Evol 60:174–182

Nakata AT, Amemura MI, Makino KO (1989) Unusual nucleotide arrangement with repeated sequences in the *Escherichia coli* K-12 chromosome. J Bacteriol 171(6):3553–3556

Nalam VJ, Alam S, Keereetaweep J, Venables B, Burdan D, Lee H, Trick HN, Sarowar S, Makandar R, Shah J (2015) Facilitation of *Fusarium graminearum* infection by 9-lipoxygenases in *Arabidopsis* and wheat. Mol Plant-Microbe Interact 28(10):1142–1152

Nalinci S (2013) Amasya ili merkez ilçedeki hane halkının et tüketim alışkanlıkları ve et tüketimini etkileyen faktörler. Master Thesis, Gaziosmanpaşa Üniversitesi (Turkish)

Neres DF, Wright C (2024) Pleiotropy, a feature or a bug? Toward co-ordinating plant growth, development, and environmental responses through engineering plant hormone signaling. Curr Opin Biotechnol 88:103151

Nigro D, Smedley MA, Camerlengo F, Hayta S (2024) Using gene editing strategies for wheat improvement. In: Ricroch A, Eriksson D, Miladinović D, Sweet J, Van Laere K, Woźniak-Gientka E (eds) A roadmap for plant genome editing. Springer, Switzerland, pp 183–201

Papikian A, Liu W, Gallego-Bartolomé J, Jacobsen SE (2019) Site-specific manipulation of *Arabidopsis* loci using CRISPR-Cas9 SunTag systems. Nat Commun 10:729

Patniak D, Khurana P (2001) Wheat biotechnology: a minireview. Electron J Biotechnol 4(2):7-8

Peng JH, Sun D, Nevo E (2011a) Wild emmer wheat, *Triticum dicoccoides*, occupies a pivotal position in wheat domestication process. Aust J Crop Sci 5:1127–1143

Peng JH, Sun D, Nevo E (2011b) Domestication evolution, genetics and genomics in wheat. Mol Breed 28:281–301

Pixley KV, Falck-Zepeda JB, Paarlberg RL, Phillips PWB, Slamet-Loedin IH, Dhugga KS, Campos H, Gutterson N (2022) Genome-edited crops for improved food security of smallholder farmers. Nat Genet 54:364–367

Pourcel C, Salvignol G, Vergnaud G (2005) CRISPR elements in *Yersinia pestis* acquire new repeats by preferential uptake of bacteriophage DNA, and provide additional tools for evolutionary studies. Microbiology 151(3):653–663

Qin L, Hao C, Hou J, Wang Y, Li T, Wang L, Ma Z, Zhang X (2014) Homologous haplotypes, expression, genetic effects and geographic distribution of the wheat yield gene *TaGW2*. BMC Plant Biol 14:107

Raffan S, Sparks C, Huttly A, Hyde L, Martignago D, Mead A, Hanley SJ, Wilkinson PA, Barker G, Edwards KJ, Curtis TY, Usher S, Kosik O, Halford NG (2021) Wheat with greatly reduced accumulation of free asparagine in the grain, produced by CRISPR/Cas9 editing of asparagine synthetase gene *TaASN2*. Plant Biotechnol J 19(8):1602–1613

Rafiei F, Wiersma J, Scofield S, Zhang C, Alizadeh H, Mohammadi M (2024) Facts, uncertainties, and opportunities in wheat molecular improvement. Heredity 133(6):371–380

Reynolds MP, Braun HJ (2022) Wheat improvement: food security in a changing climate. Springer, Switzerland

Rössner C, Lotz D, Becker A (2022) VIGS goes viral: how VIGS transforms our understanding of plant science. Annu Rev Plant Biol 73:703–728

Sánchez-León S, Gil-Humanes J, Ozuna CV, Giménez MJ, Sousa C, Voytas DF, Barro F (2018) Low-gluten, nontransgenic wheat engineered with CRISPR/Cas9. Plant Biotechnol J 16(4):902–910

Schiml S, Fauser F, Puchta H (2014) The CRISPR/Cas system can be used as a nuclease for *in planta* gene targeting and as paired nickases for directed mutagenesis in *Arabidopsis*, resulting in heritable progeny. Plant J 80(6):1139–1150

Shah S (2024) Crop improvement using CRISPR/Cas9 systems: advances and challenges. Plant Biol 1:100004

Shan Q, Wang Y, Li J, Zhang Y, Chen K, Liang Z, Zhang K, Liu J, Xie JJ, Qiu JL (2013) Targeted genome modification of crop plants using a CRISPR-Cas system. Nat Biotechnol 31(8):686–688

Shewry PR (2018) Do ancient types of wheat have health benefits compared with modern bread wheat? J Cereal Sci 79:469–476

Simmonds J, Scott P, Brinton J, Mestre TC, Bush M, Del Blanco A, Dubjovski J, Uauy C (2016) A splice acceptor site mutation in *TaGW2-A1* increases thousand grain weight in tetraploid and hexaploid wheat through wider and longer grains. Theor Appl Genet 129(6):1099–1112

Smedley MA, Hayta S, Clarke M, Harwood WA (2021) CRISPR-Cas9 based genome editing in wheat. Curr Protoc 1(3):e65

Song L, Liu J, Cao B, Liu B, Zhang X, Chen Z, Dong C, Liu X, Zhang Z, Wang W, Chai L, Liu J, Zhu J, Cui S, He F, Peng H, Hu Z, Su Z, Guo W, Xin M, Yao Y, Yan Y, Song Y, Bai G, Ni Z (2023) Reducing brassinosteroid signalling enhances grain yield in semi-dwarf wheat. Nature 617(7959):118–124

Sun Z, Zhang M, An Y, Han X, Guo B, Lv G, Zhao Y, Guo Y, Li S (2022) CRISPR/Cas9-mediated disruption of xylanase inhibitor protein (XIP) gene improved the dough quality of common wheat. Front Plant Sci 13:811668

Taj M, Sajjad M, Li M, Yasmeen A, Mubarik MS, Kaniganti S, He C (2022) Potential targets for CRISPR/Cas knockdowns to enhance genetic resistance against some diseases in wheat (*Triticum aestivum* L.). Front Genet 13:926955

Tian X, Qin Z, Zhao Y, Wen J, Lan T, Zhang L, Wang F, Qin D, Yu K, Zhao A, Hu Z, Yao Y, Ni Z, Sun Q, Smet ID, Peng H, Xin M (2022) Stress granule-associated TaMBF1c confers thermotolerance through regulating specific mRNA translation in wheat (*Triticum aestivum*). New Phytol 233(4):1719–1731

Tsakirpaloglou N, Septiningsih EM, Thomson MJ (2023) Guidelines for performing CRISPR/Cas9 genome editing for gene validation and trait improvement in crops. Plants 12(20):3564

Turnbull C, Lillemo M, Hvoslef-Eide TA (2021) Global regulation of genetically modified crops amid the gene edited crop boom–a review. Front Plant Sci 12:630396

Van der Oost J, Jore MM, Westra ER, Lundgren M, Brouns SJ (2009) CRISPR-based adaptive and heritable immunity in prokaryotes. Trends Biochem Sci 34(8):401–407

Venske E, Dos Santos RS, Busanello C, Gustafson P, Costa de Oliveira A (2019) Bread wheat: a role model for plant domestication and breeding. Hereditas 29:156:16

Verdezoto-Prado J, Chicaiza-Ortiz C, Mejía-Pérez AB, Freire-Torres C, Viteri-Yánez M, Deng L, Barba-Ostria C, Guamán LP (2025) Advances in environmental biotechnology with CRISPR/Cas9: bibliometric review and cutting-edge applications. Discov Appl Sci 7(3):167

Wang W, Akhunov E (2022) Application of CRISPR-Cas-based genome editing for precision breeding in wheat. In: Reynolds MP, Braun HJ (eds) Wheat improvement: food security in a changing climate. Springer, Switzerland, pp 539–556

Wang Y, Cheng X, Shan Q, Zhang Y, Liu J, Gao C, Qiu JL (2014) Simultaneous editing of three homoeoalleles in hexaploid bread wheat confers heritable resistance to powdery mildew. Nat Biotechnol 32(9):947–951

Wang M, Gao S, Zeng W, Yang Y, Ma J, Wang Y (2020) Plant virology delivers diverse tool-sets for biotechnology. Viruses 12:1338

Wang W, Yu Z, He F, Bai G, Trick HN, Akhunova A, Akhunov E (2022) Multiplexed promoter and gene editing in wheat using a virus-based guide RNA delivery system. Plant Biotechnol J 20:2332–2341

Wang P, Chen Z, Meng Y, Shi H, Lou C, Zheng X, Li G, Li X, Peng W, Kang G (2023) Wheat *PHT1;9* acts as one candidate arsenate absorption transporter for phytoremediation. J Hazard Mater 452:131219

Whelan AI, Lema MA (2019) Regulation of genome editing in plant biotechnology: Argentina. In: Dederer HG, Hamburger D (eds) Regulation of genome editing in plant biotechnology: a comparative analysis of regulatory frameworks of selected countries and the EU. Springer, Switzerland, pp 19–62

Wienert B, Cromer MK (2022) CRISPR nuclease off-target activity and mitigation strategies. Front Genome Editing 4:1050507

William A, Alain B, Maarten VG (2011) The world wheat book: a history of wheat breeding. Volume 2, Lavoisier, Paris

Wong AYT, Chan AWK (2016) Genetically modified foods in China and the United States: a primer of regulation and intellectual property protection. Food Sci Human Wellness 5(3):124–140

Yadav MR, Choudhary M, Singh J, Lal MK, Jha PK, Udawat P, Gupta NK, Rajput VD, Garg NK, Maheshwari C, Hasan M, Gupta S, Jatwa TK, Kumar R, Yadav AK, Prasad PVV (2022) Impacts, tolerance, adaptation, and mitigation of heat stress on wheat under changing climates. Int J Mol Sci 23(5):2838

Zhang XH, Tee LY, Wang XG, Huang QS, Yang SH (2015) Off-target Effects in CRISPR/Cas9-mediated Genome Engineering. Mol Ther Nucleic Acids 4:e264

Zhang Y, Bai Y, Wu G, Zou S, Chen Y, Gao C, Tang D (2017) Simultaneous modification of three homoeologs of *TaEDR1* by genome editing enhances powdery mildew resistance in wheat. Plant J 91(4):714–724

Zhang S, Zhang R, Song G, Gao J, Li W, Han X, Chen M, Li Y, Li G (2018a) Targeted mutagenesis using the *Agrobacterium tumefaciens*-mediated CRISPR-Cas9 system in common wheat. BMC Plant Biol 18:302

Zhang Y, Li D, Zhang D, Zhao X, Cao X, Dong L, Liu J, Chen K, Zhang H, Gao C, Wang D (2018b) Analysis of the functions of *TaGW2* homoeologs in wheat grain weight and protein content traits. Plant J 94(5):857–866

Zhang S, Zhang R, Gao J, Gu T, Song G, Li W, Li D, Li Y, Li G (2019a) Highly efficient and heritable targeted mutagenesis in wheat via the *Agrobacterium tumefaciens*-mediated CRISPR/Cas9 system. Int J Mol Sci 20(17):4257

Zhang R, Liu J, Chai Z, Chen S, Bai Y, Zong Y, Chen K, Li J, Jiang L, Gao C (2019b) Generation of herbicide tolerance traits and a new selectable marker in wheat using base editing. Nat Plants 5:480–485

Zhang J, Zhang H, Li S, Li J, Yan L, Xia L (2021a) Increasing yield potential through manipulating of an *ARE1* ortholog related to nitrogen use efficiency in wheat by CRISPR/Cas9. J Integr Plant Biol 63(9):1649–1663

Zhang F, Liu S, Zhang T et al (2021b) Construction and biological characterization of an infectious full-length cDNA clone of a Chinese isolate of wheat yellow mosaic virus. Virology 556:101–109

Zhang J, Yu H, Li J (2023) De novo domestication: retrace the history of agriculture to design future crops. Curr Opin Biotechnol 81:102946

Zhou X, Zhao Y, Ni P, Ni Z, Sun Q, Zong Y (2023) CRISPR-mediated acceleration of wheat improvement: advances and perspectives. J Genet Genomics 50:815-834

Zimny T, Sowa S, Tyczewska A, Twardowski T (2019) Certain new plant breeding techniques and their marketability in the context of EU GMO legislation: recent developments. New Biotechnol 25(51):49–56

Chapter 8
Wheat Prolamin Superfamily as a Challenging Target for Genome Editing

Oksana I. Sozinova, Natalia O. Kozub, Anatolii V. Karelov, Ihor O. Sozinov, and Yaroslav B. Blume

Abstract Gluten proteins (gliadins and glutenins) comprise nearly 80% of the total protein of wheat grain and account for the specific visco-elastic properties of dough. They, along with other groups of less abundant grain proteins such as puroindolines determining endosperm texture, α-amylase/trypsin inhibitors (ATI), farinins, purinins, and non-specific lipid transfer proteins, are attributed to the prolamin superfamily. Gluten proteins, mainly gliadins, puroindolines, and ATIs, became CRISPR/Cas9 editing targets for wheat improvement via knockout of respective genes. Prolamins can cause multiple diseases, of which the most important are celiac disease, wheat allergy, and nonceliac gluten sensitivity in predisposed patients. The immunotoxicity of prolamins is due to specific conservative amino acid motifs. Different approaches can be used to reduce immunotoxicity and thus provide people with sensitivity to particular groups of prolamins with safe food products. The traditional way to obtain wheat with reduced immunotoxicity is to select genotypes with null alleles at certain prolamin loci resulting from spontaneous or induced mutagenesis. Another approach is genetic modification (GM) for silencing prolamin genes by RNA interference. The most precise, flexible, and modern approach is gene editing by CRISPR/Cas techniques, which allows targeted knockout of certain groups of prolamin genes or even single genes. All three approaches provided wheat genotypes with reduced immunotoxicity. Still, CRISPR/Cas editing is the most effective approach for decreasing wheat immunotoxicity, as it produces non-GM

O. I. Sozinova (✉) · N. O. Kozub
Institute of Food Biotechnology and Genomics, National Academy of Sciences of Ukraine, Kyiv, Ukraine

Institute of Plant Protection, National Academy of Agrarian Sciences, Kyiv, Ukraine

A. V. Karelov · Y. B. Blume (✉)
Institute of Food Biotechnology and Genomics, National Academy of Sciences of Ukraine, Kyiv, Ukraine
e-mail: cellbio@cellbio.freenet.viaduk.net

I. O. Sozinov
Institute of Plant Protection, National Academy of Agrarian Sciences, Kyiv, Ukraine

K. A. Abd-Elsalam, A. Ahmad (eds.), *Gene Editing in Cereals*, Sustainability Sciences in Asia and Africa, https://doi.org/10.1007/978-981-95-6722-5_8

genotypes with only targeted mutations, without undesirable mutations in other genomic regions. CRISPR/Cas9 techniques were effectively employed to reduce the contents of different groups of prolamins, mostly gliadins, and thus decrease the immunotoxic properties of wheat products. In most cases, the reduction in gliadin content was associated with an increase in bread-making quality indices due to greater accumulation of glutenins. Genotypes with targeted ATI genes are promising for developing wheat cultivars with reduced allergenicity. Products from gene-edited wheat could also be considered safer prophylaxis foods for people without pronounced symptoms of wheat-induced disorders.

Keywords CRISPR/Cas · Puroindoline · Gluten · α-Amylase/trypsin inhibitors · Celiac disease · Null alleles · RNA interference

8.1 Introduction

Wheat is one of the most important crops worldwide (Willenborg and Van Acker 2008; Erenstein et al. 2022). It is a major staple food crop; its global yield reaches about 799 million metric tons (FAO 2024a). Moreover, globally, wheat provides the highest daily amount of protein per capita (17.3 g), which significantly exceeds the average contribution of other major staple food crops (10.6 g for rice and 1.4 g for potato) and even meat (7.7 g for poultry meat and 5.9 g for pig meat), milk (9.1 g) and eggs (3.3 g) (FAO 2024b) and thus remains the most reliable source of protein for human consumption (Shewry and Hey 2015). Because of this, wheat is an object of constant improvement via classical breeding, involvement of introgressive material, physical or chemical mutagenesis, and such advanced approaches as genetic transformation and, ultimately, gene editing (Kiszonas and Morris 2017; Zhang et al. 2020). The range of traits that gene editing technologies can improve includes yield traits, tolerance to both abiotic and biotic factors, and various grain quality traits, mainly through knockout of respective genes (Schenke and Cai 2020; Verma et al. 2021; Abdallah et al. 2025; Taj et al. 2022; Zhou et al. 2023; Nigro et al. 2024; Smulders et al. 2024; Jamil et al. 2025; Mukhtiar et al. 2025; Wang et al. 2025; Yao et al. 2025).

First genome-editing systems employed programmable sequence-specific nucleases (SSNs), namely transcription activator-like effector nucleases (TALENs), zinc-finger nucleases (ZFNs), and meganucleases (Gaj et al. 2013). Such nucleases produce double-strand breaks (DSBs) in DNA at certain sites, resulting in targeted genome modifications through DNA repair mechanisms (Voytas and Gao 2014). For example, ZFNs were used for producing wheat lines with the edited acetohydroxyacid synthase gene conferring imidazolinone herbicide resistance (Ran et al. 2018) and genotypes with mutations in three *TaMLO1* homoeoalleles for achieving broad-spectrum resistance to powdery mildew (Wang et al. 2014). The methods based on SSNs are quite pricy as custom enzymes are needed to be synthesized to target sites of interest; there were no reports of gene editing of prolamins via SSNs.

Gene-editing systems based on clustered regularly interspaced short palindromic repeats (CRISPR) recognition and CRISPR-associated enzymes Cas9 and Cas12a (i.e., CRISPR/Cas9 or CRISPR/12a) are much cheaper and have a wider range of applications than SSNs. Unlike RNAi silencing, they can provide repeatable results, and the progeny of transformed lines do not return to the wild type. Moreover, modern techniques using CRISPR/Cas produce transgene-free lines (He et al. 2022). Genome editing via CRISPR/Cas systems needs plasmid constructs with the genes for Cas9 from *Streptococcus pyogenes*, which recognizes the 5′-NGG-3′ protospacer adjacent motif (PAM), or Cas12a nuclease from *Lachnospiraceae* bacterium recognizing 50-T-rich PAM sequences to be introduced into wheat samples (Bandyopadhyay et al. 2020). Single guide RNA (sgRNA) for Cas9 or single CRISPR RNA (crRNA) for Cas12a complementary to target sites is used as part of CRISPR/Cas systems (Cardi et al. 2023). *Agrobacterium*-mediated, particle bombardment-mediated systems, and virus-based systems are used to deliver constructs to plants. The most progress in gene editing of wheat traits was achieved using the CRISPR/Cas9 technique (Zhou et al. 2023).

Wheat end-use quality traits undergo improvement in several directions: improvement of technological properties such as bread-making indices and grain hardness (Zhang et al. 2018, 2019, 2021; Shewry 2022; Sun et al. 2022), nutrient values (Li et al. 2021; Raffan et al. 2021), and reduction of deleterious properties of wheat product consumption for sensitive people via reduction of the content of certain groups of grain proteins (Jouanin et al. 2020; Gilissen et al. 2021; Verma et al. 2021; Smulders et al. 2024).

8.2 Grain Proteins in Wheat: Prolamin Superfamily

In wheat, the major seed storage proteins are called gluten or prolamin proteins due to their amino acid composition: glutamine and proline are predominant amino acids with proportions of 32–53% and 11–29% in different gluten protein types (Wieser et al. 2023a). Currently, gluten proteins along with other groups of wheat grain proteins whose proportion in total grain protein is much lower, indolines (puroindolines a and b, which determine grain hardness, grain softness protein (GSP)) (Morris et al. 2021), α-amylase/trypsin inhibitors (ATI), farinins, purinins, and non-specific lipid transfer proteins (LTP), are referred to as prolamin superfamily (Shewry 2019) (Fig. 8.1). It is the first three prolamin protein groups that became gene editing targets for wheat improvement (Zhou et al. 2023).

Puroindoline a and puroindoline b, small tryptophan-rich proteins are controlled by single genes at the *Hardness (Ha)* locus on 5D determining grain hardness, are easy targets for gene editing to produce hard-textured wheat from soft-textured ones. Hard texture of wheat grain can be obtained through mutations in either of the puroindoline genes in soft-textured wheat (Morris 2002). Zhang et al. (2018, 2019,

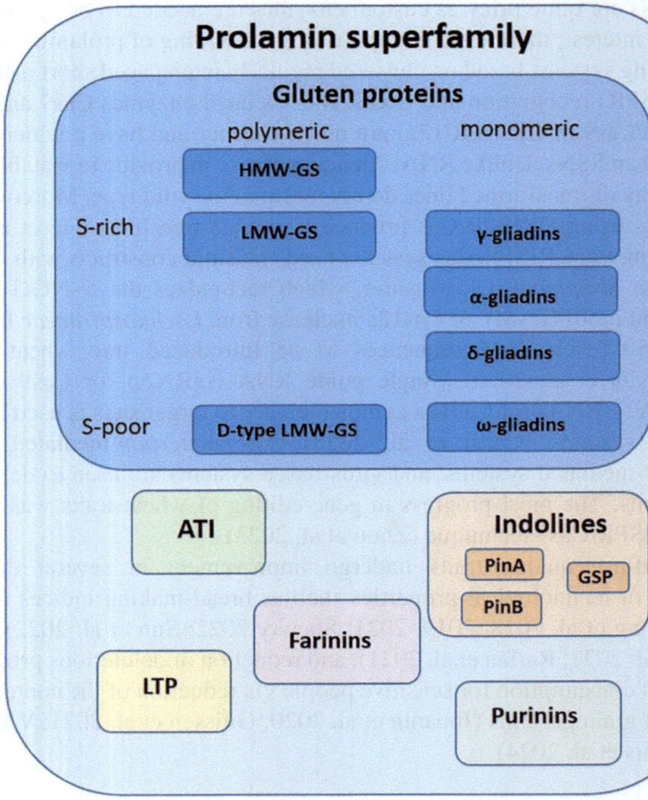

Fig. 8.1 Prolamin superfamily according to Shewry (2019). *HMW-GS* high-molecular-weight glutenin subunits, *LMW-GS* low-molecular-weight glutenin subunits, *PinA* puroindoline a, *PinB* puroindoline B, *GSP* grain softness protein, *ATI* α-amylase/trypsin inhibitors, *LTP* lipid-transfer proteins

2021) produced sgRNA for the *Pinb* gene coding sequence and transformed immature embryos. As a result, in T1 plants, 5, 7, 12, 29, or 34-bp deletions were observed (Zhang et al. 2019). The editing effectiveness rose significantly across generations and was the highest in T2 (92.26%). Plants with the *Pinb* gene carrying a 34-bp deletion causing frameshift mutations showed an almost four-fold increase in grain hardness index (Zhang et al. 2021).

A more challenging task is manipulation with other prolamin proteins, such as gluten proteins or ATIs, whose genes occur in multiple copies and variations in the wheat genome. Gluten proteins have unique biochemical properties; they establish a complex protein network in the wheat dough, which captures CO_2 released during fermentation (Wieser 2006; Delcour et al. 2012; Shewry and Belton 2024). Besides their direct application as components of traditional wheat-based products, gluten proteins are broadly used in the food sector as a thickening agent to upgrade the quality of soups, sauces, ketchup, marinades, dressings, ice cream, etc., as well as in chewing gum, chew candies, and fruit chews (Wieser et al. 2021). Deamidation and hydrolysis of gluten increase the solubility of wheat proteins and expand their use for meat products as well as in cosmetics (Burnett et al. 2018; Gabler and Scherf 2020; Chen et al. 2021).

Wheat gluten represents a complex group of proteins consisting of a mixture of monomeric gliadins and glutenin polymers, in which glutenin subunits, high-molecular-weight and low-molecular-weight ones (HMW–GS and LMW–GS, respectively), are linked via disulfide bonds (Shewry 2022) (Fig. 8.1). The ratio of gliadins and glutenins ranges from 0.75 to 2.5 (Vereijken et al. 2000; Dhaka et al. 2015; Wieser et al. 2023b); together they form 80% of the total grain protein content in the wheat kernel (Shewry 2019). The remainders are albumins and globulins, 12 and 8%, respectively (Van den Broeck et al. 2009). In particular, the α-amylase/trypsin inhibitor (ATI) family, including both albumins and globulins, makes up 4% of grain total protein (Dupont et al. 2011). Other protein groups of the prolamin superfamily are much less abundant: Dupont et al. (2011) estimated the content of purinins as 0.9%, farinins as 0.8%, and indolines (GSP and puroindolines) as 0.3%.

Woychik et al. (1961) introduced the actual gliadin classification into α-, β-, γ-, and ω-gliadins on the basis of their mobility while separated by acid gel electrophoresis, where mobility increases from ω- to α-gliadins (Fig. 8.2). According to the latest updates, α-gliadins and β-gliadins are combined into a group of α-gliadins (Shewry and Halford 2002; Shewry 2019). On the basis of their N-terminal amino acid sequence, ω-gliadins are referred to as ARE-, ARQ-, KEL- type proteins, which are referred to as ω-1 and ω-2 gliadins, and SRL-type corresponding to ω-5 gliadins (Kasarda et al. 1983). As ω-gliadins lack cysteine, they belong to S-poor proteins (Shewry and Halford 2002). Their special feature is a predominant central

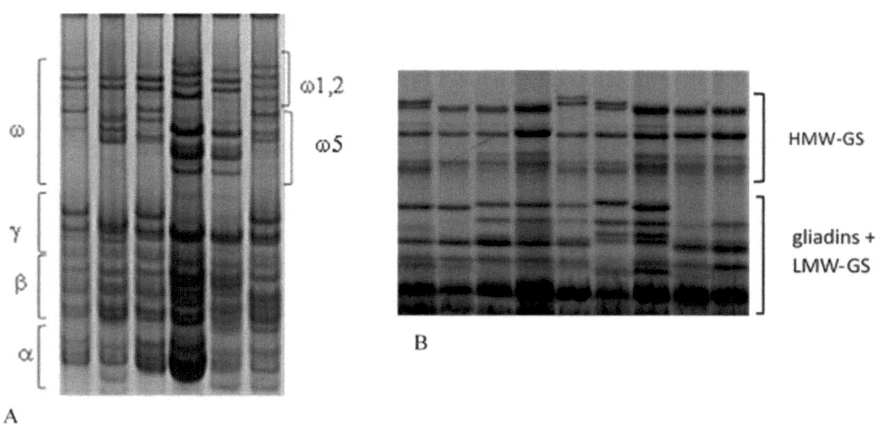

Fig. 8.2 Electrophoretic patterns of wheat gluten proteins. (**a**) zones of α-, β-, γ- and ω-gliadins after acid polyacrylamide gel electrophoresis, pH 4.2, by the procedure of Kozub et al. (2009); (**b**) SDS-electrophoresis of total grain proteins by the procedure of Laemmli (1970)

domain of repetitive units. The structure of α- and γ-gliadins, which are S-rich prolamins, differs from that of ω-gliadins (Shewry and Halford 2002).

SDS-electrophoresis of reduced glutenins (Laemmli 1970) reveals HMW–GS and LMW–GS. Whereas HMW glutenin subunits and ω-gliadins belong to two discrete groups, LMW–GS are similar to α- and γ-gliadins in their structure (Shewry and Halford 2002). The most frequent type of LMW–GS is the B-type, while rare C-type LMW–GS are α- or γ-gliadins with an additional cysteine (Shewry 2019). Similarly, rare D-type LMW–GS are ω-gliadins with a single cysteine resulting from a mutation in the respective ω-gliadin gene (Masci et al. 1993). Wieser et al. (2023a, b) refer to D-type LMW–GS as ωb-gliadins. Another gliadin group, δ-gliadins (Anderson et al. 2012; Wan et al. 2013), shows low abundance (1.2%) (Shewry 2019). A common feature of different gluten proteins is the repetitive domain with glutamine- and proline-rich motifs (Shewry and Belton 2024).

Gliadins of common wheat are controlled by the six major loci that are present in distal parts of the short arms of the homoeologous group 1 and 6 chromosomes (Payne 1987). *Gli-A1, Gli-B1,* and *Gli-D1* located on chromosomes 1A, 1B, and 1D, respectively, contain clusters of genes for ω - and γ-gliadins as well as single δ-gliadin genes, whereas at *Gli-A2, Gli-B2*, and *Gli-D2* on chromosomes 6A, 6B, and 6D there are clusters of α-gliadin genes (Payne 1987; Anderson et al. 2012). LMW-GS are encoded by gene clusters at the loci *Glu-A3, Glu-B3,* and *Glu-D3,* which are closely linked to the respective gliadin loci *Gli-A1, Gli-B1,* and *Gli-D1.* Minor gliadin loci on 1BS and 1AS (*Gli-A3, Gli-B5,* and *Gli-A5*) encode ω–gliadins, *Gli-B3/Glu-B2* encodes an ω–gliadin or, alternatively, a D-type LMW-GS (Jackson et al. 1985; Metakovsky et al. 1996). *Glu-1* loci with two genes for HMW-GS (*x*- and *y*-subunits) are on the long arms of the homoeologous group 1 chromosomes, but *Glu-A1* encodes only one subunit or none (Payne 1987). Proteomic profiling revealed 23 α-, 13 γ-, and 7 ω-gliadins in the variety Butte 86 (Dupont et al. 2011), 23 α-, 11 γ-, and 5 ω-gliadins in the Korean variety Keumkang (Cho et al. 2018). Similarly, using RNA sequencing as well as proteomics methods, the number of functional gliadin genes was assessed as 42, including 25 α-, 11 γ-, 1 δ-, and 5 ω-gliadins in the bread wheat variety Xiaoyan 81 (Wang et al. 2017). Dupont et al. (2011) revealed 22 active LMW–GS genes. In the cultivar Fielder, Liu et al. (2023) identified 0, 3, and 2 ω-gliadin genes and 3, 4, and 4 γ-gliadin genes on 1A, 1B, and 1D, respectively, as well as 8, 11, and 9 α-gliadin genes on 6A, 6B, and 6DS, respectively, whereas the annotation of the Fielder genome performed by Yu et al. (2024) revealed more functional gliadin genes: 2, 2, and 5 of ω-gliadin genes along with 5, 7, and 9 γ-gliadin genes, on 1A, 1B, and 1D, respectively, as well as 10, 15, and 9 α-gliadin genes on 6A, 6B, and 6DS, respectively.

8.3 Human Diseases Caused by Prolamins

Due to their chemical composition and structure, prolamins may cause different kinds of immunotoxicity (Shewry and Tatham 2016). Three major wheat-related diseases are currently distinguished: celiac disease (CD), wheat allergy, as well as nonceliac gluten sensitivity (Green et al. 2015; Cabanillas 2020). However, the latter disorder might also be associated with other components, such as ATIs (Cabanillas 2020). The best-known of these diseases is CD (Green et al. 2015). CD represents a chronic enteropathy, which is triggered by gluten in people carrying human leukocyte antigens HLA-DQ2 and HLA-DQ8, affecting 1% of the population in Western countries, which reaches 1.5% in Northern Europe (Mustalahti et al. 2010; Catassi et al. 2014; Ludvigsson et al. 2014). Many epitopes in gluten proteins were described (Sollid et al. 2019; Chlubnová et al. 2023). Antigenic peptides derive from repetitive domains of gluten proteins, which are rich in the amino acids proline and glutamine (Shewry and Tatham 2016). Celiac disease-relevant T-cell epitopes restricted by HLA–DQ molecules were found in all groups of gluten proteins, α-, ω-, and γ-gliadins, LMW–GS and HMW–GS (Sollid et al. 2019), except for δ-gliadins (Wang et al. 2017). The groups of gluten proteins show different epitope distribution and the HLA molecule to which they bind (DQ2.5, DQ2.2, DQ8, or DQ8.5). The α-gliadin fraction provides the most stimulatory epitopes associated with the disease (Tye-Din et al. 2010; Sollid et al. 2019). Vriz et al. (2021) proposed the following ranking of gluten proteins, in which the immunotoxicity decreases: α-gliadins > ω-gliadins > γ-gliadins > glutenins. However, gluten proteins controlled by certain homoeologous loci might differ in the occurrence of epitopes (Van Herpen et al. 2006). For example, CD epitopes are abundant in ω-gliadins encoded by loci of the A and D genomes and absent on ω-5-gliadins encoded by *Gli-B1* (Shewry and Tatham 2016). Similarly, α-gliadins encoded by the A, B, and D genome genes also differ in the presence of toxic epitopes with the order of D > A > B (van Herpen et al. 2006; Wang et al. 2017). All γ-gliadins encoded by the major gliadin loci on 1A, 1B, and 1D show a high number of toxic CD epitopes (Wang et al. 2017).

Another dangerous disorder associated with prolamins is wheat-dependent exercise-induced anaphylaxis (WDEIA), mainly caused by ω-5-gliadins (Scherf et al. 2016; Le et al. 2016; Kennard et al. 2018). WDEIA manifests itself as an allergic response in case of physical activities after ingestion of food with gluten; the physical activities can be more or less intense, the time between the ingestion and activities can be from 1 to 4 h, and 10–50 min of the activities are enough to trigger the symptoms; WDEIA manifests in such type I allergic symptoms as hives, pruritus, urticaria, gastrointestinal symptoms, hypotension, and even anaphylaxis (Scherf et al. 2016). The disease was described comparatively not long ago, in 1985, but the number of cases is growing in Europe and Japan, and more often among adult patients (Kushimoto and Aoki 1985; Le et al. 2016; Kennard et al. 2018). Seven IgE-binding epitopes, QQIPQQQ, QQFPQQQ, QQSPQQQ, QQSPEQQ, QQLPQQQ, QQYPQQQ, and PYPP, were identified in ω-5-gliadins (Matsuo et al.

2004). In addition, in 10% of cases, WDEIA is triggered by the IgE-binding epitopes QQPGQ, QQPGQGQQ, and QQSGQGQ from HMW-GS (Matsuo et al. 2005), and rare cases caused by toxic epitopes in other groups of gluten proteins were also revealed (Battais et al. 2005). Besides, ω-5-gliadins were found to cause allergies among children, manifesting a series of immediate hypersensitivity symptoms, among which are erythema, urticaria, and even the anaphylactic reaction, as well as atopic dermatitis (Palosuo et al. 2001).

Leduc et al. (2003) first described an allergy to deamidated gluten in patients with angioedema and generalized urticaria after ingesting processed meat. Denery-Papini et al. (2012) revealed that ω-2-gliadins and γ-gliadins containing the epitope QPQQPFPQ were responsible for this allergy. The same groups of proteins were also found to cause immediate hypersensitivity to hydrolyzed wheat protein (Laurière et al. 2007).

Other prolamin superfamily proteins that are considered to have increased immunotoxicity causing non-celiac wheat sensitivity (NCWS) and other allergies to wheat are ATIs and LTPs (Battais et al. 2008; Tatham and Shewry 2008; Manza et al. 2025). ATIs inhibit heterologous α-amylase and trypsin, protecting plants against insects, mites, and mammals; besides, they are involved in providing osmotic stress tolerance via precluding starch degradation (Gu et al. 2015). ATI subunits are 12–16 kDa proteins containing 10 cysteines forming 4–5 intrachain S-S bonds (Reig-Otero et al. 2018). Based on their aggregation degree, ATIs can be classified into monomeric 12-kDa WMAI-1 and WMAI-2, homodimeric 24-kDa WDAI 1–3, and heterotetrameric 60-kDa WTAI (Sanchez-Monge et al. 1989, 1992; Gomez et al. 1991; Altenbach et al. 2011). Genes for ATI subunits were mapped on a wide range of chromosomes, 7BS, 7DS, 4BS, 4D, 6AL, 6BL, 6DL, 3A, 3B, and 3D (Tatham and Shewry 2008). Among wheat grain proteins, ATI subunits 0.19, CM2, CM3, and CM16 show the highest IgE-reactive properties and are believed to be the most probable triggers for a number of wheat allergies, including bakers' asthma (Zapatero et al. 2003).

8.4 The Approaches Used to Mitigate Prolamin Immunotoxicity

Different approaches can be applied to provide wheat products with reduced immunotoxicity, namely selection of lines with null alleles, RNAi silencing, and gene editing via CRISPR/Cas systems (Shewry and Tatham 2016; Smulders et al. 2024) (Fig. 8.3).

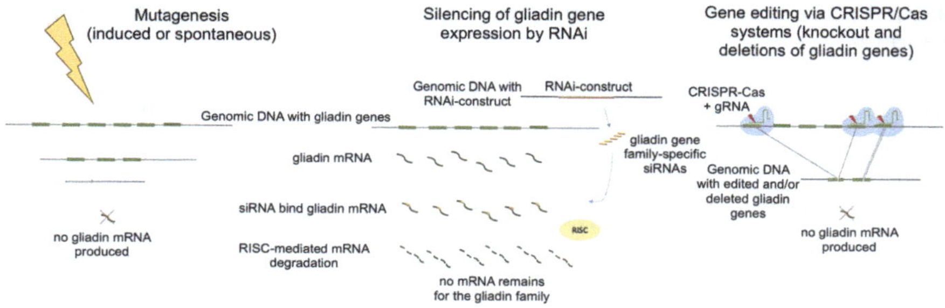

Fig. 8.3 Approaches to reduce immunotoxicity of wheat: mutagenesis, RNAi silencing, and gene editing via CRISPR/Cas systems. (Cited from Jouanin et al. (2020) with modifications)

8.4.1 Wheat 'Traditional' Mutants with Null Alleles

Wheat lines and cultivars with null alleles at different giladin loci were reported some time ago, although the authors did not highlight changes in immunotoxicity (Lafiandra et al. 1987; Branlard and Dardevet 1994). This approach is comparatively simple from a methodology standpoint, requiring only one-dimensional acid PAGE to separate gliadins. Further tests with ELISA are needed to determine immunotoxicity in wheat with null alleles (Marjanović-Balaban et al. 2021). Genotypes with null alleles can be obtained by traditional mutagenesis or screening for spontaneous mutations, and commercial cultivars with such alleles are available (Table 8.1). Genotypes obtained by this approach are not considered to be GM products and may be used without legal restrictions all over the world. The downside is low effectiveness as the target null alleles may not be found in a large sample of progeny after mutagenic treatment or have undesirable mutations in other genomic regions; moreover, even when found, the lines or cultivars with null alleles may still be immunotoxic due to compensatory effects (Galili et al. 1986).

Waga et al. (2013) reported lines showing null alleles at the *Gli-B2* or *Gli-D1* loci or with fewer proteins encoded by *Gli-B1* derived from crossing a bread wheat line with spelt wheat, in which ELISA tests showed a reduction of immunoreactivity by 18% in the case of the *Gli-B1* mutant. Waga and Skoczowski (2014) further obtained ω-gliadin-free lines showing a decrease in immunoreactivity by up to 30%. Kohno et al. (2016) developed the line 1BS-18 with the deletion in chromosome 1B that lacked ω-5-gliadins and showed the least deficiency of chromosome 1B. They demonstrated a significant reduction in its allergic properties concerning WDEIA in guinea pig and rat allergy models (Yamada et al. 2019). Moreover, Yamada et al. (2022) suggested that products from hypoallergenic 1BS-18H wheat may serve as a potential prophylaxis food against WDEIA caused by ω-5-gliadins in healthy individuals. Wheat line DH20 with the null allele at *Glu-B3* (Lee et al. 2017) was also shown to lack gliadins controlled by *Gli-B1*, resulting in markedly reduced IgE reactivity (Altenbach et al. 2018). Lines with null alleles at *Gli-A1, Gli-B1, Gli-D1 Gli-A2*, and *Gli-D2*, as well as with the combination of null-alleles at *Gli-A2* and

Table 8.1 Examples of wheat material with mutations at gliadin loci resulting in null allele formation

Null allele	S/I[a] mutation	Cultivar/line	Reference
Gli-D1	S	Darius	Branlard and Dardevet (1994)
Gli-D1 ω-gliadins	S	Biotypes of Nepal landraces MG 27081, MG 27079, MG 27080	Lafiandra et al. (1987)
Gli-B1 *Gli-B1/Gli-D1*	S	Ble-des-Vosges, Florence-Aurora Touzelle	Metakovsky et al. (2018)
Gli-D1, Gli-B1 (partial), *Gli-B2* (partial)	S	Spelt wheat (Oberkummler Rotkorn) x LAD 480	Waga et al. (2013)
ω-gliadin-free	S	B1N x Nep1	Waga and Skoczowski (2014)
Gli-1, Gli-2	I	Xiaoyan 81 deletion lines resulting from ion beam treatment	Wang et al. (2017)
Gli-B1		CS deletion line 1BS-18	Kohno et al. (2016)
Glu-B3/Gli-B1	S	Line DH20, a dihaploid line from the cross Keumkang x Olgeu	Lee et al. (2017) Altenbach et al. (2018)
Gli-A1, Gli-B1, Gli-D1 *Gli-A2, Gli-D2. Gli-A2/ Gli-D2*	S	Pegaso lines with introduced null alleles	Camerlengo et al. (2017)
Loss of α-gliadin, γ- and ω- gliadin groups	I	Paragon lines produced by γ-irradiation	Jouanin et al. (2019)
Gli-B1, Gli-D1	S	Lines from the cross B16 x OKK	Kozub et al. (2022)
Gli-A2, Gli-B2, Gli-D2	I	Gamma-irradiation of RIL143	Rottersman et al. (2024)

[a]*S*—spontaneous; *I*—induced

Gli-D2, were developed on the basis of the Italian bread wheat cultivar Pegaso by crossing it with previously detected carriers of null alleles and APAGE selection. The amount of toxic and immunogenic epitopes in the deletion line without both *Gli-A2* and *Gli-D2* loci was reduced by 85% compared with the cultivar Pegaso (Camerlengo et al. 2017). Common wheat lines with spontaneous mutations accounting for the formation of null alleles at the *Gli-B1* and *Gli-D1* loci in the Ukrainian germplasm were reported, but immunoreactivity of the lines was not studied (Kozub et al. 2022).

The frequency of M4 lines with changed gliadin patterns was assessed to be 3.9% in the Paragon γ-irradiated population (14 out of 360 screened lines): the changes included the loss of certain groups of α-gliadins, loss of γ- and ω- gliadins controlled by gliadin loci on 1AS (line P4-86), as well as loss of γ- and ω- gliadins controlled by *Gli-B1* (line P5-20) (Jouanin et al. 2019). Lines derived from gamma-irradiation of RIL143 with deletions in α-gliadin gene clusters at the *Gli-A2*, *Gli-B2,* and *Gli-D2* loci were produced, and the effects of the deletions were studied: the loss of α-gliadins controlled by *Gli-D2* resulted in improved baking quality and the reduced amount of CD-immunodominant epitopes (Rottersman et al. 2024).

8.4.2 GM Wheat with Prolamin Genes Silenced by RNA Interference

RNA silencing is a more complex approach from a technical standpoint: DNA fragments that downregulate *Gli/Glu* gene expression are to be delivered to target plants, and further screening for the transformed lines is to be performed (Guo et al. 2016). Moreover, the resulting lines are considered GM organisms, leading to certain legal restrictions for commercial usage in most countries. Plant transformation systems based on *Agrobacterium* or particle bombardment are used to transform wheat (Su et al. 2023). All GM lines with RNAi fragments affecting gliadin genes presented in Table 8.2 were developed with the use of biolistic transformation as it allows the concurrent transfer of a number of genes and does not involve introduction T-DNA border repeats; moreover, it is more suitable for multiple copy insertion (Borisjuk et al. 2019).

One of the first attempts at prolamin silencing targeted α-gliadins in common wheat, which resulted in a decrease in the α-gliadin amount by 63% and improved dough strength but somewhat reduced loaf volume (Becker et al. 2006; Wieser et al. 2007). Other research targeted γ-gliadins, stating a reduction of the target proteins by about 55–80% in Bobwhite lines (Gil-Humanes et al. 2008). Further comparative investigation of ω/α-gliadins silencing effectiveness, reporting reduction of target proteins by up to 85.6% was made; some of the lines obtained by the authors showed the reduced content of γ-gliadins and showed the significantly reduced reactivity (Gil-Humanes et al. 2010). Those results were confirmed by further research (Pistón et al. 2013). Gil-Humanes et al. (2010) observed that the reduced γ-gliadins content affected the response to the celiac disease epitopes for some lines and almost did not affect it for others. It was further explained by the increased production of ω/α-gliadins in the GM lines with downregulated γ-gliadins (Pistón et al. 2011). The compensatory effect in the lines with negative downregulation of γ-gliadins and none of such an effect in the lines with suppression of ω/α-gliadins was further confirmed while working with a wider range of plant material (Gil-Humanes et al. 2011). Baking quality for the transgenic lines with reduced contents of prolamins was additionally tested: some of them showed the deleterious effect of the dough, while others with the increased content of LMW-GS showed the same level of bread-making indices as in the initial cultivar (Gil-Humanes et al. 2014).

The reduction in ω-5-gliadin expression by RNA silencing in the cultivar Butte 86 was reported (Altenbach and Allen 2011). This cultivar was used to produce other downregulated lines with a significant reduction of gliadin production; although both lines reported by the authors showed low immunotoxicity, only one of them retained the dough quality sufficient for further commercial use (Altenbach

Table 8.2 Examples of RNAi silencing of gliadin genes

Cultivar or line used for transformation	Target gliadin group	Plasmid	Effect on prolamin content	Reference
Florida	α- gliadins	N/A	Reduction of α-gliadins by 63%	Becker et al. (2006) and Wieser et al. (2006)
BW2003, BW208	γ- gliadins	pghp8.1 + pAHC20	γ-gliadin reduction by 33–80%; total gliadin content depended on the line: From increase by 18% to reduction by 42%	Gil-Humanes et al. (2008)
BW2003, BW208	All gliadin groups	pDhp-ω/α, pGhp-ω/α, pghpg8.1	Reduction of up to 81% for ω-gliadins, 91% for α-gliadins, up to 100% for γ-gliadins, and up to 88% for total gliadins	Gil-Humanes et al. (2010)
Butte 86	ω-5-gliadins	Omega-RNAi-13 + pAHC20	Absent from the electrophoretic patterns	Altenbach and Allen (2011)
Florida	α-gliadins	pGiaRNAiSpacer + pCalneo.	Reduction of α-gliadins by >63%, increase in ω5-gliadins by 23%, γ- gliadins by 20%, HMW-GS by 16%	Becker et al. (2012)
BW208	α-, ω-, γ-gliadins, and LMW glutenin subunits	pghpg8.1, pDhp_ω/α, pDhp_α/βZR, pDhp_ω4ZR, pDhp_ω8ZR	Total gluten content decreased by 17%–66% depending on the plasmid combination	Barro et al. (2016)
Butte 86	ω-1,2-gliadins	pAHC25	Total gluten protein reduction from 6.9% to 45.7%, depending on the line	Altenbach et al. (2019)

et al. 2019). The cultivar Florida with 1BL.1RS (the wheat-rye translocation) was used to produce RNAi lines with silenced α-gliadins genes; the authors reported a reduction in α-gliadin amount in most of the RNAi lines by more than 50% and the total gliadin reduction by around 16%; the rheological tests did not show a significant difference in the dough properties (Becker et al. 2012). The findings reported by Marín-Sanz and Barro (2022) demonstrate that silencing the gliadin genes in wheat has a more significant impact on the transcriptome in leaves than in seeds. However, it is likely that the observed leaf transcriptome changes are caused by alterations that occur in the grain in the course of grain filling and no effect on leaf tanscriptome leaves is anticipated at other developmental stages. These results

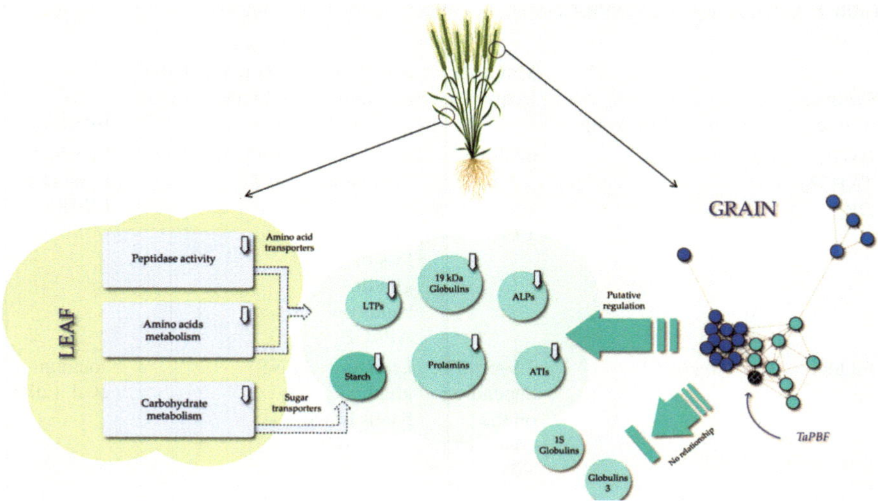

Fig. 8.4 A model depicting the role of the network of transcription factors in regulating the accumulation of seed storage proteins and starch, and the impact on processes related to the interplay between the grain and leaf in an RNA interference (RNAi) line where the expression of gliadins is downregulated. (Reprinted from Marín-Sanz and Barro 2022)

are a step towards unravelling the intricate regulatory processes in the grain during the grain-filling stage caused by the reduction in gliadin expression through RNAi. The network of transcription factors in the grain could be involved in regulation of the synthesis of starch and storage proteins, but some globulins may be not subject to this regulation, playing a primary role in protein compensation (Fig. 8.4).

8.4.3 Gene Editing of Wheat by CRISPR: A Modern Approach to Reduce Prolamin-Related Immunotoxicity

The CRISPR/Cas9 technology proved effective in reducing the contents of certain groups of prolamins, thus decreasing the immunotoxic properties of wheat products (Table 8.3). For producing mutations, sgRNAs were delivered by bombardment (Sánchez-León et al. 2018, 2024; Camerlengo et al. 2020; Yu et al. 2024) or *Agrobacterium tumefaciens*-mediated transformation (Jouanin et al. 2019; Liu et al. 2023), and further transgene-free gene-edited plants were selected. In all cases with gluten proteins, the target regions were in the sequence of the repetitive domain containing toxic epitopes.

Sánchez-León et al. (2018) were the first to comprehensively demonstrate simultaneous targeted knockdown of multiple gliadin genes via gene editing to produce low-gluten wheat. They used the bread wheat BW028 (the accession of cv. Bobwhite without the 1BL.1RS translocation) and TAH53 and the durum wheat cultivars DP

Table 8.3 Summary of CRISPR/Cas9 studies affecting prolamin contents

Cultivar or line	Target prolamin group	Mutation frequency, %	Effect on prolamin content	Effect on immunoreactivity (R5 and G12 mAbs)	Reference
BW028, TAH53, DP	α-gliadins	62.3–75.1, 5.2–6. 1,6–14,8 in T1	Reduction in gliadin content up to 69%, loss of gliadin bands on APAGE	Reduction by 85%	Sánchez-León et al. (2018)
Fielder	α-, γ- gliadins	7.9–58.3%, depending on the construct in T0	Loss of gliadin bands based on APAGE	NS	Jouanin et al. (2019)
Fielder	γ-, ω- gliadins	Not specified	Decrease in γ- and ω-gliadin contents by 75% and 64%, respectively	Reduction by 97.6%	Yu et al. (2024)
BW208	γ-, ω- gliadins	34% in T0 (up to 48.5% for γ-gliadins and up to 100% for ω-gliadins, depending on the construct)	> 50% in some lines, Loss of gliadin bands on APAGE	Reduction by up to 97%	Sánchez-León et al. (2024)
Fielder	Two γ-gliadins encoded by TraesFLD1D01G005600 and TraesFLD1B01G010600 genes	Not specified	About 10% reduction	Reduction by up to 76%	Liu et al. (2023)
Svevo	ATI CM3 and CM16 subunits	14% in T0	No gene products observed	No immunoreactivity to CM3	Camerlengo et al. (2020)

to reduce α-gliadins by CRISPR/Cas9 targeting of multiple copies of α-gliadin genes. For this, two sgRNAs were designed to target the sequence encoding a 33-mer in the repetitive domain of α-gliadins with multiple CD-epitopes, of which sgAlpha-2 proved to be more efficient. sgAlpha-2-transformed BW208 lines showed the highest frequency of mutations in α-gliadin genes, amounting to

62.3–75.1%, whereas, in the case of TAH53 and DP, they were lower (1.50–14.77% and 5.16%–7.86%, respectively). As a result, 35 out of 45 identified α-gliadin genes were edited. The decrease in α-gliadin content ranged from 32% to 82% in T_1 lines, while total gliadin content decreased by up to 69%. The authors showed the absence of off-target mutations in other prolamin genes, although in most T1 lines, γ-gliadin content was also decreased. The reduction in gliadin content was generally accompanied by increased HMW–GS content and SDS-sedimentation volume in some lines, suggesting their good bread-making performance. Moreover, ELISA tests with the R5 and G12 monoclonal antibodies (mAbs) showed a reduction of up to 85% in immunoreactivity of the T_1 lines from BW208 in comparison with the initial cultivar. Removing gene editing components after target gene editing is essential for obtaining stable nontransgenic genotypes (He et al. 2022). Sánchez-León et al. (2018) succeeded in selecting transgene-free low-gluten genotypes among T2 lines, three lines derived from BW208, and six lines from durum wheat DP.

Jouanin et al. (2019) edited α- and γ-gliadin genes of the cultivar Fielder. They designed three sgRNAs for targeting α-gliadin genes and three sgRNAs for γ-gliadin genes, and further performed *A. tumefaciens*-mediated transformation of immature embryos with constructs containing one to five sgRNAs (in four variants). As a result, among 117 T0 plants expressing Cas9, APAGE revealed changes in gliadin patterns of T1 seeds derived from 36 T0 plants (30.8%). The constructs differed in the frequency of mutations. In addition, in some cases, constructs targeting γ-gliadins caused changes in ω-gliadin bands, which the authors explained by deletions of ω-gliadin genes located between the targeted γ-gliadin genes.

In further studies, gene editing was used to target multiple γ- and ω-gliadin genes in the cultivars Fielder and BW208 (Yu et al. 2024; Sánchez-León et al. 2024). Yu et al. (2024) performed CRISPR/Cas9-based multiplex editing of all ω-gliadin genes and a portion of γ-gliadin genes in Fielder. For this, the authors used simultaneous biolistic transformation of immature embryos with seven sgRNA constructs targeting the repetitive regions of γ- and ω-gliadin genes. When designing sgRNAs, the focus was on targeting ω-gliadin genes to remove all of them. Those sgRNAs also targeted some copies of γ-gliadin genes. As a result, eight transgenic Cas + T0 plants were produced, of which one, 387, was used for further studies and was shown to be transformed with four sgRNAs. Using next-generation sequencing of PCR amplicons amplified for the ω-gliadin genes, the T1 line 387-3-6 was shown to be homozygous for the edited alleles in one, two, and five ω-gliadin genes on chromosomes 1A, 1B, and 1D, respectively. The majority of edited T1 lines derived from 387-3, including line 387-3-6, did not contain the Cas9 sequence. Gene editing in γ- and ω-gliadin genes was demonstrated by whole genome sequencing of line 387-3-6 and non-gene-edited line 387-1-8 obtained from the same T0 plant. The gene editing reduced the amount of immunotoxic peptides by 55.2% and 77.2% according to ELISA tests with mAbs R5 and G12. In the edited lines, γ-gliadin content decreased by 63.9%, the amount of ω-1,2-gliadins by 84.6%, whereas the ω-5-gliadin content decreased by only 22%. At the same time, the glutenin/gliadin ratio increased in the gene-edited lines, resulting in higher gluten strength of such lines.

To edit the same groups of genes, γ- and ω-gliadin genes, in the above-mentioned cultivar BW208, Sánchez-León et al. (2024) designed seven sgRNAs (four, two, and two sgRNAs for γ-gliadin, ω-1,2-gliadin, and ω-5-gliadin genes, respectively, of which three targeted both γ- and ω-1,2-gliadin genes). Transformation with constructs containing two sgRNAs in four combinations, as well as one variant of cotransformation with two constructs, yielded 59 transformed lines; in 20 of those lines, the authors observed significant alterations in the gliadin APAGE profiles corresponding to mutations in the γ- and ω-gliadin genes. Depending on the construct used for editing, there were lines without γ-gliadins on gels, without certain groups of γ-gliadins (without the γ-gliadin controlled by *Gli-B1* in the progeny of the line Gamma-229), and without both γ- and ω-1,2-gliadins (in the progeny of Omera-43). In the lines with altered gliadin patterns, the presence of induced InDels was testified by sequencing analysis. sgRNAs differed in the editing efficiency. For example, the efficiency of sgOmega4 ranged from 28.6% to 33.3% for γ-gliadin genes and 83.3–100% for ω-gliadin genes, whereas sgOmega2 showed 50% for both groups of genes. Noteworthy, transgene-free lines with mutations in several γ-gliadin genes were identified. Lines with edited γ- and ω-gliadin genes showed decreased content of these target groups of proteins and, in most cases, reduced α-gliadin content along with increased content of HMW–GS and LMW–GS, as judged from reverse-phase HPLC quantification. Lines with further reduced gluten content were obtained via crossing the gene-edited lines with reduced γ- and ω-gliadin content to those with edited α-gliadin genes (Sánchez-León et al. 2018); their APAG gels lacked α-γ- and ω-1,2-gliadins although ω-5-gliadins were still present. Such lines are promising materials for producing wheat cultivars with immune-safe gluten.

The same cultivar, Fielder, and the CRISPR/Cas9 technique were used in the study aimed at determining the role of certain γ-gliadins in bread-making quality (Liu et al. 2023). Using precise CRISPR/Cas9-mediated knockout of certain γ-gliadin genes, *Gli-γ1-1D* (TraesFLD1B01G010600) and *Gli-γ2-1B* (TraesFLD1D01G005600), whose expression accounted for 35% and 23% of the total γ-gliadin content, with one sgRNA, the authors produced lines with mutations of two γ-gliadin genes (KO-1 or KO-2), as well as a line with the mutation in the *Gli-γ1-1D* gene only (KO-3). Lines KO-1 or KO-2 showed decreased γ-gliadin content and increased HMW-GS content, resulting in higher SDS-sedimentation and increased bread-making quality. The inactivation of only the *Gli-γ1-1D* gene also reduced gluten content but was insufficient for improving quality indices. The knockout of the γ-gliadin genes in the edited lines caused an approximately 70% decrease in R5 CD epitopes in comparison with the initial cultivar.

Thus, the studies on gene editing of gliadin genes have shown its high potential for producing low-gluten hypoimmunogenic wheat genotypes with improved bread-making quality primarily due to the compensating rise in glutenin content. Noteworthy is the efficiency of the APAGE method for cheap and easy primary screening of mutants with altered gliadin patterns employed in different laboratories (Sánchez-León et al. 2018, 2024; Jouanin et al. 2019; Liu et al. 2023).

Multiple ATI genes are another challenging group of prolamin genes subjected to gene editing. To reduce the amount of subunits WTAI-CM3 and WTAI CM16, Camerlengo et al. (2020) performed editing of CM3 and CM16 genes with the CRISPR construct containing seven gRNAs in polycistronic tRNA/gRNA via bombardment of scutella of immature embryos of the durum wheat variety Svevo. Mutations (large and small IN/DELs and SNPs) were identified by PCR, high-resolution melting analysis, and sequencing in 14 of 97 regenerated plants. Of their progeny, four transgene-free edited plants containing large deletions were selected. ELISA tests showed no immunoreactivity related to CM3 subunits, indicating that there were no CM3 subunits in the edited lines. The lines with edited CM3 and CM16 genes are promising for developing durum wheat cultivars with reduced allergenicity.

8.5 Conclusion

CRISPR/Cas9 technology was effectively used to decrease the contents of certain groups of prolamins, mostly gliadins, and decrease the immunotoxic properties of wheat products. In addition, in most cases, the decrease in gliadin content was accompanied by an increase in bread-making indices due to a higher accumulation of HMW–GS. However, it is not ruled out that some successful ‘classical’ spontaneous or induced wheat mutants with null alleles at certain gliadin loci, for example those at *Gli-B1* (without ω-5-gliadins), could become objects for further gene editing to reduce the content of additional groups of proteins, because the relative degree of reduction in ω-gliadins reported in the CRISPR/Cas9 studies, especially ω-5-gliadins inducing WDEIA, was generally lower than that for α- and γ-gliadins (Sánchez-León et al. 2024; Yu et al. 2024). As for wheat with reduced content of CD or WDEIA epitopes, it could at least be considered safer prophylaxis food for people without pronounced symptoms of wheat-induced disorders. Despite the abundance of proline and glutamine, wheat gluten proteins provide adequate amounts of essential amino acids required for adult human nutrition, except for lysine (Shewry and Hey 2015). So, it could be expected that future achievements with gene edition will deal with this problem, for example, by the production of precise mutations in wheat prolamin-box binding factors, like it was done by the nontransgenic targeting-induced local lesions in genomes (TILLING) method (Moehs et al. 2019).

References

Abdallah NA, Elsharawy H, Abulela HA, Thilmony R, Abdelhadi AA, Elarabi NI (2025) Multiplex CRISPR/Cas9-mediated genome editing to address drought tolerance in wheat. GM Crops Food 16:1–17. https://doi.org/10.1080/21645698.2022.2120313

Altenbach SB, Allen PV (2011) Transformation of the US bread wheat 'Butte 86' and silencing of omega-5 gliadin genes. GM Crops 2:66–73. https://doi.org/10.4161/gmcr.2.1.15884

Altenbach SB, Vensel WH, Dupont FM (2011) The spectrum of low molecular weight alpha-amylase/protease inhibitor genes expressed in the US bread wheat cultivar Butte 86. BMC Res Notes 4:242. https://doi.org/10.1186/1756-0500-4-242

Altenbach SB, Chang HC, Simon-Buss A, Jang YR, Denery-Papini S, Pineau F, Gu YQ, Huo N, Lim SH, Kang CS, Lee JY (2018) Towards reducing the immunogenic potential of wheat flour: omega gliadins encoded by the D genome of hexaploid wheat may also harbor epitopes for the serious food allergy WDEIA. BMC Plant Biol 18:291. https://doi.org/10.1186/s12870-018-1506-z

Altenbach SB, Chang HC, Yu XB, Seabourn BW, Green PH, Alaedini A (2019) Elimination of omega-1,2 gliadins from bread wheat (*Triticum aestivum*) flour: effects on immunogenic potential and end-use quality. Front Plant Sci 10:580. https://doi.org/10.3389/fpls.2019.00580

Anderson OD, Dong L, Huo N, Gu YQ (2012) A new class of wheat gliadin genes and proteins. PLoS One 7:e52139. https://doi.org/10.1371/journal.pone.0052139

Bandyopadhyay A, Kancharla N, Javalkote VS, Dasgupta S, Brutnell TP (2020) CRISPR-Cas12a (Cpf1): a versatile tool in the plant genome editing tool box for agricultural advancement. Front Plant Sci 11:584151. https://doi.org/10.3389/fpls.2020.584151

Barro F, Iehisa JC, Giménez MJ, García-Molina MD, Ozuna CV, Comino I, Sousa C, Gil-Humanes J (2016) Targeting of prolamins by RNAi in bread wheat: effectiveness of seven silencing-fragment combinations for obtaining lines devoid of coeliac disease epitopes from highly immunogenic gliadins. Plant Biotechnol J 14:986–996. https://doi.org/10.1111/pbi.12455

Battais F, Mothes T, Moneret-Vautrin DA, Pineau F, Kanny G, Popineau Y, Bodinier M, Denery-Papini S (2005) Identification of IgE-binding epitopes on gliadins for patients with food allergy to wheat. Allergy 60:815–821. https://doi.org/10.1111/j.1398-9995.2005.00795.x

Battais F, Richard C, Jacquenet S, Denery-Papini S, Moneret-Vautrin DA (2008) Wheat grain allergies: an update on wheat allergens. Eur Ann Allergy Clin Immunol 40:67–76

Becker D, Folck A, Knies P, Lorz H, Wieser H (2006) Silencing the α-gliadins in hexaploid bread wheat. In: Lookhart GL, Ng PKW (eds) Gluten proteins AACC, St Paul, pp 86–89

Becker D, Wieser H, Koehler P, Folck A, Mühling KH, Zörb C (2012) Protein composition and techno-functional properties of transgenic wheat with reduced α-gliadin content obtained by RNA interference. J Appl Bot Food Qual 85:23–33

Borisjuk N, Kishchenko O, Eliby S, Schramm C, Anderson P, Jatayev S, Kurishbayev A, Shavrukov Y (2019) Genetic modification for wheat improvement: from transgenesis to genome editing. Biomed Res Int 2019:6216304. https://doi.org/10.1155/2019/6216304

Branlard G, Dardevet M (1994) A null *Gli-D1* allele with a positive effect on bread wheat quality. J Cereal Sci 20:235–244. https://doi.org/10.1006/jcrs.1994.1063

Burnett C, Bergfeld WF, Belsito DV, Hill RA, Klaassen CD, Liebler DC, Marks JG Jr, Shank RC, Slaga TJ, Snyder PW, Andersen FA, Heldreth B (2018) Safety assessment of hydrolyzed wheat protein and hydrolyzed wheat gluten as used in cosmetics. Int J Toxicol 37(1_suppl):55S–66S. https://doi.org/10.1177/1091581818776013

Cabanillas B (2020) Gluten-related disorders: celiac disease, wheat allergy, and nonceliac gluten sensitivity. Crit Rev Food Sci Nutr 60:2606–2621. https://doi.org/10.1080/10408398.2019.1651689

Camerlengo F, Sestili F, Silvestri M, Colaprico G, Margiotta B, Ruggeri R, Lupi R, Masci S, Lafiandra D (2017) Production and molecular characterization of bread wheat lines with reduced amount of α-type gliadins. BMC Plant Biol 17:248. https://doi.org/10.1186/s12870-017-1211-3

Camerlengo F, Frittelli A, Sparks C, Doherty A, Martignago D, Larré C, Lupi R, Sestili F, Masci S (2020) CRISPR-CAS9 multiplex editing of the α-amylase/trypsin inhibitor genes to reduce allergen proteins in durum wheat. Front Sustain Food Syst 4:104. https://doi.org/10.3389/fsufs.2020.00104

Cardi T, Murovec J, Bakhsh A, Boniecka J, Bruegmann T, Bull SE, Eeckhaut T, Fladung M, Galovic V, Linkiewicz A, Lukan T, Mafra I, Michalski K, Kavas M, Nicolia A, Nowakowska J, Sági L, Sarmiento C, Yıldırım K, Zlatković M, Hensel G, Van Laere K (2023) CRISPR/Cas-mediated plant genome editing: outstanding challenges a decade after implementation. Trends Plant Sci 28:1144–1165. https://doi.org/10.1016/j.tplants.2023.05.012

Catassi C, Gatti S, Fasano A (2014) The new epidemiology of celiac disease. JPGN Rep 59(Suppl 1):S7–S9. https://doi.org/10.1097/01.mpg.0000450393.23156.59

Chen X, Fu W, Luo Y, Cui C, Suppavorasatit I, Liang L (2021) Protein deamidation to produce processable ingredients and engineered colloids for emerging food applications. Compr Rev Food Sci Food Saf 20:3788–3817. https://doi.org/10.1111/1541-4337.12759

Chlubnová M, Christophersen AO, Sandve GKF, Lundin KEA, Jahnsen J, Dahal-Koirala S, Sollid LM (2023) Identification of gluten T cell epitopes driving celiac disease. Sci Adv 9:eade5800. https://doi.org/10.1126/sciadv.ade5800

Cho K, Beom HR, Jang YR, Altenbach SB, Vensel WH, Simon-Buss A, Lim SH, Kim MG, Lee JY (2018) Proteomic profiling and epitope analysis of the complex α-, γ-, and ω-gliadin families in a commercial bread wheat. Front Plant Sci 9:818. https://doi.org/10.3389/fpls.2018.00818

Delcour JA, Joye IJ, Pareyt B, Wilderjans E, Brijs K, Lagrain B (2012) Wheat gluten functionality as a quality determinant in cereal-based food products. Annu Rev Food Sci Technol 3:469–492. https://doi.org/10.1146/annurev-food-022811-101303

Denery-Papini S, Bodinier M, Larré C, Brossard C, Pineau F, Triballeau S, Pietri M, Battais F, Mothes T, Paty E, Moneret-Vautrin DA (2012) Allergy to deamidated gluten in patients tolerant to wheat: specific epitopes linked to deamidation. Allergy 67:1023–1032. https://doi.org/10.1111/j.1398-9995.2012.02860.x

Dhaka V, Khatkar B (2015) Effects of gliadin/glutenin and HMW-GS/LMW-GS ratio on dough rheological properties and bread-making potential of wheat varieties. J Food Qual 38:71–82. https://doi.org/10.1111/jfq.12122

Dupont FM, Vensel WH, Tanaka CK, Hurkman WJ, Altenbach SB (2011) Deciphering the complexities of the wheat flour proteome using quantitative two-dimensional electrophoresis, three proteases and tandem mass spectrometry. Proteome Sci 9:10. https://doi.org/10.1186/1477-5956-9-10

Erenstein O, Jaleta M, Mottaleb KA, Sonder K, Donovan J, Braun H (2022) Global trends in wheat production, consumption and trade. In: Reynolds MP, Braun HJ (eds) Wheat improvement. Springer, Cham, pp 47–66. https://doi.org/10.1007/978-3-030-90673-3_4

FAO. Crops and livestock products (2024a) License: CC BY-NC-SA 3.0IGO. 2024a Dec 20; [cited 20 Feb 2025]. Available from https://www.fao.org/faostat/en/#data/QCL

FAO. Food balance sheet (2024b) License: CC BY-NC-SA 3.0 IGO. 2024b Jul 19; [cited 20 Feb 2025]. Available from https://www.fao.org/faostat/en/#data/FBS

Gabler AM, Scherf KA (2020) Comparative characterization of gluten and hydrolyzed wheat proteins. Biomolecules 10:1227. https://doi.org/10.3390/biom10091227

Gaj T, Gersbach CA, Barbas CF 3rd (2013) ZFN, TALEN, and CRISPR/Cas-based methods for genome engineering. Trends Biotechnol 31:397–405. https://doi.org/10.1016/j.tibtech.2013.04.004

Galili G, Levy AA, Feldman M (1986) Gene-dosage compensation of endosperm proteins in hexaploid wheat *Triticum aestivum*. PNAS 83:6524–6528. https://doi.org/10.1073/pnas.83.17.6524

Gil-Humanes J, Pistón F, Hernando A, Alvarez JB, Shewry PR, Barro F (2008) Silencing of γ-gliadins by RNA interference (RNAi) in bread wheat. J Cereal Sci 48:565–568. https://doi.org/10.1016/j.jcs.2008.03.005

Gil-Humanes J, Pistón F, Tollefsen S, Sollid LM, Barro F (2010) Effective shutdown in the expression of celiac disease-related wheat gliadin T-cell epitopes by RNA interference. PNAS 107:17023–17028. https://doi.org/10.1073/pnas.1007773107

Gil-Humanes J, Pistón F, Shewry PR, Tosi P, Barro F (2011) Suppression of gliadins results in altered protein body morphology in wheat. J Exp Bot 62:4203–4213. https://doi.org/10.1093/jxb/err119

Gil-Humanes J, Pistón F, Barro F, Rosell CM (2014) The shutdown of celiac disease-related gliadin epitopes in bread wheat by RNAi provides flours with increased stability and better tolerance to over-mixing. PLoS One 9:e91931. https://doi.org/10.1371/journal.pone.0091931

Gilissen LJWJ, Smulders MJM (2021) Low gluten and coeliac-safe wheat through gene editing. In: Ricroch A, Chopra S, Kuntz M (eds) Plant biotechnology. Springer, Cham, pp 231–248. https://doi.org/10.1007/978-3-030-68345-0_16

Gomez L, Sanchez-Monge R, Lopez-Otín C, Salcedo G (1991) Wheat inhibitors of heterologous α-amylases: characterization of major components from the monomeric class. Plant Physiol 96:768–774. https://doi.org/10.1104/pp.96.3.768

Green PH, Lebwohl B, Greywoode R (2015) Celiac disease. J Allergy Clin Immunol 135:1099–1107. https://doi.org/10.1016/j.jaci.2015.01.044

Gu A, Hao P, Lv D, Zhen S, Bian Y, Ma C, Xu Y, Zhang W, Yan Y (2015) Integrated proteome analysis of the wheat embryo and endosperm reveals central metabolic changes involved in the water deficit response during grain development. J Agric Food Chem 63:8478–8487. https://doi.org/10.1021/acs.jafc.5b00575

Guo Q, Liu Q, Smith NA, Liang G, Wang MB (2016) RNA silencing in plants: mechanisms, technologies and applications in horticultural crops. Curr Genomics 17:476–489. https://doi.org/10.2174/1389202917666160520103117

He Y, Mudgett M, Zhao Y (2022) Advances in gene editing without residual transgenes in plants. Plant Physiol 188:1757–1768. https://doi.org/10.1093/plphys/kiab574

Jackson EA, Holt LM, Payne PI (1985) *Glu-B2*, a storage protein locus controlling the D group of LMW glutenin subunits in bread wheat (*Triticum aestivum*). Genet Res 46:11–17. https://doi.org/10.1017/s0016672300022412

Jamil M, Ahmad W, Sanwal M, Maqsood MF (2025) Gene editing and GWAS for digital imaging analysis of wheat grain weight, size and shape are inevitable to enhance the yield. Cereal Res Commun. https://doi.org/10.1007/s42976-025-00630-x

Jouanin A, Schaart JG, Boyd LA, Cockram J, Leigh FJ, Bates R, Wallington EJ, Visser RGF, Smulders MJM (2019) Outlook for coeliac disease patients: towards bread wheat with hypoimmunogenic gluten by gene editing of α- and γ-gliadin gene families. BMC Plant Biol 19:333. https://doi.org/10.1186/s12870-019-1889-5

Jouanin A, Gilissen LJWJ, Schaart JG, Leigh FJ, Cockram J, Wallington EJ, Boyd LA, van den Broeck HC, van der Meer IM, America AHP, Visser RGF, Smulders MJM (2020) CRISPR/Cas9 gene editing of gluten in wheat to reduce gluten content and exposure-reviewing methods to screen for coeliac safety. Front Nutr 7:51. https://doi.org/10.3389/fnut.2020.00051

Kasarda DD, Autran J, Lew EJ, Nimmo CC, Shewry PR (1983) N-terminal amino acid sequences of ω-gliadins and ω-secalins. Biochim Biophys Acta (BBA)—Protein Proteomic 747:138–150. https://doi.org/10.1016/0167-4838(83)90132-2

Kennard L, Thomas I, Rutkowski K, Azzu V, Yong PFK, Kasternow B, Hunter H, Cabdi NMO, Nakonechna A, Wagner A (2018) A multicenter evaluation of diagnosis and management of omega-5 gliadin allergy (also known as wheat-dependent exercise-induced anaphylaxis) in 132 adults. J Allergy Clin Immunol Practice 6:1892–1897. https://doi.org/10.1016/j.jaip.2018.02.013

Kiszonas AM, Morris CF (2017) Wheat breeding for quality: a historical review. Cereal Chem 95:17–34. https://doi.org/10.1094/cchem-05-17-0103-fi

Kohno K, Takahashi H, Endo TR, Matsuo H, Shiwaku K, Morita E (2016) Characterization of a hypoallergenic wheat line lacking ω-5 gliadin. Allergol Int 65:400–405. https://doi.org/10.1016/j.alit.2016.03.002

Kozub NA, Sozinov IA, Sobko TA, Kolyuchii VT, Kuptsov SV, Sozinov AA (2009) Variation at storage protein loci in winter common wheat cultivars of the Central Forest-Steppe of Ukraine. Cytol Genet 43:55–62. https://doi.org/10.3103/S0095452709010101

Kozub NO, Sozinov IO, Bidnyk HY, Demianova NA, Sozinova OI, Karelov AV, Spivak SI, Blume Y (2022) Common wheat genotypes with null alleles at gliadin loci. Factors in Experimental Evolution of Organisms 31:59–64. https://doi.org/10.7124/FEEO.v31.1485

Kushimoto H, Aoki T (1985) Masked type I wheat allergy. Relation to exercise-induced anaphylaxis. Arch Dermatol 121:355–360. https://doi.org/10.1001/archderm.1985.01660030077023

Laemmli UK (1970) Cleavage of structural proteins during the assembly of the head of bacteriophage T4. Nature 227:680–685. https://doi.org/10.1038/227680a0

Lafiandra D, Colaprico G, Kasarda DD, Porceddu E (1987) Null alleles for gliadin blocks in bread and durum wheat cultivars. Theor Appl Genet 74:610–616. https://doi.org/10.1007/BF00288860

Laurière M, Pecquet C, Boulenc E, Bouchez-Mahiout I, Snégaroff J, Choudat D, Raison-Peyron N, Vigan M, Branlard G (2007) Genetic differences in omega-gliadins involved in two different immediate food hypersensitivities to wheat. Allergy 62:890–896. https://doi.org/10.1111/j.1398-9995.2007.01456.x

Le TA, Al Kindi M, Tan JA, Smith A, Heddle RJ, Kette FE, Hissaria P, Smith WB (2016) The clinical spectrum of omega-5-gliadin allergy. Intern Med J 46:710–716. https://doi.org/10.1111/imj.13091

Leduc V, Moneret-Vautrin DA, Guerin L, Morisset M, Kanny G (2003) Anaphylaxis to wheat isolates: immunochemical study of a case proved by means of double-blind, placebo-controlled food challenge. J Allergy Clin Immunol 111:897–899. https://doi.org/10.1067/mai.2003.1345

Lee J, Kang C, Beom H, Jang Y, Altenbach SB, Lim S, Kim Y, Park C (2017) Characterization of a wheat mutant missing low-molecular-weight glutenin subunits encoded by the B-genome. J Cereal Sci 73:158–164. https://doi.org/10.1016/j.jcs.2016.12.004

Li J, Jiao G, Sun Y, Chen J, Zhong Y, Yan L, Jiang D, Ma Y, Xia L (2021) Modification of starch composition, structure and properties through editing of *TaSBEIIa* in both winter and spring wheat varieties by CRISPR/Cas9. Plant Biotechnol J 19:937–951. https://doi.org/10.1111/pbi.13519

Liu D, Yang H, Zhang Z, Chen Q, Guo W, Rossi V, Xin M, Du J, Hu Z, Liu J, Peng H, Ni Z, Sun Q, Yao Y (2023) An elite γ-gliadin allele improves end-use quality in wheat. New Phytol 239:87–101. https://doi.org/10.1111/nph.18722

Ludvigsson JF, Bai JC, Biagi F, Card TR, Ciacci C, Ciclitira PJ, Green PH, Hadjivassiliou M, Holdoway A, van Heel DA, Kaukinen K, Leffler DA, Leonard JN, Lundin KE, McGough N, Davidson M, Murray JA, Swift GL, Walker MM, Zingone F, Sanders DS, BSG coeliac disease guidelines development group, british society of gastroenterology (2014) Diagnosis and management of adult coeliac disease: guidelines from the British Society of Gastroenterology. Gut 63:1210–1228. https://doi.org/10.1136/gutjnl-2013-306578

Manza F, Lungaro L, Costanzini A, Caputo F, Carroccio A, Mansueto P, Seidita A, Raju SA, Volta U, De Giorgio R, Sanders DS, Caio G (2025) Non-celiac gluten/wheat sensitivity-state of the art: a five-year narrative review. Nutrients 17:220. https://doi.org/10.3390/nu17020220

Marín-Sanz M, Barro F (2022) RNAi silencing of wheat gliadins alters the network of transcription factors that regulate the synthesis of seed storage proteins toward maintaining grain protein levels. Front Plant Sci 13:935851. https://doi.org/10.3389/fpls.2022.935851

Marjanović-Balaban Ž, Cvjetković VG, Grujić R (2021) Gliadin proteins from wheat flour: the optimal determination conditions by ELISA. Foods and Raw Materials 9:364–370. https://doi.org/10.21603/2308-4057-2021-2-364-370

Masci S, Lafiandra D, Porceddu E, Lew EL, Tao HP, Kasarda DD (1993) D-glutenin subunits: N-terminal sequences and evidence for the presence of cysteine. Cereal Chem 70:581–585

Matsuo H, Morita E, Tatham AS, Morimoto K, Horikawa T, Osuna H, Ikezawa Z, Kaneko S, Kohno K, Dekio S (2004) Identification of the IgE-binding epitope in omega-5 gliadin, a major allergen in wheat-dependent exercise-induced anaphylaxis. J Biol Chem 279:12135–12140. https://doi.org/10.1074/jbc.M311340200

Matsuo H, Kohno K, Niihara H, Morita E (2005) Specific IgE determination to epitope peptides of omega-5 gliadin and high molecular weight glutenin subunit is a useful tool for diagnosis of wheat-dependent exercise-induced anaphylaxis. J Biol Chem 175:8116–8122. https://doi.org/10.4049/jimmunol.175.12.8116

Metakovsky EV, Chernakov VM, Upelniek VP, Redaelli R, Dardevet M, Branlard G, Pogna NE (1996) Recombination mapping of minor ω-gliadin-coding loci on chromosome 1A of common wheat: a revision. J Genetic Breed 50:277–286

Metakovsky E, Melnik V, Rodriguez-Quijano M, Upelniek V, Carrillo JM (2018) A catalog of gliadin alleles: polymorphism of 20th-century common wheat germplasm. Crop J 6:628–641. https://doi.org/10.1016/j.cj.2018.02.003

Moehs CP, Austill WJ, Holm A, Large TAG, Loeffler D, Mullenberg J, Schnable PS, Skinner W, van Boxtel J, Wu L, McGuire C (2019) Development of decreased-gluten wheat enabled by determination of the genetic basis of *lys3a* barley. Plant Physiol 179:1692–1703. https://doi.org/10.1104/pp.18.00771

Morris CF (2002) Puroindolines: the molecular genetic basis of wheat grain hardness. Plant Mol Biol 48:633–647. https://doi.org/10.1023/a:1014837431178

Morris CF, Luna J, Caffe-Treml M (2021) The vromindolines of cv. Hayden oat (*Avena sativa* L.)—a review of the Poeae and Triticeae indolines and a suggested system for harmonization of nomenclature. J Cereal Sci 97:103135. https://doi.org/10.1016/j.jcs.2020.103135

Mukhtiar A, Mahmood A, Khan MA, Ameen M, Al-Khayri JM, Qari SH (2025) Transforming field crops with CRISPR/Cas: a new era in genome editing. Rend Fis Acc Lincei 36:195–208. https://doi.org/10.1007/s12210-025-01308-6

Mustalahti K, Catassi C, Reunanen A, Fabiani E, Heier M, McMillan S, Murray L, Metzger M, Gasparin M, Bravi E, Mäki M (2010) The prevalence of celiac disease in Europe: results of a centralized, international mass screening project. Ann Med 42:587–595. https://doi.org/10.3109/07853890.2010.505931

Nigro D, Smedley MA, Camerlengo F, Hayta S (2024) Using gene editing strategies for wheat improvement. In: Ricroch A, Eriksson D, Miladinović J, Sweet K, Van Laere E, Woźniak-Gientka E (eds) A roadmap for plant genome editing. Springer, Cham, pp 183–202. https://doi.org/10.1007/978-3-031-46150-7_12

Palosuo K, Varjonen E, Kekki O, Klemola T, Kalkkinen N, Alenius H, Reunala T (2001) Wheat ω-5 gliadin is a major allergen in children with immediate allergy to ingested wheat. J Allergy Clin Immunol 108:634–638. https://doi.org/10.1067/mai.2001.118602

Payne PI (1987) Genetics of wheat storage proteins and the effect of allelic variation on bread-making quality. Annu Rev Plant Physiol 38:141–153. https://doi.org/10.1146/annurev.pp.38.060187.001041

Pistón F, Gil-Humanes J, Rodríguez-Quijano M, Barro F (2011) Down-regulating γ-gliadins in bread wheat leads to non-specific increases in other gluten proteins and has no major effect on dough gluten strength. PLoS One 6:e24754. https://doi.org/10.1371/journal.pone.0024754

Pistón F, Gil-Humanes J, Barro F (2013) Integration of promoters, inverted repeat sequences and proteomic data into a model for high silencing efficiency of coeliac disease related gliadins in bread wheat. BMC Plant Biol 13:136. https://doi.org/10.1186/1471-2229-13-136

Raffan S, Sparks C, Huttly A, Hyde L, Martignago D, Mead A, Hanley SJ, Wilkinson PA, Barker G, Edwards KJ, Curtis TY, Usher S, Kosik O, Halford NG (2021) Wheat with greatly reduced accumulation of free asparagine in the grain, produced by CRISPR/Cas9 editing of asparagine synthetase gene *TaASN2*. Plant Biotechnol J 19:1602–1613. https://doi.org/10.1111/pbi.13573

Ran Y, Patron N, Kay P, Wong D, Buchanan M, Cao YY, Sawbridge T, Davies JP, Mason J, Webb SR, Spangenberg G, Ainley WM, Walsh TA, Hayden MJ (2018) Zinc finger nuclease-mediated precision genome editing of an endogenous gene in hexaploid bread wheat (*Triticum aestivum*) using a DNA repair template. Plant Biotechnol J 16:2088–2101. https://doi.org/10.1111/pbi.12941

Reig-Otero Y, Mañes J, Manyes L (2018) Amylase-trypsin inhibitors in wheat and other cereals as potential activators of the effects of nonceliac gluten sensitivity. J Med Food 21:207–214. https://doi.org/10.1089/jmf.2017.0018

Rottersman MG, Zhang W, Zhang J, Grigorian G, Burguener G, Carter C, Vang T, Hegarty J, Zhang X, Dubcovsky J (2024) Deletion of wheat alpha-gliadins from chromosome 6D improves gluten strength and reduces immunodominant celiac disease epitopes. bioRxiv. https://doi.org/10.1101/2024.07.19.604379

Sánchez-León S, Gil-Humanes J, Ozuna CV, Giménez MJ, Sousa C, Voytas DF, Barro F (2018) Low-gluten, nontransgenic wheat engineered with CRISPR/Cas9. Plant Biotechnol J 16:902–910. https://doi.org/10.1111/pbi.12837

Sánchez-León S, Marín-Sanz M, Guzmán-López MH, Gavilán-Camacho M, Simón E, Barro F (2024) CRISPR/Cas9-mediated multiplex gene editing of gamma and omega gliadins, paving the way for gliadin-free wheat. J Exp Bot 75:7079–7095. https://doi.org/10.1093/jxb/erae376

Sanchez-Monge R, Gomez L, García-Olmedo F, Salcedo G (1989) New dimeric inhibitor of heterologous α-amylases encoded by a duplicated gene in the short arm of chromosome 3B of wheat (*Triticum aestivum* L.). Eur J Biochem 183:37–40. https://doi.org/10.1111/j.1432-1033.1989.tb14893.x

Sanchez-Monge R, Gomez L, Barber D, Lopez-Otin C, Armentia A, Salcedo G (1992) Wheat and barley allergens associated with baker's asthma. Glycosylated subunits of the α-amylase-inhibitor family have enhanced IgE-binding capacity. Biochem J 281:401–405. https://doi.org/10.1042/bj2810401

Schenke D, Cai D (2020) Applications of CRISPR/CAS to improve crop disease resistance: beyond inactivation of susceptibility factors. iScience 23:101478. https://doi.org/10.1016/j.isci.2020.101478

Scherf KA, Brockow K, Biedermann T, Koehler P, Wieser H (2016) Wheat-dependent exercise-induced anaphylaxis. Clin Exp Allergy 46:10–20. https://doi.org/10.1111/cea.12640

Shewry P (2019) What is gluten-why is it special? Front Nutr 6:101. https://doi.org/10.3389/fnut.2019.00101

Shewry P (2022) Wheat grain proteins: past, present, and future. Cereal Chem 100:9–22. https://doi.org/10.1002/cche.10585

Shewry PR, Belton PS (2024) What do we really understand about wheat gluten structure and functionality? J Cereal Sci 117:103895. https://doi.org/10.1016/j.jcs.2024.103895

Shewry PR, Halford NG (2002) Cereal seed storage proteins: structures, properties and role in grain utilization. J Exp Bot 53:947–958. https://doi.org/10.1093/jexbot/53.370.947

Shewry PR, Hey SJ (2015) The contribution of wheat to human diet and health. Food Energy Secur 4:178–202. https://doi.org/10.1002/fes3.64

Shewry PR, Tatham AS (2016) Improving wheat to remove coeliac epitopes but retain functionality. J Cereal Sci 67:12–21. https://doi.org/10.1016/j.jcs.2015.06.005

Smulders MJ, Gilissen LJ, Juranić M, Schaart JG, van de Wiel CC (2024) Gene editing of wheat to reduce coeliac disease epitopes in gluten. In: Ricroch A, Eriksson D, Miladinović D, Sweet J, Van Laere K, Woźniak-Gientka E (eds) A roadmap for plant genome editing. Springer Nature Switzerland, Cham, pp 203–222. https://doi.org/10.1007/978-3-031-46150-7_13

Sollid LM, Tye-Din JA, Qiao S, Anderson RP, Gianfrani C, Koning F (2019) Update 2020: nomenclature and listing of celiac disease-relevant gluten epitopes recognized by CD4+ T cells. Immunogenetics 72:85–88. https://doi.org/10.1007/s00251-019-01141-w

Su W, Xu M, Radani Y, Yang L (2023) Technological development and application of plant genetic transformation. Int J Mol Sci 24:10646. https://doi.org/10.3390/ijms241310646

Sun Z, Zhang M, An Y, Han X, Guo B, Lv G, Zhao Y, Guo Y, Li S (2022) CRISPR/Cas9-mediated disruption of xylanase inhibitor protein (XIP) gene improved the dough quality of common wheat. Front Plant Sci 13:811668. https://doi.org/10.3389/fpls.2022.811668

Taj M, Sajjad M, Li M, Yasmeen A, Mubarik MS, Kaniganti S, He C (2022) Potential targets for CRISPR/Cas knockdowns to enhance genetic resistance against some diseases in wheat (*Triticum aestivum* L.). Front Genet 13:926955. https://doi.org/10.3389/fgene.2022.926955

Tatham AS, Shewry PR (2008) Allergens to wheat and related cereals. Clin Exp Allergy 38:1712–1726. https://doi.org/10.1111/j.1365-2222.2008.03101.x

Tye-Din JA, Stewart JA, Dromey JA, Beissbarth T, van Heel DA, Tatham A, Henderson K, Mannering SI, Gianfrani C, Jewell DP, Hill AV, McCluskey J, Rossjohn J, Anderson RP (2010)

Comprehensive, quantitative mapping of T cell epitopes in gluten in celiac disease. Sci Transl Med 2:41ra51. https://doi.org/10.1126/scitranslmed.3001012
van den Broeck HC, America AH, Smulders MJ, Bosch D, Hamer RJ, Gilissen LJ, van der Meer IM (2009) A modified extraction protocol enables detection and quantification of celiac disease-related gluten proteins from wheat. J Chromatogr B Analyt Technol Biomed Life Sci 877:975–982. https://doi.org/10.1016/j.jchromb.2009.02.035
van Herpen TW, Goryunova SV, van der Schoot J, Mitreva M, Salentijn E, Vorst O, Schenk MF, van Veelen PA, Koning F, van Soest LJ, Vosman B, Bosch D, Hamer RJ, Gilissen LJ, Smulders MJ (2006) Alpha-gliadin genes from the A, B, and D genomes of wheat contain different sets of celiac disease epitopes. BMC Genomics 7:1. https://doi.org/10.1186/1471-2164-7-1
Vereijken J, Klostermann V, Beckers F, Spekking W, Graveland A (2000) Intercultivar variation in the proportions of wheat protein fractions and relation to mixing behaviour. J Cereal Sci 32:159–167. https://doi.org/10.1006/jcrs.2000.0333
Verma AK, Mandal S, Tiwari A, Monachesi C, Catassi GN, Srivastava A, Gatti S, Lionetti E, Catassi C (2021) Current status and perspectives on the application of CRISPR/CAS9 gene-editing system to develop a low-gluten, non-transgenic wheat variety. Foods 10:2351. https://doi.org/10.3390/foods10102351
Voytas DF, Gao C (2014) Precision genome engineering and agriculture: opportunities and regulatory challenges. PLoS Biol 12:e1001877. https://doi.org/10.1371/journal.pbio.1001877
Vriz R, Moreno FJ, Koning F, Fernandez A (2021) Ranking of immunodominant epitopes in celiac disease: identification of reliable parameters for the safety assessment of innovative food proteins. Food Chem Toxicol 157:112584. https://doi.org/10.1016/j.fct.2021.112584
Waga J, Skoczowski A (2014) Development and characteristics of ω-gliadin-free wheat genotypes. Euphytica 195:105–116. https://doi.org/10.1007/s10681-013-0984-1
Waga J, Zientarski J, Szaleniec M, Obtułowicz K, Dyga W, Skoczowski A (2013) Null alleles in gliadin coding loci and wheat allergenic properties. Am J Plant Sci 04:160–168. https://doi.org/10.4236/ajps.2013.41021
Wan Y, Shewry PR, Hawkesford MJ (2013) A novel family of γ-gliadin genes are highly regulated by nitrogen supply in developing wheat grain. J Exp Bot 64:161–168. https://doi.org/10.1093/jxb/ers318
Wang Y, Cheng X, Shan Q, Zhang Y, Liu J, Gao C, Qiu J (2014) Simultaneous editing of three homoeoalleles in hexaploid bread wheat confers heritable resistance to powdery mildew. Nat Biotechnol 32:947–951. https://doi.org/10.1038/nbt.2969
Wang DW, Li D, Wang J, Zhao Y, Wang Z, Yue G, Liu X, Qin H, Zhang K, Dong L, Wang D (2017) Genome-wide analysis of complex wheat gliadins, the dominant carriers of celiac disease epitopes. Sci Rep 7:44609. https://doi.org/10.1038/srep44609
Wang W, Pan Q, Tian B, Yu Z, Davidson D, Bai G, Akhunova A, Trick H, Akhunov E (2025) Non-additive dosage-dependent effects of *TaGS3* gene editing on grain size and weight in wheat. Theor Appl Genet 138:38. https://doi.org/10.1007/s00122-025-04827-w
Wieser H (2006) Chemistry of gluten proteins. Food Microbiol 24:115–119. https://doi.org/10.1016/j.fm.2006.07.004
Wieser H, Koehler P, Folck A, Becker D (2007) Characterization of wheat with strongly reduced α-gliadin content. In: Lookhart GL, Ng PKW (eds) Gluten proteins 2006 AACC, St Paul, pp 13–16
Wieser H, Segura V, Ruiz-Carnicer Á, Sousa C, Comino I (2021) Food safety and cross-contamination of gluten-free products: a narrative review. Nutrients 13:2244. https://doi.org/10.3390/nu13072244
Wieser H, Koehler P, Scherf KA (2023a) Chemistry of wheat gluten proteins: qualitative composition. Cereal Chem 100:23–35. https://doi.org/10.1002/cche.10572
Wieser H, Koehler P, Scherf KA (2023b) Chemistry of wheat gluten proteins: quantitative composition. Cereal Chem 100:36–55. https://doi.org/10.1002/cche.10553

Willenborg CJ, van Acker RC (2008) The biology and ecology of hexaploid wheat (*Triticum aestivum* L.) and its implications for trait confinement. Can J Plant Sci 88:997–1013. https://doi.org/10.4141/CJPS07144

Woychik JH, Boundy JA, Dimler RJ (1961) Starch gel electrophoresis of wheat gluten proteins with concentrated urea. Arch Biochem Biophys 94:477–482. https://doi.org/10.1016/0003-9861(61)90075-3

Yamada Y, Yokooji T, Ninomiya N, Taogoshi T, Morita E, Matsuo H (2019) Evaluation of the allergenicity of ω5-gliadin-deficient Hokushin wheat (1BS-18) in a wheat allergy rat model. Biochem Biophys Rep 20:100702. https://doi.org/10.1016/j.bbrep.2019.100702

Yamada Y, Yokooji T, Kunimoto K, Inoguchi K, Ogino R, Taogoshi T, Morita E, Matsuo H (2022) Hypoallergenic wheat line (1BS-18H) lacking ω5-gliadin induces oral tolerance to wheat gluten proteins in a rat model of wheat allergy. Foods 11:2181. https://doi.org/10.3390/foods11152181

Yao Y, Guo W, Gou J, Hu Z, Liu J, Ma J, Zong Y, Xin M, Chen W, Li Q, Wang Z, Zhang R, Uauy C, Baloch FS, Ni Z, Sun Q (2025) Wheat2035: Integrating pan-omics and advanced biotechnology for future wheat design. Mol Plant 18:272–297. https://doi.org/10.1016/j.molp.2025.01.005

Yu Z, Yunusbaev U, Fritz A, Tilley M, Akhunova A, Trick H, Akhunov E (2024) CRISPR-based editing of the ω- and γ-gliadin gene clusters reduces wheat immunoreactivity without affecting grain protein quality. Plant Biotechnol J 22:892–903. https://doi.org/10.1111/pbi.14231

Zapatero L, Martínez MI, Alonso E, Salcedo G, Sánchez-Monge R, Barber D, Lombardero M (2003) Oral wheat flour anaphylaxis related to wheat α-amylase inhibitor subunits CM3 and CM16. Allergy 58:956. https://doi.org/10.1034/j.1398-9995.2003.00158.x

Zhang S, Zhang R, Song G, Gao J, Li W, Han X, Chen M, Li Y, Li G (2018) Targeted mutagenesis using the *Agrobacterium tumefaciens*-mediated CRISPR-Cas9 system in common wheat. BMC Plant Biol 18:302. https://doi.org/10.1186/s12870-018-1496-x

Zhang S, Zhang R, Gao J, Gu T, Song G, Li W, Li D, Li Y, Li G (2019) Highly efficient and heritable targeted mutagenesis in wheat via the *Agrobacterium tumefaciens*-mediated CRISPR/CAS9 system. Int J Mol Sci 20:4257. https://doi.org/10.3390/ijms20174257

Zhang Y, Pribil M, Palmgren M, Gao C (2020) A CRISPR way for accelerating improvement of food crops. Nat Food 1:200–205. https://doi.org/10.1038/s43016-020-0051-8

Zhang S, Zhang R, Gao J, Song G, Li J, Li W, Qi Y, Li Y, Li G (2021) CRISPR/Cas9-mediated genome editing for wheat grain quality improvement. Plant Biotechnol J 19:1684–1686. https://doi.org/10.1111/pbi.13647

Zhou X, Zhao Y, Ni P, Ni Z, Sun Q, Zong Y (2023) CRISPR-mediated acceleration of wheat improvement: advances and perspectives. J Genet Genomics 50:815–834. https://doi.org/10.1016/j.jgg.2023.09.007

Chapter 9
Harnessing CRISPR/Cas Gene Editing for Wheat Improvement: Advancing Yield, Quality, and Resilience in a Changing Climate

Sayanti Mandal, Swati Hazra, Ramsha Syed, Ansu Chetri, Deepak Kumar Upadhyay, Enakshee Samanta, Kashish Vats, and Himani Thakur

Abstract Population growth, famines, climate change, and other issues are just some of the barriers that agriculture needs to overcome to provide food to meet world population needs while maintaining food safety and nutritional value. The wheat (*Triticum aestivum*) is enormously important for human nutrition, as a source of carbohydrates (and thus calories), fiber, protein, vitamins, and micronutrients. It is the second most-produced crop after maize (*Zea mays*) with 808 million tons of grain produced in 2021–2022. Plant breeding has played an important role in meeting the ever-growing demand for agricultural produce and will do so for decades to come. This will be aided by new technologies, and here we discuss how clustered regularly interspaced short palindromic repeats (CRISPR)-associated (Cas) nucleases technology can be harnessed for the improvement of agricultural genetics and wheat. The emerging applications and implications of the CRISPR-Cas system have raised new regulatory issues globally, from moral, ethical, social, safety, and technology points of view regarding its applications in preclinical and clinical research, biomedicine, and agriculture. Most importantly, this chapter also concentrates on the ethical and societal issues and further challenges related to the use of the CRISPR-Cas system in plant genome editing.

Keywords Genome editing · Wheat CRISPR · Crop improvement · Disease resistance · Multi-stress tolerant · *de-novo* domestication · Ethics

S. Mandal (✉) · R. Syed · A. Chetri · D. K. Upadhyay
Department of Chemistry and Biochemistry, Sharda School of Engineering and Sciences, Sharda University, Greater Noida, India

S. Hazra · E. Samanta · K. Vats · H. Thakur
Department of Life Science, Sharda School of Bio-Science and Technology, Sharda University, Greater Noida, India

K. A. Abd-Elsalam, A. Ahmad (eds.), *Gene Editing in Cereals*, Sustainability Sciences in Asia and Africa, https://doi.org/10.1007/978-981-95-6722-5_9

9.1 Introduction

One of the most widespread crops grown on earth is wheat (*Triticum aestivum*), even when the world produces too much food to ensure that there is enough food. Wheat is a major crop that is cultivated across the globe (Erenstein et al. 2022). To ensure that no one goes hungry, the world agricultural industry produces a diverse assortment of food crops. According to the Ethiopian Institute of Agriculture Research, 7.5% of the population relies on wheat. Wheat has many foods that are derived. Due to it, wheat plays a critical role in the life of people (Senbeta and Worku 2023; Mesfin et al. 2024). All around the world, there are numerous types of wheat harvests. Some of these factors, which may pose a risk to food security, include infections of wheat and weather variation. It is also revealed that 21.5% of all losses in wheat negatively affect food security. The types of wheat crops grown in the world are numerous (Farhad et al. 2023). Problems that can complicate the process of providing food security include climate change and diseases that can infect wheat. Savary et al. (2019) state that the losses of wheat globally are 21.5% and this affects food security. Wheat is grown on more than 220 million hectares of land all over the world. Wheat yields 778 million metric tonnes per annum with a value added by a factor of 10 in the agricultural sector (Grosse-Heilmann et al. 2024). It produces an average of 3.5 tons/sq. meter annually. It contains 1.8% fiber, 9.4% protein, 69% carbohydrates, and 2.5% fat. The wheat crop is even more important as the population increases and with it, the demand of food and nutrition increases (Nyaupane et al. 2024). It yields 3.5 tons per hectare annually on average. It contains 1.8% fiber, 9.4% protein, 69% carbohydrates, and 2.5% fat (Ahmed et al. 2022). Now more than ever, the wheat crop is significant because the population is growing, which implies that the demand in terms of food and nourishment is also increasing (Grosse-Heilmann et al. 2024).

Examples of abiotic stressors that complicate the task of feeding the world through wheat include heat stress and drought. In 2021–2022, the worldwide consumption of 889.1 million metric tonnes of wheat was exceeded by only 778 million metric tonnes produced. This translated to a deficit of 111.1 million metric tonnes of wheat (15%) in the market relative to the required quantity (Ko et al. 2025). That wheat would have been sufficient to feed approximately 1 billion people, which is 12.5% of the world population (FAOSTAT, 2022). The solution to the food shortage due to the lack of wheat grown might be climate-tolerant genotypes (Bhandari et al. 2023). The effects of climate change and wheat productivity remain; the temperature might increase to 1.07 °C in 2045 AD (Bhandari et al. 2023). Due to the gradual increase in population in the region, wheat demand will increase two-fifths by the year 2050. Meanwhile, climate-related challenges such as heat distress and drought will reduce wheat yields in South Asia by 44–47% (Tran et al. 2025). Thus, the market segment and wheat demand will increase yearly. Wheat is the most significant crop globally, yet we must produce more to provide to the entire globe (Erenstein et al. 2022). Although much effort has been made to breed, the production of valuable crops such as wheat has stalled (Erenstein et al. 2022). Due to the large impact of climate-related abiotic stresses on wheat production, the development of climate-resilient genotypes is crucial to ensure the safety of food and nutrition globally. The negative impact of abiotic stressors on the productivity of wheat requires the development of genotypes that are resistant to the effects of climate change (Benitez-Alfonso et al. 2023).

Plant breeding using genome editing technology has been implemented in wheat among other species (Waites et al. 2025). These technologies are potentially high. Tools such as zinc finger nucleases (ZFN) to modify the Acetohydroxyacid synthase gene (Ran et al. 2018) and transcription activator-like effector nucleases (TALENs) to modify the Mildew resistance locus O (TaMLO1) gene have enabled allohexaploid wheat to be modified to a small number of genes. This has caused the blanket resistance to powdery mildew (Ran et al. 2018; Ahmar et al. 2019). However, both approaches have their issues, such as the time and cost-intensive task of preparing protein domains to discover the exact editing locations under attack. The most popular method of editing genomes, however, is now the clustered regularly interspaced short palindromic repeat (CRISPR)/CRISPR-associated nuclease (Cas) system. This is primarily because it is highly precise, convenient, adaptable, effective and can edit up to or more than one genome simultaneously (Akram et al. 2023). Recently developed and improved CRISPR/Cas-based genome editing systems have since been used to edit common wheat. Base editing (BE), prime editing (PE), targeted gene mutagenesis, and multiplex genome editing are some of those systems (Li et al. 2021a, b). What the field needs is that these systems provide breeders of wheat with genetic tools that are more precise and effective. Regardless of these advances, the genomes of other valuable crops have been edited more frequently than the wheat genome (Brock et al. 2025). Even with these improvements, wheat genome editing has fallen behind those of other agriculturally important crops because wheat has a complex genome, polyploid, and doesn't work as well when it comes to transformation (Schaart and Smulders 2021). All of them state that the quality of the available wheat genomes sequencing data and the increasing amount of next-generation sequencing and multi-omics data are probably the primary factors that will make people begin applying CRISPR-based genome editing to wheat (Akram et al. 2025).

First, we are discussing the current work of genome editing in wheat with the help of CRISPR/Cas. In the following section, we will discuss the general processes that are involved in editing the wheat genome. We also discuss the latest applications like base edit (BE) and prime edit (PE) in wheat improvement, which are designed to increase yield and development, end-use quality, resistance to diseases, and abiotic stress. We intend to provide a comprehensive description of the recent advances in CRISPR/Cas technologies and their wider application in wheat. We will also consider the ethical issues and opportunities that await the use of CRISPR in the future.

9.2 CRISPR Working Mechanism

CRISPR functions as an adaptive immune response in Prokaryotes specifically in bacteria and archaea. When a bacterium gets invaded by exogenous DNA (such as that from viruses or plasmids), then the segments of the genomic DNA are incorporated into the host's CRISPR array, which forms a record of past encounters (Li et al. 2024). These sequences are transcribed upon successive invasions that guide the Cas proteins to recognize and cleave matching foreign DNA. In CRISPR system particularly has two main components that are Cas9 endonuclease (it is a large, multidomain protein that has different regions of recognition and cleavage, as well as a region that interfaces with PAM) and a Guide RNA (gRNA) (it is a fusion of crRNA, which determines the Target DNA) (Bhatia et al. 2023).

Cas9 in its inactive form undergoes conformational activation upon sgRNA binding leading to transitioning into a target-searching state. The Cas9–sgRNA forms a complex that scans the genomic DNA for PAM (protospacer adjacent motif). It is a short sequence that is downstream of the target region and is necessary for Cas9 binding (Bhatia et al. 2023). Once a PAM is determined, Cas9 destabilizes the DNA duplex, which allows the sgRNA to pair with the complementary strand leading to the formation of R-loop (an RNA–DNA hybrid). Upon the subsequent base pairing of sgRNA, the conformational rearrangements in Cas9 are triggered, which leads to the activation of nuclease domains. The HNH domain cleaves the complementary DNA strand, which guides the RNA, whereas the RuvC domain cleaves the non-complementary DNA strand. This cleavage basically occurs three base pairs upstream of the PAM that creates a double-strand break (DSB) with non-cohesive ends (Pacesa et al. 2022). After the formation of DSB (double-strand break), the eukaryotic cell follows the two primary repair pathways.

Nonhomologous End Joining (NHEJ) It is a repair pathway that rejoins the non-cohesive ends of the DNA through insertions or deletions, which results in gene disruption (Stinson and Loparo 2021).

Homology-Directed Repair (HDR) It is a repair mechanism that uses the template of the Homologous DNA to precisely insert or replace genetic sequences (Stinson and Loparo 2021).

The CRISPR–Cas system in prokaryotes (bacteria and archaea) consists of three core stages (Rath et al. 2015):

1. *Adaptation:* It is the first stage of the CRISPR–Cas system, where it determines and captures the short DNA sequences from invading foreign DNA (Viruses or plasmids). These are called protospacers that are incorporated into the host genome as *spacers*.
2. *crRNA Biogenesis:* During this stage, the transcription and processing of the CRISPR array occur to produce the CRISPR RNAs (crRNAs). The CRISPR array is first transcribed into pre-crRNA, and then it further gets processed into mature crRNAs.

3. *Interference:* It is the final stage in which the CRISPR–Cas effector complex uses the crRNA for identification and neutralization of DNA or RNA on the basis of sequence complementarity. This stage degrades the invader's DNA, which prevents infection or replication.

9.3 CRISPR Base Editing and Prime Editing

Among the most popular genetic tools to edit genomes is known as CRISPR-Cas. When RNA-guided Cas nucleases create a DSB in DNA, cellular repair mechanisms, NHEJ, imprecise repair, HDR, or precise repair are activated. Conversely, in most cases, DNA double-strand breaks cause unexpected changes in the DNA and mortality in certain species. There is a need to possess efficient and accurate editing tools to have bacteria and plants as useful bioproduction platforms. Therefore, the most discussed in this chapter are the non-DSB and template-free methods of genome editing that can be referred to as base editing (BE) and prime editing (PE). The methods are applied to bacteria and plants. We shall start by discussing about the base and prime editors, and by describing the research done on crop plants.

9.3.1 Base Editing in Wheat

In the absence of donor templates, double-stranded DNA breaks, or the unintended products of the NHEJ and HDR mechanisms, base editing is a paradigm shift in genome editing technology, which creates transition and transversion mutations at the single-base level (Xu et al. 2024a, b; Prakash et al. 2025) (Fig. 2.1a). Base editors (BEs) promise much in terms of applications in plant genome editing because this gene editing method has much fewer undesired changes (Chen and Liu 2022). A base editor uses a nucleotide deaminase together with a catalytically inactivated Cas protein to create a base edit at a given position in either DNA or RNA (Yoon et al. 2023). When the Cas protein denatures the double-stranded DNA, a small fragment of single-stranded DNA is exposed due to an initial interaction of the Cas protein-gRNA complex with its target locus in the DNA, forming an R-loop (Paul and Montoya 2020). That single-stranded DNA base convergence is then catalyzed by the deaminase enzyme. Finally, the target area was formed because of DNA repair and replication through the permanent insertion of single-base changes (Gong et al. 2018).

The first base editors developed were cysine base editors (CBEs), which change cytosine (C) to thymine (T) and guanine (G) to adenine (A) within the target sequence (Luo et al. 2023) (Fig. 9.1). In 2016, the David Liu lab made the first-generation base-editor (CBE1): by inserting a 16 amino XTEN linker between rATP cytidine deaminase and the N terminus of dCas9 (dead Cas9) (Komor et al. 2016). CBE1 can successfully convert C:G to T:A in vitro; however, in vivo the base

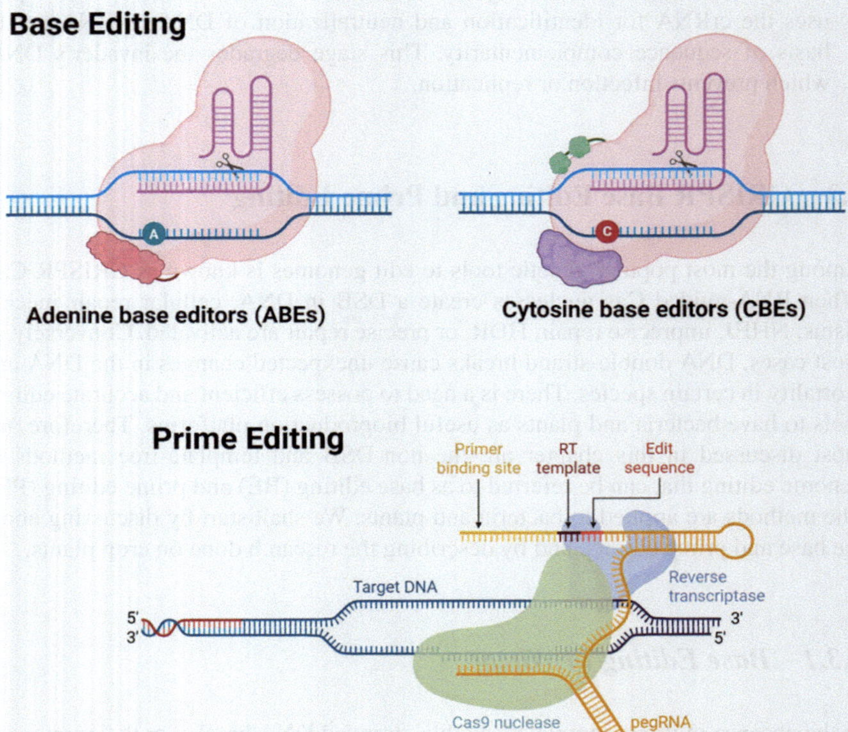

Fig. 9.1 Base editing using adenine base, cytosine base editors, and prime editing. (Images created with BioRender)

excision repair (BER) identifies any G:U base pair as a mismatch and removes the uracil's through uracil N-glycosylases (UNGs). In order to circumvent this limitation and increase productivity, CBE2 was developed to include an uracil DNA glycosylase inhibitor (UGI) covalently conjugated to the C-terminus of dCas9 through a 4-amino acid linker (Beger et al. 1995). Since UGI prevents uracil DNA glycosylase (UDG) in the organism, this novel version is three times more effective at editing than CBE1. CBE3 was developed by replacing dCas9 with a nickase-like nCas9 (H840, HNH catalytic domain) variant to enhance the editing capability. In the novel form, nCas9 would cause a nick in the DNA strand containing G, which would activate the cellular mismatch repair (MMR) process (Kantor et al. 2020). This MMR process replaces G on the damaged strand with an A creating U:A pairing with the target strand. U:A pair eventually was fixed, and the necessary T:A replacement was achieved. Although using UGI to inhibit BER in CBE2 enhanced editing efficiency up to threefold, nicking used in CBE3 enhanced efficiencies up to six times higher than in CBE2 in human cells (Jiang and Yao 2022). Plant CBE was initially applied to rice and subsequently to other plants with diverse objectives, such as the acquisition of disease and herbicide resistance, agricultural domestication, and yield and improvement of nutrient utilization efficiency in different plants (Xu et al. 2024a, b). Table 9.1 summarizes CBE-mediated genome editing across a wide range of objectives. Table 9.1 displays uses of base editors in plants with a focus on plant species, gene of interest, type of BE, purpose of the intended modification, and reagent delivery plan.

9.3.2 Prime Editing in Wheat

In 2019, David Lui et al. announced prime editing, a form of search and replace that can make any possible edits, including all 12 possible base-to-base replacement or insertion or deletion mutations, without breaking a strand or creating a donor DNA template (Komor et al. 2016; Wanikar et al. 2025) (Fig. 9.1). Prime editor is a versatile and precise genome editing tool comprised of a Cas9 nick (nCas9; H840A mutation) conjugated to an altered reverse transcriptase (RT) and a prime editing guide RNA (pegRNA) comprising a primer binding site (PBS) and an RT template. When pegRNA is introduced to a cell, a nick (nCas9; H840A mutation) recognizes the noncomplementary strand of DNA and cleaves it three bases above the PAM site (Li et al. 2025). The PBS is hybridized with the bases above the nick created by the nCas9 (H840A), and the RT template contains the necessary alterations and undergoes reverse transcription. The desired modification is inserted in the new DNA, and the strand that has not been modified is also adjusted to match the modified strand through a cellular DNA repair process (Chen et al. 2022).

In the original publication about the topic, Anzalone et al. presented three major versions of editing systems. The first prime editor (PE1) uses wild-type reverse transcriptase of the commercial Moloney murine leukemia virus (M-MLV), a pegRNA, and Cas9 (H840A) nickase (Lu et al. 2022). The effectiveness of prime editing has been explored in a range of RT mutations that alter RNaseH activity, thermostability, processivity, and affinity of DNA to RNA substrates. PE1 activity was increased by the presence of the targeted M-MLV-RT pentamutant (M-MLV RT (D200N/L603W/T330P/T306K/W313F)) in PE1; hence the name PE2, prime editor 2. PE2 raised on average 1.6 to 5.1 times the editing efficiency to harbor point mutations (Huang and Liu 2023). Furthermore, it suggested that it could work with shorter PBS sequences and that it had a higher efficiency in editing in indels. With an additional sgRNA to direct Cas9 (H840A) nickaze to make a nick in the uncut DNA strand, a better prime editor named PE3 increased the editing efficiency by approximately three times over PE2 (Huang and Liu 2023). The researchers were able to avoid this problem because the position of the extra sgRNA could cause a large number of indels, so the extra sgRNA was designed to complement the modified, but not original, strand. Since it preserved editing efficiency but reduced the

Table 9.1 Applications of prime and base editing in wheat genes

S. No	Target gene	Function	Trait	Editing tool	References
1.	*TaEPSPS*	Glyphosate-resistant	Herbicide resistance	Prime editing	Chen et al. (2025)
2.	*TaGW2*	Modulate weight of the grain	Yield	Prime editing	Ni et al. (2023), Zong et al. (2022) and Lin et al. (2020)
3.	*TaLOX2*	Lipoxygenase gene affecting the storability of wheat grains	Yield	Prime editing	Ni et al. (2023), Zong et al. (2022) and Lin et al. (2020)
4.	*TaRht1*	Regulates plant height	Yield	Prime editing	Ni et al. (2023)
5.	*TaSBElla*	Codes for starch-branching enzyme	End-use quality	Prime editing	Ni et al. (2023) and Zong et al. (2022)
6.	*TcLCYe*	Codes for lycopene epsilon cyclase, also connected with carotenoid biosynthesis	End-use quality	Prime editing	Ni et al. (2023)
7.	*TaBCH2*	Codes for beta-carotene hydroxylase and interlinked with carotenoid biosynthesis	End-use quality	Prime editing	Ni et al. (2023)
8.	*TaMLO1*		Biotic stress	Cytosine base editing	Gaillochet et al. (2023)
9.	*TaWTK3*	Codes for a protein named tandem kinase	Biotic stress	Prime editing	Ni et al. (2023)
10.	*TaDIR*	Detrimental role in the defense response against fusarium crown rot	Biotic stress	Prime editing	Ni et al. (2023)
11.	*TaSINA*	Codes for a protein named E3 ubiquitin ligase protein, associated to heat stress and root growth	Abiotic stress	Prime editing	Ni et al. (2023)
12.	*TaPDS*	Phytoene dehydrogenase, also involved in carotenoid biosynthesis	Others	Prime editing	Ni et al. (2023)
13.	*TaDME*	5-methylcytosine DNA glycosylases	Others	Prime editing	Ni et al. (2023), Zong et al. (2022) and Lin et al. (2020)
14.	*TaVRN3*	MADS box gene FLC in the vernalization pathway	Others	Prime editing	Ni et al. (2023)

(continued)

Table 9.1 (continued)

S. No	Target gene	Function	Trait	Editing tool	References
15.	*TaGW2*	Grain weight modulation	Yield	Adenosine base editing	Han et al. (2025)
16.	*TaDEP1*	Oversees dense particles, erect panicle and high grain number per panicle	Yield	Adenosine base editing	Han et al. (2025)
17.	*TaNAC2*	Transcription factor corresponding to nitrate response and disease resistance	Others	Prime editing	Han et al. (2025)
18.	*TaGASR7*	Oversee grain weight and length	Yield	Prime editing	Lin et al. (2020)
19.	*TaMLO1*		Biotic stress	Prime editing	Lin et al. (2020)
20.	*TaUBI10*	Similar to ubiquitin protein	Others	Prime editing	Lin et al. (2020)
21.	*TaALS*	Encodes for Acetolactate synthase, it plays a crucial role in biosynthesis of branched chain amino acids	Herbicide resistance	Cytosine base editing	Zhang et al. (2019a, b) and Zong et al. (2018)
22.	*TaACCase*	Codes for Acetylase-CoA carboxylase	Herbicide resistance	Cytosine base editing	Zhang et al. (2019a, b)
23.	*TaGW2*	Controls grain weight	Yield	Adenosine base editing	Li et al. (2018a)
24.	*TaHPPD*	Codes for 4-hydroxyphenylpyaunate dioxygenase	Herbicide resistance	Cytosine base editing	Zong et al. (2018)
25.	*TaEPSPS*	Codes for the sixth and the second last enzyme in the shikimate pathway	Herbicide resistance	Adenosine base editing	Li et al. (2018a)
26.	*TaPLA*	Codes for pollen-specific phospholipase	Haploid induction	Cytosine base editing	Zong et al. (2018)
27	*TaLOX2*	Codes for lipoxygenase, which affects the storage of wheat grains	Yield	Cytosine base editing	Zong et al. (2017)

average frequency of indel in human cell lines by 13-fold, this variation of PE3 system is called PE3b (Taha et al. 2022) (Fig. 9.1).

In conclusion, the advent of base editing and prime editing technologies has revolutionized the field of genome editing in plants as they provide fast and precise means of specific genetic modification (Table 9.1). Moreover, single-nucleotide alterations without breaking of the double-stranded DNA have opened new possibilities in crop modification, resistance to diseases, and modification of traits.

Despite the enormous advancements made in this field, there are still issues, such as improving the efficiency and specificity of the editing process, the optimization of delivery to customers, and the off-target effects. However, as base, and prime editing continue to be studied and developed, these two tools have massive potential to grow plant biotechnology and agricultural improvement.

9.4 CRISPR/Cas for Improvement of Wheat

Wheat being the rich source of carbohydrates, essential amino acids, vitamins, and minerals, is one of the major cereals cultivated all over the world. It is consumed by 33% of the global population as their staple food (Grote et al. 2021). Wheat serves more than 20% of the daily calorie and protein intakes of human population worldwide (Gupta et al. 2022; Elsharawy and Refat 2023). As the population continues to grow and arable land becomes scarcer, improvement of wheat productivity has become a critical issue for global food security (Zhou et al. 2023). Although traditional breeding techniques, for example, hybridization, mutation breeding, and marker-assisted selection have traditionally increased yields, they have certain drawbacks of them being time-consuming, labor-intensive, and vulnerable to linkage drag. Therefore, techniques for precise genome-editing have been necessary to modify specific gene of interest in wheat to upgrade their traits to enhance the yield and stress response, including disease resistance and drought tolerance (Fig. 9.2).

9.4.1 Increase in Wheat Yield and Productivity

The development of CRISPR/Cas tool provides a revolutionary method for directly altering the wheat genome to improve agronomic performance and productivity (Kim and Budak 2018; Zhou et al. 2023). The capability of CRISPR/Cas nucleases to alter various alleles is the main advantage of this technology. The precise site-directed "double-strand breaks" (DSBs) are created in DNA at the targeted genomic location by Cas9 endonuclease (or its alternative) with the help of a "single-guide RNA" (sgRNA) (Manghwar et al. 2020). After the induction of DSBs, repair process is turned on by the cell through either "homology-directed repair" (HDR), or "non-homologous end joining" (NHEJ) methods. The HDR, which allows for precise changes through targeted gene insertions or replacements, needs a DNA template ("donor") to repair the breaks in dsDNA during homologous recombination (Molla et al. 2021). This has multiple possibilities, for examples, single-base alterations can be designed; a localized sequence can be differentiated; new protein variants can be developed; and evolution of certain proteins can be sped up to create agricultural breeding lines that can be resilient to both biotic and abiotic stressors (Shimatani et al. 2017). On the other hand, NHEJ, which is often imprecise, usually results in the tiny insertions or deletions (indels), and thus typically accomplishes

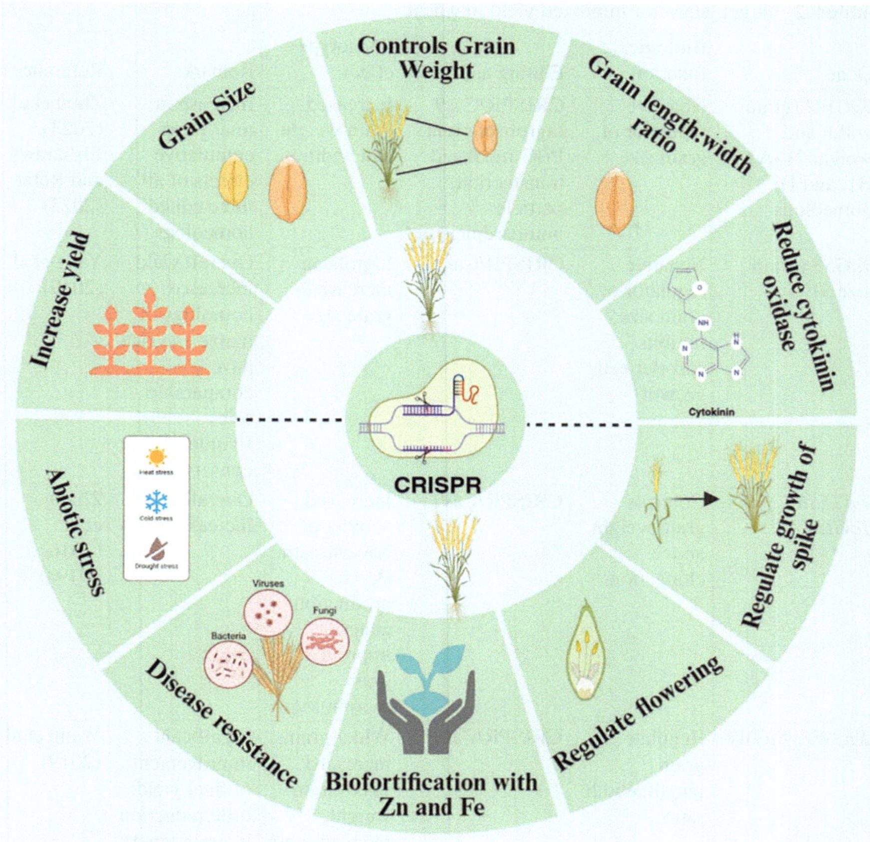

Fig. 9.2 CRISPR-based wheat improvement. (Images created with BioRender)

gene knockouts (Kim and Budak 2018; Elsharawy and Refat 2023). This enables the creation of loss-of-function. Since many of the genes in wheat's allohexaploid genome (AABBDD) have three homologs, editing all copies at once is sometimes necessary to achieve the full gene knockouts. Despite this complexity, more than 80 of wheat are associated with the agronomic traits particularly those affecting yield have been successfully edited using CRISPR/Cas9 technique (Zhou et al. 2023). Various genes that directly or indirectly control agronomic yield and yield parameters (grain size, length, width, grain weight, etc.) of wheat have been targeted by this advanced genome-editing technique (Table 9.2).

Table 9.2 Target genes for improved yield in wheat

Gene	Biological function	Editing approach	Phenotype effect	Remarks	References
TaGW2 (grain width and weight 2) -A1, B1, and D1 homeologs	Negative regulator of grain size	CRISPR/Cas9 (agrobacterium, PEG-mediated transfection, particle bombardment)	Increased grain weight and width	Increase in total yield; cumulative effects of all three edited homeologs	Zhou et al. (2023), Elsharawy and Refat (2023)
TaGS3 (grain size 3)	Negative regulator of grain size; inhibits spikelet cell growth	CRISPR/Cas12a	Significant increase in grain size	Overall yield increased; homologs in maize and rice have comparable roles—Evolutionary conservation	Yang et al. (2019)
TaGASR7, TaGLW7	Controls grain weight and elongation	CRISPR/Cas9	Increased activity of downstream starch production genes; improved grain morphology	Overall yield increase	Zhang et al. (2016, 2019a)
TaGW7, TaGW8	Regulate grain length:Width ratio	CRISPR/Cas9	Wider grains, increased 1000-grain weight; reduced grain length slightly	Significant improvement of final yield; little reduction in grain length	Wang et al. (2019)
TaCKX2-D1 (Cytokinin oxidase/ dehydrogenase)	Reduced cytokinin oxidase (CKX) which breaks down cytokinins	CRISPR/Cas9	Elevated cytokinin levels, enhanced grain filling	More grains per spike; higher yield	Zhang et al. (2019a)
TaDA1	Negative regulator of seed size	CRISPR/Cas9	Larger grains and organs	Increased grain yield	Liu et al. (2020)
TaARF12, TaARF4 (Auxin response factors)	Regulate plant height and growth of spikes	CRISPR-mediated	Optimized biomass partitioning, reduced plant height	Increased spike toughness and grain number	Xu et al. (2020)

(continued)

Table 9.2 (continued)

Gene	Biological function	Editing approach	Phenotype effect	Remarks	References
TaSPL13, *TaSPL16*	Regulation of grain shape, tillering, and flowering time	CRISPR-mediated	Larger grains, more tillers, early blooming	Enhanced yield potential	Li et al. (2022a, b)

9.4.2 Wheat Quality Improvement and Biofortification

The CRISPR/Cas technique plays a vital role in the enhancement of wheat nutritional composition. Micronutrient deficiency or "hidden hunger" among women and children has emerged as a serious health concern of current time (Prasad 1996; WHO 2025). The rising atmospheric CO_2 level has further complicated the situation by imparting the "dilution effect" on grain micronutrients, especially on iron (Fe) and zinc (Zn) concentration (Wang and Liu 2021). The CRISPR/Cas9-mediated disruption of TaIPK1 significantly reduced the levels of phytic acid while increased Fe and Zn accumulations in wheat grains (Ibrahim et al. 2022). This might be useful in tackling the "dilution" of grain Fe and Zn. A significant amount of protein is present in wheat grains in the form of gluten, which can induce moderate to severe allergic reactions among the people having gluten disorder (Cabanillas 2020). Therefore, CRISPR/Cas9 technique has been deployed to edit the genes encoding α-gliadin, a constituent of gluten, to develop "hypoimmunogenic" wheat varieties with reduced glutens (Sánchez-León et al. 2018; Jouanin et al. 2019; Verma et al. 2022). Nevertheless, CRISPR/Cas9-mediated targeted mutagenesis of *TaSBEII* a gene has been achieved to modify the starch composition of wheat varieties with high-amylose supporting modern human diet requirements (Li et al. 2021a, b).

9.4.3 Improvement of Wheat Stress Resilience

Climatic changes put a negative impact on the growth of the plants, heavily affecting 20% of wheat productivity due to the various biotic and abiotic factors (Savary et al. 2019). Abiotic factors like heat and drought cause a major wheat yield loss across the globe (Pequeno et al. 2024). Additionally, presence of heavy metals like lead (Pb) and mercury (Hg), salts (NaCl), contaminants from agrochemicals (pesticides and herbicides) in soil puts the plants under stress. These abiotic factors mainly affect the photosynthetic efficiency of plants by disrupting chlorophyll synthesis, performance of photosystems, transport of electrons, and exchange of gases (Sharma et al. 2020). Due to the changing climate, plants are also introduced to several biotic stresses caused by specific living organisms, such as viruses, weeds, ascarids, spiders, and insects (Kumar and Nautiyal 2022). These stress agents dispossess the hosts of proper nutrition leading to yield losses due to reduced plant vitality and, in intense situations, even can cause death of the host plants (Shlibak et al. 2021). Several common fungal diseases of wheat such as rust diseases, Fusarium head blight, blast, and powdery mildew cause significant damages to the crop. These diseases reduce the overall productivity of the wheat, which poses socioeconomic challenges to the farmers (Randhawa et al. 2019). Other diseases like stem, leaf and stripe rusts, bacterial and viral diseases also disrupt the wheat production by causing a major impact on the plant growth and quality of grains.

9.4.3.1 Abiotic Stress Tolerance

The CRISPR/Cas9-mediated engineering of TaDREB2 ("dehydration responsive element-binding protein 2" has helped in imparting drought tolerance successfully in wheat (Kim and Budak 2018). "Ethylene response factors" (ERFs) and AP2/ERF homologous protein family with a DNA-binding segment contribute tolerance power toward numerous abiotic stressors, such as heat and cold, drought, salt, etc. (Yu et al. 2022). Therefore, ERF (e.g., TaERF3) was frequently targeted by the CRISPR/Cas9 system to enhance wheat tolerance to several abiotic factors simultaneously (Kim and Budak 2018). The product of the stress-granule-associated gene TaMBF1c is a member of the family of multiprotein bridging factors (MBF), a type of transcriptional coactivator protein regulating stress-responsive genes expression and thus plants' responses to drought, salt, and temperature changes (Ram et al. 2020). The overexpression of the gene TaMBF1c in wheat plants boosts their tolerance to these abiotic factors, especially to the heat, catering to better growth and productivity even under severe conditions (Yadav et al. 2022).

9.4.3.2 Enhanced Disease Resistance

Disease resistance potential of plants depends on two major kinds of plant-specific genes—R (resistance) and S (susceptibility) genes. The combination of these S and R genes together is able to offer a great resistance power against any plant disease. It might also help to plan for the future resistance against the newly evolving pathogenic diseases, which might drastically damage the crop in the next few years. Conventional breeding played a crucial role in amplifying the host disease resistance by using both qualitative and quantitative resistance potentials against the diseases. A group of major effective R genes regulates the qualitative traits, whereas a group of minor effective genes controls the quantitative traits (Merrick et al. 2021). Introduction of a combination of multiple cloned R genes into wheat through the process of transgenesis provided enhanced resistance against some diseases. It helped in introducing particular traits into wheat. Indicatively, the gene, NLR Yr10 has been overexpressed to confer resistance against stripe rust, which is a serious fungal infection of wheat (Liu et al. 2014). Under the conventional breeding approach, markers-associated selection was used to insert the R genes into the better cultivars. As an example, the cultivars Yr5, Yr10, and Yr15 resist the wheat stripe rust disease more efficiently (Haider et al. 2023). The QTL mapping also contributed to the identification of the genetic locations that enable plants to be resistant to some diseases. Another method to make wheat more disease-resistant is called transgenic RNA interference (RNAi). It has been demonstrated that the RNAi technique can be used to effectively silence the genes required by pathogens to survive (Chen et al. 2014; Tatineni et al. 2020).

Recently, CRISPR-mediated genome editing has become highly significant in stabilizing wheat production and increasing its resistance to stress and disease. This will assist farmers to avoid making huge financial losses (Cowger et al. 2022). As an example, deoxynivalenol (DON), a mycotoxin virulence factor, facilitates the growth of the pathogen *Fusarium graminearum* in floral tissues and the translocation of the pathogen among different tissues through a series of cell-damaging effects (necrosis, tissue bleaching, and other cell-damaging processes) and connected to the defense (Brauer et al. 2020). CRISPR/Cas system could be used to activate certain genes, such as TaNFXLI, by direct treatment of tissues with DON. This may assist fungus to develop resistance to it. Crops have also been made more resistant to virus diseases using the CRISPR/Cas nucleases to target the genomes of viruses (Yuan et al. 2024). Therefore, the approach has been reported to be very effective in providing a broad spectrum of resistances to several diseases caused by different types of pathogens.

Table 9.3 indicates that the most common applications of the genome-editing method include selective knockout of various plant genes which are weak. This aids in enhancing the resistance of plants to many of the major diseases of wheat. This approach has also provided a new research domain to investigate the effectiveness of various gene knockout combinations in acting against either a single pathogen or multiple pathogens. This is to use them in breeding of plants.

9.5 CRISPR Accelerates *de-novo* Domestication for Developing Resilient and Sustainable Wheat Crops

To ensure food security, de novo domestication is considered as a viable and alternative approach to crop development and the maintenance of agricultural yields using limited resources (Huang et al. 2022). Stress-adapted crop wild relatives (CWRs) of numerous crop species provide better genetic tools in de novo domestication of numerous crop species (Bohra et al. 2022). De novo domestication is the addition of domestication genes to the wild species in order to produce new crops. De novo domestication requires imitation of the natural breeding process, rather than inserting domesticated genes of CRWs into cultivated species, which is how a traditional process is carried out (Jian et al. 2023). It perhaps can be possible to rapidly and precisely transform CWRs into cultivated crops using a technique known as de-novo domestication. In this method, most of the positive alleles controlling

Table 9.3 Target genes for improved disease resistance in wheat

Gene	Pathogen(s)	Disease symptoms	Remarks	Reference(s)
Fusarium head blight S genes: *TaLAC4* and *TaUGT6*	Fungus Fusarium	Infected heads appear discolored with pink spores visible on them	Improved resistance in wheat plants	Cuthbert et al. (2006)
Powdery mildew S genes: *TaMLO* and *TaEDR1*	Fungus like *Erisiphe, Podosphaera*	White, powdery growth on the surface of stem, leaves	Knockouts are resistant to the disease with reduced crop losses and also require less pesticides	Wang et al. (2014)
Rust disease S genes: *Lr10, Lr13; Sr2; Yr*	Fungus *Puccinia* sps.	Leaf rust, stem rust, and stripe rust	Knockouts provide resistance to leaf rust, stem rust, and stripe rust diseases, respectively	He et al. (2023) and Liu et al. (2024)
NLR ("nucleotide binding domain, leucine-rich repeats") resistance genes				Tamborski and Krasileva (2020)

resilience to stress may not be lost during the breeding process, and neither are many other agronomic and nutritional traits (Razzaq et al. 2021).

De novo domestication may be extremely significant to navigate certain genetic issues, such as adapting the facilities of a diverse natural ecosystem, establishing robust relations between certain ecological niches and crops, and assisting crops to endure extremely adverse or stressful climate. It may also serve to enhance the diversity and genetic benefits of farming systems (Bohra et al. 2022). The knowledge gained through the study of the process of domestication in the model species could be critical to the analysis of CWRs and the development of an in-depth understanding of their genomes (Pisias et al. 2022). Domestication of monogenic traits is relatively easy, but abiotic stress tolerance traits are polygenic and very difficult to access (Mohd Saad et al. 2022). Furthermore, with the help of a complex set of genome editors to target the monogenic genes in CWRs with stress resilience features, it becomes possible to domesticate them at the beginning (Tahakik et al. 2024).

The objective of de novo domestication is to study and alter stress-related genes, resulting in the development of new types of crops that have better adaptability and are more resistant to changes in climatic conditions (Razzaq et al. 2021). Crops and model plants with limited use have been sequenced in addition to well-investigated agricultural species (Pisias et al. 2022). As genomes of an increasing number of organisms are sequenced, it is more possible to understand the genetic composition of these organisms, and these datasets can be used to determine genes controlling de novo domestication. The initial step in realizing successful de-novo domestication is the ability to identify stress-resistant genes in CWRs by using high-throughput genomics and transcriptomics. The development of next-generation sequencing (NGS) technology has made it possible to read the complete genome of any given species reliably, precisely, and economically (Curtin et al. 2021). The discovery of several domesticated genes, as well as the availability of modern genome editing tools, already offers an attractive path toward the creation of climate-resistant crops under de novo domestication (Bohra et al. 2022).

De novo domestication can positively impact crop wild relatives that are efficient sources of abiotic stress tolerance and are rich sources of nutritional value (Kapazoglou et al. 2022). With the emergence of CRISPR/Cas technology, there has been a paradigm shift in the approach to improving abiotic stress tolerance of crops. Hundreds of years are often required to domesticate crops, but with the help of CRISPR technology, plant breeders can accelerate the process by altering only a few specific loci (Li et al. 2022a, b). With the introduction of multiplex genome editing (MGE) technologies, the process of de novo domestication has become significantly faster. As an example, wild African rice landraces are very resistant to biotic and abiotic stress and are more nutritious than cultivated rice. Four genes, GN1A, GW2, GS3, and High-Tillering Dwarf1 (HTD1), were changed simultaneously in the MGE platform to accelerate domestication of African rice landraces (Kumar et al. 2023). Moreover, recently a MGE system based on CRISPR/Cas was employed to convert the wild rice species *Oryza alta* to cultivated rice. Systems were directed at A1An-1, An-2/LAB, Bh4, GAD1/RAE2, GW5/GSE5/qSW, OsSD1, OsLG1, OsGS3, PROG1, Rc, and qSH1, Sh4. The mutant plants had remarkable improvements in length of awn, panicle shape, hull color, pericarp color, shattering, breadth and size of grain, height of the plant, size of the grain, and tiller angle, which is a new finding (Yu et al. 2021; Abdullah et al. 2022). These findings highlight the importance of CRISPR-mediated de novo domestication of agronomically valuable crops and their wild counterparts to guarantee food security across the world.

9.6 Future Challenges and Perspectives of Plant CRISPR Technologies

There remain gaps in our understanding of the CRISPR-Cas genome editing framework, particularly concerning the mechanism by which Cas9 protein separates from the sgRNA and its subsequent reuse. Additionally, challenges in plant genome editing include ensuring effective transmission of mutations to subsequent generations. Future large-scale studies should focus on the frequency of off-target mutations in plant genomes and strategies to minimize them (Zischewski et al. 2017). Furthermore, additional research on Cas9 could lead to the development of new variants with novel PAM specificities. Using various sgRNA in a delivery systems could also provide opportunities for deleting gene clusters. An accessible database containing comprehensive information on CSRIPR-Cas system structures, supported by bioinformatics and AI, could facilitate discoveries. Currently, most research focuses on applying CRISPR-Cas system to enhance improved agricultural traits in crops. Advances in Cas proteins are expected to broaden the scope of their application and improve efficiency. Recent studies on Cas13 proteins suggest that these systems could offer increased precision and effectiveness. However, future development is needed to apply these technologies to food crops. The recent decision by the Court of Justice of the European Union (ECJ) has brought attention to the use of CRISPR in crops (Callaway 2018). The European Court of Justice said that crops modified with CRISPR fall under the 2001 directive, which was originally designed to address older methods of breeding and requires a long approval process. This ruling has been heavily criticized by many scientists who believe that this would have a huge impact on hindering EU investment in CRISPR research. Conversely, environmental groups such as the Amsterdam-based Friends of the Earth believe that the strict laws are necessary because they are crucial in promoting the safety of crops. The decision of the ECJ and the positive feedback of the scientific community prove that, among other technical challenges, there are also political ones that should be overcome to create this promising technology.

9.7 Conclusion

CRISPR-Cas systems are widely used to engineer agriculturally crops for various applications ranging from disease resistance to functional genomics. Despite their precision, these techniques can result in off-target editing, on-target effects, and interfere with gene regulation, potentially causing environmental concerns and unexpected interactions. Furthermore, gene drives could have biosafety concerns due to their irreversible nature in population. Scientific studies revealing off-target editing in CRISPR-edited plants highlight the need for government agencies to adopt precautionary measures for approving and marketing these CRISPR-edited crops. Crops and foods developed through CRISPR-Cas genome editing system should undergo independent evaluation for environmental and food safety before being approved for consumption. If scientific, regulatory, and political authorities fail to create an explicit regulatory framework, the advancement of genome-edited crops will be significantly hindered. Additionally, a global regulatory system is needed to address the social issues of individual countries regarding CRISPR-edited crops. It is also essential to actively involve the public in the development, regulation, and political aspects of genome-edited organisms to facilitate the better integration of CRISPR-edited crops in the society. Therefore, to ensure that CRISPR genome-edited crops contribute to food security and sustainable agriculture, we need a robust universal regulatory system to build public trust.

References

Abdullah M, Okemo P, Furtado A, Henry R (2022) Potential of genome editing to capture diversity from Australian wild rice relatives. Front Genom Ed 4:875243. https://doi.org/10.3389/fgeed.2022.875243

Ahmar S, Saeed S, Khan MH, Ullah Khan S, Kamran M, Faheem A, Maqsood A, Rauf M, Saleem S, Hong W, Jung K (2019) A revolution toward gene-editing technology and its application to crop improvement. Int J Mol Sci 21(16):5665. https://doi.org/10.3390/ijms21165665

Ahmed HG, Zeng Y, Shah AN, Yar MM, Ullah A, Ali M (2022) Conferring of drought tolerance in wheat (Triticum aestivum L.) genotypes using seedling indices. Front Plant Sci 13:961049. https://doi.org/10.3389/fpls.2022.961049

Akram F, Sahreen S, Aamir F et al (2023) An insight into modern targeted genome-editing technologies with a special focus on CRISPR/Cas9 and its applications. Mol Biotechnol 65:227–242. https://doi.org/10.1007/s12033-022-00501-4

Akram MU, Uğuz AB, Uygun U, Salantur A, Yilmaz R (2025) Investigation of Phenolic Composition and Antioxidant Capacities in Selected Turkish Indigenous Wheat Varieties. Food Sci Nutr 13(2):e4614. https://doi.org/10.1002/fsn3.4614

Beger RD, Balasubramanian S, Bennett SE, Mosbaugh DW, Bolton PH (1995) Tertiary structure of uracil-DNA glycosylase inhibitor protein. J Biol Chem 270(28):16840–16847. https://doi.org/10.1074/jbc.270.28.16840

Benitez-Alfonso, Y., Soanes, B. K., Zimba, S., Sinanaj, B., German, L., Sharma, V., Bohra, A., Kolesnikova, A., Dunn, J. A., Martin, A. C., Khashi u Rahman, M., Saati-Santamaría, Z., García-Fraile, P., Ferreira, E. A., Frazão, L. A., Cowling, W. A., Siddique, K. H., Pandey, M. K., Farooq, M., . . . Foyer, C. H. (2023). Enhancing climate change resilience in agricultural crops. Curr Biol, 33(23), R1246-R1261. doi:https://doi.org/10.1016/j.cub.2023.10.028

Bhandari RK, Nyaupane S, Poudel MR (2023) Expression and association of quantitative traits of wheat (*Triticum aestivum* L.) genotypes under different wheat growing environments. Cogent Food Agric 9(2). https://doi.org/10.1080/23311932.2023.2288394

Bhatia S, Yadav SK (2023) CRISPR-Cas for genome editing: classification, mechanism, designing and applications. Int J Biol Macromol 238:124054. https://doi.org/10.1016/j.ijbiomac.2023.124054

Bohra A, Kilian B, Sivasankar S, Caccamo M, Mba C, McCouch SR, Varshney RK (2022) Reap the crop wild relatives for breeding future crops. Trends Biotechnol 40(4):412–431. https://doi.org/10.1016/j.tibtech.2021.08.009

Brauer EK, Balcerzak M, Rocheleau H, Leung W, Schernthaner J, Subramaniam R, Ouellet T (2020) Genome editing of a deoxynivalenol-induced transcription factor confers resistance to *Fusarium graminearum* in wheat. Mol Plant-Microbe Interact 33(3):553–560. https://doi.org/10.1094/MPMI-11-19-0332-R

Brock N, Kaur N, Halford NG (2025) Advances in genome editing in plants within an evolving regulatory landscape, with a focus on its application in wheat breeding. J Plant Biochem Biotechnol 34:599–614. https://doi.org/10.1007/s13562-025-00981-w

Cabanillas B (2020) Gluten-related disorders: celiac disease, wheat allergy, and nonceliac gluten sensitivity. Crit Rev Food Sci Nutr 60(15):2606–2621. https://doi.org/10.1080/10408398.2019.1651689

Callaway E (2018) EU law deals blow to CRISPR crops. Nature 560(7716):16–16

Chen PJ, Liu DR (2022) Prime editing for precise and highly versatile genome manipulation. Nat Rev Genet 24(3):161. https://doi.org/10.1038/s41576-022-00541-1

Chen M, Sun L, Wu H, Chen J, Ma Y, Zhang X, Du L, Cheng S, Zhang B, Ye X, Pang J (2014) Durable field resistance to wheat yellow mosaic virus in transgenic wheat containing the antisense virus polymerase gene. Plant Biotechnol J 12(4):447–456. https://doi.org/10.1111/pbi.12151

Chen T, Li Y, Qiu Z, Zou C, Zhang B (2025) Prime editing of *TaEPSPS* homeologs enhances wheat resistance to glyphosate. J Agric Food Chem 73(34):21516–21525. https://doi.org/10.1021/acs.jafc.5c05769

Cowger C, Bulli P, Chen F, Lillemo M, Maccaferri M (2022) Advances in breeding for wheat disease resistance. Front Plant Sci 13:1034506. https://doi.org/10.3389/fpls.2022.1034506

Curtin S, Qi Y, P Peres LE, Fernie AR, Zsögön A (2021) Pathways to de novo domestication of crop wild relatives. Plant Physiol 188(4):1746. https://doi.org/10.1093/plphys/kiab554

Cuthbert PA, Somers DJ, Thomas J, Cloutier S, Brulé-Babel A (2006) Fine mapping Fhb1, a major gene controlling fusarium head blight resistance in bread wheat (*Triticum aestivum* L.). Theor Appl Genet 112(8):1465–1472. https://doi.org/10.1007/s00122-006-0249-7

Elsharawy H, Refat M (2023) CRISPR/Cas9 genome editing in wheat: enhancing quality and productivity for global food security—a review. Funct Integr Genomics 23(3):265. https://doi.org/10.1007/s10142-023-01190-1

Erenstein O, Jaleta M, Mottaleb KA, Sonder K, Donovan J, Braun HJ (2022) Global trends in wheat production, consumption and trade. In: Reynolds MP, Braun HJ (eds) Wheat improvement. Springer, Cham. https://doi.org/10.1007/978-3-030-90673-3_4

Farhad M, Kumar U, Tomar V, Bhati PK, Barek V, Brestic M, Hossain A (2023) Heat stress in wheat: a global challenge to feed billions in the current era of the changing climate. Front Sustain Food Syst 7:1203721. https://doi.org/10.3389/fsufs.2023.1203721

Gaillochet C, Pena Fernandez A, Goossens V, D'Halluin K, Drozdzecki A, Shafie M, Van Duyse J, Van Isterdael G, Gonzalez C, Vermeersch M et al (2023) Systematic optimization of Cas12a base editors in wheat and maize using the ITER platform. Genome Biol 24:6

Gong S, Yu HH, Johnson KA, Taylor DW (2018) DNA unwinding is the primary determinant of CRISPR-Cas9 activity. Cell Rep 22(2):359–371. https://doi.org/10.1016/j.celrep.2017.12.041

Grosse-Heilmann M, Cristiano E, Deidda R, Viola F (2024) Durum wheat productivity today and tomorrow: A review of influencing factors and climate change effects. Resources, Environment and Sustainability 17:100170. https://doi.org/10.1016/j.resenv.2024.100170

Grote U, Fasse A, Nguyen TT, Erenstein O (2021) Food security and the dynamics of wheat and maize value chains in Africa and Asia. Front Sustain Food Syst 4:617009. https://doi.org/10.3389/fsufs.2020.617009

Gupta OP, Singh AK, Singh A, Singh GP, Bansal KC, Datta SK (2022) Wheat biofortification: utilizing natural genetic diversity, genome-wide association mapping, genomic selection, and genome editing technologies. Front Nutr 9:826131. https://doi.org/10.3389/fnut.2022.826131

Haider MW, Kaur J, Bala R, Singh S, Srivastava P, Sharma A, Singh R, Kumari J (2023) Stripe rust resistance gene (s) postulation in wheat germplasm with the help of differentials and tagged molecular markers. Sci Rep 13(1):9007. https://doi.org/10.1038/s41598-023-36197-y

Han M, Zhou J, Wang J, Ivanistau A, Yang Q, Feng B (2025) Wheat improves the quality of coarse cereals steamed bread by changing the gluten network. Appl Food Res 5(1):101040. https://doi.org/10.1016/j.afres.2025.101040

He F, Wang C, Sun H, Tian S, Zhao G, Liu C, Wan C, Guo J, Huang X, Zhan G, Yu X (2023) Simultaneous editing of three homoeologues of TaCIPK14 confers broad-spectrum resistance to stripe rust in wheat. Plant Biotechnol J 21(2):354–368. https://doi.org/10.1111/pbi.13956

Huang Z, Liu G (2023) Current advancement in the application of prime editing. Front Bioeng Biotechnol 11:1039315. https://doi.org/10.3389/fbioe.2023.1039315

Huang X, Huang S, Han B, Li J (2022) The integrated genomics of crop domestication and breeding. Cell 185(15):2828–2839. https://doi.org/10.1016/j.cell.2022.04.036

Ibrahim S, Saleem B, Rehman N, Zafar SA, Naeem MK, Khan MR (2022) CRISPR/Cas9 mediated disruption of inositol Pentakisphosphate 2-kinase 1 (TaIPK1) reduces phytic acid and improves iron and zinc accumulation in wheat grains. J Adv Res 37:33–41. https://doi.org/10.1016/j.jare.2021.07.006

Jian LM, Xiao YJ, Yan JB (2023) *De novo* domestication: a new way for crop design and breeding. Yi chuan = Hereditas 45(9):741–753. https://doi.org/10.16288/j.yczz.23-194

Jiang L, Yao S (2022) Enhancing prime editing via inhibition of mismatch repair pathway. Mol Biomed 3:7. https://doi.org/10.1186/s43556-022-00072-5

Jouanin A, Schaart JG, Boyd LA, Cockram J, Leigh FJ, Bates R, Wallington EJ, Visser RG, Smulders MJ (2019) Outlook for coeliac disease patients: towards bread wheat with hypoimmunogenic gluten by gene editing of α-and γ-gliadin gene families. BMC Plant Biol 19(1):333. https://doi.org/10.1186/s12870-019-1889-5

Kantor A, McClements ME, MacLaren RE (2020) CRISPR-Cas9 DNA Base-editing and prime-editing. Int J Mol Sci 21(17):6240. https://doi.org/10.3390/ijms21176240

Kapazoglou A, Gerakari M, Lazaridi E, Kleftogianni K, Sarri E, Tani E, Bebeli PJ (2022) Crop wild relatives: A valuable source of tolerance to various abiotic stresses. Plants 12(2):328. https://doi.org/10.3390/plants12020328

Kim D, Alptekin B, Budak H (2018) CRISPR/Cas9 genome editing in wheat. Funct Integr Genomics 18(1):31–41. https://doi.org/10.1007/s10142-017-0572-x

Ko I, Chapman T, Nazarov T, Uwugiaren R, Smertenko A, Chandrakanth NN, Oates D (2025) Grain under pressure: harnessing biochemical pathways to beat drought and heat in wheat. Plant J 122(6):e70253. https://doi.org/10.1111/tpj.70253

Komor AC, Kim YB, Packer MS, Zuris JA, Liu DR (2016) Programmable editing of a target base in genomic DNA without double-stranded DNA cleavage. Nature 533(7603):420. https://doi.org/10.1038/nature17946

Kumar V, Nautiyal CS (2022) Plant abiotic and biotic stress alleviation: from an endophytic microbial perspective. Curr Microbiol 79(10):311. https://doi.org/10.1007/s00284-022-03012-2

Kumar M, Prusty MR, Pandey MK, Singh PK, Bohra A, Guo B, Varshney RK (2023) Application of CRISPR/Cas9-mediated gene editing for abiotic stress management in crop plants. Front Plant Sci 14:1157678. https://doi.org/10.3389/fpls.2023.1157678

Li C, Zong Y, Wang Y, Jin S, Zhang D, Song Q, Zhang R, Gao C (2018) Expanded base editing in rice and wheat using a Cas9-adenosine deaminase fusion. Genome Biol 19:59

Li J, Jiao G, Sun Y, Chen J, Zhong Y, Yan L, Jiang D, Ma Y, Xia L (2021a) Modification of starch composition, structure and properties through editing of TaSBEIIa in both winter and spring wheat varieties by CRISPR/Cas9. Plant Biotechnol J 19(5):937–951. https://doi.org/10.1111/pbi.13519

Li J, Li Y, Ma L (2021b) Recent advances in CRISPR/Cas9 and applications for wheat functional genomics and breeding. aBIOTECH 2(4):375. https://doi.org/10.1007/s42994-021-00042-5

Li J, Xu X, Lin G, Wang Y, Liu Y, Zhang M, Zhou J, Wang Z, Zhang Y (2018a) Micro-irrigation improves grain yield and resource use efficiency by co-locating the roots and N-fertilizer distribution of winter wheat in the North China Plain. Sci Total Environ 643:367–377. https://doi.org/10.1016/j.scitotenv.2018.06.157.

Li L, Shi F, Wang G, Guan Y, Zhang Y, Chen M, Chang J, Yang G, He G, Wang Y, Li Y (2022a) Conservation and divergence of Squamosa-Promoter Binding Protein-Like (SPL) gene family between wheat and rice. Int J Mol Sci 23(4):2099. https://doi.org/10.3390/ijms23042099

Li Y, Wu X, Zhang Y, Zhang Q (2022b) CRISPR/Cas genome editing improves abiotic and biotic stress tolerance of crops. Front Genome Ed 4:987817. https://doi.org/10.3389/fgeed.2022.987817

Li J, Wu S, Zhang K, Sun X, Lin W, Wang C, Lin S (2024) Clustered regularly interspaced short palindromic repeat/CRISPR-associated protein and its utility all at sea: status, challenges, and prospects. Microorganisms 12(1):118. https://doi.org/10.3390/microorganisms12010118

Li M, Lin Y, Cheng Q, Wei T (2025) Prime editing: A revolutionary technology for precise treatment of genetic disorders. Cell Prolif 58(4):e13808. https://doi.org/10.1111/cpr.13808

Lin Q, Zong Y, Xue C, Wang S, Jin S, Zhu Z, Wang Y, Anzalone AV, Raguram A, Doman JL et al (2020) Prime genome editing in rice and wheat. Nat Biotechnol 38:582e585

Liu W, Frick M, Huel R, Nykiforuk CL, Wang X, Gaudet DA, Eudes F, Conner RL, Kuzyk A, Chen Q, Kang Z (2014) The stripe rust resistance gene Yr10 encodes an evolutionary-conserved and unique CC–NBS–LRR sequence in wheat. Mol Plant 7(12):1740–1755. https://doi.org/10.1093/mp/ssu112

Liu H, Li H, Hao C, Wang K, Wang Y, Qin L, An D, Li T, Zhang X (2020) TaDA1, a conserved negative regulator of kernel size, has an additive effect with TaGW2 in common wheat (*Triticum aestivum* L.). Plant Biotechnol J 18(5):1330–1342. https://doi.org/10.1111/pbi.13298

Liu S, Zhang F, Su J, Fang A, Tian B, Yu Y, Bi C, Ma D, Xiao S, Yang Y (2024) CRISPR-targeted mutagenesis of mitogen-activated protein kinase phosphatase 1 improves both immunity and yield in wheat. Plant Biotechnol J 22(7):1929–1941. https://doi.org/10.1111/pbi.14312

Lu C, Kuang J, Shao T, Xie S, Li M, Zhu L, Zhu L (2022) Prime editing: an all-rounder for genome editing. Int J Mol Sci 23(17):9862. https://doi.org/10.3390/ijms23179862

Luo J, Abid M, Tu J, Cai X, Zhang Y, Gao P, Huang H (2023) Cytosine base editors (CBEs) for inducing targeted DNA base editing in *Nicotiana benthamiana*. BMC Plant Biol 23:305. https://doi.org/10.1186/s12870-023-04322-8

Manghwar H, Li B, Ding X, Hussain A, Lindsey K, Zhang X, Jin S (2020) CRISPR/Cas systems in genome editing: methodologies and tools for sgRNA design, off-target evaluation, and strategies to mitigate off-target effects. Adv Sci 7(6):1902312. https://doi.org/10.1002/advs.201902312

Merrick LF, Burke AB, Chen X, Carter AH (2021) Breeding with major and minor genes: genomic selection for quantitative disease resistance. Front Plant Sci 12:713667. https://doi.org/10.3389/fpls.2021.713667

Mesfin AH, Gebremedhin KG, Tefera T, Chimsa MD, Vonk R (2024) Pathways and policy options for food and nutrition gaps in arid agricultural farming systems in the Tigray region, North Ethiopia. Heliyon 10(7):e28213. https://doi.org/10.1016/j.heliyon.2024.e28213

Mohd Saad NS, Neik TX, Thomas WJ, Amas JC, Cantila AY, Craig RJ, Edwards D, Batley J (2022) Advancing designer crops for climate resilience through an integrated genomics approach. Curr Opin Plant Biol 67:102220. https://doi.org/10.1016/j.pbi.2022.102220

Molla KA, Sretenovic S, Bansal KC, Qi Y (2021) Precise plant genome editing using base editors and prime editors. Nat Plants 7(9):1166–1187. https://doi.org/10.1038/s41477-021-00991-1

Ni P, Zhao Y, Zhou X et al (2023) Efficient and versatile multiplex prime editing in hexaploid wheat. Genome Biol 24:156. https://doi.org/10.1186/s13059-023-02990-1

Nyaupane S, Poudel MR, Panthi B, Dhakal A, Paudel H, Bhandari R (2024) Drought stress effect, tolerance, and management in wheat—a review. Cogent Food Agric 10(1). https://doi.org/10.1080/23311932.2023.2296094

Pacesa M, Loeff L, Querques I et al (2022) R-loop formation and conformational activation mechanisms of Cas9. Nature 609:191–196. https://doi.org/10.1038/s41586-022-05114-0

Paul B, Montoya G (2020) CRISPR-Cas12a: functional overview and applications. Biom J 43(1):8. https://doi.org/10.1016/j.bj.2019.10.005

Pequeno DN, Ferreira TB, Fernandes JM, Singh PK, Pavan W, Sonder K, Robertson R, Krupnik TJ, Erenstein O, Asseng S (2024) Production vulnerability to wheat blast disease under climate change. Nat Clim Chang 14(2):178–183. https://doi.org/10.1038/s41558-023-01902-2

Pisias MT, Bakala HS, McAlvay AC, Mabry ME, Birchler JA, Yang B, Pires JC (2022) Prospects of Feral Crop *De Novo* Redomestication. Plant Cell Physiol 63(11):1641–1653. https://doi.org/10.1093/pcp/pcac072

Prakash A, Bahadur Singh R, Kumari N, Tiwari P, Singh S, Ranjan S, Mandal S (2025) CRISPR/Cas-mediated base editing in plants. In: Gene-Edited Crops, pp 108–121. https://doi.org/10.1201/9781003500933-12

Prasad AS (1996) Zinc deficiency in women, infants and children. J Am Coll Nutr 15(2):113–120. https://doi.org/10.1080/07315724.1996.10718575

Ram H, Kaur A, Gandass N, Katoch M, Roy S, Kushwaha NK, Soni P (2020) Stomatal adaptive response in plants under drought stress. In: Kuila A (ed) Plant stress biology. Apple Academic Press, pp 167–183. https://doi.org/10.1201/9781003055358-8

Ran Y, Patron N, Kay P, Wong D, Buchanan M, Cao Y, Sawbridge T, Davies JP, Mason J, Webb SR, Spangenberg G, Ainley WM, Walsh TA, Hayden MJ (2018) Zinc finger nuclease-mediated precision genome editing of an endogenous gene in hexaploid bread wheat (*Triticum aesti-*

vum) using a DNA repair template. Plant Biotechnol J 16(12):2088. https://doi.org/10.1111/pbi.12941

Randhawa MS, Bhavani S, Singh PK, Huerta-Espino J, Singh RP (2019) Disease resistance in wheat: present status and future prospects. In: Wani SH (ed) Disease resistance in crop plants. Springer, Cham, pp 61–81. https://doi.org/10.1007/978-3-030-20728-1_4

Rath D, Amlinger L, Rath A, Lundgren M (2015) The CRISPR-Cas immune system: biology, mechanisms and applications. Biochimie 117:119–128. https://doi.org/10.1016/j.biochi.2015.03.025

Razzaq A, Saleem F, Wani SH, Abdelmohsen SA, Alyousef HA, Abdelbacki AM, Alkallas FH, Tamam N, Elansary HO (2021) De-novo domestication for improving salt tolerance in crops. Front Plant Sci 12:681367. https://doi.org/10.3389/fpls.2021.681367

Sánchez-León S, Gil-Humanes J, Ozuna CV, Giménez MJ, Sousa C, Voytas DF, Barro F (2018) Low-gluten, nontransgenic wheat engineered with CRISPR/Cas9. Plant Biotechnol J 16(4):902–910. https://doi.org/10.1111/pbi.12837

Savary S, Willocquet L, Pethybridge SJ, Esker P, McRoberts N, Nelson A (2019) The global burden of pathogens and pests on major food crops. Nat Ecol Evol 3(3):430–439. https://doi.org/10.1038/s41559-018-0793-y

Schaart JG, Smulders MJ (2021) Genome editing of polyploid crops: prospects, achievements and bottlenecks. Transgenic Res 30(4):337. https://doi.org/10.1007/s11248-021-00251-0

Senbeta AF, Worku W (2023) Ethiopia's wheat production pathways to self-sufficiency through land area expansion, irrigation advance, and yield gap closure. Heliyon 9(10):e20720. https://doi.org/10.1016/j.heliyon.2023.e20720

Sharma A, Kumar V, Shahzad B, Ramakrishnan M, Singh Sidhu GP, Bali AS, Handa N, Kapoor D, Yadav P, Khanna K, Bakshi P (2020) Photosynthetic response of plants under different abiotic stresses: a review. J Plant Growth Regul 39(2):509–531. https://doi.org/10.1007/s00344-019-10018-x

Shimatani Z, Kashojiya S, Takayama M, Terada R, Arazoe T, Ishii H, Teramura H, Yamamoto T, Komatsu H, Miura K, Ezura H (2017) Targeted base editing in rice and tomato using a CRISPR-Cas9 cytidine deaminase fusion. Nat Biotechnol 35(5):441–443. https://doi.org/10.1038/nbt.3833

Shlibak AA, Örgeç M, Zencirci N (2021) Wheat landraces versus resistance to biotic and abiotic stresses. In: Zencirci N, Baloch FS, Habyarimana E, Chung G (eds) Wheat landraces. Springer, Cham, pp 193–214. https://doi.org/10.1007/978-3-030-77388-5_10

Stinson BM, Loparo JJ (2021) Repair of DNA double-Strand breaks by the non-homologous end joining pathway. Annu Rev Biochem 90:137. https://doi.org/10.1146/annurev-biochem-080320-110356

Taha EA, Lee J, Hotta A (2022) Delivery of CRISPR-Cas tools for in vivo genome editing therapy: trends and challenges. J Control Release 342:345–361. https://doi.org/10.1016/j.jconrel.2022.01.013

Tahakik RR, Deshmukh AG, Moharil MP, Jadhav PV, Kogade VT, More KD, Shinde VP (2024) Transitioning from the Green Revolution to the Gene Revolution: strengthening nutritional security using climate resilient traditional crops. Bull Natl Res Cent 48(1):1–20. https://doi.org/10.1186/s42269-024-01281-4

Tamborski J, Krasileva KV (2020) Evolution of plant NLRs: from natural history to precise modifications. Annu Rev Plant Biol 71(1):355–378. https://doi.org/10.1146/annurev-arplant-081519-035901

Tatineni S, Sato S, Nersesian N, Alexander J, Quach T, Graybosch RA, Clemente TE (2020) Transgenic wheat harboring an RNAi element confers dual resistance against synergistically interacting wheat streak mosaic virus and Triticum mosaic virus. Mol Plant-Microbe Interact 33(1):108–122. https://doi.org/10.1094/MPMI-10-19-0275-R

Tran B, Tseng W, Chen C (2025) Climate change impacts on crop yields across temperature rise thresholds and climate zones. Sci Rep 15(1):1–17. https://doi.org/10.1038/s41598-025-07405-8

Verma AK, Mandal S, Serin Y, Ali MH, Tiwari A, Lionetti E, Catassi C (2022) CRISPR/Cas9 as a promising genome-editing technology for generating low-immunogenic wheat variety: recent legislation and global implementation of genetically engineered crops. CAB Rev Perspect Agric Vet Sci Nutr Nat Resour 2022. https://doi.org/10.1079/CABIREVIEWS202217039;WEBSITE:WEBSITE:CABI-SITE

Waites J, Murali Achary VM, Syombua ED, Hearne SJ, Bandyopadhyay A (2025) CRISPR-mediated genome editing of wheat for enhancing disease resistance. Front Genome Ed 7:1542487. https://doi.org/10.3389/fgeed.2025.1542487

Wang X, Liu F (2021) Effects of elevated CO2 and heat on wheat grain quality. Plants 10(5):1027. https://doi.org/10.3390/plants10051027

Wang Y, Cheng X, Shan Q, Zhang Y, Liu J, Gao C, Qiu JL (2014) Simultaneous editing of three homoeoalleles in hexaploid bread wheat confers heritable resistance to powdery mildew. Nat Biotechnol 32(9):947–951. https://doi.org/10.1038/nbt.2969

Wang W, Pan Q, Tian B, He F, Chen Y, Bai G, Akhunova A, Trick HN, Akhunov E (2019) Gene editing of the wheat homologs of TONNEAU 1-recruiting motif encoding gene affects grain shape and weight in wheat. Plant J 100(2):251–264. https://doi.org/10.1111/tpj.14440

Wanikar R, Dixit P, Khater M, Dange M, Yevale R, Mandal S (2025) Prime editing through CRISPR-Cas and its applications in plants. In: Gene-Edited Crops, pp 122–141. https://doi.org/10.1201/9781003500933-13

WHO (2025). WHO Global Anaemia estimates, 2025 Edition. https://www.who.int/data/gho/data/themes/topics/anaemia_in_women_and_children. Accessed 24 Aug 2025

Xu L, Wang D, Liu S, Fang Z, Su S, Guo C, Zhao C, Tang Y (2020) Comprehensive atlas of wheat (*Triticum aestivum* L.) Auxin Response Factor expression during male reproductive development and abiotic stress. Front Plant Sci 11:586144. https://doi.org/10.3389/fpls.2020.586144

Xu P, Qin Y, Ma M, Liu T, Ruan F, Xue L, Cao J, Xiao G, Chen Y, Fu H, Zhou G, Xie Y, Xia D (2024a) Genome-wide association study reveals the genetic basis of rice resistance to three herbicides. Front Plant Sci 15:1476829. https://doi.org/10.3389/fpls.2024.1476829

Xu W, Zhang S, Qin H et al (2024b) From bench to bedside: cutting-edge applications of base editing and prime editing in precision medicine. J Transl Med 22:1133. https://doi.org/10.1186/s12967-024-05957-3

Yadav MR, Choudhary M, Singh J, Lal MK, Jha PK, Udawat P, Gupta NK, Rajput VD, Garg NK, Maheshwari C, Hasan M (2022) Impacts, tolerance, adaptation, and mitigation of heat stress on wheat under changing climates. Int J Mol Sci 23(5):2838. https://doi.org/10.3390/ijms23052838

Yang J, Zhou Y, Zhang YE, Hu W, Wu Q, Chen Y, Wang X, Guo G, Liu Z, Cao T, Zhao H (2019) Cloning, characterization of TaGS3 and identification of allelic variation associated with kernel traits in wheat (*Triticum aestivum* L.). BMC Genet 20(1):98. https://doi.org/10.1186/s12863-019-0800-6

Yoon DE, Kim R, Park J, Jeong TY, Eun B, Cho Y, Lim Y, Lee H, Seong JK, Kim K (2023) Precise base editing without unintended indels in human cells and mouse primary myoblasts. Exp Mol Med 55(12):2586. https://doi.org/10.1038/s12276-023-01128-4

Yu H, Lin T, Meng X, Du H, Zhang J, Liu G, Chen M, Jing Y, Kou L, Li X, Gao Q, Liang Y, Liu X, Fan Z, Liang Y, Cheng Z, Chen M, Tian Z, Wang Y et al (2021) A route to de novo domestication of wild allotetraploid rice. Cell 184(5):1156–1170.e14. https://doi.org/10.1016/j.cell.2021.01.013

Yu Y, Yu M, Zhang S, Song T, Zhang M, Zhou H, Wang Y, Xiang J, Zhang X (2022) Transcriptomic identification of wheat AP2/ERF transcription factors and functional characterization of TaERF-6-3A in response to drought and salinity stresses. Int J Mol Sci 23(6):3272. https://doi.org/10.3390/ijms23063272

Yuan X, Xu K, Yan F, Liu Z, Spetz C, Zhou H, Wang X, Jin H, Wang X, Liu Y (2024) CRISPR/Cas9-mediated resistance to wheat dwarf virus in hexaploid wheat (*Triticum aestivum* L.). Viruses 16(9):1382. https://doi.org/10.3390/v16091382

Zhang Y, Liang Z, Zong Y, Wang Y, Liu J, Chen K, Qiu JL, Gao C (2016) Efficient and transgene-free genome editing in wheat through transient expression of CRISPR/Cas9 DNA or RNA. Nat Commun 7(1):12617. https://doi.org/10.1038/ncomms12617

Zhang Z, Hua L, Gupta A, Tricoli D, Edwards KJ, Yang B, Li W (2019a) Development of an agrobacterium-delivered CRISPR/Cas9 system for wheat genome editing. Plant Biotechnol J 17(8):1623–1635. https://doi.org/10.1111/pbi.13088

Zhang R, Liu J, Chai Z, Chen S, Bai Y, Zong Y, Chen K, Li J, Jiang L, Gao C (2019b) Generation of herbicide tolerance traits and a new selectable marker in wheat using base editing. Nat Plants 5:480e485

Zhou X, Zhao Y, Ni P, Ni Z, Sun Q, Zong Y (2023) CRISPR-mediated acceleration of wheat improvement: advances and perspectives. J Genet Genomics 50(9):815–834. https://doi.org/10.1016/j.jgg.2023.09.007

Zischewski J, Fischer R, Bortesi L (2017) Detection of on-target and off-target mutations generated by CRISPR/Cas9 and other sequence-specific nucleases. Biotechnol Adv 35(1):95–91

Zong Y, Wang Y, Li C, Zhang R, Chen K, Ran Y, Qiu JL, Wang D, Gao C (2017) Precise base editing in rice, wheat and maize with a Cas9-cytidine deaminase fusion. Nat Biotechnol 35:438e440

Zong Y, Song Q, Li C, Jin S, Zhang D, Wang Y, Qiu JL, Gao C (2018) Efficient C-to-T base editing in plants using a fusion of nCas9 and human APOBEC3A. Nat Biotechnol 36:950e953

Zong Y, Liu Y, Xue C, Li B, Li X, Wang Y, Li J, Liu G, Huang X, Cao X, Gao C (2022) An engineered prime editor with enhanced editing efficiency in plants. Nat Biotechnol 40(9):1394–1402. https://doi.org/10.1038/s41587-022-01254-w

Chapter 10
CRISPR-Cas-Based Gene Editing in Barley for Trait Enhancement

Sohini Jash, Ansu Chetri, Ramsha Syed, Deepak Kumar Upadhyay, and Sayanti Mandal

Abstract Barley (*Hordeum vulgare* Linn.) is one of the ancient, cultivated grains, widely utilized for food, feed, and commercial applications, including malting and brewing industries. Its adaptability to changing climates has made it a staple crop; however, barley production faces numerous challenges, including abiotic and biotic stresses, low soil fertility, and climate variability, all of which affect yield and quality. Enhancing barley's quality is crucial to improving nutritional value, disease resistance, and industrial suitability. Traditional breeding methods have contributed to genetic improvement, but they are time-consuming and lack precision. The emergence of CRISPR-Cas gene editing has transformed barley enhancement by enabling precise modifications to genes responsible for grain quality, stress tolerance, and resistance to pathogens. Despite its efficiency, CRISPR-Cas genome editing faces limitations, including off-target mutations, regulatory constraints, and challenges in achieving high-frequency homology-directed repair (HDR) in plants. Future advancements aim to optimize Cas variants, improve editing efficiency, and integrate new techniques such as base and prime editing to overcome current barriers. As research progresses, CRISPR-Cas holds immense potential for developing superior barley varieties, ensuring higher yields, improved quality, and resilience to challenging environmental stressors, ultimately contributing to world hunger and sustainable agriculture.

Keywords Genome editing · Barley (*Hordeum vulgare* Linn.) · Cas9 · Cas12a · Climate-smart crops · Food security

S. Jash
JMS School of Life Sciences, University of Sussex, Falmer, Brighton, UK

A. Chetri · R. Syed · D. K. Upadhyay · S. Mandal (✉)
Department of Chemistry and Biochemistry, Sharda School of Engineering and Sciences, Sharda University, Greater Noida, India

K. A. Abd-Elsalam, A. Ahmad (eds.), *Gene Editing in Cereals*, Sustainability Sciences in Asia and Africa, https://doi.org/10.1007/978-981-95-6722-5_10

10.1 Introduction

Barley (*Hordeum vulgare* Linn.) is a diploid (2n) plant species, with 14 chromosomes and 5.1 gigabases (gb) genome classified under the tribe *Triticeae* within the genus *Hordeum* and the *Poaceae* family, also referred to as *Gramineae* (Brown et al. 1999; Sato 2020). Archaeological evidence suggests that barley (*Hordeum vulgare* L.) was first domesticated in the Fertile Crescent approximately 10,000 years ago from its wild parent, *Hordeum spontaneum* (Badr et al. 2000; Sato 2020). Over time, domestication led to the development of traits that improved its suitability for agriculture (Alvi and Saleem 2024; Sato 2020; Raza et al. 2025). Key characteristics such as the non-brittle rachis, six-rowed spike, and naked caryopsis were essential in transforming wild barley into its cultivated form (Alvi and Saleem 2024). Additionally, adaptations like reduced vernalization requirements and photoperiod insensitivity enabled barley to spread far beyond its original habitat (Pourkheirandish and Komatsuda 2007). The transition from wild to cultivated barley is believed to have taken thousands of years (Newman 2006).

While the Fertile Crescent is generally regarded as the primary region of barley domestication, recent evidence supports the possibility of multiple origins. *H. spontaneum* has been identified in several regions outside the Fertile Crescent, including North Africa (Algeria, Egypt, Morocco, and Libya), Crete, Tibet, and Ethiopia. Genetic studies suggest distinct differences between barley populations from Southern and Asia Eastern compared to those from North Africa, Western Asia, and Europe (Alvi and Saleem 2024). Molecular research further proposes Central Asia, particularly along the eastern Iranian Plateau, as another potential domestication site (Wang et al. 2015). Other suggested regions include the Himalayas, Morocco, and Ethiopia (Badr et al. 2000). Genetic analyses of wild and cultivated barley varieties indicate that Tibet may have been a center for barley domestication, reinforcing the idea of multiple domestication events (Dai et al. 2012, 2014; Wang et al. 2015).

Existing breeding objectives for barley are very variable and dependent on the characteristics of the local environment. It is necessary for barley breeding to concentrate on the development of high-yielding and stress-tolerant varieties that can thrive even in challenging climatic conditions to meet the growing demand for livestock feed, starch, and a variety of alcoholic and nonalcoholic beverages. Some examples of these beverages include beer, whisky, and ethanol. Other examples include barley tea, barley coffee, and more. There are certain common objectives that are made for breeding and agricultural crop production (Garcia-Gimenez and Jobling 2021; Verma et al. 2022; Czembor and Czembor 2024). These objectives are developed as a reaction to changes that are occurring on a global scale, such as the predicted increase in the pace of human population growth and diet. When it comes to guaranteeing food security and fulfilling the ever-increasing demand, high yields, breeding for disease resistance, and stability are crucial components that are necessary all over the world. These components are required in almost every country on the planet. Barley breeding has been pushed to speed up the process to combine several features in new varieties because of the accelerated evolution of the global environment. Therefore, the aim of this book chapter is to present an overview of successful genome editing applications in barley, with a particular emphasis on the use of Clustered Regularly Interspaced Short Palindromic Repeats-CRISPR associated protein 9 (CRISPR-Cas9) technology for the purpose of manipulating attributes related to grain quality and stress tolerance. Following the invention of programmable nucleases for gene editing, the chapter then moves on to an analysis of genome editing experiments in barley and ethical challenges associated with CRISPR. We end with a discussion of the possible prospects that may be realized via the use of CRISPR-Cas technology, as well as regulatory considerations and the ways in which the gene editing technique could potentially develop to broaden the scope of present barley breeding.

10.2 Barley Grain Structure

10.2.1 Structural Composition

Barley grain is primarily composed of several structural components, including the endosperm (which makes up about 75% of the grain), the aleurone layer, the germ, and the pericarp. The pericarp, along with the seed coat, serves as a protective barrier, shielding the seed from physical damage and external environmental factors by completely enclosing it (Shaveta et al. 2019).

The endosperm is the most critical part of the barley grain, as its chemical composition directly influences malt quality. This tissue acts as the primary storage site for starch granules, which are essential for various food and brewing applications. The aleurone layer, another significant component of the barley grain, is composed of specialized cells that also contain starch granules. It is typically a few cell layers thick and surrounds the entire endosperm. However, its thickness varies significantly in different regions of the grain, being a single-cell layer near the germ while displaying considerable variation in the crease region (Shaveta et al. 2019).

The aleurone layer plays a crucial role in the structural integrity and biochemical functions of barley grain. It is rich in non-starch polysaccharides, specifically arabinoxylan and β-glucan, which are primarily located within the aleurone cell walls. The proportion of these compounds differs between the aleurone and endosperm cell walls. In the aleurone, the β-glucan-to-arabinoxylan ratio is approximately 25:75, whereas in the endosperm, this ratio is reversed (Kaur et al. 2019). This compositional difference influences the physical properties of the grain, including cell wall thickness, which varies among different barley cultivars. Research indicates that cultivars with a higher β-glucan content tend to have thicker cell walls (Šimić et al. 2019). Beyond its structural role, the aleurone layer is functionally significant in barley germination, as it is actively involved in the production and regulation of enzymes responsible for breaking down endosperm components. These enzymes include:

1. β-glucanases—Essential for the degradation of β-glucan, facilitating the mobilization of stored nutrients.
2. Limit Dextrinase—A key enzyme that breaks down complex carbohydrates into simpler forms (Sissons et al. 1992).
3. α-amylase—Plays a crucial role in starch degradation, converting starch into fermentable sugars.
4. Proteinases and Peptidases—Responsible for protein hydrolysis, aiding in the release of amino acids required for seedling growth (Jones 2005).
5. α-glucosidase—Catalyzes the breakdown of glucose-containing polysaccharides into simple sugars.
6. Protease inhibitors—Regulate protease activity to ensure controlled protein breakdown (Jones 2005).
7. Limit dextrinase and Trypsin—Involved in carbohydrate and protein metabolism, supporting seed germination and early plant development (MacGregor et al. 2002; Stahl et al. 2007).

The composition and functional attributes of the barley grain, particularly the aleurone layer, are integral to both its nutritional value and industrial applications, influencing processes such as malting, brewing, and food production.

10.2.2 Chemical Composition

Barley grain is composed of approximately 65% to 68% starch, 10% to 17% protein, 4% to 9% beta-glucan, 2% to 3% free lipids, and 1.5% to 2.5% minerals (Kaur et al. 2024). In hulled barley, the caryopsis remains enclosed within the hull, whereas hull-less barley lacks this outer covering. Nutritionally, hull-less barley contains higher amounts of β-glucan, essential amino acids, starch, and total dietary fiber compared to hulled varieties (Cyran 1996). Due to its increased starch content and lower fiber concentration, hull-less barley provides greater digestible energy (Sohn et al. 2007). Its total dietary fiber content varies between 11% and 34%, with soluble dietary fiber making up 3% to 20% of the total (Raj et al. 2023). As contrary to this, hull-less or dehulled barley has a total dietary fiber content that ranges from 11% to 20%. This includes 11% to 14% insoluble fiber and 3% to 10% soluble fiber. The pearling process, which involves the gradual removal of the outer layers of the grain, significantly alters its composition. This process strips away the hull (palea and lemma), bran layers (pericarp and testa), and the germ (embryo), all of which contain higher concentrations of insoluble fiber, protein, minerals (ash), and lipids. As a result, pearling increases the starch and beta-glucan content while reducing the levels of insoluble fiber, protein, ash, and free lipids. Since these outer layers are rich in essential nutrients but lower in starch and beta-glucan, their removal impacts both the nutritional value and functional properties of barley-based food products (Geng et al. 2021; Raj et al. 2023).

The structural and compositional attributes of barley grain play an important role in determining its processing characteristics and the quality of food products derived from pearled barley or barley flour (Raj et al. 2023). Due to its minimal hull attachment, hull-less barley requires little to no additional processing to separate the hull during threshing or milling. This makes it a more efficient and preferable option for food applications compared to hulled barley, as it reduces processing costs and retains a higher proportion of its dietary fiber and essential nutrients. Consequently, hull-less barley is considered more suitable for human consumption and various food processing industries (Geng et al. 2021; Raj et al. 2023).

10.3 Barley as a Fundamental Crop to Promote Sustainable Agriculture

In North Africa and across the world, cereals are among the most widely grown crops. They are also the main source of carbohydrates for the majority of people (Geng et al. 2021; Raj et al. 2023). In addition to being one of the oldest crops to be cultivated, barley is an essential component of both the agricultural sector and the economy of the whole world (Langridge 2018). Although its production has not increased as significantly as that of other staple cereals, barley continues to be a widely adaptable crop grown in diverse climates (Verstegen et al. 2014).

Barley (*Hordeum vulgare* L.) ranks as the fourth most significant cereal crop, following wheat (*Triticum aestivum* L., *Triticum durum*), maize (*Zea mays* L.), and rice (*Oryza sativa* L.). While barley is grown for both human consumption and livestock feed, wheat remains the dominant staple food in Egypt, leading to barley being primarily used as animal fodder (Visioni et al. 2023). The majority of barley production is allocated to animal feed (70%), with smaller portions used in malting and brewing (21%) and human consumption (6%) (Tricase et al. 2018).

Barley (*Hordeum vulgare* L.) has gained attention from numerous health experts due to its high fiber content, particularly β-glucan, which plays a role in lowering blood pressure, weight management, reducing cholesterol levels (Shaveta et al. 2019; Visioni et al. 2023). Among dietary fibers, β-glucans, along with other indigestible food components such as lactosucrose, soybean oligosaccharides, and isomalto-oligosaccharides, are currently being studied for their potential prebiotic effects (Kaur et al. 2019). Additionally, it helps in preventing colon cancer. This grain is widely available and affordable, making it an accessible source of essential nutrients. Barley contains both soluble and insoluble fiber, along with protein, B and E vitamins, minerals like selenium, magnesium, iron, and copper, as well as beneficial plant compounds such as flavonoids and anthocyanins (Visioni et al. 2023).

Barley is known to be a great source of silicon, a mineral that is found in all bodily tissues but is most abundant in bones and connective tissues including skin, arteries, hair, and nails (Jugdaohsingh 2007). Silicon plays a crucial role in forming bonds with glycosaminoglycans and contributes significantly to establishing cross-links between proteoglycans and collagen (Carlisle 1981). Although the exact process of mineralization remains unclear, Carlisle suggested that silicon likely enhances bone calcification by making the bone matrix more receptive to calcium deposition (Carlisle 1981). Research indicates that the silicon concentration in osteoid tissue is 25 times higher than in surrounding areas, gradually decreasing as calcification progresses (Carlisle 1970). Elevated silicon levels promote bone formation, reduce calcium excretion through urine, and enhance bone mineral density.

The soluble fiber in barley, particularly β-glucan, binds with bile acids in the intestines, which helps lower plasma cholesterol levels (Wang et al. 2016). Once absorbed, this fiber also reduces cholesterol production in the liver and helps cleanse blood vessels (Mio et al. 2020). Insoluble fiber adds bulk to the digestive system, promoting satiety, improving bowel movements, alleviating constipation, and reducing harmful bacteria in the colon—thereby lowering the risk of colon cancer. Furthermore, barley is rich in niacin, which aids in decreasing LDL (bad cholesterol) and increasing HDL (good cholesterol) (Kaur et al. 2019). Its selenium and vitamin E content provide powerful antioxidant benefits, while magnesium, an important co-factor in carbohydrate metabolism, works alongside fiber to support better blood sugar control in Type 2 diabetes. Barley also has diuretic properties, making it beneficial for urinary tract health, preventing gallstone formation, and exhibiting an attenuating effect by reducing calcium deposits in the kidney (Al-Razzuqi and Al-Ebady 2011).

10.4 Challenges Associated with Barley Yield

The rising demand for cereals, coupled with insufficient supply, highlights the necessity for increased cereal production, particularly for crops that can withstand drought and require minimal water. Such crops are better suited for arid and semi-arid regions in Egypt and other parts of North Africa (El Ansari et al. 2023). Barley is being reconsidered as a sustainable alternative to water-intensive cereals like wheat, maize, and rice due to its resilience to drought and environmental stresses, low water requirements, and minimal input needs, which enhance its economic viability (Raj et al. 2023; Verstegen et al. 2014).

Although barley is considered relatively tolerant to abiotic and biotic stresses, it remains vulnerable to environmental challenges that adversely affect its growth, yield, and overall productivity. With climate change posing a severe threat to global agriculture, these stressors contribute to significant yield losses and jeopardize food security (Shaheen et al. 2023). Key abiotic stress factors include drought, salinity, sodicity, soil acidity, waterlogging, extreme temperatures (both heat and freezing stress), nutrient imbalances (either deficiencies or toxicities), heavy metal contamination, and salt stress. These factors influence plant metabolism, disrupt physiological processes, and ultimately reduce grain yield (Rani and Chaudhary 2018). The severity of the impact depends on the duration of the environmental stress and the growth stage at which it occurs, with early stage stress often leading to more significant yield losses.

Barley is susceptible to a wide variety of biotic challenges, the majority of which come in the form of diseases that are caused by fungus, viruses, and other pathogens. Among the most prevalent diseases affecting barley production are net blotch, rusts, spot blotch, stripe disease, Septoria speckled leaf blotch, molya disease, barley yellow dwarf, and powdery mildew (Gangwar et al. 2019). These diseases weaken the plant, lower its resistance to environmental stressors, and significantly reduce grain yield. Furthermore, other fungal infections, such as black point and smuts, are particularly problematic for the brewing and malting industries as they degrade the quality of malt and beer, making them unsuitable for commercial processing.

To sustain and enhance barley production, it is essential to develop strategies that improve stress tolerance, such as breeding stress-resistant varieties, optimizing agronomic practices, and implementing integrated disease management approaches. Understanding how different stress factors interact and affect barley at various growth stages is crucial for minimizing their adverse effects and ensuring stable production (Rani and Chaudhary 2018). Ethiopia has significant potential for barley production; however, farmers face challenges due to inadequate modern storage facilities and substantial postharvest losses (Kifle 2016). Studies have identified several key constraints affecting the barley value chain (Rashid et al. 2019). These include a limited supply of high-yielding and disease-resistant varieties, infestations of weeds and pests, declining soil fertility and low soil pH, poor soil drainage, adverse weather conditions such as frost and drought, and various biotic stresses. Additionally, rising transportation costs and fluctuations in market prices further hinder the profitability and sustainability of barley farming in the country (Kifle 2016).

10.5 Quality Enhancement Techniques

Enhancing wheat quality is essential for improving human nutrition, meeting industrial processing needs, and ensuring agricultural sustainability. Due to ongoing changes, sustaining current barley production levels while ensuring future increases remains a significant challenge (Zhou et al. 2023). Effective quality enhancement strategies include developing resistant crop varieties and adopting proper agronomic practices. Resistant varieties enhance resilience, seed treatments protect young plants, and cultivation methods like crop rotation and soil management reduce disease risk (Verstegen et al. 2014). Advancements in molecular genetics, such as RFLPs, PCR-based markers, and third-generation SNP markers, have revolutionized breeding by enabling high-throughput genotyping. This allows precise mapping of disease resistance traits, accelerating crop improvement. Integrating these modern tools with traditional strategies enhances crop resilience, boosts yields, and strengthens food security against agricultural challenges (Salvi et al. 2014).

The CRISPR system is a cutting-edge genome-editing tool that plays an important role in improving barley quality by allowing precise genetic modifications. This technology enhances essential traits such as nutritional content, resistance to diseases, and adaptability to environmental conditions. By modifying specific genes, CRISPR enables the development of barley varieties with superior grain composition, increased protein levels, and enhanced malting properties, making it highly valuable for the food and beverage industries. Furthermore, it strengthens resistance against both biotic and abiotic stresses, reducing reliance on chemical treatments and supporting sustainable farming practices. By accelerating genetic improvements, CRISPR contributes to higher yields and better-quality barley, meeting the increasing global demand. Scientists have refined CRISPR-Cas9 and Cas12a toolkits to enhance genome editing precision in barley and wheat, achieving near-perfect efficiency in modifying specific genes, with reported success rates reaching up to 100% (Lawrenson et al. 2024).

10.6 CRISPR-Cas9 Overview

The CRISPR-Cas system was first recognized as a prokaryotic adaptive immune mechanism that protects bacterial cells from invading genetic elements such as plasmids and viruses. When foreign DNA enters the bacterial cell, CRISPR-Cas detects and cleaves the genetic material. Additionally, it retains fragments of viral DNA as molecular memory, allowing the organism to recognize and neutralize similar threats upon future encounters (Datsenko et al. 2012). More recently, scientists have repurposed the CRISPR-Cas system as a highly precise, efficient, and user-friendly tool for targeted genome editing. This breakthrough technology is transforming multiple fields, including biotechnology, medicine, and life sciences. In prokaryotic organisms such as bacteria and archaea, the CRISPR-Cas system functions as an RNA-guided defense mechanism. It integrates short fragments of foreign DNA, known as spacer sequences, into the CRISPR loci of the host genome. These sequences are transcribed into precursor CRISPR RNAs (pre-crRNAs), which are then processed into mature CRISPR RNAs (crRNAs). In conjunction with a trans-activating CRISPR RNA (tracrRNA), these crRNAs serve as a guide for Cas endonuclease proteins, allowing them to locate and breakdown DNA components that are identical to viral or plasmid DNA. Bacteria and archaea are able to establish an adaptive immune response via this mechanism, which enables them to resist future genetic invasions that are comparable to those that have occurred in the past (Kalinina et al. 2020).

10.6.1 Key Components of CRISPR-Cas9

I. Protospacer Adjacent Motif (PAM): The PAM sequence is a short, highly conserved DNA motif (2–4 base pairs) located next to the target DNA region. It plays a critical role in guiding Cas proteins to the correct site, ensuring specificity and preventing unintended DNA cleavage (Koonin and Makarova 2019). The PAM sequence differs across Cas proteins and influences the efficiency of DNA recognition and cleavage (Anders et al. 2014; Leenay and Beisel 2017).

II. CRISPR Array: This region of the bacterial genome consists of repeating DNA sequences interspersed with unique spacer sequences acquired from previous viral or plasmid infections. These stored sequences act as a genetic memory, enabling future immune responses against familiar threats (Bolotin et al. 2005; Pourcel et al. 2005).

III. CRISPR RNA (crRNA): The crRNA carries spacer sequences complementary to the target DNA, allowing sequence-specific targeting. The CRISPR array is transcribed into a long precursor CRISPR RNA (pre-crRNA), which is processed by Cas proteins into mature crRNAs that guide Cas enzymes for DNA cleavage (Koonin and Makarova 2019).

IV. Trans-activating CRISPR RNA (tracrRNA) (specific to Cas9): tracrRNA plays a key role in crRNA processing and function. It pairs with crRNA to form a stable single-guide RNA (sgRNA), which directs Cas9 to the target DNA sequence for precise cleavage (Liao and Beisel 2021; Liu et al. 2022).

V. Direct Repeats (DR) (specific to Cas12a): Direct repeats are sequences within the CRISPR array that contribute to pre-crRNA processing and crRNA maturation. They also help in forming R-loop structures that are essential for rRNA binding and function (Stella et al. 2017).

VI. Single Guide RNA (sgRNA) / Guide RNA (gRNA): A synthetic chimeric RNA formed by fusing tracrRNA and crRNA. The guide RNA (gRNA) contains a 20-nucleotide sequence that is complementary to the target DNA, directing Cas9 to its specific genomic location. The seed region, located at the 3′ end of the gRNA, is essential for accurate target recognition (Ran et al. 2013; Wong et al. 2015).

VII. Cas Proteins: These are CRISPR-associated endonucleases that use gRNA to identify and cleave the target DNA sequence. Cas9, for example, introduces double-stranded breaks, leading to gene disruption or targeted modifications (Szczepankowska 2012).

Through these components, the CRISPR-Cas system has evolved into a revolutionary genome editing tool, enabling precise genetic modifications with applications in agriculture, medicine, and biotechnology.

10.6.2 Classification of the CRISPR-Cas System

The CRISPR-Cas system is a diverse family of genome-editing tools, classified based on their mechanisms of action and effector proteins. It operates in three main stages: adaptation, expression and maturation, and interference. The adaptation module facilitates spacer acquisition by incorporating foreign genetic sequences from viruses or plasmids into the CRISPR array. The effector module is responsible for processing pre-crRNA, target recognition, and cleavage. In both Type I and Type III systems, these activities are carried out by the Cascade complex, also known as the CRISPR-associated complex for antiviral defence. This complex collaborates with Cas3 to degrade target sequences. Furthermore, in order to execute its role, Cascade works in conjunction with the Cmr complex and the Csm complex, respectively, in Type III-A and III-B systems (Chaudhuri et al. 2022).

10.6.3 Types of CRISPR-Cas Systems

Based on the effector proteins that they include, CRISPR-Cas systems may be classified into two primary categories.

Class I Systems
Class I systems utilize *multi-protein complexes* to process crRNA and recognize target sequences (Makarova et al. 2020). The effector module in Class I consists of multiple Cas proteins, including Cas5, Cas6, Cas8, and Cas10, which play distinct roles in crRNA processing, target cleavage, and spacer integration. Class I is further divided into three types:

- *Type I:* Cleaves DNA
- *Type III:* Cleaves both DNA and RNA
- *Type IV:* Its activity is still being investigated (Nidhi et al. 2021).

Class II Systems
Class II systems, in contrast to Class I systems, are dependent on a *single multidomain protein* for the processing of crRNA and the cleavage of targets (Makarova et al. 2020). Class II consists of the following types.

- *Type II:* Includes Cas9, which facilitates double-stranded DNA cleavage. It has three subtypes (II-A, II-B, and II-C).
- *Type V:* Characterized by *Cas12*, this type has 10 subtypes (V-A to V-K) and is capable of both DNA cleavage and self-processing of pre-crRNA.
- *Type VI:* Comprised of Cas13 proteins, which function as RNases and specifically target RNA. This type is further divided into four subtypes (VI-A to VI-D).

10.6.4 Cas9: Structure and Function

Streptococcus pyogenes Cas9 (SpyCas9) is a DNA endonuclease responsible for precise genome editing. It is a large protein composed of 1368 amino acids and features multiple domains with specialized functions. Cas9 facilitates the cleavage of double-stranded DNA (dsDNA) at a specific site, occurring three base pairs upstream of the protospacer adjacent motif (PAM). Structurally, Cas9 consists of two primary lobes: the recognition (REC) lobe and the nuclease (NUC) lobe, which includes conserved HNH and RuvC nuclease domains along with a variable C-terminal domain (CTD). The HNH domain is responsible for cleaving the DNA strand that is complementary to the guide RNA (gRNA), referred to as the target strand, while the RuvC domain cuts the opposite nontarget strand (Jiang and Doudna 2017) (Fig. 10.1).

10.6.5 Cas9's Working Mechanism

Cas9 remains in an inactive state unless it is bound to a guide RNA (gRNA). The gRNA serves as a molecular blueprint, directing Cas9 to the specific DNA sequence it needs to edit. It forms a T-shaped secondary structure, consisting of one tetra-loop and three stem-loops. The structure of the 5′ end of the gRNA is engineered to be complementary to the sequence of the target DNA, which provides accurate recognition (Fig. 10.1). Once gRNA binds to Cas9, it induces conformational changes in the protein, transitioning it from an inactive to an active state. The activated Cas9 then scans the genome, searching for a PAM sequence (5′-NGG-3′), a short motif required for DNA binding and cleavage. When a PAM site is located, Cas9 unwinds the adjacent DNA, allowing the gRNA to hybridize with the complementary strand.

Cas9 cleaves both strands of DNA three base pairs upstream of the PAM region once it has successfully bound to the DNA molecule by both strands. Both of its nuclease domains are responsible for carrying out the cleavage process: The HNH domain is responsible for cutting the target DNA strand, which is complementary to the sequence of the gRNA. The RuvC domain is responsible for cleaving the non-target strand, which ultimately results in a double-stranded break (DSB) with blunt ends (Hillary and Ceasar 2023). Immediately after the DNA breaks, the cell begins to start repair methods such as nonhomologous end joining (NHEJ) or homology-directed repair (HDR) to either damage the function of the gene or introduce certain genetic alterations.

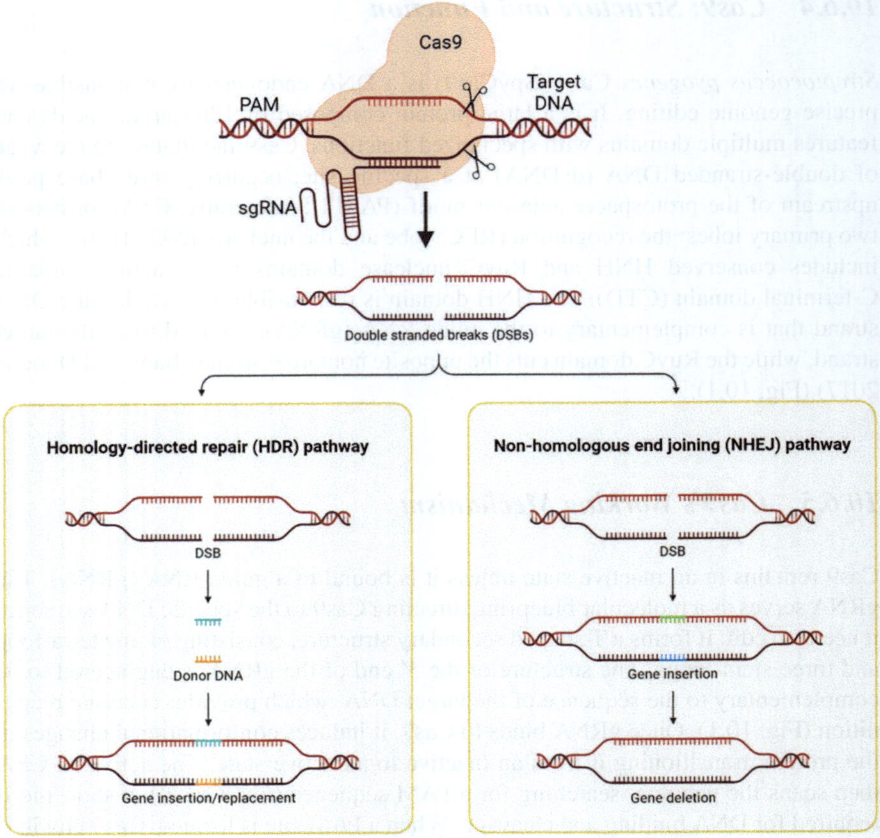

Fig. 10.1 A schematic illustration of gene knockout via CRISPR-Cas9. (Images created with BioRender)

10.6.6 Significance of Cas9 in Genome Editing

Cas9's ability to generate targeted double-stranded breaks has revolutionized genome engineering across various fields, including agriculture, medicine, and biotechnology. Its precision, efficiency, and adaptability make it an essential tool for gene knockout studies, genetic disease correction, crop improvement, and synthetic biology applications. The ability to program Cas9 using customizable gRNAs further enhances its utility, allowing scientists to target virtually any gene of interest with remarkable accuracy (Karunarathne et al. 2022; Lin et al. 2021).

10.6.7 Cas12a (Also Known as Cpf1)

Cas12a, formerly known as Cpf1, is a CRISPR-associated endonuclease identified in bacteria such as *Lachnospiraceae* and *Acidaminococcus* species, where it serves as a defense mechanism against viral invasions (Jianwei et al. 2023). Cas12a, in contrast to Cas9, does not need a trans-activating crRNA (tracrRNA) to activate itself. Instead, it makes use of a conserved pseudoknot structure that is located within the repeat-derived portion of its CRISPR RNA (crRNA) in order to process its own guides (Jedrzejczyk et al. 2022). This characteristic has been leveraged for multiplexed genome editing in vivo. Cas12a contains a single RuvC domain responsible for its catalytic activity and targets DNA sequences adjacent to a 5′-TTTV PAM. It cleaves both DNA strands distal to the PAM site, resulting in 5′ overhangs. After creating these DSBs, Cas12a remains catalytically active and can further degrade single-stranded DNA (ssDNA) in trans, a property that has been harnessed for precise gene editing and sequence-specific nucleic acid detection (Swarts and Jinek 2018) (Fig. 10.2).

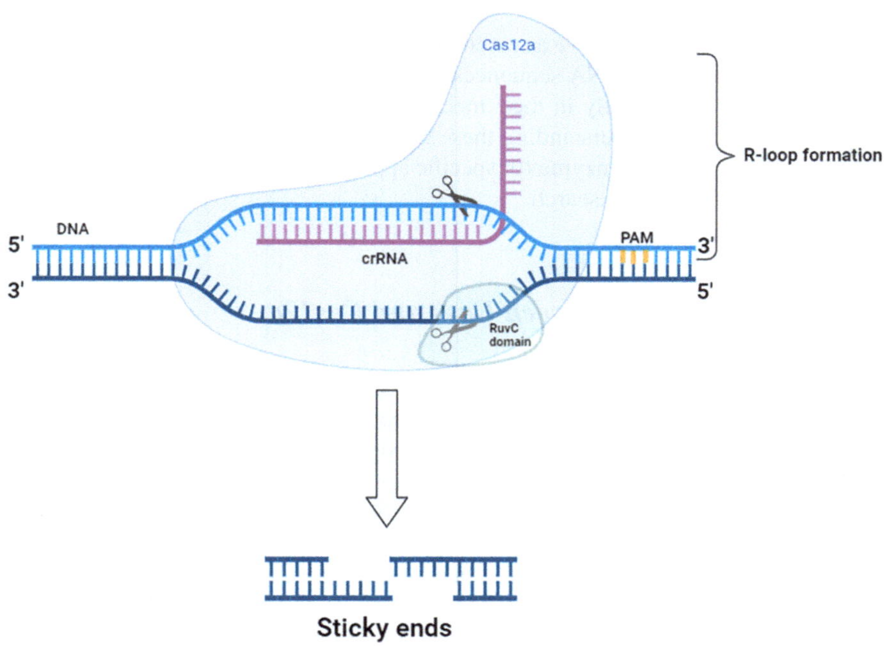

Fig. 10.2 Depiction of the CRISPR-Cas12a system, which generates double-stranded breaks (DSBs) with cohesive (sticky) ends. (Images created with BioRender)

10.6.8 Working Mechanism of Cas12a (Cpf1)

Cas12a, which is often referred to as Cpf1, is an RNA-guided endonuclease that is capable of effectively cleaving both single-stranded DNA (ssDNA) and double-stranded DNA (dsDNA) under the supervision of CRISPR RNAs (crRNAs) (Chen et al. 2018). The protein comprises a RuvC-like endonuclease domain and a nuclease (NUC) lobe, which are essential for its DNA cleavage activity (Fig. 10.2).

Upon encountering a potential target site adjacent to a PAM sequence, Cas12a initiates the formation of an R-loop by hybridizing the crRNA with the complementary target DNA strand. This process involves the alignment of the first 17 base pairs of the target sequence, resulting in the establishment of the R-loop structure (Dronina et al. 2022). Following R-loop formation, the active RuvC domain of Cas12a cleaves the nontarget DNA strand using the PAM sequence as a reference point. However, the precise mechanism by which the RuvC domain cleaves the target DNA strand remains to be fully elucidated. Notably, Cas12a generates staggered double-stranded breaks with 5′ overhangs, producing sticky ends that can facilitate specific genetic modifications during DNA repair processes (Hillary and Ceasar 2023).

Both Cas9 and Cas12a serve as essential tools in genetic engineering, facilitating precise modifications in DNA sequences. Despite their shared role in genome editing, they differ significantly in their mechanisms of action, target specificity, and structural properties. Understanding these similarities and differences is crucial for selecting the appropriate enzyme for specific applications in molecular biology, biotechnology, and medical research.

10.6.9 Similarities Between Cas9 and Cas12a

Cas9 and Cas12a both belong to the Class 2 CRISPR-Cas system, meaning they function as single, multi-domain proteins without requiring additional Cas proteins for their activity (Makarova et al. 2015). Cas9 operates using a duplex consisting of crRNA and tracrRNA, whereas Cas12a functions with only crRNA. In both cases, these RNAs form a ribonucleoprotein complex with their respective enzymes, positioning the crRNA to act as a guide RNA (gRNA) for target DNA recognition (Jinek et al. 2012). A crucial feature of both systems is their dependence on a short sequence in the target DNA known as the protospacer adjacent motif (PAM) (Mojica et al. 2009). Cas9 recognizes an "NGG" PAM, while Cas12a identifies a "TTTV" PAM, ensuring precise target selection. Once bound to the correct DNA sequence, both enzymes induce targeted DSBs, triggering repair mechanisms that can introduce genetic modifications (Jinek et al. 2012). These capabilities have made Cas9 and Cas12a widely used tools in genetic engineering, allowing for targeted gene knockouts, insertions, and other modifications in a variety of organisms, including plants, animals, and human cells (Zhang 2019).

10.6.10 Differences Between Cas9 and Cas12a

Despite their functional similarities, Cas9 and Cas12a exhibit several key differences. Structurally, Cas9 belongs to the Type II CRISPR-Cas system, while Cas12a is classified under Type V (Khan et al. 2022). One of the most fundamental distinctions between the two enzymes is that Cas9 requires both crRNA and tracrRNA for activation, whereas Cas12a operates solely with crRNA, simplifying the system. Additionally, Cas12a has an intrinsic ability to process multiple guide RNAs, allowing for the simultaneous targeting of multiple genes, making it particularly useful for multiplex genome editing (Lee et al. 2019).

Cas9 and Cas12a also differ in their DNA cleavage mechanisms. Cas9 cuts DNA 3–4 base pairs upstream of the PAM site, generating blunt-ended DSBs without overhangs (Ge et al. 2023) (Jinek et al. 2012). On the other hand, Cas12a cleaves DNA at a location that is 18–20 bp downstream of the PAM. This results in staggered cuts that bring about 5′ overhangs inside the DNA (Mishra et al. 2024). This distinction significantly impacts DNA repair pathways and the efficiency of gene insertions or deletions. The blunt-ended DSBs created by Cas9 are typically repaired through NHEJ, which can introduce small insertions or deletions. On the other hand, the staggered cuts produced by Cas12a may facilitate HDR, which can be advantageous for precise genetic modifications (de Pater et al. 2024; Pawelczak et al. 2018).

Another key difference is how these enzymes interact with DNA post-cleavage. After inducing DSBs, Cas9 remains attached to the DNA, which can hinder the recruitment of repair proteins and slow down the repair process. However, Helicase-like transcription factor (HLTF) translocase plays a crucial role in removing Cas9 from the damaged DNA ends, facilitating efficient repair. In contrast, Cas12a detaches from the DNA immediately after cleavage, allowing repair mechanisms to initiate more efficiently (Reginato et al. 2024). This characteristic of Cas12a may lead to faster and more precise genetic modifications compared to Cas9.

Overall, while both Cas9 and Cas12a serve as powerful genome-editing tools, their structural and functional differences make them suited for distinct applications. Cas9's blunt-ended cuts are useful for gene knockouts, while Cas12a's staggered cuts and multiplexing capabilities make it particularly effective for large-scale genetic modifications and precise insertions (Swarts and Jinek 2018).

10.7 Application of CRISPR in Barley Trait Enhancement

The development of barley varieties that can tolerate abiotic stressors, such as salt, drought, severe temperatures, and nutrient deficits, is very necessary in order to guarantee consistent crop yields and ensure the safety of food supplies that are available worldwide. As a result of its high degree of natural adaptation to a wide range of growing circumstances, barley is a crop that shows great potential for future production in a climate that is going through significant changes (Kopecká et al. 2023) (Table 10.1). Enhancing barley's resistance to these environmental challenges ensures consistent production, even under adverse conditions (Newton et al. 2011). For instance, drought-tolerant barley varieties can maintain yields with limited water availability, which is crucial for arid and semi-arid regions. Similarly, salt-tolerant strains can thrive in soils affected by salinization, a growing concern due to irrigation practices and climate change. Moreover, barley's resilience to harsh climates contributes to its performance under various abiotic stress factors (Kopecká et al. 2023).

Compared to other gene-editing techniques such as Sequence Specific Nucleases (SSNs), Transcription Activator-Like Effector Nucleases (TALEN), and Zinc Finger Nucleases (ZFN), the CRISPR-Cas9 system offers several advantages, including simpler target design, the ability to introduce multiple mutations simultaneously, greater versatility, and higher efficiency (Leong et al. 2018). As a result, the development of more precise gene-editing technologies, such as CRISPR-Cas9, base editing (BE), and prime editing (PE), is crucial for creating improved barley varieties with enhanced traits and increased resilience to environmental stress (Karunarathne et al. 2022; Lin et al. 2021). Gene editing in barley is mainly conducted by delivering the It is possible to modify genes using the CRISPR-Cas9 complex by either particle bombardment or Agrobacterium-mediated transformation (Han et al. 2021). A previous study utilizing *Agrobacterium*-mediated transfer of the Cas9 construct in barley reported a mutation frequency of 10–23% in the first generation due to Cas9 activity (Lawrenson et al. 2015) (Fig. 10.3). Among barley cultivars, "Golden Promise" is the most used for *Agrobacterium tumefaciens*-mediated CRISPR-Cas9 delivery due to its superior efficiency in shoot regeneration and transformation success. This preference is due to its higher shoot regeneration efficiency compared to other spring and winter barley varieties (Han et al. 2021; Hensel et al. 2008). An alternative approach to gene editing in barley involves using a virus-based vector, specifically Sonchus yellow net rhabdovirus (SYNV), to deliver the entire CRISPR-Cas9 cassette. This method eliminates the need for tissue culture, allowing for direct delivery into intact plants without requiring the isolation of plant cells or tissues, making the process more efficient and less labor-intensive (Kapusi et al. 2017).

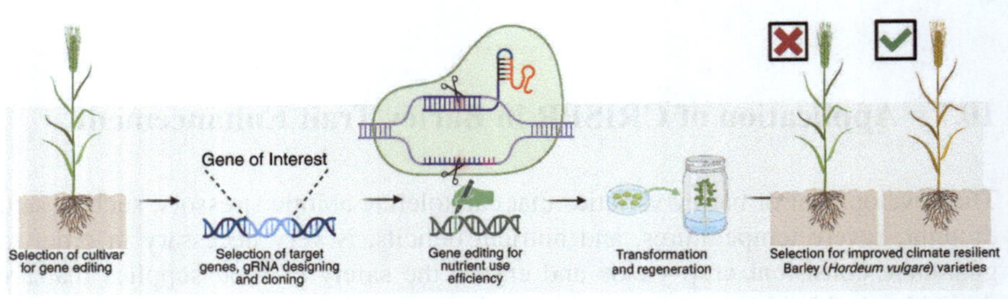

Fig. 10.3 The stages involved in the CRISPR-Cas9 genetic transformation and generation of climate-resilient barley variety are shown in a schematic diagram. (Images created with BioRender)

Table 10.1 Application of CRISPR in improving of Barley

S. No.	Gene name	Improved traits	Mutational frequency	References
1	*Ethylene response factor (HvERF62)*	Tolerant to waterlogging	76.5%	Zhu et al. (2025)
2	*Chymotrypsin inhibitor 1A (CI-1A)*	Hydrolysis of storage proteins	83.5%	Panting et al. (2025)
3	*Grain Width and Weight 2 (HvGW2.1)*	Improves grain yield; increase spike numbers	72.4%	Kis et al. (2024)
4	*GSK3/SHAGGY-like kinase (HvGSK1.1)*	Salt tolerance and improve yield	Data not available	Kloc et al. (2024)
5	*Gibberellin 3-oxidase 1 (HvGA3ox1)*	Increment in coleoptile length	78.6%	Chen et al. (2023)
6	*Albostrains (HvCMF7), Axis localised HORMA domain protein (HvMUS81), Transverse filament protein of the synaptonemal complex (HvZYP1)*	Resistance against viral infections	Up to 94%	Tamilselvan-Nattar-Amutha et al. (2023)
7	*Ovule pectin modifier 1(HvOPM1), Ovule pectin modifier 2 (HvOPM2)*	Increase in grain size and alter the ovule shape and size	Data not available	Yang et al. (2023)
8	*Fasciated ear (HvFEA2; HvFEA3; HvFEA4)*	Inflorescence meristem dilation and reduced spikes number and size also decreases grain yield	Data not available	Wang et al. (2023)
9	*Microrchidia (HvMORC1; HvMORC6a)*	Resistance against fungi that are plant pathogenic	58%	Galli et al. (2022)
10	*Nudum gene (HvNud)*	Naked barley grain	57.9%	Zang et al. (2022)
11	*Abnormal cytokinin response1 repressor 1 (HvARE1)*	Increased yield and plant height	50–90%	Karunarathne et al. (2022)
12	*Eukaryotic translation initiation factor 4E (HveIF4E)*	Resistance against viral infections	Data not available	Kershanskaya et al. (2022)
13	*Starch synthase IIa (HvSSIIa)*	Inflate the amylose and resistance starch content	18.2–21.7%	Yang et al. (2022)
14	*Phytoene desaturase (HvPDS)*	Surge in number of spikes per plant	53%	Han et al. (2021)

(continued)

Table 10.1 (continued)

S. No.	Gene name	Improved traits	Mutational frequency	References
15	*Mitogen-Activated Protein Kinase 6 (HvMPK6)*	Embryologically preestablished shoot	Data not available	Křenek et al. (2021)
16	*Caffeic acid O-methyltransferase 1 (HvCOMT1)*	Decrease in lignin content and increment in fermentable glucose content	42%	Lee et al. (2021)
17	*Non-adherent hull (HvNUD)*	Naked barley grain	57%	Gerasimova et al. (2020)
18	*D hordein gene (HvHor)*	Enhanced malt quality	Data not available	Li et al. (2020)
19	*Homogentisate phytyltransferase (HvHPT)*	Impact on barley grain development	46.2–64.7%	Zeng et al. (2020)
20	*Inositol trisphosphate 5/6 kinases (HvITPK1)*	Increment in phosphate content in barley grain	25%	Vlćko and Ohnoutkova (2020)
21	*HvdsRED*	Resistance toward wheat dwarf virus	Data not available	Kis et al. (2019)
22	*Cytokinin oxidase/ dehydrogenase gene (HvCKX; HvCKX3)*	Naked grains and protective hull are absent	88%	Gasparis et al. (2018)
23	*Microrchidia gene (HvMORC1)*	Resistance against powdery mildew	77.7%	Kumar et al. (2018)
24	*Cytokinin dehydrogenase (HvCKX1)*	Increment in grain yield	45–93%	Holubova et al. (2018)
25	*PM19 (HvPM19)*	Stunting	10–23%	Lawrenson et al. (2015)

While numerous proof-of-concept studies have demonstrated the potential of CRISPR-Cas9 for genetic improvements in barley (Lawrenson et al. 2015), only a limited number of these advancements have progressed to commercial application (Nonaka et al. 2017). Additionally, a speed breeding protocol for barley has recently been developed using CRISPR-Cas9 technology, which accelerates the growth cycle and enhances the efficiency of barley breeding programs (Watson et al. 2018). In recent investigations, the efficiency and benefits of using LbCas12a, a variation of Cas12a, for genome editing in barley have been emphasized. These studies have shown that LbCas12a is superior than Cas9 in terms of both its overall efficiency and its advantages. Two novel Cas12a variants (ttHsCas12a and ttAtCas12a + int) outperformed existing versions, achieving mutation rates of up to 87%. Unlike Cas9, which creates blunt-ended DNA breaks, Cas12a generates sticky-ended breaks, enhancing gene-targeting efficiency. It also recognizes a TTTV PAM sequence, making it effective in GC-poor regions. The study tested different guide RNA architectures (V1 and V2), with V2 showing higher efficiency. Additionally, mutations introduced by Cas12a were heritable, confirming stable gene edits. These findings suggest that optimized Cas12a-based editing could be highly effective in barley and other monocot crops like wheat and rice (Lawrenson et al. 2022). Further research and refinement of gene-editing approaches are necessary to translate these experimental breakthroughs into viable agricultural solutions that can benefit large-scale crop production.

Genetic modification allows for the introduction of specific genes into elite barley cultivars to enhance traits such as drought tolerance, salinity resistance, and nitrogen use efficiency (NUE). The use of genetic modification to introduce desirable traits into elite barley cultivars has certain limitations, primarily due to safety and health concerns associated with genetically modified organisms (GMOs) (Pellegrino et al. 2018). Nevertheless, a number of barley genes have been investigated and used to improve resistance to a variety of abiotic stressors. This presents a viable option for the purpose of enhancing crop resilience.

Specific genes have been identified for their role in combating different environmental stress factors (Hazzouri et al. 2018; Karunarathne et al. 2020; Mwando et al. 2020; Zhou et al. 2016). For instance, *Hordeum vulgare* aleurone 1 (*HVA1*) has been linked to improved drought tolerance, while high-affinity K^+ transporter 1;1 (*HKT1;1*) plays a crucial role in salinity tolerance by regulating ion homeostasis. Additionally, abnormal cytokinin response 1 repressor 1 (*ARE1*) has been found to enhance nitrogen use efficiency (NUE), contributing to better nutrient absorption and growth under nitrogen-limited conditions (Han et al. 2018; Karunarathne et al. 2022; Sivamani et al. 2000).

Beyond genetic modifications, researchers have also identified elite barley genotypes through extensive phenotype screening. Some notable stress-tolerant cultivars include "Golshan" and "Oxin," which exhibit high salinity resistance, as well as "GrangeR" and "Bridge," recognized for their superior tolerance to low nitrogen environments (Abbasipour Bahrani et al. 2023; Karunarathne et al. 2020). These naturally stress-adapted genotypes serve as valuable resources for plant breeding programs aimed at developing resilient barley varieties without relying solely on genetic engineering.

By integrating genetic studies with selective breeding approaches, scientists can enhance barley's adaptability to environmental stress while addressing concerns related to genetically modified crops. This combined strategy supports sustainable agriculture by improving crop performance in challenging conditions while maintaining consumer and regulatory acceptance.

The *HvAKT1* gene is essential for maintaining ion homeostasis, particularly in potassium (K^+) transport, and its overexpression has been linked to enhanced drought tolerance in barley. By improving K^+ uptake in the roots, this gene helps regulate osmotic balance and water retention in plant cells. As a result, barley plants with increased *HvAKT1* expression demonstrate greater drought resistance, better survival rates, and higher biomass production under water-deficient conditions (Feng et al. 2020).

Likewise, *HvHKT1;5* plays a crucial role in managing sodium (Na^+) transport, which is vital for salinity tolerance. When this gene is silenced, it prevents excessive Na^+ accumulation in leaves while maintaining a stable K^+/Na^+ ratio, thereby reducing ion toxicity. This balance allows plants to sustain normal growth and physiological functions even in high-salinity environments, making them more resilient to salinity stress (Huang et al. 2020).

Another key gene, *HvPap-1*, encodes an aquaporin protein that facilitates efficient water movement across cell membranes. Under drought conditions, its overexpression enhances water-use efficiency, minimizes excessive water loss, and helps maintain leaf hydration. Consequently, plants with increased *HvPap-1* expression exhibit improved growth, greater drought resistance, and better adaptability to arid conditions (Gomez-Sanchez et al. 2019).

In terms of micronutrient regulation, *HvZIP8* is involved in zinc (Zn) transport, ensuring adequate Zn uptake and distribution within the plant. When overexpressed, this gene promotes better Zn translocation, which is essential for enzymatic activity and overall plant health. In Zn-deficient conditions, plants with higher HvZIP8 expression show improved biomass accumulation and increased Zn content in their grains, making them nutritionally superior (Sun et al. 2015).

Similarly, *HvCBF4* functions as a cold-responsive transcription factor, activating genes that help the plant endure freezing stress. When overexpressed, it triggers the accumulation of protective compounds such as antioxidants and osmoprotectants, which mitigate the negative effects of low temperatures. Consequently, plants with increased HvCBF4 expression demonstrate better cold tolerance, reduced frost damage, and an improved ability to survive harsh winter conditions (Jeknić et al. 2014).

HvRAP76, an ERF transcription factor, plays a role in waterlogging tolerance. When silenced using RNAi (RNA interference) in lines 23, 25, and 55, a reduction in growth and survival was observed under waterlogging stress. The silencing led to a decrease in fluorescence, indicating lower HvRAP76 expression. Additionally, gene co-expression analysis revealed a significant reduction in hypoxia-responsive gene expression, confirming the involvement of HvRAP76 in stress adaptation. Furthermore, germination efficiency was found to be lower in the RNAi lines compared to the null control, emphasizing its role in waterlogging tolerance (Mendiondo et al. 2016).

10.8 Limitations of CRISPR-Cas9

10.8.1 Off-Target Effects

It is possible for the combination of programmable gRNA and Cas9 nucleases to result in unintended modifications due to the tolerance of mismatches in the gRNA sequence, particularly up to five base pairs in the 5′ region. This is despite the fact that the combination has significantly simplified and improved the efficiency of gene editing in plants (Zhao and Wolt 2017). Moreover, these off-target effects may introduce unpredictable changes in the genome, leading to unintended consequences that are difficult to assess (Liu et al. 2017). A greater level of specificity of the

CRISPR system in plant genomes may be responsible for the higher frequency of off-target edits in human and animal genomes compared to plant genomes, as shown by research (Endo et al. 2015; Feng et al. 2016; Peterson et al. 2016). Additionally, the lower occurrence of off-target modifications in plants may also be attributed to the technical limitations in detecting such mutations (Ahmad et al. 2020).

It is standard practice to use several techniques to identify changes that are caused by CRISPR. These techniques include the loss of restriction site test, the surveyor nuclease and T7 endonuclease I (T7EI) assay, high-resolution melting analysis (HRMA), and PAGE-based genotyping (Shan et al. 2014; Thomas et al. 2014; Vouillot et al. 2015; Zhu et al. 2014). Moreover, each of these techniques has its limitations. For instance, the restriction site loss assay is dependent on the presence of specific restriction sites at the target region, meaning only mutations that disrupt these sites can be detected. Similarly, T7EI and surveyor nuclease assays, along with PAGE-based genotyping, are unable to identify homozygous mutations. While HRMA is a highly sensitive method, its accuracy is affected when the difference in melting temperature (Tm) between the mutant and wild-type sequences is minimal, and it requires specialized, costly equipment (Ahmad et al. 2020).

Unlike in animals, off-target effects in plants can be minimized through selective breeding during tissue culture. Any undetected mutations present in edited plant lines that pass the selection process can be gradually eliminated through repeated backcrossing with the parental genotype to remove unwanted traits. However, this process, known as transgene outbreeding, presents challenges in certain plant species, making the elimination of undesirable phenotypes more complex (Ahmad et al. 2020).

10.8.2 Continuous Cas9 Activity

There have been several studies that have shown that the CRISPR-Cas machinery continues to function even after multiple generations have passed (Char et al. 2017; Feng et al. 2014; Wang et al. 2018). This prolonged nuclease activity can lead to unintended genetic modifications, potentially introducing mutations in otherwise stable plant lines. Since Cas9 remains functional in successive generations, it is crucial to obtain Cas9-free plants before genotyping to prevent late chimerism caused by continuous Cas9 expression (Char et al. 2017). To mitigate the issue of persistent CRISPR-Cas activity, researchers have investigated alternative methods that avoid plasmid integration during gene editing. One such approach is transient expression, where CRISPR components are introduced into plant cells temporarily without being incorporated into the nuclear genome (Ahmad et al. 2020). Another promising strategy involves using viral-based plant expression vectors, which facilitate gene editing without requiring integration into the host genome. This method not only reduces the likelihood of off-target mutations but also offers a rapid solution for generating Cas9-free plants at early developmental stages, improving the precision and efficiency of gene editing in plants (Ahmad et al. 2020).

10.9 Ethical Concerns Regarding CRISPR

The ethics of CRISPR technology revolves around several crucial problems, especially its ability to alter human genetics, modification of species, and its influence on ecology. One of the major ethical concerns with CRISPR is its use in human germline editing, which includes the alternation of the DNA of embryos or germ cells that could be passed to future generations. The unintended consequences of the alternation of the human genome are not totally understood, and there is concern over the long-term effects of these genetic modifications on human evolution (Nuffield Council on Bioethics 2020).

The use of CRISPR in agriculture, particularly for gene editing of crops and animals, raises concerns about its environmental impact. For instance, gene-edited crops, if they crossbreed with wild species, could disrupt local ecosystems, leading to a loss of biodiversity or the emergence of invasive species. The application of CRISPR to alter animals or insects, such as mosquitoes to reduce the transmission of diseases like malaria, could also pose risks to ecosystems. If CRISPR is not properly regulated, it could unintentionally introduce genetic changes that have adverse effects on the environment (O'Neill et al. 2021). Genetically modified (GM) crops are produced by altering the genetic material of the crop, i.e., DNA of the crop using different genetic engineering methods. They help in increasing the resistance of crops against several biotic and abiotic stress also increasing the life span of the crop at the same time enhancing the nutritional value of the crops in many cases. These characteristics of GM crops are quite similar to that of CRISPR-modified crops because they both involve manipulation or alteration of the genetic material. But there are several differences which might seem very minute but create a vast difference between CRISPR and GMOs. GMOs' are created by transferring a foreign DNA, i.e., gene from another non-crossable or crossable species, but on the other hand, CRISPR-modified crops do not necessarily involve transformation of any foreign DNA (Zhang et al. 2020). In contrast to GMOs, CRISPR results in accurate and small targeted modifications at a particular locus. Many CRISPR-modified crops are transgene-free and closely bear a resemblance of natural crops.

There is a debate going on whether the existing regulations on GMOs should be applied on CRISPR or not, different groups have different opinions on it. Traducers of CRISPR modified crops argue that the procedure involved in production of CRISPR modified plants is like that of the conventional GMO's and hence existing GMO regulations should be implied on CRISPR modified crops too. The main concern for both GMOs and CRISPR-edited crops is the use of marker genes for the selection of engineered crops. One of the most discussed and significant drawbacks in the case of CRISPR-edited crops is the ability of CRISPR to alter the genotype of future generations of the crop, giving rise to ethical concerns. CRISPR-Cas9 is a very efficient genetic editing tool that can be used to cure various genetic disorders, but it can be used to generate negative results and can be exploited for own benefits. Another ethical concern is the potential use of CRISPR in bioterrorism. The ease with which genetic material can be altered raises the possibility of weaponizing this technology. The creation of genetically modified organisms for malicious purposes presents a real risk, particularly with the increasing accessibility of gene-editing tools (Sandel 2020). Ensuring that CRISPR technology is not misused requires rigorous monitoring and adherence to international security standards.

In the European Union (EU), all these factors are major concerns, and the citizens are more prudent about the different drawbacks of this genetic engineering tool (Table 10.2). Due to which, in European Union, CRISPR-edited crops are considered Genetically Engineered Organisms (GMOs) and hence the strict GMO regulations are implied on CRISPR-edited crops, which causes difficulties in the commercialization of various CRISPR-modified crops (Marone et al. 2023). However, this is not the case everywhere, in countries like Argentina and Brazil, the rules and regulations are comparatively more flexible. But that is not all, the inclusion of CRISPR technology in global agriculture sector causes socioeconomic challenges. The patent landscape is predominated by US and EU regions and restricts access to small and developing countries. India and several other African countries, face challenges in ensuring the benefits of CRISPR are accessible to all farmers irrespective of their production scale (Munawar et al. 2024).

Table 10.2 Regulatory policies of different nations for CRISPR genome editing (modified from Abd-Elsalam and Ahmad 2023)

Country	Policy	References
USA	Policy does not match the regulations that are followed for GMOs.	Kleter et al. (2019)
Australia	If the final product is non-transgene, GMO regulations will not be implied.	Mallapaty (2019), Tchetvertakov (2019)
Japan	If the final product is non-transgene, policy will not follow GMO regulations.	Zannoni (2019)
Brazil	If the final product is non-transgene, GMO regulations will not be implied.	Eriksson et al. (2019)
Chile	If new amalgamation of genetic material is observed, GMO regulations will be implied.	Eriksson et al. (2019)
European Union	GMO regulations are implied, irrespective of any factor.	Jouanin et al. (2018)
Argentina	If final product is non-transgenic, GMO regulations will not be implied.	Gao et al. (2018)
China	If final product us proved to be transgene, policy will follow GMO regulations.	Gao et al. (2018)
New Zealand	GMO regulations are followed, in all cases.	Fritsche et al. (2018)
France	If the exogenous genes are not removed, GMO regulations are implied.	Pfeiffer et al. (2018)

10.10 Conclusion and Future Perspective for Efficient CRISPR Editing

The future of Cas12a in agricultural biotechnology is highly promising, with ongoing advancements aimed at optimizing its efficiency and expanding its applications in crop improvement. One key area of focus is the development of engineered Cas12a variants with improved functionality, including relaxed PAM requirements. Unlike the standard Cas12a, which relies on TTTV PAM sequences, novel versions with expanded target recognition would allow gene editing in a wider range of genomic locations, making it more versatile for modifying crop traits (Chen et al. 2020). This advancement would be particularly beneficial for multiplex genome editing, where multiple genes are edited simultaneously to enhance traits such as stress tolerance, disease resistance, and yield improvement. Another important direction is the enhancement of Cas12a's temperature stability to improve its performance in a broader range of plant species. Since Cas12a exhibits reduced activity at lower temperatures compared to Cas9, optimizing its efficiency for crops grown in diverse climatic conditions is essential. By fine-tuning Cas12a's protein structure or using synthetic biology approaches, researchers aim to enhance its editing capacity for crops cultivated in temperate and cold regions, where efficient gene editing remains a challenge (Bandyopadhyay et al. 2020).

Moreover, the integration of Cas12a with next-generation gene-editing tools such as PE and BE holds great potential for precise genetic modifications. Unlike conventional CRISPR methods that rely on DSBs, these advanced techniques allow single-base changes or targeted insertions, reducing unintended mutations and improving the precision of crop trait engineering. This would be particularly useful for modifying nutritional content, shelf life, and stress resilience without disrupting other essential genes. Another emerging application is the use of Cas12a in HDR strategies, which enable precise gene replacement rather than simple knockouts. The efficiency of HDR in plants is significantly low due to several factors, including the limited availability of donor templates and the length of these templates, which affects the repair process (Puchta 2005). Improvements in repair pathway modulation could make it a viable approach for introducing beneficial alleles, such as those associated with higher yields, improved grain quality, and pest resistance. Additionally, as synthetic biology advances, Cas12a could be harnessed for epigenome editing, allowing scientists to regulate gene expression without changing the DNA sequence itself. As regulatory frameworks evolve, Cas12a-based genome editing has the potential to transform modern agriculture by accelerating the development of climate-resilient, high-yielding, and disease-resistant crops. With continued innovation, Cas12a could play a pivotal role in sustainable agriculture, reducing reliance on chemical inputs, enhancing food security, and meeting the demands of a growing global population.

References

Abbasipour Bahrani H, Ghazvini H, Amiri B, Bazrafshan F, Nikkhah H (2023) Responses of barley (*Hordeum vulgare* L.) genotypes to salinity stress under controlled and field conditions. Gesunde Pflanzen 75:499–513. https://doi.org/10.1007/s10343-022-00711-5
Abd-Elsalam KA, Ahmad A (eds) (2023) Global regulatory outlook for CRISPRized plants. Elsevier. https://doi.org/10.1016/B978-0-443-18444-4.00024-7
Ahmad N, Rahman M, Mukhtar Z, Zafar Y, Zhang B (2020) A critical look on CRISPR-based genome editing in plants. J Cell Physiol 235:666–682. https://doi.org/10.1002/jcp.29052
Al-Razzuqi RAM, Al-Ebady Z (2011) Role of Barley in the improvement of Nephrocalcinosis in experimental rabbit model. Int J Nutr Pharmacol Neurol Dis 1:69. https://doi.org/10.4103/2231-0738.77535
Alvi M, Saleem MN (2024) Importance of Barley (*Hordeum vulgare* Linn.) as food and medicine from past to present: a brief overview. J Phytopharmacol 13:402–406. https://doi.org/10.31254/phyto.2024.13509
Anders C, Niewoehner O, Duerst A, Jinek M (2014) Structural basis of PAM-dependent target DNA recognition by the Cas9 endonuclease. Nature 513:569–573. https://doi.org/10.1038/nature13579
Badr A, M K, Sch R, Rabey HE, Effgen S, Ibrahim HH, Pozzi C, Rohde W, Salamini F (2000) On the origin and domestication history of barley (*Hordeum vulgare)*. Mol Biol Evol 17:499–510. https://doi.org/10.1093/oxfordjournals.molbev.a026330
Bandyopadhyay A, Kancharla N, Javalkote VS, Dasgupta S, Brutnell TP (2020) CRISPR-Cas12a (Cpf1): a versatile tool in the plant genome editing tool box for agricultural advancement. Front Plant Sci 11. https://doi.org/10.3389/fpls.2020.584151
Bolotin A, Quinquis B, Sorokin A, Ehrlich SD (2005) Clustered regularly interspaced short palindrome repeats (CRISPRs) have spacers of extrachromosomal origin. Microbiology (Reading) 151:2551–2561. https://doi.org/10.1099/mic.0.28048-0
Brown SE, Stephens JL, Lapitan NL, Knudson DL (1999) FISH landmarks for barley chromosomes (*Hordeum vulgare* L.). Genome 42:274–281
Carlisle EM (1970) Silicon: a possible factor in bone calcification. Science 167:279–280. https://doi.org/10.1126/science.167.3916.279
Carlisle EM (1981) Silicon: a requirement in bone formation independent of vitamin D1. Calcif Tissue Int 33:27–34. https://doi.org/10.1007/BF02409409
Char SN, Neelakandan AK, Nahampun H, Frame B, Main M, Spalding MH, Becraft PW, Meyers BC, Walbot V, Wang K, Yang B (2017) An *Agrobacterium*-delivered CRISPR/Cas9 system for high-frequency targeted mutagenesis in maize. Plant Biotechnol J 15:257–268. https://doi.org/10.1111/pbi.12611
Chaudhuri A, Halder K, Datta A (2022) Classification of CRISPR/Cas system and its application in tomato breeding. Theor Appl Genet 135:367–387. https://doi.org/10.1007/s00122-021-03984-y
Chen JS, Ma E, Harrington LB, Da Costa M, Tian X, Palefsky JM, Doudna JA (2018) CRISPR-Cas12a target binding unleashes indiscriminate single-stranded DNase activity. Science 360:436–439. https://doi.org/10.1126/science.aar6245
Chen P, Zhou J, Wan Y, Liu H, Li Y, Liu Z, Wang H, Lei J, Zhao K, Zhang Y, Wang Y, Zhang X, Yin L (2020) A Cas12a ortholog with stringent PAM recognition followed by low off-target editing rates for genome editing. Genome Biol 21:78. https://doi.org/10.1186/s13059-020-01989-2
Chen W, Zheng Y, Wang J, Wang Z, Yang Z, Chi X, Dai L, Lu G, Yang Y, Sun B (2023) Ethylene-responsive SbWRKY50 suppresses leaf senescence by inhibition of chlorophyll degradation in sorghum. New Phytol 238:1129–1145
Cyran M (1996) A note on the composition and nutritional value of hulless barley. J Anim Feed Sci
Czembor JH, Czembor E (2024) Barley genetic resources: advancing conservation and applications for breeding—series II. Agronomy 15(1):31. https://doi.org/10.3390/agronomy15010031

Dai F, Nevo E, Wu D, Comadran J, Zhou M, Qiu L, Chen Z, Beiles A, Chen G, Zhang G (2012) Tibet is one of the centers of domestication of cultivated barley. Proc Natl Acad Sci 109:16969–16973. https://doi.org/10.1073/pnas.1215265109

Dai F, Chen Z-H, Wang X, Li Z, Jin G, Wu D, Cai S, Wang N, Wu F, Nevo E, Zhang G (2014) Transcriptome profiling reveals mosaic genomic origins of modern cultivated barley. Proc Natl Acad Sci 111:13403–13408. https://doi.org/10.1073/pnas.1414335111

Datsenko KA, Pougach K, Tikhonov A, Wanner BL, Severinov K, Semenova E (2012) Molecular memory of prior infections activates the CRISPR/Cas adaptive bacterial immunity system. Nat Commun 3:945. https://doi.org/10.1038/ncomms1937

de Pater S, Kamoen L, van Schendel R, Hooykaas PJJ, Tijsterman M (2024) Profiling Cas9- and Cas12a-induced mutagenesis in *Arabidopsis thaliana*. Plant J 119:2706–2717. https://doi.org/10.1111/tpj.16943

Dronina J, Samukaite-Bubniene U, Ramanavicius A (2022) Towards application of CRISPR-Cas12a in the design of modern viral DNA detection tools (Review). J Nanobiotechnol 20:41. https://doi.org/10.1186/s12951-022-01246-7

El Ansari L, Chenoune R, Yigezu YA, Komarek AM, Gary C, Belhouchette H (2023) Intensification options in cereal-legume production systems generate trade-offs between sustainability pillars for farm households in northern Morocco. Agric Syst 212(103):769. https://doi.org/10.1016/j.agsy.2023.103769

Endo M, Mikami M, Toki S (2015) Multigene knockout utilizing off-target mutations of the CRISPR/Cas9 system in rice. Plant Cell Physiol 56:41–47. https://doi.org/10.1093/pcp/pcu154

Eriksson D, Kershen D, Nepomuceno A, Pogson BJ, Prieto H, Purnhagen K, Smyth S et al (2019) A comparison of the EU regulatory approach to directed mutagenesis with that of other jurisdictions, consequences for international trade and potential steps forward. New Phytol 222:1673–1684

Feng Z, Mao Y, Xu N, Zhang B, Wei P, Yang D-L, Wang Z, Zhang Z, Zheng R, Yang L, Zeng L, Liu X, Zhu J-K (2014) Multigeneration analysis reveals the inheritance, specificity, and patterns of CRISPR/Cas-induced gene modifications in Arabidopsis. Proc Natl Acad Sci 111:4632–4637. https://doi.org/10.1073/pnas.1400822111

Feng C, Yuan J, Wang R, Liu Y, Birchler JA, Han F (2016) Efficient targeted genome modification in maize using CRISPR/Cas9 system. J Genet Genom 43:37–43. https://doi.org/10.1016/j.jgg.2015.10.002

Feng X, Liu W, Cao F, Wang Y, Zhang G, Chen Z-H, Wu F (2020) Overexpression of *HvAKT1* improves drought tolerance in barley by regulating root ion homeostasis and ROS and NO signaling. J Exp Bot 71:6587–6600. https://doi.org/10.1093/jxb/eraa354

Fritsche S, Poovaiah C, MacRae E, Thorlby G (2018) A New Zealand perspective on the application and regulation of gene editing. Front Plant Sci 9:1323

Galli M, Martiny E, Imani J, Kumar N, Koch A, Steinbrenner J, Kogel KH (2022) CRISPR/SpCas9-mediated double knockout of barley Microrchidia *MORC1* and *MORC6a* reveals their strong involvement in plant immunity, transcriptional gene silencing and plant growth. Plant Biotechnol J 20(1):89–102

Gangwar OP, Chander Bhardwaj S, Singh GP, Prasad P, Kumar S (2019) Barley diseases and their management- an Indian perspective. Wheat Barley Res 10. https://doi.org/10.25174/2249-4065/2018/83844

Gao W, Xu W-T, Huang K-L, Guo M-Z, Luo YB (2018) Risk analysis for genome editing-derived food safety in China. Food Control 84:128–137

Garcia-Gimenez G, Jobling SA (2021) Gene editing for barley grain quality improvement. J Cereal Sci 103(103):394. https://doi.org/10.1016/j.jcs.2021.103394

Gasparis S, Kała M, Przyborowski M, Łyznik LA, Orczyk W, Nadolska-Orczyk A (2018) A simple and efficient CRISPR/Cas9 platform for induction of single and multiple, heritable mutations in barley (*Hordeum vulgare* L.). Plant Methods 14(1):111

Ge N, Liu M, Li R, Allen NM, Galvin J, Shen S, O'Brien T, Prendiville TW (2023) Using Ribonucleoprotein-based CRISPR/Cas9 to edit single nucleotide on human induced pluripotent

stem cells to model type 3 long QT Syndrome (SCN5A±). Stem Cell Rev Rep 19:2774–2789. https://doi.org/10.1007/s12015-023-10602-5
Geng L, Li M, Zhang G, Ye L (2021) Barley: a potential cereal for producing healthy and functional foods. Food Qual Saf 6. https://doi.org/10.1093/fqsafe/fyac012
Gerasimova SV, Hertig C, Korotkova AM, Kolosovskaya EV, Otto I, Hiekel S, Kochetov AV, Khlestkina EK, Kumlehn J (2020) Conversion of hulled into naked barley by Cas endonuclease-mediated knockout of the NUD gene. BMC Plant Biol 20(1):255
Gomez-Sanchez A, Gonzalez-Melendi P, Santamaria ME, Arbona V, Lopez-Gonzalvez A, Garcia A, Hensel G, Kumlehn J, Martinez M, Diaz I (2019) Repression of drought-induced cysteine-protease genes alters barley leaf structure and responses to abiotic and biotic stresses. J Exp Bot 70:2143–2155. https://doi.org/10.1093/jxb/ery410
Han Y, Yin S, Huang L, Wu X, Zeng J, Liu X, Qiu L, Munns R, Chen Z-H, Zhang G (2018) A sodium transporter *HvHKT1;1* confers salt tolerance in barley via regulating tissue and cell ion homeostasis. Plant Cell Physiol 59:1976–1989. https://doi.org/10.1093/pcp/pcy116
Han Y, Broughton S, Liu L, Zhang X-Q, Zeng J, He X, Li C (2021) Highly efficient and genotype-independent barley gene editing based on anther culture. Plant Comm 2. https://doi.org/10.1016/j.xplc.2020.100082
Hazzouri KM, Khraiwesh B, Amiri KMA, Pauli D, Blake T, Shahid M, Mullath SK, Nelson D, Mansour AL, Salehi-Ashtiani K, Purugganan M, Masmoudi K (2018) Mapping of *HKT1;5* gene in barley using GWAS approach and its implication in salt tolerance mechanism. Front Plant Sci 9. https://doi.org/10.3389/fpls.2018.00156
Hensel G, Valkov V, Middlefell-Williams J, Kumlehn J (2008) Efficient generation of transgenic barley: the way forward to modulate plant–microbe interactions. J Plant Physiol 165:71–82. https://doi.org/10.1016/j.jplph.2007.06.015
Hillary VE, Ceasar SA (2023) A review on the mechanism and applications of CRISPR/Cas9/Cas12/Cas13/Cas14 proteins utilized for genome engineering. Mol Biotechnol 65:311–325. https://doi.org/10.1007/s12033-022-00567-0
Holubova K, Hensel G, Vojta P, Tarkowski P, Bergougnoux V, Galuszka P (2018) Modification of barley plant productivity through regulation of cytokinin content by reverse-genetics approaches. Front Plant Sci 27(9):1676
Huang L, Kuang L, Wu L, Shen Q, Han Y, Jiang L, Wu D, Zhang G (2020) The HKT transporter *HvHKT1;5* negatively regulates salt tolerance1. Plant Physiol 182:584–596. https://doi.org/10.1104/pp.19.00882
Jedrzejczyk DJ, Poulsen LD, Mohr M, Damas ND, Schoffelen S, Barghetti A, Baumgartner R, Weinert BT, Warnecke T, Gill RT (2022) CRISPR-Cas12a nucleases function with structurally engineered crRNAs: SynThetic trAcrRNA. Sci Rep 12:12193. https://doi.org/10.1038/s41598-022-15388-z
Jeknić Z, Pillman KA, Dhillon T, Skinner JS, Veisz O, Cuesta-Marcos A, Hayes PM, Jacobs AK, Chen THH, Stockinger EJ (2014) *Hv-CBF2A* overexpression in barley accelerates COR gene transcript accumulation and acquisition of freezing tolerance during cold acclimation. Plant Mol Biol 84:67–82. https://doi.org/10.1007/s11103-013-0119-z
Jiang F, Doudna JA (2017) CRISPR–Cas9 structures and mechanisms. Annu Rev Biophys 46:505–529. https://doi.org/10.1146/annurev-biophys-062215-010822
Jianwei L, Jobichen C, Machida S, Meng S, Read RJ, Hongying C, Jian S, Yuan YA, Sivaraman J (2023) Structures of apo Cas12a and its complex with crRNA and DNA reveal the dynamics of ternary complex formation and target DNA cleavage. PLoS Biol 21:e3002023. https://doi.org/10.1371/journal.pbio.3002023
Jinek M, Chylinski K, Fonfara I, Hauer M, Doudna JA, Charpentier E (2012) A programmable dual-RNA–guided DNA endonuclease in adaptive bacterial immunity. Science 337:816–821. https://doi.org/10.1126/science.1225829
Jones BL (2005) The endogenous endoprotease inhibitors of barley and malt and their roles in malting and brewing. J Cereal Sci 42:271–280. https://doi.org/10.1016/j.jcs.2005.06.002

Jouanin A, Boyd LA, Visser RGF, Smulders MJ (2018) Development of wheat with hypoimmunogenic gluten obstructed by the gene editing policy in Europe. Front Plant Sci 9:1523
Jugdaohsingh R (2007) Silicon and bone health. J Nutr Health Aging 11:99–110
Kalinina NO, Khromov A, Love AJ, Taliansky ME (2020) CRISPR applications in plant virology: virus resistance and beyond. Phytopathology 110:18–28. https://doi.org/10.1094/PHYTO-07-19-0267-IA
Kapusi E, Corcuera-Gómez M, Melnik S, Stoger E (2017) Heritable genomic fragment deletions and small indels in the putative ENGase gene induced by CRISPR/Cas9 in barley. Front Plant Sci 8. https://doi.org/10.3389/fpls.2017.00540
Karunarathne SD, Han Y, Zhang X-Q, Zhou G, Hill CB, Chen K, Angessa T, Li C (2020) Genome-wide association study and identification of candidate genes for nitrogen use efficiency in barley (*Hordeum vulgare* L.). Front Plant Sci 11. https://doi.org/10.3389/fpls.2020.571912
Karunarathne SD, Han Y, Zhang X-Q, Li C (2022) CRISPR/Cas9 gene editing and natural variation analysis demonstrate the potential for *HvARE1* in improvement of nitrogen use efficiency in barley. J Integr Plant Biol 64:756–770. https://doi.org/10.1111/jipb.13214
Kaur H, Shaveta S, Kaur S, Sharma V, Kaur K (2019) Hulless Barley: a new era of research for food purposes. Wheat Barley Res 11. https://doi.org/10.25174/2249-4065/2019/83719
Kaur A, Purewal SS, Phimolsiripol Y, Punia Bangar S (2024) Unraveling the hidden potential of barley (*Hordeum vulgare*): an important review. Plants 13:2421. https://doi.org/10.3390/plants13172421
Kershanskaya OI, Yessenbaeva GL, Nelidova DS, Karabekova AN, Sadullaeva ZN (2022) CRISPR/Cas genome editing perspectives for barley breeding. Physiol Plant 174(3):e13686
Khan Z, Ali Z, Khan AA, Sattar T, Zeshan A, Saboor T, Binyamin B (2022) History and classification of CRISPR/Cas system. In: Ahmad A, Khan SH, Khan Z (eds) The CRISPR/Cas tool kit for genome editing. Springer, Singapore, pp 29–52. https://doi.org/10.1007/978-981-16-6305-5_2
Kifle SW (2016) Review on barley production and marketing in Ethiopia. J Econ Sustain Dev 7(9):91–100
Kis A, Hamar E, Tholt G, Ban R, Havelda Z (2019) Creating highly efficient resistance against wheat dwarf virus in barley by employing CRISPR/Cas9 system. Plant Biotechnol J 17(6):1004
Kis A, Polgári D, Dalmadi Á, Ahmad I, Rakszegi M, Sági L et al (2024) Targeted mutations in the *GW2.1* gene modulate grain traits and induce yield loss in barley. Plant Sci 340:111968. https://doi.org/10.1016/j.plantsci.2023.111968
Kleter GA, Kuiper HA, Kok EJ (2019) Gene-edited crops: towards a harmonized safety assessment. Trends Biotechnol 37:443–447
Kloc Y, Łaczmański Ł, Orczyk W (2024) *HvGSK1.1* controls salt tolerance and yield through the brassinosteroid signaling pathway in barley. Int J Mol Sci 25(2):998. https://doi.org/10.3390/ijms25020998
Koonin EV, Makarova KS (2019) Origins and evolution of CRISPR-Cas systems. Philos Trans R Soc B: Biol Sci 374:20180087. https://doi.org/10.1098/rstb.2018.0087
Kopecká R, Kameniarová M, Černý M, Brzobohatý B, Novák J (2023) Abiotic stress in crop production. Int J Mol Sci 24:6603. https://doi.org/10.3390/ijms24076603
Křenek P, Chubar E, Vadovič P, Ohnoutková L, Vlčko T, Bergougnoux V, Cápal P, Ovečka M, Šamaj J (2021) CRISPR/Cas9-induced loss-of-function mutation in the barley mitogen-activated protein kinase 6 gene causes abnormal embryo development leading to severely reduced grain germination and seedling shootless phenotype. Front Plant Sci 12(670):302. https://doi.org/10.3389/fpls.2021.670302
Kumar N, Galli M, Ordon J, Stuttmann J, Kogel KH, Imani J (2018) Further analysis of barley *MORC1* using a highly efficient RNA-guided Cas9 gene-editing system. Plant Biotechnol J 16(11):1892–1903
Langridge P (2018) Economic and academic importance of barley. In: Stein N, Muehlbauer GJ (eds) The barley genome. Springer International Publishing, Cham, pp 1–10. https://doi.org/10.1007/978-3-319-92528-8_1

Lawrenson T, Shorinola O, Stacey N, Li C, Østergaard L, Patron N, Uauy C, Harwood W (2015) Induction of targeted, heritable mutations in barley and *Brassica oleracea* using RNA-guided Cas9 nuclease. Genome Biol 16:258. https://doi.org/10.1186/s13059-015-0826-7

Lawrenson T, Hinchliffe A, Forner M, Harwood W (2022) Highly efficient genome editing in barley using novel LbCas12a variants and impact of sgRNA architecture. https://doi.org/10.1101/2022.0d4.28.489853

Lawrenson T, Clarke M, Kirby R, Forner M, Steuernagel B, Brown JKM, Harwood W (2024) An optimised CRISPR Cas9 and Cas12a mutagenesis toolkit for Barley and Wheat. Plant Methods 20:123. https://doi.org/10.1186/s13007-024-01234-y

Lee K, Zhang Y, Kleinstiver BP, Guo JA, Aryee MJ, Miller J, Malzahn A, Zarecor S, Lawrence-Dill CJ, Joung JK, Qi Y, Wang K (2019) Activities and specificities of CRISPR/Cas9 and Cas12a nucleases for targeted mutagenesis in maize. Plant Biotechnol J 17:362–372. https://doi.org/10.1111/pbi.12982

Lee JH, Won HJ, Hoang Nguyen Tran P, Lee S, Kim HY, Jung JH (2021) Improving lignocellulosic biofuel production by CRISPR/Cas9-mediated lignin modification in barley. GCB Bioenergy 13(4):742–752

Leenay RT, Beisel CL (2017) Deciphering, communicating, and engineering the CRISPR PAM. J Mol Biol 429:177–191. https://doi.org/10.1016/j.jmb.2016.11.024

Leong K-YB, Chan Y-H, Abdullah WMANW, Lim S-HE, Lai K-S, Leong K-YB, Chan Y-H, Abdullah WMANW, Lim S-HE, Lai K-S (2018) The CRISPR/Cas9 system for crop improvement: progress and prospects. In: Next generation plant breeding. IntechOpen. https://doi.org/10.5772/intechopen.75024

Li H, Li J, Chen J, Yan L, Xia L (2020) Precise modifications of both exogenous and endogenous genes in rice by prime editing. Mol Plant 13:671–674

Liao C, Beisel CL (2021) The tracrRNA in CRISPR biology and technologies. Annu Rev Genet 55:161–181. https://doi.org/10.1146/annurev-genet-071719-022559

Lin Q, Jin S, Zong Y, Yu H, Zhu Z, Liu G, Kou L, Wang Y, Qiu J-L, Li J, Gao C (2021) High-efficiency prime editing with optimized, paired pegRNAs in plants. Nat Biotechnol 39:923–927. https://doi.org/10.1038/s41587-021-00868-w

Liu X, Xie C, Si H, Yang J (2017) CRISPR/Cas9-mediated genome editing in plants. In: Methods, CRISPR-Cas systems for Genome engineering and investigation, vol 121–122, pp 94–102. https://doi.org/10.1016/j.ymeth.2017.03.009

Liu Y, Pinto F, Wan X, Yang Z, Peng S, Li M, Cooper JM, Xie Z, French CE, Wang B (2022) Reprogrammed tracrRNAs enable repurposing of RNAs as crRNAs and sequence-specific RNA biosensors. Nat Commun 13. https://doi.org/10.1038/s41467-022-29604-x

MacGregor AW, Bazin SL, Schroeder SW (2002) Effect of starch hydrolysis products on the determination of limit dextrinase and limit dextrinase inhibitors in barley and malt. J Cereal Sci 35:17–28. https://doi.org/10.1006/jcrs.2001.0408

Makarova KS, Wolf YI, Alkhnbashi OS, Costa F, Shah SA, Saunders SJ, Barrangou R, Brouns SJJ, Charpentier E, Haft DH, Horvath P, Moineau S, Mojica FJM, Terns RM, Terns MP, White MF, Yakunin AF, Garrett RA, van der Oost J, Backofen R, Koonin EV (2015) An updated evolutionary classification of CRISPR–Cas systems. Nat Rev Microbiol 13:722–736. https://doi.org/10.1038/nrmicro3569

Makarova KS, Wolf YI, Iranzo J, Shmakov SA, Alkhnbashi OS, Brouns SJJ, Charpentier E, Cheng D, Haft DH, Horvath P, Moineau S, Mojica FJM, Scott D, Shah SA, Siksnys V, Terns MP, Venclovas Č, White MF, Yakunin AF, Yan W, Zhang F, Garrett RA, Backofen R, van der Oost J, Barrangou R, Koonin EV (2020) Evolutionary classification of CRISPR–Cas systems: a burst of class 2 and derived variants. Nat Rev Microbiol 18:67–83. https://doi.org/10.1038/s41579-019-0299-x

Mallapaty S (2019) Australian gene-editing rules adopt 'middle ground'. Nature News. https://www.nature.com/articles/d41586-019-01282-8

Marone D, Mastrangelo AM, Borrelli GM (2023) From transgenesis to genome editing in crop improvement: applications, marketing, and legal issues. Int J Mol Sci 24:7122. https://doi.org/10.3390/ijms24087122

Mendiondo GM, Gibbs DJ, Szurman-Zubrzycka M, Korn A, Marquez J, Szarejko I, Maluszynski M, King J, Axcell B, Smart K, Corbineau F, Holdsworth MJ (2016) Enhanced waterlogging tolerance in barley by manipulation of expression of the N-end rule pathway E3 ligase 6. Plant Biotechnol J 14:40–50. https://doi.org/10.1111/pbi.12334

Mio K, Yamanaka C, Ichinose Y, Kohyama N, Yanagisawa T, Aoe S (2020) Effects of barley β-glucan with various molecular weights partially hydrolyzed by endogenous β-glucanase on glucose tolerance and lipid metabolism in mice. Cereal Chem 97:1056–1065. https://doi.org/10.1002/cche.10328

Mishra S, Nayak S, Tuteja N, Poosapati S, Swain DM, Sahoo RK (2024) CRISPR/Cas-mediated genome engineering in plants: application and prospectives. Plants 13:1884. https://doi.org/10.3390/plants13141884

Mojica FJM, Díez-Villaseñor C, García-Martínez J, Almendros C (2009) Short motif sequences determine the targets of the prokaryotic CRISPR defence system. Microbiology 155:733–740. https://doi.org/10.1099/mic.0.023960-0

Munawar N, Ahsan K, Ahmad A (2024) CRISPR-edited plants' social, ethical, policy, and governance issues. In: Abd-Elsalam KA, Ahmad A (eds) Global regulatory outlook for CRISPRized plants. Academic Press, pp 367–396

Mwando E, Han Y, Angessa TT, Zhou G, Hill CB, Zhang X-Q, Li C (2020) Genome-wide association study of salinity tolerance during germination in barley (*Hordeum vulgare* L.). Front Plant Sci 11. https://doi.org/10.3389/fpls.2020.00118

Newman (2006) A brief history of barley foods. CFW. https://doi.org/10.1094/CFW-51-0004

Newton AC, Flavell AJ, George TS, Leat P, Mullholland B, Ramsay L, Revoredo-Giha C, Russell J, Steffenson BJ, Swanston JS, Thomas WTB, Waugh R, White PJ, Bingham IJ (2011) Crops that feed the world 4. Barley: a resilient crop? Strengths and weaknesses in the context of food security. Food Sec 3:141–178. https://doi.org/10.1007/s12571-011-0126-3

Nidhi S, Anand U, Oleksak P, Tripathi P, Lal JA, Thomas G, Kuca K, Tripathi V (2021) Novel CRISPR–Cas systems: an updated review of the current achievements, applications, and future research perspectives. Int J Mol Sci 22:3327. https://doi.org/10.3390/ijms22073327

Nonaka S, Arai C, Takayama M, Matsukura C, Ezura H (2017) Efficient increase of ɣ-aminobutyric acid (GABA) content in tomato fruits by targeted mutagenesis. Sci Rep 7:7057. https://doi.org/10.1038/s41598-017-06400-y

Nuffield Council on Bioethics (2020) Genome editing and human reproduction: Social and ethical issues. Retrieved from https://www.nuffieldbioethics.org/

O'Neill MA et al (2021) Ecological consequences of CRISPR gene editing. Nat Ecol Evol 5(4):456–463. https://doi.org/10.1038/s41559-021-01456-5

Panting M, Holme IB, Dionisio G, Brinch-Pedersen H (2025) Simplex and multiplex CRISPR/Cas9-mediated knockout of grain protease inhibitors in model and commercial barley improves hydrolysis of barley and soy storage proteins. Plant Biotechnol J. https://doi.org/10.1111/pbi.70065

Pawelczak KS, Gavande NS, VanderVere-Carozza PS, Turchi JJ (2018) Modulating DNA repair pathways to improve precision genome engineering. ACS Chem Biol 13:389–396. https://doi.org/10.1021/acschembio.7b00777

Pellegrino E, Bedini S, Nuti M, Ercoli L (2018) Author Correction: impact of genetically engineered maize on agronomic, environmental and toxicological traits: a meta-analysis of 21 years of field data. Sci Rep 8:6485. https://doi.org/10.1038/s41598-018-24840-y

Peterson BA, Haak DC, Nishimura MT, Teixeira PJPL, James SR, Dangl JL, Nimchuk ZL (2016) Genome-wide assessment of efficiency and specificity in CRISPR/Cas9 mediated multiple site targeting in *Arabidopsis*. PLoS One 11:e0162169. https://doi.org/10.1371/journal.pone.0162169

Pfeiffer M, Quetier F, Ricroch A (2018) Genome editing in agricultural biotechnology. Adv Bot Res 86:245–286
Pourcel C, Salvignol G, Vergnaud G (2005) CRISPR elements in Yersinia pestis acquire new repeats by preferential uptake of bacteriophage DNA, and provide additional tools for evolutionary studies. Microbiology 151:653–663. https://doi.org/10.1099/mic.0.27437-0
Pourkheirandish M, Komatsuda T (2007) The importance of barley genetics and domestication in a global perspective. Ann Bot 100:999–1008. https://doi.org/10.1093/aob/mcm139
Puchta H (2005) The repair of double-strand breaks in plants: mechanisms and consequences for genome evolution. J Exp Bot 56:1–14. https://doi.org/10.1093/jxb/eri025
Raj R, Shams R, Pandey VK, Dash KK, Singh P, Bashir O (2023) Barley phytochemicals and health promoting benefits: a comprehensive review. J Agric Food Res 14:100677. https://doi.org/10.1016/j.jafr.2023.100677
Ran FA, Hsu PD, Wright J, Agarwala V, Scott DA, Zhang F (2013) Genome engineering using the CRISPR-Cas9 system. Nat Protoc 8:2281–2308. https://doi.org/10.1038/nprot.2013.143
Rani S, Chaudhary A (2018) Management strategies for abiotic stresses in barley. Wheat and barley research. J Cereal Res 10:151–165
Rashid S, Abate GT, Lemma S, Warner J, Kasa L, Minot N (2019) The barley value chain in Ethiopia. Gates Open Res 3:169. https://doi.org/10.21955/gatesopenres.1115011.1
Raza A, Khare T, Zhang X, Rahman MM, Hussain M, Gill SS, Chen H, Zhou M, Hu Z, Varshney RK (2025) Novel strategies for designing climate-smart crops to ensure sustainable agriculture and future food security. J Sustain Agric Environ 4(2):e70048. https://doi.org/10.1002/sae2.70048
Reginato G, Dello Stritto MR, Wang Y, Hao J, Pavani R, Schmitz M, Halder S, Morin V, Cannavo E, Ceppi I, Braunshier S, Acharya A, Ropars V, Charbonnier J-B, Jinek M, Nussenzweig A, Ha T, Cejka P (2024) HLTF disrupts Cas9-DNA post-cleavage complexes to allow DNA break processing. Nat Commun 15:5789. https://doi.org/10.1038/s41467-024-50080-y
Salvi S, Druka A, Milner SG, Gruszka D (2014) Induced genetic variation, TILLING and NGS-based cloning. In: Kumlehn J, Stein N (eds) Biotechnological approaches to barley improvement. Springer, Berlin, Heidelberg, pp 287–310. https://doi.org/10.1007/978-3-662-44406-1_15
Sandel MJ (2020) The ethical limits of gene editing: a critique of CRISPR technology. Harvard Law Review 133(7):1219–1235. https://harvardlawreview.org/2020/05/the-ethical-limits-of-gene-editing/
Sato K (2020) History and future perspectives of barley genomics. DNA Res 27:dsaa023. https://doi.org/10.1093/dnares/dsaa023
Shaheen N, Ahmad S, Alghamdi SS, Rehman HM, Javed MA, Tabassum J, Shao G (2023) CRISPR-Cas system, a possible "Savior" of rice threatened by climate change: an updated review. Rice 16:39. https://doi.org/10.1186/s12284-023-00652-1
Shan Q, Wang Y, Li J, Gao C (2014) Genome editing in rice and wheat using the CRISPR/Cas system. Nat Protoc 9:2395–2410. https://doi.org/10.1038/nprot.2014.157
Shaveta S, Kaur H, Kaur S (2019) Hulless Barley: a new era of research for food purposes. J Cereal Res 11:114
Šimić G, Horvat D, Lalić A, Koceva Komlenić D, Abičić I, Zdunić Z (2019) Distribution of β-glucan, phenolic acids, and proteins as functional phytonutrients of hull-less barley grain. Foods 8:680. https://doi.org/10.3390/foods8120680
Sissons MJ, Lance RCM, Sparrow DHB (1992) Studies on limit dextrinase in barley I. Purification of malt limit dextrinase and production of monospecific antibodies. J Cereal Sci 16:107–116. https://doi.org/10.1016/S0733-5210(09)80143-4
Sivamani E, Bahieldin A, Wraith JM, Al-Niemi T, Dyer WE, Ho T-HD, Qu R (2000) Improved biomass productivity and water use efficiency under water deficit conditions in transgenic wheat constitutively expressing the barley *HVA*1 gene. Plant Sci 155:1–9. https://doi.org/10.1016/S0168-9452(99)00247-2

Sohn M, Himmelsbach DS, Barton FE, Griffey CA, Brooks W, Hicks KB (2007) Near-infrared analysis of ground barley for use as a feedstock for fuel ethanol production. Appl Spectrosc 61:1178–1183. https://doi.org/10.1366/000370207782597148

Stahl Y, Alexander RD, Coates S, Bryce JH, Jenkinson HR, Morris PC (2007) The barley limit dextrinase inhibitor: gene expression, protein location and interaction with 14-3-3 protein. Plant Sci 172:452–461. https://doi.org/10.1016/j.plantsci.2006.10.008

Stella S, Alcón P, Montoya G (2017) Erratum: structure of the Cpf1 endonuclease R-loop complex after target DNA cleavage. Nature 547:476–476. https://doi.org/10.1038/nature23300

Sun H, Chen Z-H, Chen F, Xie L, Zhang G, Vincze E, Wu F (2015) DNA microarray revealed and RNAi plants confirmed key genes conferring low Cd accumulation in barley grains. BMC Plant Biol 15:259. https://doi.org/10.1186/s12870-015-0648-5

Swarts DC, Jinek M (2018) Cas9 versus Cas12a/Cpf1: Structure–function comparisons and implications for genome editing. WIREs RNA 9:e1481. https://doi.org/10.1002/wrna.1481

Szczepankowska A (2012) Role of CRISPR/*cas* system in the development of bacteriophage resistance. In: Łobocka M, Szybalski WT (eds) Advances in virus research, bacteriophages, Part A. Academic Press, pp 289–338. https://doi.org/10.1016/B978-0-12-394621-8.00011-X

Tamilselvan-Nattar-Amutha S, Hiekel S, Hartmann F, Lorenz J, Dabhi RV, Dreissig S, Hensel G, Kumlehn J, Heckmann S (2023) Barley stripe mosaic virus mediated somatic and heritable gene editing in barley (*Hordeum vulgare* L.). Front Plant Sci 14:1201446

Tchetvertakov G (2019) Australia approves cutting-edge CRISPR gene editing technology. https://smallcaps.com.au/australia-approves-cutting-edge-crispr-gene-editing-technology/

Thomas HR, Percival SM, Yoder BK, Parant JM (2014) High-throughput genome editing and phenotyping facilitated by high resolution melting curve analysis. PLoS One 9:e114632. https://doi.org/10.1371/journal.pone.0114632

Tricase C, Amicarelli V, Lamonaca E, Rana RL, Tricase C, Amicarelli V, Lamonaca E, Rana RL (2018) Economic analysis of the barley market and related uses. In: Grasses as food and feed. IntechOpen. https://doi.org/10.5772/intechopen.78967

Verma RPS, Lal C, Malik R, Kharub AS, Kumar L, Kumar D (2022) Barley improvement: current status and future prospects in changing scenario. In: Kashyap PL et al (eds) New Horizons in Wheat and Barley Research. Springer, Singapore. https://doi.org/10.1007/978-981-16-4449-8_6

Verstegen H, Köneke O, Korzun V, Von Broock R (2014) The world importance of barley and challenges to further improvements. In: Kumlehn J, Stein N (eds) Biotechnology in agriculture and forestry. Springer Berlin Heidelberg, Berlin, Heidelberg, pp 3–19. https://doi.org/10.1007/978-3-662-44406-1_1

Visioni A, Basile B, Amri A, Sanchez-Garcia M, Corrado G (2023) Advancing the conservation and utilization of barley genetic resources: insights into germplasm management and breeding for sustainable agriculture. Plants 12(18):3186. https://doi.org/10.3390/plants12183186

Vlćko T, Ohnoutkova L (2020) Allelic variants of CRISPR/Cas9 induced mutation in an inositol trisphosphate 5/6 kinase gene manifest different phenotypes in barley. Plants 9(2):195

Vouillot L, Thélie A, Pollet N (2015) Comparison of T7E1 and surveyor mismatch cleavage assays to detect mutations triggered by engineered nucleases. G3 Genes|Genomes|Genetics 5:407–415. https://doi.org/10.1534/g3.114.015834

Wang Y, Ren X, Sun D, Sun G (2015) Origin of worldwide cultivated barley revealed by NAM-1 gene and grain protein content. Front Plant Sci 6. https://doi.org/10.3389/fpls.2015.00803

Wang Y, Harding SV, Eck P, Thandapilly SJ, Gamel TH, Abdel-Aal E-SM, Crow GH, Tosh SM, Jones PJ, Ames NP (2016) High-molecular-weight β-glucan decreases serum cholesterol differentially based on the *CYP7A1* rs3808607 polymorphism in mildly hypercholesterolemic adults. J Nutr 146:720–727. https://doi.org/10.3945/jn.115.223206

Wang P, Zhang J, Sun L, Ma Y, Xu J, Liang S, Deng J, Tan J, Zhang Q, Tu L, Daniell H, Jin S, Zhang X (2018) High efficient multisites genome editing in allotetraploid cotton (*Gossypium hirsutum*) using CRISPR/Cas9 system. Plant Biotechnol J 16:137–150. https://doi.org/10.1111/pbi.12755

Wang C, Yang X, Zhang Y, Shen C, Shi J, Xia C, Fang T, Tu Q, Li L, Zhou X, Zhang D, Li G (2023) Barley FASCIATED EAR genes determine inflorescence meristem size and yield traits. Crop J 11:679–691

Watson A, Ghosh S, Williams MJ, Cuddy WS, Simmonds J, Rey M-D, Asyraf Md Hatta M, Hinchliffe A, Steed A, Reynolds D, Adamski NM, Breakspear A, Korolev A, Rayner T, Dixon LE, Riaz A, Martin W, Ryan M, Edwards D, Batley J, Raman H, Carter J, Rogers C, Domoney C, Moore G, Harwood W, Nicholson P, Dieters MJ, DeLacy IH, Zhou J, Uauy C, Boden SA, Park RF, Wulff BBH, Hickey LT (2018) Speed breeding is a powerful tool to accelerate crop research and breeding. Nat Plants 4:23–29. https://doi.org/10.1038/s41477-017-0083-8

Wong N, Liu W, Wang X (2015) WU-CRISPR: characteristics of functional guide RNAs for the CRISPR/Cas9 system. Genome Biol 16:218. https://doi.org/10.1186/s13059-015-0784-0

Yang Q, Ding J, Feng X, Zhong X, Lan J, Tang H, Harwood W, Li Z, Guzman C, Xu Q, Zhang Y, Jiang Y, Qi P, Deng M, Ma J, Wang J, Chen G, Lan X, Wei Y, Zheng Y, Jiang Q (2022) Editing of the starch synthase IIa gene led to transcriptomic and metabolomic changes and high amylose starch in barley. Carbohydr Polym 285:119238

Yang X, Wilkinson LG, Aubert MK, Houston K, Shirley NJ, Tucker MR (2023) Ovule cell wall composition is a maternal determinant of grain size in barley. New Phytol 237:2136–2147

Zang Y, Gong Q, Xu Y, Liu H, Bai H, Li N, Du L, Ye X, Lan C, Wang K (2022) Production of conjoined transgenic and edited barley and wheat plants for Nud genes using the CRISPR/SpCas9 system. Front Genet 13(5):873850

Zannoni L (2019) Evolving regulatory landscape for genome-edited plants. CRISPR J 2:3–8

Zeng Z, Han N, Liu C, Buerte B, Zhou C, Chen J, Wang M, Zhang Y, Tang Y, Zhu M, Wang J, Yang Y, Bian H (2020) Functional dissection of HGGT and HPT in barley vitamin E biosynthesis via CRISPR/Cas9-enabled genome editing. Ann Bot 126(5):929–942

Zhang F (2019) Development of CRISPR-Cas systems for genome editing and beyond. Q Rev Biophys 52:e6. https://doi.org/10.1017/S0033583519000052

Zhang Y, Pribil M, Palmgren M, Gao C (2020) A CRISPR way for accelerating improvement of food crops. Nat Food 1:200–205

Zhao H, Wolt JD (2017) Risk associated with off-target plant genome editing and methods for its limitation. Emerg Top Life Sci 1:231–240. https://doi.org/10.1042/ETLS20170037

Zhou G, Broughton S, Zhang X-Q, Ma Y, Zhou M, Li C (2016) Genome-wide association mapping of acid soil resistance in barley (*Hordeum vulgare* L.). Front Plant Sci 7. https://doi.org/10.3389/fpls.2016.00406

Zhou X, Zhao Y, Ni P, Ni Z, Sun Q, Zong Y (2023) CRISPR-mediated acceleration of wheat improvement: advances and perspectives. J Genet Genom Wheat Biol 50:815–834. https://doi.org/10.1016/j.jgg.2023.09.007

Zhu X, Xu Y, Yu S, Lu L, Ding M, Cheng J, Song G, Gao X, Yao L, Fan D, Meng S, Zhang X, Hu S, Tian Y (2014) An efficient genotyping method for genome-modified animals and human cells generated with CRISPR/Cas9 system. Sci Rep 4:6420. https://doi.org/10.1038/srep06420

Zhu J, Zhang Y, Zhang M, Hong Y, Sun C, Guo Y, Yin H, Lv C, Guo B, Wang F, Xu R (2025) Natural variation and CRISPR/Cas9 gene editing demonstrate the role of a group VII ethylene response factor, *HvERF62*, in regulation of barley waterlogging tolerance. J Exp Bot. https://doi.org/10.1093/jxb/eraf101

Chapter 11
CRISPR-Cas-Mediated Gene Editing in Sorghum: Recent Advances, Challenges, and Prospects

Zhifang Wang, Karen Massel, Jinping Wang, Zhiyin Jiao, Ian Godwin, Peng Lv, and Guoquan Liu

Abstract Sorghum has historically been considered one of the most challenging plant species for tissue culture, genetic transformation, and gene editing among cereals. Consequently, research and application of these biotechnologies have lagged those of its counterparts, such as rice and maize. Among modern cutting-edge biotechnologies, the Clustered Regularly Interspaced Short Palindromic Repeats (CRISPR)-Cas system has garnered worldwide attention for plant genome editing. This system efficiently and precisely targets modifications in genome sequences, accelerating modern plant breeding processes and compensating for limitations of traditional breeding methods in sorghum improvement. Significant progress in sorghum biotechnologies has been made, especially in the twenty-first century, through continuous and persistent optimization and improvement. Universality, efficiency, and stability are key factors in overcoming bottlenecks in sorghum genetic transformation and enhancing gene editing efficiency. In recent years, the development of sorghum genetic transformation system, completion of sorghum genome sequencing, identification of important agronomic trait genes, and optimization of genome editing system have led to significant progress and advancement. CRISPR-Cas-mediated gene editing has resulted in the cultivation of sorghum varieties with enhanced resistance, increased productivity, and improved quality. This chapter reviews the status of sorghum genetic transformation and genome editing tools, their application in improving superior qualities, and the challenges and prospects for future sorghum development. It aims to provide a systematic elaboration on sorghum gene editing and serve as a reference for further improvement.

Z. Wang · J. Wang · Z. Jiao · P. Lv
Institute of Millet Crops, Hebei Academy of Agriculture & Forestry Sciences/Hebei Branch of China National Sorghum Improvement Centre, Shijiazhuang, China

K. Massel · I. Godwin · G. Liu (✉)
Centre for Crop Science, Queensland Alliance for Agriculture and Food Innovation, The University of Queensland, St Lucia, Australia
e-mail: g.liu2@uq.edu.au

K. A. Abd-Elsalam, A. Ahmad (eds.), *Gene Editing in Cereals*, Sustainability Sciences in Asia and Africa, https://doi.org/10.1007/978-981-95-6722-5_11

Keywords Sorghum · Gene editing · Genetic transformation · Tissue culture · CRISPR-Cas · Plant biotechnology

11.1 Introduction

As an important C4 model plant, sorghum is the fifth-largest cereal crop in the world. It has been widely cultivated for its high yield and high level stress resistance. Based on its primary application, sorghum is generally classified into three categories (grain sorghum for food, forage sorghum for animal feed, and sweet sorghum for bioenergy) (Berenji et al. 2011). In general, sorghum as a multipurpose crop serves as staple food for more than half billion people, as an important source of silage feed for animals, and as a brewing or bioenergy source for industry (Liu et al. 2021). Moreover, sorghum can be used as an additive ingredient in the food processing industry, being incorporated into various forms of food (Ducksbury and Stefoska-Needham 2022). The productivity of sorghum grain as fodder is higher than that of barley and oats, and comparable to that of corn (Dahlberg 2019). Sorghum grain especially red sorghum, containing antioxidant compounds, is beneficial to human health (de Oliveira and de Alencar Figueiredo 2024). Furthermore, sweet sorghum with high biomass can be used as an energy and processing crop to produce biofuels, vegetable oils, and paper, etc. (Dahlberg et al. 2011; Dahlberg 2019). With the rapid development of living standard, increasing global population, and intensified climate change, the demand of sorghum productivities has been skyrocketing. Therefore, to meet the demand, it is of great significance to cultivate sorghum varieties with high resistance, high yield, and superior quality.

Currently, crop enhancement mainly relies on crossbreeding, mutation breeding, haploid breeding, and polyploid breeding in modern agriculture (Miflin 2000). However, each of these breeding methods has its own advantages and drawbacks. Hybrid breeding can improve grain yield and grain quality but requires a long time to develop new varieties and relies on the screening of desirable traits based on existing genetic germplasm. Mutagenesis breeding could generate a vast amount of variation; however, it necessitates extensive handling of materials, and the direction of mutations cannot be controlled, resulting in few beneficial mutations through selection. Haploid breeding can significantly shorten the breeding cycle, but the technique is relatively complex and needs to be combined with hybrid breeding. Polyploid breeding generally leads to low seed-setting rates. Modern breeding technologies have shed lights on crop improvement. Genetic engineering can directly breed desired traits through gene transformation, but its application is limited by lengthy and costly regulatory assessment processes, as well as concerns over genetically modified organisms (GMOs). Genome editing technology, such as CRISPR-Cas9, can precisely modify specific sites in the plant genome, streamlining the breeding process. The cutting-edge technology of CRISPR-Cas9 is worldwide used by plant breeders to achieve breeding goals in a shorter time, avoiding the introduction of unfavorable genetic backgrounds into new varieties (Pacesa et al. 2024).

Multiple strategies have been developed to create precisely mutation without any transgene residual (He et al. 2022). Plants edited in this manner, named transgene-free genome editing, are not considered as GMOs in some countries or regions, such as Australia and the United States (Zhang et al. 2021).

With the completion of genome sequencing, pan-genome, telomere-to-telomere genome assembly, and the cloning of important agronomic trait genes in sorghum (Paterson et al. 2009; McCormick et al. 2018; Tao et al. 2021; Wang et al. 2025), genome editing technology plays a significant role in the gene functional identification and the cultivation of high-yield and high-quality varieties. Since sorghum is a diploid plant with a relatively small genome, it also serves as an important model crop for studying C4 cereal crop family plants (Calviño and Messing 2012). For a long time, sorghum has been regarded as one of the most challenging crops in terms of gene delivery, tissue culture, regeneration, and genetic transformation, which has constrained the application of gene editing in sorghum (Che et al. 2018). In recent years, there have been many breakthrough achievements in overcoming genotype dependence, expanding the sources of explants, optimizing parameters for genetic transformation, and improving the efficiency and stability of transformation in sorghum (Fig. 11.1) (Liu and Godwin 2012; Liu et al. 2015, 2020; Che et al. 2022; Wang et al. 2023; Li et al. 2024). In this chapter, we review the progress of sorghum tissue culture and genetic transformation techniques, as well as genome editing technologies, particularly the application of CRISPR-Cas9 on improvement of sorghum traits. We also discuss the challenges and prospects faced by gene editing in future sorghum production, revealing the tremendous potential of genome editing in

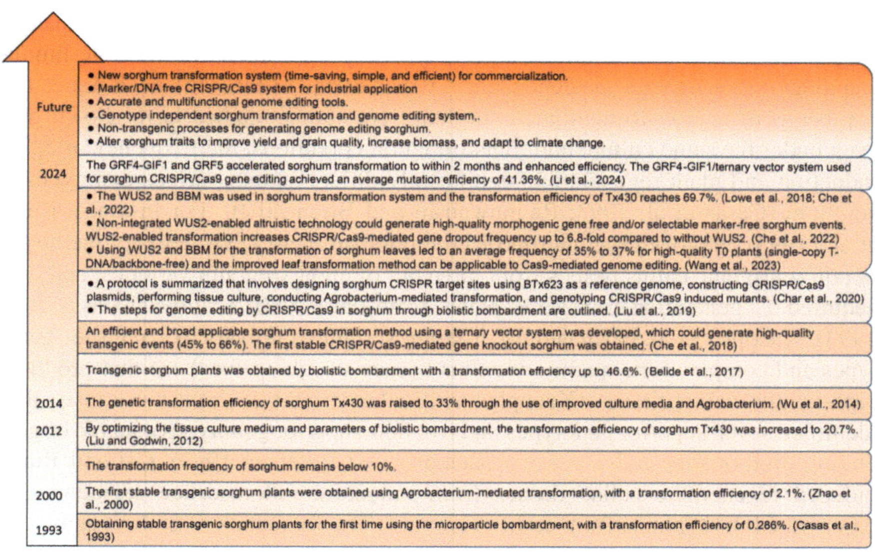

Fig. 11.1 Timeline of the main historical events of sorghum transformation

fundamental research and industrial application, which holds great promise to sorghum research and industry in future.

11.2 Sorghum Tissue Culture System

The establishment of a highly efficient sorghum tissue culture system is one of the most important prerequisites for sorghum genetic transformation and genome editing process. The genotype of sorghum, the source of explants, and optimal media are crucial for successful tissue culture and regeneration of sorghum.

11.2.1 Genotype

Up to date, sorghum tissue culture is constrained by genotypes. The induction of callus and plant regeneration in sorghum is limited to specific genotypes of sorghum. High-tannin sorghum varieties could produce more phenolic compounds during the tissue culture process, which not only have toxicity to *Agrobacterium* but also have negative effects on cell differentiation and tissue regeneration (Cai and Butler 1990; Kaeppler and Pedersen 1997; Gao et al. 2005; Nguyen et al. 2007). Therefore, transformation methods, and in particular *Agrobacterium*-mediated methodologies, may not be suitable for the genetic transformation of high-tannin sorghum varieties.

The traditional tissue culture process includes dedifferentiation, callus induction, callus proliferation, callus maturation, callus regeneration, and rooting induction stages. The callus proliferation stage is typically time-intensive and rate-limited step due to strong genotype dependency, which is one of the bottleneck issues that hamper genetic transformation in sorghum. Numerous studies have shown that, even under the same culture and transformation conditions, the frequency of callus induction and plant regeneration significantly differs among various genotypes of sorghum (Casas et al. 1993; Raghuwanshi and Birch 2010). For instance, Casas et al. (1993) employed the microprojectile bombardment DNA delivery method to obtain transgenic callus tissues from three out of eight sorghum varieties, and only acquired T_0 generation transgenic plants from the P898012 cultivar (Casas et al. 1993). Kaeppler et al. evaluated 41 different sorghum cultivars to screen for genotypes and compared the quality of callus tissues searching for those that produced high-quality callus tissues (Kaeppler and Pedersen 1997). From these tests, they identified seven superior genotypes and five non-superior genotypes capable of generating high-quality callus tissues (Kaeppler and Pedersen 1997). Cai and Butler induced regenerable callus tissues from immature inflorescence segments and regenerated 3600 plants from eight genotypes. The frequency of embryogenic callus formation ranged from 8% to 70%, depending on the genotype (Cai and Butler 1990). Whereas Raghuwanshi and Birch (2010) utilized 32 varieties of sweet sorghum to optimize the culture medium for genetic transformation experiments, achieving an overall transformation efficiency of 0.09% with a repeatable experimental protocol. They also reported that certain genotypes, including Ramada and R9188, were promising candidates for further tissue culture and genetic transformation studies (Raghuwanshi and Birch 2010). However, recent research statistics show clear favoritism for the genotypes P898012 and Tx430 as the most reliable genotypes frequently used for sorghum tissue culture and genetic transformation (Casas et al. 1993; Zhu et al. 1998; Emani et al. 2002; Jeoung et al. 2002; Gurel et al. 2009; Grootboom et al. 2010; Kumar et al. 2011; Liu and Godwin 2012; Wu et al. 2014; Do et al. 2016; Belide et al. 2017; Mookkan et al. 2017; Che et al. 2018, 2022; Nelson-Vasilchik et al. 2018; Sato-Izawa et al. 2018; Chou et al. 2020; Wang et al. 2023).

11.2.2 Source of Explants

Selecting appropriate explants for genetic transformation constitutes a crucial element in the successful regeneration of transgenic plants. The explants reported for genetic transformation of sorghum include immature embryos, immature inflorescences, shoot tips, mature embryos, leaf bases, leaves, and mature seeds. However, the most widely used and most effective explant is the immature embryo. In the early stages of protocol developments, protoplasts and cells suspension cultures combined with electroporation were used for the research in genetic transformation of sorghum, but stable transgenic sorghum plants were not successfully obtained (Battraw and Hall 1991; Hagio et al. 1991). Casas et al. (1993) successfully generated the first fertile transgenic sorghum plants by employing callus derived from the immature embryos of sorghum line P898012 (Casas et al. 1993). Subsequently, various tissues such as immature embryos, mature embryos, shoot tips, young inflorescences, leaf tissues, and embryo-derived callus were utilized in attempts to the genetic transformation of sorghum, and these various explants show differences in frequencies of callus induction and regeneration (Jambagi et al. 2010; Chou et al. 2020; Wang et al. 2021, 2023).

Based on the published reports, the use of immature embryos as explants remains predominant, and the ideal period for collection of immature embryos is between 12 and 15 days post-pollination. However, the use of morphogenic regulators (see Sect. 11.5 in this chapter for more detail) has provided significant breakthroughs in transformation based on immature embryos, but also expanded the range of explants available for genetic transformation (Lowe et al. 2016). For instance, fertile transgenic plants can be regenerated by using maize mature embryo slices and leaf base tissue segments (Lowe et al. 2016). More recently, the expression of morphogenic regulators shows the ability to directly stimulate the rapid formation of somatic embryos and plant regeneration from immature leaf whorls (Wang et al. 2023). The improved leaf transformation method could be a universal method for leaf transformation and gene editing in gramineous crops (Wang et al. 2023). This is a breakthrough in research that would allow for a more consistent supply of explants to be grown for only 12–18 days in vitro cultures rather than waiting the 60–70 days for immature embryos from staggered plantings.

11.2.3 Sorghum Tissue Culture Media

Tissue culture media have played a critical role in providing nutrition and phytohormones to plant cells. Sorghum tissue culture has encountered five major challenges, including genotype dependency, narrow regeneration timeframe, low regeneration efficiency, toxic pigments (phenolics), and tissue culture contamination (Seetharama and Godwin 2004; Pola et al. 2008; Raghuwanshi and Birch 2010). Despite those challenges, remarkable progress has been made through optimization and innovation. The robust improved tissue culture systems have been established in many research groups (Liu et al. 2015; Belide et al. 2017; Che et al. 2022). The tissue culture system has been involved in multiple aspects, including genotypes, explants, different stages for tissue culture, transformation methods, different phytohormones, various concentration of phytohormones, combinations of phytohormones, and additional compositions (Liu et al. 2014). For example, by adding increased amounts of copper sulphate, the callus induction, callus regeneration, and root induction rates are significantly improved for sorghum tissue culture (Liu et al. 2013). Besides, the interaction between those aspects makes tissue culture more complicated. Therefore, identifying the optimal standardized medium is very challenging for plant tissue culture.

Tissue culture is essential for transformation even for most genome editing by CRISPR/Cas9. In recent years, great progress has been made in sorghum transformation and genome editing. Based on the particle bombardment system, a robust sorghum tissue culture system, highly efficient transformation, great genome editing system, and transgene-free genome editing system have been established (Liu and Godwin 2012; Liu et al. 2015; Massel et al. 2022; Zhang et al. 2025). Meanwhile, based on the *Agrobacterium*-mediated transformation system, from tissue culture to genome editing has been well-established with high efficiency (Che et al. 2022; Fontanet-Manzaneque et al. 2024).

11.3 Sorghum Genetic Transformation Methods

The progress on sorghum genetic transformation has been slower compared to other cereal crops such as rice and maize, primarily due to the low success rate of genetic transformation and the absence of effective transformation techniques. The biolistic bombardment and *Agrobacterium*-mediated transformation are two commonly used techniques for sorghum genetic transformation (Fig. 11.2a and b). Protoplast transfection is also a commonly used method to deliver transient gene expression (Meng

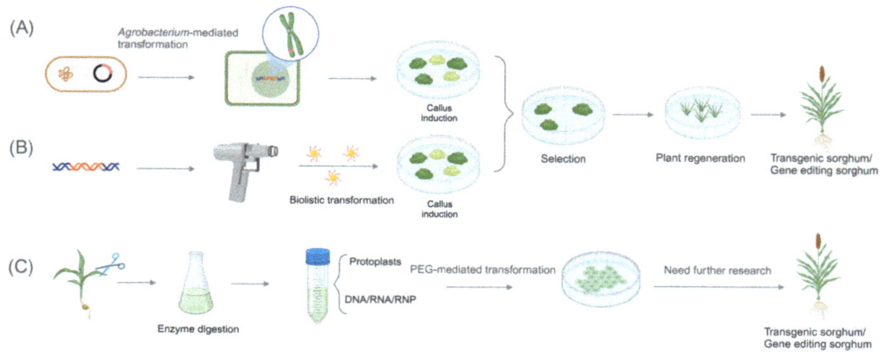

Fig. 11.2 General illustrations of delivery systems for sorghum genetic transformation. (**a**) Transgenic or gene editing sorghum generated by *Agrobacterium*-mediated DNA delivery. (**b**) Conventional methods for biolistic bombardment-mediated DNA delivery. (**c**) DNA, RNA, or CRISPR-Cas-sgRNA ribonucleoprotein (RNP) can be delivered into plant cells by PEG-mediated transformation of plant protoplasts. (Created in BioRender. Wang, Z. (2025) https://BioRender.com/a9gnsxj)

et al. 2020); however, further exploration is needed to achieve stable plant regeneration (Fig. 11.2c). Other less popular methods, including electroporation and pollen-mediated transformation, have been explored with advantages and disadvantages (Wang et al. 2007).

11.3.1 Biolistic Bombardment

Biolistic bombardment transformation refers to the method in which micro-particles wrapped with exogenous genes are accelerated through the protective cell wall of plants, resulting in the integration of exogenous genes into the host genome, with the advantage of not being limited by host specificity (Hagio et al. 1991; Finer et al. 1992; Altpeter et al. 2005; Rech et al. 2008); therefore, it is widely used for the genetic transformation of gramineous crops that are insensitive to *Agrobacterium* (Tzfira and Citovsky 2006). The first successful transformation of sorghum in 1993 was achieved through the biolistic bombardment method, with a transformation efficiency of 0.286% (Casas et al. 1993). Over the subsequent nearly 20 years, although there were multiple reports on obtaining transgenic sorghum using the gene gun, the transformation efficiency has consistently been very low (Casas et al. 1997; Zhu et al. 1998; Able et al. 2001; Emani et al. 2002; Jeoung et al. 2002; Tadesse et al. 2003; Girijashankar et al. 2005; Grootboom et al. 2010; Raghuwanshi and Birch 2010). Liu and Godwin improved the genetic transformation efficiency of sorghum to 20.7% by optimizing the tissue culture medium and parameters of biolistic bombardment (Liu and Godwin 2012). Belide et al. (2017) reported that differentiating embryogenic callus derived from immature embryos of sorghum

Tx430 could maintain its regenerative capacity for more than one year when cultured on modified MS medium (Belide et al. 2017). By using biolistic bombardment on these callus, transgenic sorghum plants could be obtained, with a transformation efficiency as high as 46.6% (Belide et al. 2017).

Although the biolistic bombardment has been successfully applied to the genetic transformation of grains such as rice, corn, wheat, barley, and sorghum (Shou et al. 2004; Yao et al. 2006; He et al. 2008; Liu and Godwin 2012), the technology also has certain drawbacks. For example, the insertion of multiple transgene copies may result in complex rearrangements and/or recombination; the bombardment with high-velocity and micro-particles may damage the DNA of host cells, leading to genomic instability; and the randomness of exogenous gene insertion might lead to unpredictable changes in gene expression or gene silencing, etc. (Zhu et al. 1998; Emani et al. 2002). Despite the existing disadvantages of the biolistic bombardment, studies have shown that using 2.5 ng of DNA per bombardment can increase the probability of single-copy insertion events (Lowe et al. 2009). Through co-transformation, the selective marker genes can be selectively removed from the transgenic progeny (Hare and Chua 2002; Wang et al. 2006).

11.3.2 **Agrobacterium-*Mediated Transformation***

Agrobacterium-mediated transformation is widely used due to its advantages such as being cost-effective, low gene copy number, capability to transform large gene fragments, and genetic stability of the exogenous genes. Because monocotyledonous plants were not within the host range of *Agrobacterium* infection, it was initially believed that *Agrobacterium* was not suitable for transforming monocots, and *Agrobacterium*-mediated transformation was limited to dicotyledonous plants (De Cleene and De Ley 1976; Rahman et al. 2024). However, with the advancement of tissue culture techniques, the transformation of major monocotyledonous plants has also been successfully demonstrated such as rice (*Oryza sativa L.*) (Zhang et al. 2016), maize (*Zea mays L.*) (Jiao et al. 2022), wheat (*Triticum aestivum L.*) (Hayta et al. 2019), and barley (*Hordeum vulgare L.*) (Ismagul et al. 2014).

The first report of obtaining transgenic sorghum plants using *Agrobacterium*-mediated transformation was published in 2000, with an average transformation efficiency of 2.1% (Zhao et al. 2000). Since then, numerous studies aiming to optimize the *Agrobacterium*-mediated method have been conducted. Although optimizations were performed in the components of the culture medium, embryo pretreatment, and various other factors, the transformation frequency of sorghum has increased but still remains below 10% (Gao et al. 2005; Gurel et al. 2009; Kumar et al. 2011).

The particular strain of *Agrobacterium* being used has been shown to impact transformation efficiencies. Wu et al. (2014) utilized immature embryos of sorghum Tx430 as explants and infected them with the *Agrobacterium* strain LBA4404. Whereas studies utilizing *Agrobacterium* AGL1 strain demonstrated transformation efficiencies as high as 33% due to its increased aggressiveness for T-DNA transfer (Wu et al. 2014). This signifies that sorghum has joined the ranks of cereals capable of effective transformation through *Agrobacterium*-mediated methods. Do et al. (2016) optimized the regenerative transformation of the public sorghum genotype P898012 using a standard binary vector and *Agrobacterium* strain AGL1, achieving an average transformation efficiency of 14%, with 40–50% of the transformed progeny containing a single intact T-DNA copy (Do et al. 2016).

In addition to strain of *Agrobacterium*, modifying the infection media can also play a role in transformation efficiencies. Based on previously reported *Agrobacterium* transformation protocols (Wu et al. 2014), Che et al. (2018) found that the application of 0.005% Silwet-70 could significantly enhance the transformation efficiency (Che et al. 2018). There is also evidence showing how ternary vector system could enhance the transformation efficiency from 15.1% to 29% as compared to using a binary vector system (Che et al. 2018; Fontanet-Manzaneque et al. 2024).

11.3.3 Protoplast Transfection

Transient protoplast transfection is an efficient method for rapid and high-throughput studies of gene expression, protein subcellular localization, and protein interactions. Moreover, the use of genome editing tools to transfect protoplasts has been successfully conducted, with efficient genome editing already achieved in sorghum. Meng et al. established an efficient method for isolating sorghum protoplasts and utilized a PEG-mediated transformation method to transfect two CRISPR/Cas9 plasmids into sorghum protoplasts (Meng et al. 2020). They demonstrated that protoplast assays can be applied for transient gene expression and single-guide RNA (sgRNA) screening in the process of CRISPR/Cas9 gene editing (Meng et al. 2020). Lee et al. (2023) optimized the plasmid transformation for sorghum protoplasts and also explored a ribonucleoprotein (RNP)-based protoplast delivery system (Lee et al. 2023). They successfully edited target genes in protoplasts using Cas9-sgRNA RNPs, which reduced off-target efficiency and did not introduce exogenous DNA, indicating that Cas9-sgRNA RNPs are an effective genome editing tool in sorghum (Lee et al. 2023). Protoplast regeneration ability varies significantly among plant species and tissue types, influenced by genotype (Jeong et al. 2021). There is an absence of efficient regeneration methods for sorghum protoplasts at present. The improvement of the regeneration potential of sorghum protoplasts is challenging but crucial for the successful production of regenerated plants, which requires ongoing research and exploration.

11.3.4 Other Transformation Methods

Electroporation and the pollen tube pathway are methods used in the early sorghum transformation, belonging to the direct DNA delivery methods. The explants used in electroporation are primarily protoplasts or cell suspensions. Ou-lee et al. (1986) were the early adopters to introduce exogenous genes into sorghum protoplasts through electroporation (Ou-Lee et al. 1986; Battraw and Hall 1991), and subsequently, Hagio et al. (1991) also successfully introduced exogenous genes into a sorghum cell suspension using electroporation, but no stable regenerating plants were obtained (Hagio et al. 1991). Wang et al. (2007) sonicated plasmids and pollen from sorghum A(2)V4B in a 0.3 mol/l sucrose solution and then pollinated the treated pollen onto the male-sterile line A(2)V4A, resulting in transgenic plants. This provides a method for obtaining transgenic sorghum plants through pollen-mediated transformation (Wang et al. 2007). Although this method circumvents the complex process of in vitro culture, it displays some disadvantages. For example, ultrasonic waves can destroy pollen and affect seed production. Additionally, this method could result in a high number of gene copies, which can lead to gene silencing and cause extensive trait segregation in the progeny.

11.4 Gene Editing in Sorghum

There is a range of gene editing tools available to researchers, which allow for the targeted changes in the DNA by site-directed mutations. These include Meganucleases, Zinc Finger Nucleases (ZFNs), and Transcription activator-like effector nucleases (TALENs). However, the field of gene editing experienced significant advancement due to the emergence of Clustered Regularly Interspaced Short Palindromic Repeats (CRISPR) technology (Jinek et al. 2012; Shan et al. 2013). This new tool offers a straightforward and user-friendly methodology for precise and highly efficient genome editing, with the ability for it to be tailored for a wide range of outcomes. The rest of this chapter will focus on CRISPR/Cas systems and their adoption and usefulness for sorghum breeding.

One commonality between all gene editing tools is that they all lead to the creation of double-stranded breaks (DSBs) at precise location in the genome. From there, the plant undergoes endogenous repair mechanisms through two pathways: (1) Nonhomologous End Joining (NHEJ) and/or (2) Homology-driven repair (HDR). In plants, NHEJ is highly favored, often resulting in erroneous repair that leads to insertions or deletions at the break site (Cong et al. 2013). Consequently, this tool is the most commonly used to target the coding regions of genes, leading to frame-shift mutations and mistranslated proteins, ultimately resulting in a functional knockout of the targeted gene.

11.4.1 CRISPR Systems

CRISPR tools were developed from the immune systems of bacteria, which have been modified for its use as a biotechnological tool (Mali et al. 2013; Shan et al. 2013). Classically, the immune system uses a Cas nuclease, a CRISPR RNA (crRNA), and transactivating RNA (tracrRNA) in combination to break up invading genetic elements. For it to be utilized by as a biotechnological tool, optimizations and tweaks have been performed to allow for function. For example, for it to be used in eukaryotic cells, nuclear localization signals were added to the Cas nucleases for proximity to the host genome.

There is a range of CRISPR-based systems that can be broadly divided into two main classes (Classes I and II), each with three subclasses (Type I to VI) (Xu and Li 2020). Class I systems, containing Types I, III, and IV, require the use of multiple different Cas proteins, making them more complex and less useful to be modified into a biotechnological tool (Makarova et al. 2017). In contrast, Class II systems, containing Type II, V, and VI, rely on a single Cas nuclease, and this simplicity has allowed them to be the most prevalent nuclease utilized for gene editing tools (Chylinski et al. 2014; Tang and Fu 2018). Within the class II systems, the most well-known and widely used is the Type II CRISPR system, which utilizes the Cas9 enzyme (Jinek et al. 2012; Shan et al. 2013). This is followed by Type V using Cas12a (or Cpf1) as the effector protein (Bandyopadhyay et al. 2020). Additionally, class II also includes Type VI employs Cas13 which targets RNA instead of DNA, making it useful for RNA editing applications but lacks the ability to create heritable changes in the genome (Cook et al. 2019; O'Connell 2019).

Streptococcus pyogenes Cas9 (SpCas9) was the first Cas9 variant used in plant gene editing. It encodes a 160 kD protein (1368 amino acid), forming a multidomain DNA endonuclease (Jiang et al. 2013). For its activity, it requires the expression of SpCas9 nuclease alongside a guide RNA (gRNA). The gRNA is actually the fusion of the original transactivating RNA (tracrRNA) and CRISPR RNA (crRNA) as a gRNA duplex, allowing for effective gene editing, which is easier to modify and tweak for target genes (Jinek et al. 2012). The gRNA 20-nt complementary that binds to specific regions of the genome, provided this complementary region is directly upstream of a Protospacer Adjacent Motif (PAM), which in the case of Cas9 is 5′-NGG-3′. The Cas9 protein itself consists of two main lobes: a Recognition (REC) lobe with REC1 and REC2 domains responsible for guide RNA binding, as well as a Nuclease (NUC) lobe (Nishimasu et al. 2014). The NUC lobe contains three domains: the RuvC domain for noncomplementary strand cleavage, the HNH domain for complementary strand cleavage, and lastly the PAM-interacting domains that facilitate binding to the genome (Nishimasu et al. 2014). This structure allows the gRNA-Cas9 complex to specifically target precise regions of the genome based on the guide RNA's specificity and create a blunt cut in both strands of DNA through endonuclease activity.

Class II Type V CRISPR systems include Cas12a (formerly known as Cpf1), which shares features of the type II Cas9 systems (Paul and Montoya 2020). For example, it also requires a single multidomain Cas protein (Cas12a), which also has a RuvC domain for noncomplementary strand cleavage. This nuclease lacks HNH domain, and instead it features a zinc finger-like domain. Also, unlike Cas9, it lacks the requirement of tracrRNA and relies on only the smaller crRNA, consisting of 42–44 nucleotides that largely simplifies the gRNA design and delivery. For targeting potential, it requires a T-rich PAM motif of 5′-TTTV-3′ and creates staggered cut in the double-stranded DNA (Bandyopadhyay et al. 2020). These cuts are made approximately 18-19 nt from the 3′ end of the PAM site on the complementary strand and 23 nt on the noncomplementary strand, resulting in 5′ overhang of 4–5 bases. Moreover, Cas12a can self-process the maturation of pre-crRNAs into mature crRNAs for simplified multiplexing when the user wants to target multiple sites simultaneously (Charpentier et al. 2015). In plants, several orthologs have been tested from bacteria such as *Acidaminococcus* sp. *BV3L6* (AsCas12a), *Francisella novicida* (FnCas12a) and *Lachnospiraceae bacterium* (LbCas12a) and shown immense potential (Malzahn et al. 2019).

Within the class II Type V systems, there is a range of Cas12a-like proteins that are termed “miniCRISPR” systems due to their small size (type V-F) and often share a unified RuvC-like endonuclease domain near their C-terminus (Harrington et al. 2018; Karvelis et al. 2020; Pausch et al. 2020). Their compact size is notable as they could be packaged into viral vectors, which often have a very limited cargo capacity, allowing for alternative approaches for transgene-free editing. Although some work has been achieved to make strides in utilizing miniCRISPRs in plants, there is room for improvement in crops, and in particular monocots that have limited grafting potential.

The CRISPR-Cas13 nucleases fall within class II Type VI, which exhibit unique characteristics due to their RNA-targeting capabilities and are often grouped into four different subtypes (a-d) (Yoon et al. 2024). Similar to other CRISPR systems, it has a REC lobe for recognition of the crRNA and an NUC domain, but in this instance contains two Higher Eukaryotic and Prokaryotic Nucleotide-binding HEPN catalytic sites rather than DNA endonucleases. They are known to have a higher specificity threshold in comparison to the DNA-targeting CRISPR systems, requiring a near-perfect complementarity between crRNA and RNA target.

11.4.2 Multiplexed Genome Editing

The CRISPR-Cas system is a useful tool for multiplexed genome editing. For the target specificity is determined by the short gRNA, multiplexed genome editing typically requires the parallel expression of multiple sgRNAs sequences to conveniently modify several genes at once. RNA polymerase III (Pol III)-driven and Pol II–driven systems have been reported for multiplexed genome editing (Zhu et al. 2020). Multiple Pol III promoters (U3 and U6) are used in a single construct to drive the expression of multiple sgRNAs (Xing et al. 2014; Ma and Liu 2016). While in Pol II-driven systems, multiple sgRNAs are controlled by a single promoter, and multiple Csy-type ribonuclease 4 s (Csy4s) are interspersed between the sgRNAs under the regulation of the Poly II promoter (Čermák et al. 2017). When Csy4 is expressed, it is capable of cleaving the 20-nucleotide sequences that flank the sgRNAs (Čermák et al. 2017). Currently, multiplexed targeting mutagenesis in the P898012 genotype of sorghum using a Pol III-driven system has been reported (Char et al. 2020). In the future, multiplexed genome editing could be combined with morphogenetic regulators such as WUS2 or BBM to enhance both transformation efficiency and gene-editing efficiency (Che et al. 2022).

11.4.3 Cas Fusion Systems

The CRISPR-Cas system provides opportunities for precise manipulation without the need for double-strand DNA breaks, donor templates, or HDR (Anzalone et al. 2019, 2020). The capability of highly specific targeting of the genome has been exploited with enzyme fusions to Cas nucleases. In these scenarios, the endonuclease activity of the Cas enzymes is deactivated in one or both nuclease domains to create a nickase (nCas) or dead Cas (dCas), respectively. These nCas or dCas systems are then fused to other enzymes via a linker to make specific changes to targeted regions in the host genome.

The most widely used example in plants is the use of base editors, where specific nucleotide changes can be achieved (Gaudelli et al. 2017). At present, base editors are being employed for accurate and efficient single-base alterations in the genes of several crops, such as rice, tomato, wheat, maize, Arabidopsis, rapeseed, and potato (Shimatani et al. 2017; Zong et al. 2017; Kang et al. 2018; Li et al. 2018b; Zong et al. 2018; Veillet et al. 2019; Parikh et al. 2021). Cytosine base editors (CBEs) often utilize APOBEC systems that have a cytidine deaminase domain and an uracil glycosylase inhibitor (UGI), allowing for the conversion of cytosine to uracil within a specific editing window. The uracil can then be converted to thymine through natural DNA repair mechanisms. Adenine base editors (ABEs) allow for the conversion of adenine-to-guanine following a similar mechanism with an engineered TadA (tRNA adenine deaminase domain) that converts the nucleotide to an inosine within a specific editing window, and DNA replication and repair mechanisms will naturally convert inosine to guanine. The major limitation to base editing systems is the window of conversion, where if it's too large, it stops being specific to the key targets of the amino acids, and if it is too narrow, it loses targeting potential, which is limited by the PAM motif. Similarly, due to the complexities of the PAM recognition site by diverse Cas nucleases, fewer alternative Cas proteins have been adopted in the plant world (Koonin et al. 2023).

Similarly, other enzymes can be fused to a deactivated Cas nuclease, such as transcriptional activators, transcriptional repressors, and methyltransferases (Gilbert et al. 2013; Lowder et al. 2015, 2018; Li et al. 2017; Shariati et al. 2019; Ghoshal et al. 2021; Rahman et al. 2024). However, using this method requires continuous expression of the transgenic components, and therefore cannot be designated as non-GM outcomes.

Prime editing is a term to classify a catalytically impaired Cas nuclease alongside an engineered reverse transcriptase enzyme (Molla et al. 2021). In this system, rather than a gRNA or crRNA, it uses a prime editing gRNA (pegRNA) that identifies the target sequence and encodes the desired modification. As such, it's almost a modified version of the HDR editing, but the 'template' repair is within the deployed guide RNA and doesn't create a double-stranded break. This system relies on the creation of a single-stranded break (or nick), where the pegRNA hybridizes to the nicked DNA strand, where the RT uses the pegRNA template to synthesize the new DNA strand containing the edit, and the repair mechanisms ligate it back into the DNA. Although an amazing achievement that could be deployed to 'rewrite' sections of the gene, in plants, the levels of prime editing are considerably lower, and the outcomes are likely to be regulated as a "GMO" due to the use of a template in the repair (Vats et al. 2024). Prime editing has been implemented in only a limited number of crops because of its comparatively low editing efficiency (Anzalone et al. 2019; Lin et al. 2020; Xu et al. 2020; Zhu et al. 2020). However, there are few reports on the application of base editing and prime editing in sorghum, and the effective methods of using it in sorghum need to be explored.

11.4.4 DNA-Free Editing Approaches

Although many countries have decided to regulate specific types of gene editing outcomes as non-GM (see Sect. 11.6 in this chapter for more details), many of the methodologies used to create the gene-edited varieties still have an intermediary GMO step. This can cause issues with confidence of non-GM status and release of material. Tools have been developed that allow for transgene-free gene editing that avoid or limit the complications of approaches that have a stable transformation step within the pipeline.

Haploid Induction Editing (HI-Edit) is a novel genome editing technique developed by Syngenta that utilizes the natural process of haploid induction and combines it with the genetic engineering capabilities of genome editing. It allows for direct modification of commercial crop varieties that are often not amenable to tissue culture, which remains a significant challenge. The haploid inducer line is transformed with the gene editing construct, and the transgenic line is used as a pollen donor for crossing with elite material. During fertilization, the paternal genome from the HI line is gradually eliminated in the first cell divisions, but before the complete elimination the CRISPR-Cas9 machinery is expressed and induces the targeted mutations. It results in a haploid offspring with an edited genome and lack both the haploid-inducer parental DNA and the editing machinery and can undergo well-established tools for chromosome doubling using chemicals. This idea provides the benefit of directly modifying commercial crop varieties and significantly reduces the time required for introgression the gene-edited trait into the elite germplasm. Although this does still require the male donor line to be stably transformed, the gene editing outcome in the elite material remains transgene-free with an easier path for commercialization.

Rather than using DNA to transform the plants with the CRISRPR machinery for integration, groups have also focused on the uses of RNP complexes into cells that allow for non-integrative solutions of gene editing. These have been deployed through protoplast transfection and biolistic transformation depending on the crop. Due to the instability of the RNP complexes and the delivery load required to facilitate gene editing, the rates are significantly lower than if provided as DNA, which allows for robust expression. In these systems, purified Cas nuclease is formed into the complex with gRNAs in vitro, which can be introduced via polyethylene glycol (PEG)-mediated protoplast transfection. However, as discussed previously, protoplast regeneration systems are not well-established for most crops with limited examples in sorghum. Whereas binding RNP complexes for biolistic delivery may prove a more realistic approach in sorghum, where groups have bombarded immature embryos with RNPs in wheat and maize with some success (Svitashev et al. 2016; Liang et al. 2019). However, these RNP systems do not allow for any selective pressure to be placed on the outcomes and therefore the downstream burden of screening for gene-edited progenies is high. Similarly, researchers have attempted to use delivery methods such as carbon nanoparticles to carry the CRISPR cargo and applied with a foliar spray (Lew et al. 2020; Ali et al. 2022; Dunbar et al. 2022). In these examples, single-walled carbon nanotubes (CNTs) to edit a variety of tissues, which then allows for direct production of offspring without tissue culture, which avoid some of the issues surrounding the cell wall barriers of most nanotechnology tools when delivered a spray.

Virus-induced genome editing (VIGE) has also emerged as a powerful tool for transgene-free gene editing. The use of plant viruses can be engineered to be deployed as vectors for editing, and these include the potato virus X (PVX), tobacco rattle virus (TRV), barley stripe mosaic virus (BSMV). There are two strategies for VIGE in crops: the first is to have a Cas nuclease-positive transgenic plant, and the guides are placed within the viral vector and delivered. This allows for the creation of a single transgenic plant, and user-friendly and expedited strategy where only the gRNAs need to be tweaked for various targets. Similarly, transgenic Cas9 positive plants have been used with a grafting strategy but rely on the addition of mRNA/protein signal tags to target specific tissue types, for example, tRNAs and FT mobile elements may allow for movement and infection of the meristem of developing tissues (Yang et al. 2023). There have been attempts to place the Cas nuclease and guides within a single viral vector, but the cargo capacity of these vectors is often quite limited, which is why there is such a strong push to develop miniature Cas nucleases.

11.4.5 Optimizations of Gene Editing in Sorghum

Each aspect of the CRISPR/Cas9 system can be optimized for each crop species and cultivar. One way to ensure strong gene editing efficiencies is to perform guide testing. Numerous groups have developed in silico methodologies for gRNA design and prediction of their efficacy. Unfortunately, many of the training populations were not designed in plants, and with the high variance in gene editing methodologies, the outcomes are not always correlative with findings in vivo. Therefore, tools to verify guide efficacy can be important for gRNA design tools. As such, Meng et al. (2020) developed a protoplast assay to test gRNA efficiency providing in vivo context for editing.

Once designed the appropriate gRNAs for specific and high efficacy, the expression of these elements is of the utmost importance for effective gene editing. For Cas9 expression, typically a constitutive promoter is selected such as the maize *ubi-1* promoter, or specific use of egg-cell specific promoters has also shown to work for high levels of expression in calli tissues. In addition, to ensure optimal translational efficiency, codon bias optimization has been performed, and most groups are utilizing the maize codon bias optimized versions. Since more research has gone into understanding the expression patterns of protein-coding genes, there was a need to optimize U3/U6 promoters for expression of these small noncoding RNAs. Most works in sorghum have used rice or maize U3 or U6 promoters, but studies have shown promise of opting to use endogenous U6 promoters to drive expression with improved rates of gene editing efficiencies (Massel et al. 2022). Following this study, the transgene-free genome editing system has been established, and transgene-free gene-edited plants were achieved within one generation (T_0) in sorghum, demonstrating the feasibility with *phytoene desaturase* (*PDS*) gene (Zhang et al. 2025).

Gene editing has been successfully demonstrated to be an effective approach to target and modify a range of genes in sorghum for crop improvement. Early testing of different CRISPR approaches targeted a transgenic plant to knockout a reporter gene (Jiang et al. 2013) or targeted the *PDS* gene as a reportable knockout (Liu et al. 2019). Then flowering time has been studied through targeted modifications to FT gene (Char et al. 2020; Lee et al. 2023), leaf inclination angle via ligueless-1 gene targeting (Brant et al. 2021), and brown midrib mutants created through targeted CAD genes (Peng et al. 2018). For grain quality, editing of the kafirin genes has also been explored (Li et al. 2018a; Massel et al. 2021, 2022; Elkonin et al. 2023; Hurst et al. 2023). As well as targeting genes involved in phytohormone synthesis for striga resistance (Hao et al. 2023), and seeing if haploid induction could occur through knocking out CENH3, which was shown to be lethal in sorghum (Che et al. 2018). With further improvements and optimizations being performed in gene editing, along with the increased efficacy of Cas-fusion constructs, there is scope to target a range of more genes for sorghum crop improvement.

11.5 The Role of Plant Developmental Regulators

The CRISPR-Cas technology utilization for crops, including sorghum, is constrained not only by the limitations of delivering gene editing components into plant cells but also by the capability to regenerate fertile plants through tissue culture. Plant cell division, somatic embryogenesis, and plant regeneration are controlled by a complex regulatory network that is associated with a range of developmental regulators, including Wuschel (WUS), Wound-Induced Dedifferentiation 1 (WIND1), Auxin Response Factor (ARF), Growth-Regulating Factors, Baby-Boom (BBM), and so on (Passarinho et al. 2008; Bouchabké-Coussa et al. 2013; Lian et al. 2022). At present, numerous studies have shown that WUS2 and BBM are crucial regulators for overcoming genotype and explant selection bottlenecks in the genetic transformation of various plants.

11.5.1 Wuschel 2 (WUS2) and Baby-Boom (BBM)

Wuschel (*WUS*) gene is a class of transcription factors that maintain a pluripotent stem cell pool in the shoot apical meristem, and overexpression of the *WUS* gene can promote somatic embryogenesis (Bouchabké-Coussa et al. 2013; Kadri et al. 2021; Lopes et al. 2021). BBM encodes an AP2/ERF transcription factor, which is involved in the development of roots, seeds, basal embryos, and shoot meristems (Passarinho et al. 2008). Overexpression of *BBM* in could also promote somatic embryogenesis (Passarinho et al. 2008).

The earliest report indicated that WUS2 and BBM improved the transformation efficiency of several stubborn maize inbred lines by utilizing immature embryos, achieving a significant breakthrough in genetic transformation (Lowe et al. 2016). They demonstrated that the genotype dependence can be mitigated when the plants had additional combined overexpression of *BBM* and *WUS2*. These morphogenic regulators also allowed for broadening the range of explant selection to some extent. Since using a strong promoter to drive the expression can lead to the formation of chimeric and necrotic callus tissue, the experiment conducted by Lowe et al. (2016) only used the maize Ubiquitin promoter one to drive the expression *BBM,* whereas a weak promoter was selected to drive the expression of *WUS2*. In addition, since these morphogenic genes can impact plant development, to enable plant regeneration without pleiotropic impacts the CRE-loxP system was utilized to excise WUS2 and BBM (Delneri et al. 2000; Hoa et al. 2002). The loxP sites were inserted at both ends of CRE, WUS, and BBM expression cassette, and the expression of *CRE* was driven by the drought-inducible promoter *RAB17*, through desiccation to excise the sequences flanked by the loxP sites (Lowe et al. 2016). Within the four maize inbred lines PH581, PHP38, PHN46, and PHH5G, the frequencies of obtaining T_0 plants with single-copy, excised transformants and without the *Agrobacterium* plasmid backbone (outside the T-DNA) were 3.2%, 4.2%, 7.7%, and 11.0%, respectively.

Furthermore, the combination of BBM and WUS2 can also enhance transformation of sorghum, sugarcane, and rice (Lowe et al. 2016). Based on the research of Lowe et al. (2016), Mookkan et al. (2017) conducted genetic transformation of maize B73 and sorghum P898012 by co-expressing the maize *WUS2* and *BBM* genes, demonstrating that this approach could increase the frequency of genetic transformation in recalcitrant varieties. Nelson-Vasilchik et al. (2018) summarized the *Agrobacterium*-mediated genetic transformation method for sorghum using the vector PHP78891 (Mookkan et al. 2017), which was designed to promote somatic embryo formation (Nelson-Vasilchik et al. 2018). This method is applicable to P898012 and BTx623, as well as other recalcitrant sorghum varieties (Nelson-Vasilchik et al. 2018). Lowe et al. (2018) conducted further investigations into the promoters driving *WUS2* and *BBM*. They utilized a maize auxin-inducible promoter (*Zm-Axig1*$_{pro}$) to drive *WUS* in combination with a maize phospholipid transfer protein gene promoter (*Zm-PLTP*$_{pro}$) to drive *BBM*, establishing a callus-free transformation without the need to excise WUS2 and BBM (Lowe et al. 2018). Che et al. (2018) developed an efficient and broad applicable sorghum transformation system using a ternary vector system with the pVIR accessory plasmid, which improved *Agrobacterium*-mediated sorghum transformation and generated high-quality transgenic events. Moreover, by integrating CRISPR/Cas gene-editing technology, they achieved the first stable gene knockout in sorghum (Che et al. 2018).

The use of BBM and WUS2 has enabled more efficient transformation of various varieties. However, adding auxiliary components for transformation and the element for excising these components not only increases the size of the T-DNA but also adds complexity to the transformation system. A new "altruistic transformation" system has been developed and characterized by a small T-DNA size and a simple structure. Studies have shown that WUS protein can migrate between cells via plasmodesmata (Yadav and Reddy 2012; Negin et al. 2017). The movement of WUS protein can stimulate embryogenesis in neighboring cells (Yadav et al. 2011). Hoerster et al. (2020) carried out co-infection using *Agrobacterium* LBA4404 thymidine auxotrophic strain (LBA4404Thy^{-}), separately harboring WUS2 expression cassettes (driven by maize Pltp promoter *Zm-Pltp*$_{pro}$) and traditional binary T-DNA plasmids (with selectable and visual markers). The method was optimized by altering the ratios of two *Agrobacterium* strains and using enhancers to regulate the expression of *WUS2*, thereby increasing the frequency of obtaining T_0 maize plants that did not contain WUS2 but contain selectable marker (Hoerster et al. 2020). Che et al. (2022) conducted experiments by utilizing "altruistic transformation" to test whether this altruistic method can be expanded to sorghum. They co-transformed *Agrobacterium* carrying T-DNA vectors with selectable markers and visual marker genes along with *Agrobacterium* carrying altruistic T-DNA binary plasmids (with a *WUS2* expression cassette) at a ratio of 9:1 (Che et al. 2022). Compared to traditional transformation methods, this not only significantly improved transformation efficiency but also led to a higher frequency of high-quality events (Che et al. 2022). It was also shown that a mixture of *Agrobacterium* containing 10% altruistic plasmids was sufficient to deliver WUS2 T-DNA into cells, inducing somatic embryo formation. And the probability of the altruistic and conventional plasmids co-integrating into a single transgenic event is low (Hoerster et al. 2020; Che et al. 2022). Nonintegrated WUS2-enabled altruistic transformation was also utilized for CRISPR/Cas-mediated genome editing, to compare editing frequency with that of conventional transformation and integrated morphogenic gene-enabled transformation. The frequency of gene editing and gene deletion was highest in the case of integrated morphogenic gene-enabled transformation (Che et al. 2022). The gene-editing frequencies mediated by nonintegrated WUS2-enabled altruistic transformation and conventional transformation were equivalent, but the frequency of gene deletion was lowest when mediated by conventional transformation (Che et al. 2022). The results suggested a significant correlation between the frequency of gene editing and the WUS2 protein level in transformed cells, but how WUS2 contributes to the enhancement of CRISPR/Cas-mediated genome editing remains unknown (Che et al. 2022).

Immature embryos are the main explants used in the studies on genetic transformation and gene editing in sorghum. However, in order to obtain immature embryos, it is necessary to plant ahead in the greenhouse and to sample at specific times, which is time-consuming and labor-intensive. If it were possible to select more easily obtainable explants, it would significantly improve the efficiency of genetic transformation or gene editing. Although other explants, such as leaf tissues, have also been studied, their success rates are relatively low, which limited the widespread application. Lowe et al. (2016) directly transformed BBM and WUS2 into embryo slices from leaf segments and mature seeds, resulting in normal fertile T_0 plants (Lowe et al. 2016). Wang et al. (2023) utilized a novel promoter combination to regulate the expression of WUS2 and BBM, after the *Agrobacterium* infection of early leaf tissue derived from maize and sorghum seedlings, directly stimulated the rapid formation of somatic embryos and the regeneration of T_0 plants (Wang et al. 2023). This improved leaf transformation method could also be used for the implementation of CRISPR/Cas9-mediated genome modifications in maize, and this method has also been applied to the leaf transformation of other grass crops, including rice, millet, and barley (Wang et al. 2023). Mature embryos are easier to obtain compared to immature embryos. However, in sorghum, there is limited research on the application of morphogenic regulators-mediated genetic transformation methods in mature embryos and their influence on genome editing efficiency.

11.5.2 Growth-Regulating Factor (GRF) and GRF-Interacting Factor (GIF)

The growth-regulating factor (GRF) family is a small, plant-specific transcription factor family that plays an important role in controlling flowering, root development, growth under stress conditions, and plant longevity regulation (Omidbakhshfard et al. 2015). GRF4 is also involved in the embryonic development of cotyledons and the shoot apical meristem (SAM) (Omidbakhshfard et al. 2015). GRFs can form complexes with GRF-interacting factors (GIFs), thereby enhancing the activity of GRFs (He et al. 2017; Zhang et al. 2018a). It has been reported that the GRF4-GIF1 complex can improve the regeneration rate and efficiency of transgenic wheat in the absence of exogenous cytokinins, resulting in transgenic plants that are fertile and show no significant developmental defects (Debernardi et al. 2020). In watermelon, the use of ClGRF4-ClGIF1 increased the transformation efficiency by approximately ninefold (Feng et al. 2021). Li et al. (2024) improved the transformation efficiency of sorghum to 38.28% by combining GRF4-GIF1 with the auxiliary plasmid pVS1-VIR2. Compared to BBM-WUS, the overexpression of GRF4-GIF1 did not induce significant growth defects in sorghum (Li et al. 2024). Furthermore, Li et al. (2024) combined the GRF4-GIF1/ternary vector system with CRISPR/Cas9 to establish an efficient gene editing tool for sorghum (Li et al. 2024). The GRF-GIF chimeric method is a promising alternative to BBM-WUS for it can be driven by a single constitutive promoter to produce fertile plants without excising the GRF-GIF chimeric construct, and it can also enhance the transformation frequency of both monocots and dicots. Vandeputte et al. (2024) consider that the GRF-GIF chimeric has an effect on the maturation of somatic embryos during the plant regeneration process, which works in different regeneration processes with BBM-WUS. Therefore, it may be promising to combine these two systems to explore better genetic transformation methods or to investigate the potential for co-expression of GRF-GIF with nonintegrated WUS (Vandeputte et al. 2024).

11.5.3 WUSCHEL-Related Homeobox (WOX) Genes

The WUS-related homeobox (WOX) genes encode plant-specific homeobox transcription factors, which play a crucial role in the maintenance of stem cells homeostasis, embryonic or tissue development, organ formation, and the response to stress, etc. (van der Graaff et al. 2009; Fambrini et al. 2022). Numerous studies have shown the function of the WOX genes in the induction of somatic embryogenesis. Research indicates that WOX2 and WOX9 are essential in regulating tobacco embryonic patterns (Zhou et al. 2018). Overexpression of the *MtWOX9* gene in *M. truncatula* can stimulate embryogenesis by altering the expression levels of various somatic embryogenesis-related genes (Tvorogova et al. 2019). Auxin and cytokinin can induce the specific expression of the endogenous *TaWOX5* in wheat roots or callus tissue, indicating that TaWOX5 is related to hormone regulation in somatic embryogenesis (Zhao et al. 2014). Subsequently, WOX5 was shown that is related to organ regeneration by promoting the production of auxin and enhancing the sensitivity to cytokinins, which is essential for the acquisition of pluripotency (Zhai and Xu 2021). Wang et al. (2022) reported that during the genetic transformation process in wheat (as well as in *Triticum monococcum*, triticale, rye, barley, and maize), the overexpression of the wheat gene *TaWOX5* could significantly increase the transformation frequency and reduce genotype dependence. Furthermore, overexpression of *WOX5* does not exhibit a noticeable negative impact on callus differentiation or plant phenotype, but transgenic plants exhibit wide, short flag leaves and thick stems (Wang et al. 2022). Based on this phenotype, it can help to identify unmarked transgenic or non-transgenic edited plants (Wang et al. 2022). Furthermore, Chang et al. (2024) established an effective *Agrobacterium*-mediated transformation system using TaWOX5 for the genetic transformation of cultivated *T. monococcum*, hexaploid triticale, and rye (Chang et al. 2024). However, there are few reports on the study of *WOX* genes in genetic transformation and gene editing in sorghum, which needs further research in the future.

11.6 Challenges and Perspectives for Future Sorghum Production

11.6.1 Challenges in Applying Developmental Regulators

Sorghum gene editing is based on an efficient genetic transformation system, and the transformation efficiency, in turn, can affect the success of genome editing. Since the first transgenic sorghum was obtained using biolistic bombardment method in 1993 (Casas et al. 1993) and transgenic sorghum plants were produced using *Agrobacterium*-mediated transformation in 2000 (Zhao et al. 2000), there have been a number of reports on sorghum genetic transformation research, and the transformation efficiency of sorghum has gradually improved. Subsequently, the genetic transformation efficiency of sorghum experienced a significant improvement with a transformation efficiency of 20.7% achieved using biolistic bombardment in 2012 (Liu and Godwin 2012), and an efficiency of 33% obtained through *Agrobacterium*-mediated transformation in 2014 (Wu et al. 2014). However, these transformation methods are still limited by genotype. Currently, studies on genetic transformation in sorghum mainly utilize immature embryos as explants. The process of preparing these explants involves cultivating sorghum plants ahead of time and collecting immature embryos at designated stages, making it a time-consuming and labor-intensive endeavor. It is essential to devise genetic transformation techniques that employ more readily available explants to enhance the efficiency of gene editing in sorghum. In recent years, the application of plant developmental regulatory factors such as WUS2, BBM in sorghum genetic transformation has significantly improved the transformation efficiency and can overcome genotype dependence, as well as broaden the range of explant selection (Lowe et al. 2016, 2018; Mookkan et al. 2017; Che et al. 2018; Nelson-Vasilchik et al. 2018; Wang et al. 2023). However, the use of WUS2 is constrained by patents in plant transformation, and the application of GRF4-GIF1 in plant regeneration is also protected by patents, which greatly limits the application of these regulators for genetic transformation in the breeding research of sorghum. In summary, significant progress has been made in the genetic transformation of sorghum, greatly advancing the development of gene editing in sorghum; however, these technologies have not yet been widely applied on a large scale.

11.6.2 The Regulatory for Genome-Edited Crops

Genome editing technology showcases significant promise for agriculture, as it has become a mature tool for improving abiotic stress, biotic stress, yield, and nutritional quality of crops as illustrated in Fig. 11.3. This versatile tool has broken new ground in the pursuit of crop improvement by enabling precise, efficient, and relatively cost-effective manipulation of plant genomes. The future of this technology

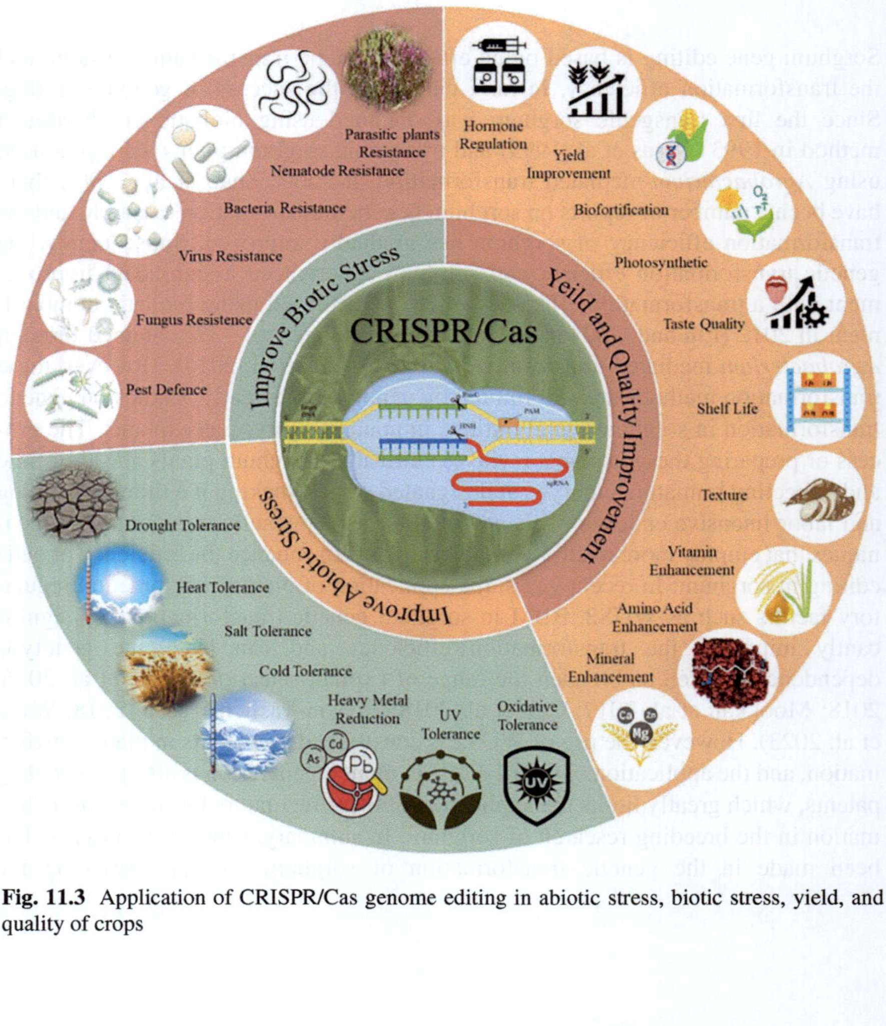

Fig. 11.3 Application of CRISPR/Cas genome editing in abiotic stress, biotic stress, yield, and quality of crops

promises further exploration and more innovative applications, potentially revolutionizing our approach to sustainable agriculture.

Biotic and abiotic stress significantly reduce agricultural productivity and crop yields. By targeting specific genes and pathways, the CRISPR/Cas technology provides a precise tool for enhancing the tolerance of grain crops. Furthermore, by strategically managing hormone pathways, augmenting photosynthesis, and elevating the quality of nutritional substances, etc., the yield and quality of crops can be improved, thus ensuring a healthier and more abundant food supply (Chen et al. 2024).

Nonetheless, the regulatory landscape for genome-edited crops is intricate, encompassing various factors such as scientific, ethical, legal, and social dimensions. And laws and regulations concerning genetically edited crops differ across various countries and regions. In the United States, gene editing is viewed as comparable to conventional breeding in some instances. Consequently, modifications that do not include genes associated with herbicide and pesticide resistance are exempt from regulation under the U.S. regulatory framework (Waltz 2016a, b). Currently, in Australia, Japan, Argentina, Chile, and Brazil, products developed using gene editing technology can be exempt from regulation if the final product does not contain genetically modified components (Whelan and Lema 2015; Duensing et al. 2018; South et al. 2019; Van Vu et al. 2019; El-Mounadi et al. 2020; Zhang et al. 2021). In Canada, New Zealand, and European Union countries, products developed through gene editing technology require strict regulation (Whelan and Lema 2015; Waltz 2016a, b; Shimatani et al. 2017; Smyth 2017; Eckerstorfer et al. 2019; El-Mounadi et al. 2020).

11.6.3 Concerns About Genome-Edited Sorghum

During the utilization process of gene editing, it is necessary to ensure that the application of gene editing technologies is safe and does not have negative impacts on the environment and human health. This includes assessing the long-term effects of gene-edited crops and ensuring the precision and controllability of the gene editing process.

Concerns about gene editing technology mainly focus on the possibility for off-target mutations to lead to unintended genetic alterations in plants (Pineda et al. 2019). To tackle this issue, various improvements have been implemented in the gene editing process, including enhancing gRNA specificity, utilizing transformed CRISPR-Cas9 ribonucleoprotein complexes instead of nucleotide sequences, applying inducible promoters for targeted regulation of gRNA and/or Cas9 expression, and modifying deaminase activity to decrease the ability of Cas9 to bind DNA, all aimed at boosting the precision of gene editing and minimizing off-target effects (Svitashev et al. 2016; Shimatani et al. 2017; Liang et al. 2018; El-Mounadi et al. 2020).

Concerns about CRISPR-Cas9 technology also include the biosafety risks associated with the use of the Cas9 protein in gene editing applications; therefore, the Cas9 and its homologous genes were analyzed. In many bacteria and archaea, the CRISPR-Cas system provides protection by targeting specific sequences (Horvath and Barrangou 2010). *Streptococcus thermophilus* and *Lactobacillus plantarum* are two widely used bacteria in the production of food such as milk, yogurt, and cheese (Lecomte et al. 2016; Behera et al. 2018), which showed the 23% ~ 58% and 35% similarity with the Cas9 from *S. pyogenes* used in plant genome editing (El-Mounadi et al. 2020). *Lactobacillus plantarum* is considered a kind of beneficial bacteria which present in the human gastrointestinal tract that contributes to the balance of the beneficial microbiota (Nagpal et al. 2012; Kassayová et al. 2014). These indicate that humans have already been exposed to the Cas9 protein for a long time before the appearance of gene editing technology. Therefore, after the application of Cas9 in gene editing technology, the biosafety risks to human need to be further evaluated.

Societal concerns about genomic editing stem from a lack of clear understanding of the principles, benefits, risks, and safety of genomic edited crops, as well as the degree to which this information is accepted by the public (Eckerstorfer et al. 2019; El-Mounadi et al. 2020). To prevent the spread of misinformation and corrections, it is necessary to strengthen public education on the principles of genome editing, especially regarding crops could be obtained that utilize gene editing technology to achieve desired traits without containing transgenic components.

In China, sorghum is primarily used for brewing liquor, and during this process, it undergoes fermentation, distillation, meaning that sorghum raw material does not directly enter the human body. In the future, it is anticipated that varieties of brewing sorghum obtained through gene editing, which do not contain transgenic components, may be easier to pass transgenic regulation and gain public acceptance compared to other crops.

11.6.4 CRISPR-Cas System Technological Advancements

The CRISPR-Cas9 system operates by creating double-strand DNA breaks (DSB) at target sites, inducing insertions, deletions, or base substitutions for the purpose of gene editing. The mutations produced through this editing process involve a certain level of unpredictability. Homology-directed repair (HDR) mediated by DSBs could be used for accurate DNA alterations, and this method depends on external donor DNA repair templates; however, its low efficiency in higher plants is the main reason limiting its application (Huang and Puchta 2019).

One limitation of modulating expression patterns with Cas-fusions is that it requires continuous expression of dCas9 and sgRNA in the target genome, and it is not possible to remove the transgenic components, leading to regulatory challenges. Therefore, utilizing Cas9 rather than dCas9 allows for the editing of cis-regulatory elements in the promoter of the target gene, facilitating the regulation of gene expression (Rodríguez-Leal et al. 2017). Therefore, utilizing Cas9 rather than dCas9 allows for the editing of cis-regulatory elements in the promoter of the target gene, facilitating the regulation of gene expression (Rodríguez-Leal et al. 2017). Many RNA-targeting CRISPR-Cas systems, such as Cas13a and Cas13b, have been developed to cleave RNA, allowing for more specific downregulation of individual transcripts (Abudayyeh et al. 2017). CRISPR-Cas can also be used to regulate pre-mRNA splicing by targeting key splicing motifs GU-AG or the 5′ untranslated region to influence gene expression (Li et al. 2019). Upstream open reading frame (uORF) is a special structure located upstream of the start codon (AUG) that can regulate protein synthesis by affecting the translation efficiency of mRNA, thereby participating in the regulation of gene expression (Zhang et al. 2018b). For instance, modifying the uORF of the gene responsible for a key enzyme in vitamin C biosynthesis led to an increase in translation levels, which boosted ascorbic acid content in lettuce by approximately 150% (Zhang et al. 2018b). Moreover, by altering the uORF of *bZIP1.1*, the translation of the primary ORF was improved, resulting in sweeter strawberries (Xing et al. 2020). In addition, editing other noncoding regions, such as noncoding RNA genes, which include long noncoding RNAs (lncRNAs) and various small RNAs (including miRNAs, siRNAs, and piRNAs), can also regulate gene expression (Chung et al. 2020; Hong et al. 2021). In sorghum, research has indicated that some small RNAs are linked to beneficial traits like drought tolerance and sugar accumulation (Calviño et al. 2011; Hamza et al. 2016), while studies on the editing of noncoding RNA genes in plants, including sorghum, remain relatively limited. In summary, the application of gene editing in sorghum is relatively limited; the above-mentioned technology, which regulates gene expression by editing noncoding regions through gene editing, provides valuable insights for the development of gene editing in sorghum.

Currently, other reported gene editing experimental methods based on the CRISPR-Cas system include conditional CRISPR-Cas systems, virus-assisted gene editing, de novo meristem induction method, and haploid-inducer mediated genome editing. In plants, approximately 10% of gene knockouts can lead to functional diversity or lethality (Lloyd et al. 2015). When studying the functions of these genes, in addition to downregulating their expression using RNAi or CRISPR-Cas-based regulation mentioned in 6.4.2, a method known as conditional CRISPR-Cas systems has been developed (Decaestecker et al. 2019). Conditional CRISPR-Cas systems primarily utilize tissue-specific expression promoters to drive the expression of Cas9 in specific tissues, thereby editing target genes within those tissues (Decaestecker et al. 2019). Additionally, inducible promoters (inducers, specific light sources, etc.) can be combined with CRISPR-Cas system to enhance the flexibility and compatibility of genome editing (Ochoa-Fernandez et al. 2020; Wang et al. 2020). Using plant virus-assisted gene editing to obtain gene-edited plants without the need for tissue culture is a very promising approach. Ellison et al. (2020) used initial Cas9 transgenic plants and a tobacco rattle virus (TRV) vector that expresses movable sgRNA (where a gene is linked to the 3′ end of sgRNA to assist its movement to the shoot apical meristem in plants) through *Agrobacterium* infiltration to obtain gene-edited regenerated plants without the need for tissue culture

(Ellison et al. 2020). Due to the limited cargo capacity of currently developed plant virus vector systems, these viruses are unable to co-encode Cas9 along with sgRNA. This technical challenge has recently been addressed by utilizing *sonchus yellow net rhabdovirus* (SYNV). Ma et al. (2020) constructed Cas9 and gRNA sequences into a vector containing the SYNV genome and achieved the editing of single or multiple genes in tobacco plants through a DNA-free genome editing method via *Agrobacterium* infiltration, without relying on tissue culture (Ma et al. 2020). However, this method is currently limited by the host range of available viral species. Furthermore, de novo meristem could be induced by infiltrating *Agrobacterium* containing morphogenetic regulators such as WUS2, IPT, STM, and sgRNA into Cas9-overexpressing plants, thereby producing gene-edited plants with heritability. (Maher et al. 2020). Gene editing can also be combined with haploid induction technology, which can circumvent the challenges of transformation and produce homozygous non-transgenic gene-edited plants (Kelliher et al. 2019; Wang et al. 2019).

In recent years, gene editing technology has developed rapidly, leading to the emergence of a series of gene editing tools to meet various editing needs. However, these multiple tools still do not fully address all the requirements for plant genome manipulation. Further development will be crucial for the application of CRISPR-Cas in plants, particularly for the development of time-saving, simple, and efficient methods suitable for sorghum genome editing.

11.6.5 Discovering Targets of Agronomic Importance

Recent studies have revealed various genes and quantitative trait loci (QTLs) that regulate essential agronomic traits in sorghum. For example, Han et al. (2015) localized qGW1 to a 101 kb region on the short arm of chromosome 1, which was an important locus related to grain weight. SSR markers on qGW1 can be used for marker-assisted selection to improve grain yield in sorghum (Han et al. 2015). Mace et al. (2019) identified approximately 340 QTLs associated with yield (Mace et al. 2019). Tao et al. (2020) conducted a large-scale GWAS on 837 sorghum materials and 1421 BC-NAM populations, resulting in the identification of 81 QTLs associated with grain size (Tao et al. 2020). In terms of sorghum grain quality, Wu et al. (2012) found that the nucleotide polymorphisms in the *TANNIN1* gene, which encodes a WD40 protein, regulate the biosynthesis of tannins in sorghum (Wu et al. 2012). The two recessive alleles *Tan1-a* and *Tan1-b* result in a nonfunctional *TANNIN1*, thereby affecting tannin synthesis (Wu et al. 2012). Kimani et al. (2020) analyzed the natural variation in grain quality of 196 sorghum varieties and found that there were 14 genetic loci associated with sorghum tannins, 14 associated with starch, and 492 associated with amino acids (Kimani et al. 2020). The progress of genetic transformation and gene editing in sorghum has been relatively slow, resulting in many identified genetic loci and genes linked to desirable traits lacking strong validation through the use of transgenic or gene editing materials. Recently, Zhang et al. (2023) conducted a genome-wide association analysis by utilizing resequencing data from 352 sorghum germplasm combined with alkaline tolerance phenotypes (Zhang et al. 2023). They identified an alkaline tolerance gene, *Alkaline Tolerance 1* (*AT1*), and validated its function using rice, sorghum mutants (obtained through CRISPR-Cas9 technology), and overexpression materials (Zhang et al. 2023). This provided a significant reference haplotype for breeding alkaline-tolerant varieties and markedly improved the research on stress resistance in sorghum (Zhang et al. 2023). In the present and future, the development of sorghum genetic transformation and gene editing provides significant support for sorghum research, and the application prospects are promising. By utilizing these key technologies to accelerate breeding, sorghum will become an important crop for promoting the achievement of sustainable development.

11.7 Conclusion

Sorghum holds immense potential as a versatile crop, addressing global needs for food, feed, and bioenergy. The advent of genome editing technologies, particularly CRISPR-Cas9, has revolutionized sorghum research, enabling precise modifications to enhance yield, stress resistance, and nutritional quality. However, challenges such as genotype dependency, limited explant availability, and regulatory hurdles remain significant barriers to broad application. To fully harness the benefits of genome editing, further advancements in tissue culture and transformation techniques are essential. Expanding the range of explants, optimizing media formulations, and leveraging developmental regulators like WUS2, BBM, and GRF4-GIF1 can improve transformation efficiency and reduce genotype limitations. Additionally, addressing public concerns and establishing clear regulatory frameworks for transgene-free genome-edited crops will be crucial for their acceptance and commercialization. As research progresses, the integration of CRISPR-Cas9 with other innovative approaches, such as DNA-free editing and virus-assisted methods, promises to streamline the development of high-performing sorghum elite lines. By combining these cutting-edge technologies with traditional breeding efforts, sorghum can play a pivotal role in achieving sustainable agriculture and meeting the challenges posed by a growing global population and climate change. The future of sorghum lies in the continued collaboration between scientists, breeders, and policymakers to unlock its full potential as a resilient and resource-efficient crop.

References

Able JA, Rathus C, Godwin ID (2001) The investigation of optimal bombardment parameters for transient and stable transgene expression in sorghum. In Vitro Cell Dev Biol - Plant (3):37

Abudayyeh OO, Gootenberg JS, Essletzbichler P, Han S, Joung J, Belanto JJ, Verdine V, Cox DBT, Kellner MJ, Regev A, Lander ES, Voytas DF, Ting AY, Zhang F (2017) RNA targeting with CRISPR-Cas13. Nature 550(7675):280–284

Ali Z, Serag MF, Demirer GS, Torre B, di Fabrizio E, Landry MP, Habuchi S, Mahfouz M (2022) DNA-carbon nanotube binding mode determines the efficiency of carbon nanotube-mediated DNA delivery to intact plants. ACS Appl Nano Mater 5(4):4663–4676

Altpeter F, Baisakh N, Beachy RN, Visser R (2005) Particle bombardment and the genetic enhancement of crops: myths and realities. Mol Breed 15(3):305–327

Anzalone AV, Randolph PB, Davis JR, Sousa AA, Koblan LW, Levy JM, Chen PJ, Wilson C, Newby GA, Raguram A, Liu DR (2019) Search-and-replace genome editing without double-strand breaks or donor DNA. Nature 576(7785):149–157

Anzalone AV, Koblan LW, Liu DR (2020) Genome editing with CRISPR-Cas nucleases, base editors, transposases and prime editors. Nat Biotechnol 38(7):824–844

Bandyopadhyay A, Kancharla N, Javalkote VS, Dasgupta S, Brutnell TP (2020) CRISPR-Cas12a (Cpf1): a versatile tool in the plant genome editing tool box for agricultural advancement. Front Plant Sci 11

Battraw M, Hall TC (1991) Stable transformation of Sorghum bicolor protoplasts with chimeric neomycin phosphotransferase II and β-glucuronidase genes. Theor Appl Genet 82(2):161–168

Behera SS, Ray RC, Zdolec N (2018) Lactobacillus plantarum with functional properties: an approach to increase safety and shelf-life of fermented foods. Biomed Res Int 2018:9361614

Belide S, Vanhercke T, Petrie JR, Singh SP (2017) Robust genetic transformation of sorghum (Sorghum bicolor L.) using differentiating embryogenic callus induced from immature embryos. Plant Methods 13:109

Berenji J, Dahlberg J, Sikora V, Latkovic D (2011) Origin, history, morphology, production, improvement, and utilization of broomcorn [*Sorghum bicolor* (L.) Moench] in Serbia. Econ Bot 65(2):190–208

Bouchabké-Coussa O, Obellianne M, Linderme D, Montes E, Maia-Grondard A, Vilaine F, Pannetier C (2013) Wuschel overexpression promotes somatic embryogenesis and induces organogenesis in cotton (Gossypium hirsutum L.) tissues cultured in vitro. Plant Cell Rep 32(5):675–686

Brant EJ, Baloglu MC, Parikh A, Altpeter F (2021) CRISPR/Cas9 mediated targeted mutagenesis of LIGULELESS-1 in sorghum provides a rapidly scorable phenotype by altering leaf inclination angle. Biotechnol J 16(11)

Cai T, Butler L (1990) Plant regeneration from embryogenic callus initiated from immature inflorescences of several high-tannin sorghums. Plant Cell Tissue Org Cult 20(2):101–110

Calviño M, Messing J (2012) Sweet sorghum as a model system for bioenergy crops. Curr Opin Biotechnol 23(3):323–329

Calviño M, Bruggmann R, Messing J (2011) Characterization of the small RNA component of the transcriptome from grain and sweet sorghum stems. BMC Genomics 12:356

Casas AM, Kononowicz AK, Zehr UB, Tomes DT, Axtell JD, Butler LG, Bressan RA, Hasegawa PM (1993) Transgenic sorghum plants via microprojectile bombardment. Proc Natl Acad Sci USA 90(23):11212–11216

Casas AM, Kononowicz AK, Haan TG, Zhang L, Tomes DT, Hasegawa BPM (1997) Transgenic sorghum plants obtained after microprojectile bombardment of immature inflorescences. In Vitro Cell Dev Biol Plant 33(2):92–100

Čermák T, Curtin SJ, Gil-Humanes J, Čegan R, Kono TJY, Konečná E, Belanto JJ, Starker CG, Mathre JW, Greenstein RL, Voytas DF (2017) A multipurpose toolkit to enable advanced genome engineering in plants. Plant Cell 29(6):1196–1217

Chang Y, Liu J, Liu C, Liu H, Tang H, Qiu Y, Lin Z, Wang K, Yan Y, Ye X (2024) Establishment of a transformation system in close relatives of wheat under the assistance of TaWOX5. J Integr Agric 23(6):1839–1849

Char SN, Wei J, Mu Q, Li X, Zhang ZJ, Yu J, Yang B (2020) An *Agrobacterium*-delivered CRISPR/Cas9 system for targeted mutagenesis in sorghum. Plant Biotechnol J 18(2):319–321

Charpentier E, Richter H, van der Oost J, White MF (2015) Biogenesis pathways of RNA guides in archaeal and bacterial CRISPR-Cas adaptive immunity. FEMS Microbiol Rev 39(3):428–441
Che P, Anand A, Wu E, Sander JD, Simon MK, Zhu W, Sigmund AL, Zastrow-Hayes G, Miller M, Liu D, Lawit SJ, Zhao ZY, Albertsen MC, Jones TJ (2018) Developing a flexible, high-efficiency *Agrobacterium*-mediated sorghum transformation system with broad application. Plant Biotechnol J 16(7):1388–1395
Che P, Wu E, Simon MK, Anand A, Lowe K, Gao H, Sigmund AL, Yang M, Albertsen MC, Gordon-Kamm W, Jones TJ (2022) Wuschel2 enables highly efficient CRISPR/Cas-targeted genome editing during rapid de novo shoot regeneration in sorghum. Commun Biol 5(1):344
Chen F, Chen L, Yan Z, Xu J, Feng L, He N, Guo M, Zhao J, Chen Z, Chen H, Yao G, Liu C (2024) Recent advances of CRISPR-based genome editing for enhancing staple crops. Front Plant Sci 15:1478398
Chou J, Huang J, Huang Y (2020) Simple and efficient genetic transformation of sorghum using immature inflorescences. Acta Physiol Plant 42(3):41
Chung PJ, Chung H, Oh N, Choi J, Bang SW, Jung SE, Jung H, Shim JS, Kim JK (2020) Efficiency of recombinant CRISPR/rCas9-mediated miRNA gene editing in rice. Int J Mol Sci 21(24):9606
Chylinski K, Makarova KS, Charpentier E, Koonin EV (2014) Classification and evolution of type II CRISPR-Cas systems. Nucleic Acids Res 42(10):6091–6105
Cong L, Ran FA, Cox D, Lin SL, Barretto R, Habib N, Hsu PD, Wu XB, Jiang WY, Marraffini LA, Zhang F (2013) Multiplex genome engineering using CRISPR/Cas systems. Science 339(6121):819–823
Cook DE, Sharma V, Zheng W, Huang J, Zhang W (2019) Developing CRISPR-Cas13 for multi-use RNA-targeting in plants. Mol Plant-Microbe Interact 32(10):214–214
Dahlberg J (2019) The role of sorghum in renewables and biofuels. Methods Mol Biol 1931:269–277
Dahlberg J, Berenji J, Sikora V, Latkovi D (2011) Assessing sorghum (Sorghum bicolor (L) Moench) germplasm for new traits: food, fuels & unique uses. Istituto Sperimentale per la Cerealicoltura(2)
De Cleene M, De Ley J (1976) The host range of crown gall. Bot Rev 42(4):389–466
de Oliveira LL, de Alencar Figueiredo LF (2024) Sorghum phytonutrients and their health benefits: a systematic review from cell to clinical trials. J Food Sci 89(S1):A5–A29
Debernardi JM, Tricoli DM, Ercoli MF, Hayta S, Ronald P, Palatnik JF, Dubcovsky J (2020) A GRF-GIF chimeric protein improves the regeneration efficiency of transgenic plants. Nat Biotechnol 38(11):1274–1279
Decaestecker W, Buono RA, Pfeiffer ML, Vangheluwe N, Jourquin J, Karimi M, Van Isterdael G, Beeckman T, Nowack MK, Jacobs TB (2019) CRISPR-TSKO: a technique for efficient mutagenesis in specific cell types, tissues, or organs in Arabidopsis. Plant Cell 31(12):2868–2887
Delneri D, Tomlin GC, Wixon JL, Hutter A, Sefton M, Louis EJ, Oliver SG (2000) Exploring redundancy in the yeast genome: an improved strategy for use of the cre-loxP system. Gene 252(1–2):127–135
Do PT, Lee H, Mookkan M, Folk WR, Zhang ZJ (2016) Rapid and efficient *Agrobacterium*-mediated transformation of sorghum (Sorghum bicolor) employing standard binary vectors and bar gene as a selectable marker. Plant Cell Rep 35(10):2065–2076
Ducksbury C, Stefoska-Needham A (2022) A cross-sectional audit of sorghum in selected cereal food products in Australian supermarkets. Nutrients 14(9):1821
Duensing N, Sprink T, Parrott WA, Fedorova M, Lema MA, Wolt JD, Bartsch D (2018) Novel features and considerations for ERA and regulation of crops produced by genome editing. Front Bioeng Biotechnol 6:79
Dunbar T, Tsakirpaloglou N, Septiningsih EM, Thomson MJ (2022) Carbon nanotube-mediated plasmid DNA delivery in rice leaves and seeds. Int J Mol Sci 23(8):4081
Eckerstorfer MF, Engelhard M, Heissenberger A, Simon S, Teichmann H (2019) Plants developed by new genetic modification techniques-comparison of existing regulatory frameworks in the EU and non-EU countries. Front Bioeng Biotechnol 7:26

Elkonin LA, Gerashchenkov GA, Borisenko N, Kenzhegulov OA, Sarsenova SK, Rozhnova NA, Panin VM (2023) Development of sorghum mutants with improved protein digestibility by CRISPR/Cas9 editing of kafirin genes. Crop Journal 11(5):1411–1418

Ellison EE, Nagalakshmi U, Gamo ME, Huang PJ, Dinesh-Kumar S, Voytas DF (2020) Multiplexed heritable gene editing using RNA viruses and mobile single guide RNAs. Nat Plants 6(6):620–624

El-Mounadi K, Morales-Floriano ML, Garcia-Ruiz H (2020) Principles, applications, and biosafety of plant genome editing using CRISPR-Cas9. Front Plant Sci 11:56

Emani C, Sunilkumar G, Rathore KS (2002) Transgene silencing and reactivation in sorghum. Plant Sci 162(2):181–192

Fambrini M, Usai G, Pugliesi C (2022) Induction of somatic embryogenesis in plants: different players and focus on WUSCHEL and WUS-RELATED HOMEOBOX (WOX) transcription factors. Int J Mol Sci 23(24):15950

Feng Q, Xiao L, He Y, Liu M, Wang J, Tian S, Zhang X, Yuan L (2021) Highly efficient, genotype-independent transformation and gene editing in watermelon (Citrullus lanatus) using a chimeric ClGRF4-GIF1 gene. J Integr Plant Biol 63(12):2038–2042

Finer JJ, Vain P, Jones MW, McMullen MD (1992) Development of the particle inflow gun for DNA delivery to plant cells. Plant Cell Rep 11(7):323–328

Fontanet-Manzaneque JB, Haeghebaert J, Aesaert S, Coussens G, Pauwels L, Cano-Delgado AI (2024) Efficient sorghum and maize transformation using a ternary vector system combined with morphogenic regulators. Plant J 120(5):2076–2088

Gao Z, Xie X, Ling Y, Muthukrishnan S, Liang GH (2005) *Agrobacterium tumefaciens*-mediated sorghum transformation using a mannose selection system. Plant Biotechnol J 3(6):591–599

Gaudelli NM, Komor AC, Rees HA, Packer MS, Badran AH, Bryson DI, Liu DR (2017) Programmable base editing of A.T to G.C in genomic DNA without DNA cleavage. Nature 551(7681):464-+

Ghoshal B, Picard CL, Vong B, Feng SH, Jacobsen SE (2021) CRISPR-based targeting of DNA methylation in by a bacterial CG-specific DNA methyltransferase. Proc Natl Acad Sci U S A 118(23):e2125016118

Gilbert LA, Larson MH, Morsut L, Liu Z, Brar GA, Torres SE, Stern-Ginossar N, Brandman O, Whitehead EH, Doudna JA, Lim WA, Weissman JS, Qi LS (2013) CRISPR-mediated modular RNA-guided regulation of transcription in eukaryotes. Cell 154(2):442–451

Girijashankar V, Sharma HC, Sharma KK, Swathisree V, Prasad LS, Bhat BV, Royer M, Secundo BS, Narasu ML, Altosaar I, Seetharama N (2005) Development of transgenic sorghum for insect resistance against the spotted stem borer (Chilo partellus). Plant Cell Rep 24(9):513–522

Grootboom AW, Mkhonza NL, O'Kennedy MM, Chakauya E, Kunert KJ, Chikwamba RK (2010) Biolistic mediated sorghum (Sorghum bicolor L. Moench) transformation via mannose and bialaphos based selection systems. Int J Bot 6(2)

Gurel S, Gurel E, Kaur R, Wong J, Meng L, Tan HQ, Lemaux PG (2009) Efficient, reproducible *Agrobacterium*-mediated transformation of sorghum using heat treatment of immature embryos. Plant Cell Rep 28(3):429–444

Hagio T, Blowers AD, Earle ED (1991) Stable transformation of sorghum cell cultures after bombardment with DNA-coated microprojectiles. Plant Cell Rep 10(5):260–264

Hamza NB, Sharma N, Tripathi A, Sanan-Mishra N (2016) MicroRNA expression profiles in response to drought stress in Sorghum bicolor. Gene Expr Patterns 20(2):88–98

Han L, Chen J, Mace ES, Liu Y, Zhu M, Yuyama N, Jordan DR, Cai H (2015) Fine mapping of qGW1, a major QTL for grain weight in sorghum. Theor Appl Genet 128(9):1813–1825

Hao JJ, Yang Y, Futrell S, Kelly EA, Lorts CM, Nebie B, Runo S, Yang JL, Alvarez S, Lasky JR, Schachtman DP (2023) CRISPR/Cas9-mediated mutagenesis of *Carotenoid Cleavage Dioxygenase* (CCD) genes in sorghum alters strigolactone biosynthesis and plant biotic interactions. Phytobiomes J 7(3):339–351

Hare PD, Chua NH (2002) Excision of selectable marker genes from transgenic plants. Nat Biotechnol 20(6):575–580

Harrington LB, Burstein D, Chen JS, Paez-Espino D, Ma E, Witte IP, Cofsky JC, Kyrpides NC, Banfield JF, Doudna JA (2018) Programmed DNA destruction by miniature CRISPR-Cas14 enzymes. Science 362(6416):839-+

Hayta S, Smedley MA, Demir SU, Blundell R, Hinchliffe A, Atkinson N, Harwood WA (2019) An efficient and reproducible *Agrobacterium*-mediated transformation method for hexaploid wheat (Triticum aestivum L.). Plant Methods 15:121

He X, Miyasaka SC, Fitch MM, Moore PH, Zhu YJ (2008) Agrobacterium tumefaciens-mediated transformation of taro (Colocasia esculenta (L.) Schott) with a rice chitinase gene for improved tolerance to a fungal pathogen Sclerotium rolfsii. Plant Cell Rep 27(5):903–909

He Z, Zeng J, Ren Y, Chen D, Li W, Gao F, Cao Y, Luo T, Yuan G, Wu X, Liang Y, Deng Q, Wang S, Zheng A, Zhu J, Liu H, Wang L, Li P, Li S (2017) OsGIF1 positively regulates the sizes of stems, leaves, and grains in rice. Front Plant Sci 8:1730

He YB, Mudgett M, Zhao YD (2022) Advances in gene editing without residual transgenes in plants. Plant Physiol 188(4):1757–1768

Hoa TT, Bong BB, Huq E, Hodges TK (2002) Cre/ lox site-specific recombination controls the excision of a transgene from the rice genome. Theor Appl Genet 104(4):518–525

Hoerster G, Wang N, Ryan L, Wu E, Gordon-Kamm B (2020) Use of non-integrating Zm-Wus2 vectors to enhance maize transformation: non-integrating WUS2 enhances transformation. In Vitro Cell Dev Biol - Plant 56(6):265–279

Hong Y, Meng J, He X, Zhang Y, Liu Y, Zhang C, Qi H, Luan Y (2021) Editing miR482b and miR482c simultaneously by CRISPR/Cas9 enhanced tomato resistance to *Phytophthora infestans*. Phytopathology 111(6):1008–1016

Horvath P, Barrangou R (2010) CRISPR/Cas, the immune system of bacteria and archaea. Science 327(5962):167–170

Huang TK, Puchta H (2019) CRISPR/Cas-mediated gene targeting in plants: finally a turn for the better for homologous recombination. Plant Cell Rep 38(4):443–453

Hurst JP, Yobi A, Li AX, Sato S, Clemente TE, Angelovici R, Holding DR (2023) Large and stable genome edits at the sorghum alpha kafirin locus result in changes in chromatin accessibility and globally increased expression of genes encoding lysine enrichment. Front Plant Sci 14:1116886

Ismagul A, Mazonka I, Callegari C, Eliby S (2014) *Agrobacterium*-mediated transformation of barley (*Hordeum vulgare* L.). Methods Mol Biol 1145:203–211

Jambagi S, Bhat R, Sumangala S, Kuruvinashetti M (2010) Agrobacterium-mediated transformation studies in sorghum using an improved gfp reporter gene. J SAT Agric Res

Jeong YY, Lee HY, Kim SW, Noh YS, Seo PJ (2021) Optimization of protoplast regeneration in the model plant Arabidopsis thaliana. Plant Methods 17(1):21

Jeoung JM, Krishnaveni S, Muthukrishnan S, Trick HN, Liang GH (2002) Optimization of sorghum transformation parameters using genes for green fluorescent protein and beta-glucuronidase as visual markers. Hereditas 137(1):20–28

Jiang WZ, Zhou HB, Bi HH, Fromm M, Yang B, Weeks DP (2013) Demonstration of CRISPR/Cas9/sgRNA-mediated targeted gene modification in Arabidopsis, tobacco, sorghum and rice. Nucleic Acids Res 41(20):e188

Jiao P, Jin S, Chen N, Wang C, Liu S, Qu J, Guan S, Ma Y (2022) Improvement of cold tolerance in maize (Zea mays L.) using Agrobacterium-mediated transformation of ZmSAMDC gene. GM Crops Food 13(1):131–141

Jinek M, Chylinski K, Fonfara I, Hauer M, Doudna JA, Charpentier E (2012) A programmable dual-RNA-guided DNA endonuclease in adaptive bacterial immunity. Science 337(6096):816–821

Kadri A, Grenier De March G, Guerineau F, Cosson V, Ratet P (2021) WUSCHEL overexpression promotes callogenesis and somatic embryogenesis in Medicago truncatula Gaertn. Plants (Basel) 10(4):715

Kaeppler HF, Pedersen JF (1997) Evaluation of 41 elite and exotic inbred Sorghum genotypes for high quality callus production. Plant Cell Tissue Organ Cult 48(1):71–75

Kang BC, Yun JY, Kim ST, Shin Y, Ryu J, Choi M, Woo JW, Kim JS (2018) Precision genome engineering through adenine base editing in plants. Nat Plants 4(7):427–431

Karvelis T, Bigelyte G, Young JK, Hou ZL, Zedaveinyte R, Budre K, Paulraj S, Djukanovic V, Gasior S, Silanskas A, Venclovas C, Siksnys V (2020) PAM recognition by miniature CRISPR-Cas12f nucleases triggers programmable double-stranded DNA target cleavage. Nucleic Acids Res 48(9):5016–5023

Kassayová M, Bobrov N, Strojný L, Kisková T, Mikeš J, Demečková V, Orendáš P, Bojková B, Péč M, Kubatka P, Bomba A (2014) Preventive effects of probiotic bacteria Lactobacillus plantarum and dietary fiber in chemically-induced mammary carcinogenesis. Anticancer Res 34(9):4969–4975

Kelliher T, Starr D, Su X, Tang G, Chen Z, Carter J, Wittich PE, Dong S, Green J, Burch E, McCuiston J, Gu W, Sun Y, Strebe T, Roberts J, Bate NJ, Que Q (2019) One-step genome editing of elite crop germplasm during haploid induction. Nat Biotechnol 37(3):287–292

Kimani W, Zhang LM, Wu XY, Hao HQ, Jing HC (2020) Genome-wide association study reveals that different pathways contribute to grain quality variation in sorghum (Sorghum bicolor). BMC Genomics 21(1):112

Koonin EV, Gootenberg JS, Abudayyeh OO (2023) Discovery of diverse CRISPR-Cas systems and expansion of the genome engineering toolbox. Biochemistry 62(24):3465–3487

Kumar V, Campbell LAM, Rathore KS (2011) Rapid recovery- and characterization of transformants following Agrobacterium-mediated T-DNA transfer to sorghum. Plant Cell Tissue Organ Cult 104(2):137–146

Lecomte X, Gagnaire V, Lortal S, Dary A, Genay M (2016) Streptococcus thermophilus, an emerging and promising tool for heterologous expression: advantages and future trends. Food Microbiol 53(Pt A):2–9

Lee JS, Bae SJ, Kim JS, Kim C, Kang BC (2023) A streamlined guide RNA screening system for genome editing in Sorghum bicolor. Plant Methods 19(1):90

Lew TTS, Park M, Wang YZ, Gordiichuk P, Yeap WC, Rais SKM, Kulaveerasingam H, Strano MS (2020) Nanocarriers for transgene expression in pollen as a plant biotechnology tool. Acs Mater Lett 2(9):1057–1066

Li Z, Zhang D, Xiong X, Yan B, Xie W, Sheen J, Li JF (2017) A potent Cas9-derived gene activator for plant and mammalian cells. Nat Plants 3(12):930–936

Li AX, Jia SG, Yobi A, Ge ZX, Sato SJ, Zhang C, Angelovici R, Clemente TE, Holding DR (2018a) Editing of an alpha-kafirin gene family increases digestibility and protein quality in sorghum. Plant Physiol 177(4):1425–1438

Li C, Zong Y, Wang Y, Jin S, Zhang D, Song Q, Zhang R, Gao C (2018b) Expanded base editing in rice and wheat using a Cas9-adenosine deaminase fusion. Genome Biol 19(1):59

Li Z, Xiong X, Wang F, Liang J, Li JF (2019) Gene disruption through base editing-induced messenger RNA missplicing in plants. New Phytol 222(2):1139–1148

Li J, Pan W, Zhang S, Ma G, Li A, Zhang H, Liu L (2024) A rapid and highly efficient sorghum transformation strategy using GRF4-GIF1/ternary vector system. Plant J 117(5):1604–1613

Lian Z, Nguyen CD, Liu L, Wang G, Chen J, Wang S, Yi G, Wilson S, Ozias-Akins P, Gong H, Huo H (2022) Application of developmental regulators to improve in planta or in vitro transformation in plants. Plant Biotechnol J 20(8):1622–1635

Liang Z, Chen K, Zhang Y, Liu J, Yin K, Qiu JL, Gao C (2018) Genome editing of bread wheat using biolistic delivery of CRISPR/Cas9 in vitro transcripts or ribonucleoproteins. Nat Protoc 13(3):413–430

Liang Z, Chen KL, Gao CX (2019) Biolistic delivery of CRISPR/Cas9 with ribonucleoprotein complex in wheat. Methods Mol Biol 1917:327–335

Lin Q, Zong Y, Xue C, Wang S, Jin S, Zhu Z, Wang Y, Anzalone AV, Raguram A, Doman JL, Liu DR, Gao C (2020) Prime genome editing in rice and wheat. Nat Biotechnol 38(5):582–585

Liu G, Godwin ID (2012) Highly efficient sorghum transformation. Plant Cell Rep 31(6):999–1007

Liu GQ, Gilding EK, Godwin ID (2013) Additive effects of three auxins and copper on sorghum *in vitro* root induction. In Vitro Cell Dev Biol-Plant 49(2):191–197

Liu G, Gilding EK, Godwin ID (2014) Genetic transformation and breeding. In: Wang YH, Upadhyaya HD, Kole C (eds) Genetics, genomics and breeding of sorghum. CRC Press, pp 251–283
Liu G, Gilding EK, Godwin ID (2015) A robust tissue culture system for sorghum [*Sorghum bicolor* (L.) Moench]. S Afr J Bot 98:157–160
Liu G, Massel K, Tabet B, Godwin ID (2020) Biolistic DNA delivery and its applications in sorghum bicolor. Methods Mol Biol 2124:197–215
Liu G, Zhang Y, Gong H, Li S, Pan Y, Davis C, Jing HC, Wu L, Godwin ID (2021) Stem vacuole-targetted sucrose isomerase enhances sugar content in sorghum. Biotechnol Biofuels 14(1):53
Lloyd JP, Seddon AE, Moghe GD, Simenc MC, Shiu SH (2015) Characteristics of plant essential genes allow for within- and between-species prediction of lethal mutant phenotypes. Plant Cell 27(8):2133–2147
Lopes FL, Galvan-Ampudia C, Landrein B (2021) WUSCHEL in the shoot apical meristem: old player, new tricks. J Exp Bot 72(5):1527–1535
Lowder LG, Zhang D, Baltes NJ, Paul JW 3rd, Tang X, Zheng X, Voytas DF, Hsieh TF, Zhang Y, Qi Y (2015) A CRISPR/Cas9 toolbox for multiplexed plant genome editing and transcriptional regulation. Plant Physiol 169(2):971–985
Lowder LG, Zhou J, Zhang Y, Malzahn A, Zhong Z, Hsieh TF, Voytas DF, Zhang Y, Qi Y (2018) Robust transcriptional activation in plants using multiplexed CRISPR-Act2.0 and mTALE-act systems. Mol Plant 11(2):245–256
Lowe BA, Shiva Prakash N, Way M, Mann MT, Spencer TM, Boddupalli RS (2009) Enhanced single copy integration events in corn via particle bombardment using low quantities of DNA. Transgenic Res 18(6):831–840
Lowe K, Wu E, Wang N, Hoerster G, Hastings C, Cho MJ, Scelonge C, Lenderts B, Chamberlin M, Cushatt J, Wang L, Ryan L, Khan T, Chow-Yiu J, Hua W, Yu M, Banh J, Bao Z, Brink K, Igo E, Rudrappa B, Shamseer PM, Bruce W, Newman L, Shen B, Zheng P, Bidney D, Falco C, Register J, Zhao ZY, Xu D, Jones T, Gordon-Kamm W (2016) Morphogenic regulators Baby boom and Wuschel improve monocot transformation. Plant Cell 28(9):1998–2015
Lowe K, La Rota M, Hoerster G, Hastings C, Wang N, Chamberlin M, Wu E, Jones T, Gordon-Kamm W (2018) Rapid genotype "independent" Zea mays L. (maize) transformation via direct somatic embryogenesis. In Vitro Cell Dev Biol Plant 54(3):240–252
Ma X, Liu YG (2016) CRISPR/Cas9-based multiplex genome editing in monocot and dicot plants. Curr Protoc Mol Biol 115:31.36.31–31.36.21
Ma X, Zhang X, Liu H, Li Z (2020) Highly efficient DNA-free plant genome editing using virally delivered CRISPR-Cas9. Nat Plants 6(7):773–779
Mace E, Innes D, Hunt C, Wang X, Tao Y, Baxter J, Hassall M, Hathorn A, Jordan D (2019) The sorghum QTL atlas: a powerful tool for trait dissection, comparative genomics and crop improvement. Theor Appl Genet 132(3):751–766
Maher MF, Nasti RA, Vollbrecht M, Starker CG, Clark MD, Voytas DF (2020) Plant gene editing through de novo induction of meristems. Nat Biotechnol 38(1):84–89
Makarova KS, Zhang F, Koonin EV (2017) SnapShot: class 1 CRISPR-Cas systems. Cell 168(5):946-+
Mali P, Yang LH, Esvelt KM, Aach J, Guell M, DiCarlo JE, Norville JE, Church GM (2013) RNA-guided human genome engineering via Cas9. Science 339(6121):823–826
Malzahn AA, Tang X, Lee K, Ren QR, Sretenovic S, Zhang YX, Chen HQ, Kang M, Bao Y, Zheng XL, Deng KJ, Zhang T, Salcedo V, Wang K, Zhang Y, Qi YP (2019) Application of CRISPR-Cas12a temperature sensitivity for improved genome editing in rice, maize, and Arabidopsis. BMC Biol 17:9
Massel K, Lam Y, Wong ACS, Hickey LT, Borrell AK, Godwin ID (2021) Hotter, drier, CRISPR: the latest edit on climate change. Theor Appl Genet 134(6):1691–1709
Massel K, Lam Y, Hintzsche J, Lester N, Botella JR, Godwin ID (2022) Endogenous U6 promoters improve CRISPR/Cas9 editing efficiencies in Sorghum bicolor and show potential for applications in other cereals. Plant Cell Rep 41(2):489–492

McCormick RF, Truong SK, Sreedasyam A, Jenkins J, Shu S, Sims D, Kennedy M, Amirebrahimi M, Weers BD, McKinley B, Mattison A, Morishige DT, Grimwood J, Schmutz J, Mullet JE (2018) The Sorghum bicolor reference genome: improved assembly, gene annotations, a transcriptome atlas, and signatures of genome organization. Plant J 93(2):338–354

Meng R, Wang C, Wang L, Liu Y, Zhan Q, Zheng J, Li J (2020) An efficient sorghum protoplast assay for transient gene expression and gene editing by CRISPR/Cas9. PeerJ 8:e10077

Miflin B (2000) Crop improvement in the 21st century. J Exp Bot 51(342):1–8

Molla KA, Sretenovic S, Bansal KC, Qi YP (2021) Precise plant genome editing using base editors and prime editors. Nat Plants 7(9):1166–1187

Mookkan M, Nelson-Vasilchik K, Hague J, Zhang ZJ, Kausch AP (2017) Selectable marker independent transformation of recalcitrant maize inbred B73 and sorghum P898012 mediated by morphogenic regulators BABY BOOM and WUSCHEL2. Plant Cell Rep 36(9):1477–1491

Nagpal R, Kumar A, Kumar M, Behare PV, Jain S, Yadav H (2012) Probiotics, their health benefits and applications for developing healthier foods: a review. FEMS Microbiol Lett 334(1):1–15

Negin B, Shemer O, Sorek Y, Eshed Williams L (2017) Shoot stem cell specification in roots by the WUSCHEL transcription factor. PLoS One 12(4):e0176093

Nelson-Vasilchik K, Hague J, Mookkan M, Zhang ZJ, Kausch A (2018) Transformation of recalcitrant sorghum varieties facilitated by Baby Boom and Wuschel2. Curr Protoc Plant Biol 3(4):e20076

Nguyen TV, Thu TT, Claeys M, Angenon G (2007) Agrobacterium-mediated transformation of sorghum (Sorghum bicolor (L.) Moench) using an improved in vitro regeneration system. Plant Cell Tissue Organ Cult 91(2):155–164

Nishimasu H, Ran FA, Hsu PD, Konermann S, Shehata SI, Dohmae N, Ishitani R, Zhang F, Nureki O (2014) Crystal structure of Cas9 in complex with guide RNA and target DNA. Cell 156(5):935–949

Ochoa-Fernandez R, Abel NB, Wieland FG, Schlegel J, Koch LA, Miller JB, Engesser R, Giuriani G, Brandl SM, Timmer J, Weber W, Ott T, Simon R, Zurbriggen MD (2020) Optogenetic control of gene expression in plants in the presence of ambient white light. Nat Methods 17(7):717–725

O'Connell MR (2019) Molecular mechanisms of RNA targeting by Cas13-containing type VI CRISPR-Cas systems. J Mol Biol 431(1):66–87

Omidbakhshfard MA, Proost S, Fujikura U, Mueller-Roeber B (2015) Growth-regulating factors (GRFs): a small transcription factor family with important functions in plant biology. Mol Plant 8(7):998–1010

Ou-Lee TM, Turgeon R, Wu R (1986) Expression of a foreign gene linked to either a plant-virus or a Drosophila promoter, after electroporation of protoplasts of rice, wheat, and sorghum. Proc Natl Acad Sci USA 83(18):6815–6819

Pacesa M, Pelea O, Jinek M (2024) Past, present, and future of CRISPR genome editing technologies. Cell 187(5):1076–1100

Parikh A, Brant EJ, Baloglu MC, Altpeter F (2021) CRISPR/Cas-mediated genome editing in sorghum—recent progress, challenges and prospects. In Vitro Cell Dev Biol – Plant 57(4):720–730

Passarinho P, Ketelaar T, Xing M, van Arkel J, Maliepaard C, Hendriks MW, Joosen R, Lammers M, Herdies L, den Boer B, van der Geest L, Boutilier K (2008) BABY BOOM target genes provide diverse entry points into cell proliferation and cell growth pathways. Plant Mol Biol 68(3):225–237

Paterson AH, Bowers JE, Bruggmann R, Dubchak I, Grimwood J, Gundlach H, Haberer G, Hellsten U, Mitros T, Poliakov A, Schmutz J, Spannagl M, Tang H, Wang X, Wicker T, Bharti AK, Chapman J, Feltus FA, Gowik U, Grigoriev IV, Lyons E, Maher CA, Martis M, Narechania A, Otillar RP, Penning BW, Salamov AA, Wang Y, Zhang L, Carpita NC, Freeling M, Gingle AR, Hash CT, Keller B, Klein P, Kresovich S, McCann MC, Ming R, Peterson DG, Mehboob ur R, Ware D, Westhoff P, Mayer KF, Messing J, Rokhsar DS (2009) The Sorghum bicolor genome and the diversification of grasses. Nature 457(7229):551–556

Paul B, Montoya G (2020) CRISPR-Cas12a: functional overview and applications. Biom J 43(1):8–17

Pausch P, Al-Shayeb B, Bisom-Rapp E, Tsuchida CA, Li Z, Cress BF, Knott GJ, Jacobsen SE, Banfield JF, Doudna JA (2020) CRISPR-CasΦ from huge phages is a hypercompact genome editor. Science 369(6501):333-+

Peng C, Wang H, Xu XL, Wang XF, Chen XY, Wei W, Lai YM, Liu GQ, Godwin ID, Li JQ, Zhang L, Xu JF (2018) High-throughput detection and screening of plants modified by gene editing using quantitative real-time polymerase chain reaction. Plant J 95(3):557–567

Pineda M, Lear A, Collins JP, Kiani S (2019) Safe CRISPR: challenges and possible solutions. Trends Biotechnol 37(4):389–401

Pola S, Mani NS, Ramana T (2008) Plant tissue culture studies in *Sorghum bicolor*: immature embryo explants as the source material. Int J Plant Prod 2(1):1–14

Raghuwanshi A, Birch RG (2010) Genetic transformation of sweet sorghum. Plant Cell Rep 29(9):997–1005

Rahman SU, Khan MO, Ullah R, Ahmad F, Raza G (2024) Agrobacterium-mediated transformation for the development of transgenic crops; present and future prospects. Mol Biotechnol 66(8):1836–1852

Rech EL, Vianna GR, Aragão FJ (2008) High-efficiency transformation by biolistics of soybean, common bean and cotton transgenic plants. Nat Protoc 3(3):410–418

Rodríguez-Leal D, Lemmon ZH, Man J, Bartlett ME, Lippman ZB (2017) Engineering quantitative trait variation for crop improvement by genome editing. Cell 171(2):470–480.e478

Sato-Izawa K, Tokue K, Ezura H (2018) Development of a stable Agrobacterium-mediated transformation protocol for Sorghum bicolor Tx430. Plant Biotechnol (Tokyo) 35(2):181–185

Seetharama N, Godwin I (2004) Sorghum tissue culture and transformation. In: Sorghum tissue culture and transformation: viii + 168 pp

Shan QW, Wang YP, Li J, Zhang Y, Chen KL, Liang Z, Zhang K, Liu JX, Xi JJ, Qiu JL, Gao CX (2013) Targeted genome modification of crop plants using a CRISPR-Cas system. Nat Biotechnol 31(8):686–688

Shariati SA, Dominguez A, Xie S, Wernig M, Qi LS, Skotheim JM (2019) Reversible disruption of specific transcription factor-DNA interactions using CRISPR/Cas9. Mol Cell 74(3):622–633.e624

Shimatani Z, Kashojiya S, Takayama M, Terada R, Arazoe T, Ishii H, Teramura H, Yamamoto T, Komatsu H, Miura K, Ezura H, Nishida K, Ariizumi T, Kondo A (2017) Targeted base editing in rice and tomato using a CRISPR-Cas9 cytidine deaminase fusion. Nat Biotechnol 35(5):441–443

Shou H, Frame BR, Whitham SA, Wang K (2004) Assessment of transgenic maize events produced by particle bombardment or Agrobacterium-mediated transformation. Mol Breed 13(2):201–208

Smyth SJ (2017) Canadian regulatory perspectives on genome engineered crops. GM Crops Food 8(1):35–43

South PF, Cavanagh AP, Liu HW, Ort DR (2019) Synthetic glycolate metabolism pathways stimulate crop growth and productivity in the field. Science 363(6422):eaat9077

Svitashev S, Schwartz C, Lenderts B, Young JK, Mark Cigan A (2016) Genome editing in maize directed by CRISPR-Cas9 ribonucleoprotein complexes. Nat Commun 7:13274

Tadesse Y, Sági L, Swennen R, Jacobs M (2003) Optimisation of transformation conditions and production of transgenic sorghum (Sorghum bicolor) via microparticle bombardment. Plant Cell Tissue Organ Cult 75(1):1–18

Tang YY, Fu Y (2018) Class 2 CRISPR/Cas: an expanding biotechnology toolbox for and beyond genome editing. Cell Biosci 8

Tao Y, Zhao X, Wang X, Hathorn A, Hunt C, Cruickshank AW, van Oosterom EJ, Godwin ID, Mace ES, Jordan DR (2020) Large-scale GWAS in sorghum reveals common genetic control of grain size among cereals. Plant Biotechnol J 18(4):1093–1105

Tao Y, Luo H, Xu J, Cruickshank A, Zhao X, Teng F, Hathorn A, Wu X, Liu Y, Shatte T, Jordan D, Jing H, Mace E (2021) Extensive variation within the pan-genome of cultivated and wild sorghum. Nat Plants 7(6):766–773

Tvorogova VE, Fedorova YA, Potsenkovskaya EA, Kudriashov AA, Lutova LA (2019) The WUSCHEL-related homeobox transcription factor MtWOX9-1 stimulates somatic embryogenesis in Medicago truncatula. Plant Cell Tissue Organ Cult (PCTOC) 138(3):517–527

Tzfira T, Citovsky V (2006) Agrobacterium-mediated genetic transformation of plants: biology and biotechnology. Curr Opin Biotechnol 17(2):147–154

van der Graaff E, Laux T, Rensing SA (2009) The WUS homeobox-containing (WOX) protein family. Genome Biol 10(12):248

Van Vu T, Sung YW, Kim J, Doan DTH, Tran MT, Kim JY (2019) Challenges and perspectives in homology-directed gene targeting in monocot plants. Rice (N Y) 12(1):95

Vandeputte W, Coussens G, Aesaert S, Haeghebaert J, Impens L, Karimi M, Debernardi JM, Pauwels L (2024) Use of GRF-GIF chimeras and a ternary vector system to improve maize (Zea mays L.) transformation frequency. Plant J 119(4):2116–2132

Vats S, Kumar J, Sonah H, Zhang F, Deshmukh R (2024) Prime editing in plants: prospects and challenges. J Exp Bot 75(17):5344–5356

Veillet F, Perrot L, Chauvin L, Kermarrec MP, Guyon-Debast A, Chauvin JE, Nogue F, Mazier M (2019) Transgene-free genome editing in tomato and potato plants using agrobacterium-mediated delivery of a CRISPR/Cas9 cytidine base editor. Int J Mol Sci 20(2)

Waltz E (2016a) CRISPR-edited crops free to enter market, skip regulation. Nat Biotechnol 34(6):582

Waltz E (2016b) Gene-edited CRISPR mushroom escapes US regulation. Nature 532(7599):293

Wang D, Zhao Q, Zhu D, Ao G, Yu J (2006) Particle-bombardment-mediated co-transformation of maize with a lysine rich protein gene (sb401) from potato. Euphytica 150(1–2):75–85

Wang W, Wang J, Yang C, Li Y, Liu L, Xu J (2007) Pollen-mediated transformation of *Sorghum bicolor* plants. Biotechnol Appl Biochem 48(Pt 2):79–83

Wang B, Zhu L, Zhao B, Zhao Y, Xie Y, Zheng Z, Li Y, Sun J, Wang H (2019) Development of a haploid-inducer mediated genome editing system for accelerating maize breeding. Mol Plant 12(4):597–602

Wang X, Ye L, Lyu M, Ursache R, Löytynoja A, Mähönen AP (2020) An inducible genome editing system for plants. Nat Plants 6(7):766–772

Wang L, Gao L, Liu G, Meng R, Liu Y, Li J (2021) An efficient sorghum transformation system using embryogenic calli derived from mature seeds. PeerJ 9:e11849

Wang K, Shi L, Liang X, Zhao P, Wang W, Liu J, Chang Y, Hiei Y, Yanagihara C, Du L, Ishida Y, Ye X (2022) The gene TaWOX5 overcomes genotype dependency in wheat genetic transformation. Nat Plants 8(2):110–117

Wang N, Ryan L, Sardesai N, Wu E, Lenderts B, Lowe K, Che P, Anand A, Worden A, van Dyk D, Barone P, Svitashev S, Jones T, Gordon-Kamm W (2023) Leaf transformation for efficient random integration and targeted genome modification in maize and sorghum. Nat Plants 9(2):255–270

Wang F, Bao J, Zhang H, Zhai G, Song T, Liu Z, Han Y, Yu F, Zou G, Zhu Y (2025) A telomere-to-telomere genome assembly of Chinese grain sorghum 654. Sci Data 12(1):460

Whelan AI, Lema MA (2015) Regulatory framework for gene editing and other new breeding techniques (NBTs) in Argentina. GM Crops Food 6(4):253–265

Wu Y, Li X, Xiang W, Zhu C, Lin Z, Wu Y, Li J, Pandravada S, Ridder DD, Bai G, Wang ML, Trick HN, Bean SR, Tuinstra MR, Tesso TT, Yu J (2012) Presence of tannins in sorghum grains is conditioned by different natural alleles of Tannin1. Proc Natl Acad Sci USA 109(26):10281–10286

Wu E, Lenderts B, Glassman K, Berezowska-Kaniewska M, Christensen H, Asmus T, Zhen S, Chu U, Cho MJ, Zhao ZY (2014) Optimized Agrobacterium-mediated sorghum transformation protocol and molecular data of transgenic sorghum plants. In Vitro Cell Dev Biol Plant 50(1):9–18

Xing HL, Dong L, Wang ZP, Zhang HY, Han CY, Liu B, Wang XC, Chen QJ (2014) A CRISPR/Cas9 toolkit for multiplex genome editing in plants. BMC Plant Biol 14:327

Xing S, Chen K, Zhu H, Zhang R, Zhang H, Li B, Gao C (2020) Fine-tuning sugar content in strawberry. Genome Biol 21(1):230

Xu YY, Li ZJ (2020) CRISPR-Cas systems: overview, innovations and applications in human disease research and gene therapy. Comput Struct Biotechnol J 18:2401–2415

Xu R, Li J, Liu X, Shan T, Qin R, Wei P (2020) Development of plant prime-editing systems for precise genome editing. Plant Commun 1(3):100043

Yadav RK, Reddy GV (2012) WUSCHEL protein movement and stem cell homeostasis. Plant Signal Behav 7(5):592–594

Yadav RK, Perales M, Gruel J, Girke T, Jönsson H, Reddy GV (2011) WUSCHEL protein movement mediates stem cell homeostasis in the Arabidopsis shoot apex. Genes Dev 25(19):2025–2030

Yang L, Machin F, Wang SF, Saplaoura E, Kragler F (2023) Heritable transgene-free genome editing in plants by grafting of wild-type shoots to transgenic donor rootstocks. Nat Biotechnol 41(7):958-+

Yao Q, Cong L, Chang JL, Li KX, Yang GX, He GY (2006) Low copy number gene transfer and stable expression in a commercial wheat cultivar via particle bombardment. J Exp Bot 57(14):3737–3746

Yoon PH, Zhang ZY, Loi KJ, Adler BA, Lahiri A, Vohra K, Shi HL, Rabelo DB, Trinidad M, Boger RS, Al-Shimary MJ, Doudna JA (2024) Structure-guided discovery of ancestral CRISPR-Cas13 ribonucleases. Science 385(6708):538–543

Zhai N, Xu L (2021) Pluripotency acquisition in the middle cell layer of callus is required for organ regeneration. Nat Plants 7(11):1453–1460

Zhang Y, Li J, Gao C (2016) Generation of stable transgenic rice (Oryza sativa L.) by Agrobacterium-mediated transformation. Curr Protoc Plant Biol 1(2):235–246

Zhang D, Sun W, Singh R, Zheng Y, Cao Z, Li M, Lunde C, Hake S, Zhang Z (2018a) GRF-interacting factor1 regulates shoot architecture and meristem determinacy in maize. Plant Cell 30(2):360–374

Zhang H, Si X, Ji X, Fan R, Liu J, Chen K, Wang D, Gao C (2018b) Genome editing of upstream open reading frames enables translational control in plants. Nat Biotechnol 36(9):894–898

Zhang Y, Restall J, Crisp P, Godwin I, Liu G (2021) Current status and prospects of plant genome editing in Australia. In Vitro Cell Dev Biol Plant 57(4):574–583

Zhang H, Yu F, Xie P, Sun S, Qiao X, Tang S, Chen C, Yang S, Mei C, Yang D, Wu Y, Xia R, Li X, Lu J, Liu Y, Xie X, Ma D, Xu X, Liang Z, Feng Z, Huang X, Yu H, Liu G, Wang Y, Li J, Zhang Q, Chen C, Ouyang Y, Xie Q (2023) A Gγ protein regulates alkaline sensitivity in crops. Science 379(6638):eade8416

Zhang Y, Cheng M, Massel K, Godwin ID, Liu G (2025) An accelerated transgene-free genome editing system using microparticle bombardment of sorghum immature embryos. aBIOTECH

Zhao ZY, Cai T, Tagliani L, Miller M, Wang N, Pang H, Rudert M, Schroeder S, Hondred D, Seltzer J, Pierce D (2000) *Agrobacterium*-mediated sorghum transformation. Plant Mol Biol 44(6):789–798

Zhao S, Jiang QT, Ma J, Zhang XW, Zhao QZ, Wang XY, Wang CS, Cao X, Lu ZX, Zheng YL, Wei YM (2014) Characterization and expression analysis of WOX5 genes from wheat and its relatives. Gene 537(1):63–69

Zhou X, Guo Y, Zhao P, Sun MX (2018) Comparative analysis of WUSCHEL-related homeobox genes revealed their parent-of-origin and cell type-specific expression pattern during early embryogenesis in tobacco. Front Plant Sci 9:311

Zhu H, Muthukrishnan SS, Krishnaveni S, Wilde G, Liang GH (1998) Biolistic transformation of sorghum using a rice chitinase gene. J Genet Breed 52(3):243–252

Zhu H, Li C, Gao C (2020) Applications of CRISPR-Cas in agriculture and plant biotechnology. Nat Rev Mol Cell Biol 21(11):661–677

Zong Y, Wang Y, Li C, Zhang R, Chen K, Ran Y, Qiu JL, Wang D, Gao C (2017) Precise base editing in rice, wheat and maize with a Cas9-cytidine deaminase fusion. Nat Biotechnol 35(5):438–440
Zong Y, Song Q, Li C, Jin S, Zhang D, Wang Y, Qiu JL, Gao C (2018) Efficient C-to-T base editing in plants using a fusion of nCas9 and human APOBEC3A. Nat Biotechnol 36:950–953

Chapter 12
Gene Editing in Maize: Progress and Challenges

Samia Hassan, Nayla Munawar, Arshiba Khawar, Paulo Zaini, Ahmed Al-Harrasi, Sameer H. Qari, and Aftab Ahmad

Abstract The deployment of genome editing, notably CRISPR/Cas9, in maize breeding catalyzed the meticulous introgression of desirable agronomic traits, thereby conferring enhanced disease resistance, herbicide tolerance, and abiotic stress tolerance. Multiplex genome editing allowed the modification of several genes at the same time, accelerating the development of elite maize varieties with optimized phenotypic profiles. In addition, genome editing has been used in waxy maize to improve its quality and functioning. Latest advancements in transformation-free genome-editing approaches, such as CRISPR/Cas9 ribonucleoprotein (RNP) delivery, have reduced the need for transgenic intermediates, which enables the fast production of genetically modified crops. The base and prime editing have accelerated the gene editing potential by specific trait improvement with exceptional accuracy due to their precision and effectiveness. However, challenges related to genetically engineered maize for practical implementation include environmental risks and potential off-target effects. Ethical concerns for CRISPR-edited maize have evolved, particularly the discussions for how to categorize and label these

Samia Hassan and Nayla Munawar contributed equally with all other contributors.

S. Hassan · A. Khawar
Department of Biochemistry, The University of Agriculture, Faisalabad, Pakistan

N. Munawar
Department of Chemistry, College of Science, United Arab Emirates University, Al-Ain, United Arab Emirates

P. Zaini
Department of Plant Sciences, University of California, Davis, CA, USA

A. Al-Harrasi · A. Ahmad (✉)
Natural and Medical Sciences Research Center (NMSRC), University of Nizwa, Nizwa, Oman
e-mail: aftab.ahmad@unizwa.edu.om

S. H. Qari
Department of Biology, AlJumum University College, Umm Al-Qura University, Makkah, Saudi Arabia

K. A. Abd-Elsalam, A. Ahmad (eds.), *Gene Editing in Cereals*, Sustainability Sciences in Asia and Africa, https://doi.org/10.1007/978-981-95-6722-5_12

genome modified crops. The incorporation of genome editing and double haploid technology, and also progresses in HI-Edit technology, has revolutionized maize breeding, speed up the developments of elite lines with precise traits. By addressing the challenges related with stable integration and transgene-free techniques, genome editing has significantly improved the maize productivity and resilience, playing a major contribution to the world's food security and promoting green agriculture.

12.1 Introduction

Zea is an old Greek term for "sustaining life," while *Mays* is a Taino word for "life giver." The plant is also known as zea, silk maize, makka, and barajovar. Maize, full of nutrients and a major staple crop, is being the primary target of genome editing programs due to its high impact on the world's hunger and nutritional status (Alam et al. 2020; Sangare et al. 2023). Maize breeding gives a chance for targeted genetic changes that focus on improving its nutritional value, increasing crop yields, and inducing desired agronimic traits for its rich variety of essential vitamins and minerals, including carbohydrates, fibers, vitamin C, vitamin E, vitamin K, vitamin B2 (niacin), vitamin B1 (thiamine), vitamin B3 (riboflavin), B6 (pyridoxine), N-p-coumaryl tryptamine, vitamin B5 (pantothenic acid), selenium, folic acid, and N-ferrulyl tryptamine (Saritha et al. 2020). Table 12.1 presents the composition of maize kernel, the edible nutritive component of the plant. Maize has important phytochemicals that include phytosterols, phenolic compounds, and carotenoids. It is known as the "queen of cereals" globally due to its sufficient yield potential (Rani et al. 2021; Kaur et al. 2021, 2022; Marangoni et al. 2020; Ding et al. 2020; Mallakpour et al. 2021; Guo et al. 2022a, b; Kaur and Singh 2022; Alsharairi 2022); (Mrowicka et al. 2022; Widomska et al. 2020; Gulati et al. 2023; Bhushan et al. 2021; Kaur et al. 2021; Kim et al. 2023; Burlini et al. 2020).

The advancements in genome editing caused a paradigm change in maize breeding, permitting the precise and targeted editing of crop's genome with exceptional specificity. The CRISPR/Cas9 system acts as a strong tool for introducing desirable traits into maize (*Zea mays* L.), the cereal crop of the Poaceae family (Voytas and Gao 2014; Zhu et al. 2021). With the help of CRISPR/Cas9, studies have effectively improved various traits such as an improved resistance to pathogens, herbicide tolerance, and abiotic stress tolerance in maize, ultimately accelerating its resilience and yield (Liu et al. 2020; Wang et al. 2020). Multiplex genome editing enabled the targeting of multiple genes at once to improve the production of top maize varieties with optimal phenotypic profiles and agronomic performance (Li et al. 2020a, b). Latest advancements in transformation-free genome editing approaches, such as CRISPR/Cas9 ribonucleoprotein (RNP) delivery, have reduced the need for transgenic intermediates, which enables the fast production of genetically modified crops.

The base and prime editing have accelerated the gene editing potential by specific trait improvement with exceptional accuracy due to their precision and effectiveness (Anzalone et al. 2019). The incorporation of genome editing and double haploid technology, and also progresses in HI-Edit technology, has revolutionized maize breeding, speed up the developments of elite lines with precise traits. (Liu et al. 2020; Li et al. 2022). On the other hand, the testing of genome transferred maize in fields cause many problems as well, such as environmental risks and potential off-target effects, require complete research and mitigation measures (Kuzma and Grieger 2020). Ethical concerns for CRISPR-edited maize have evolved, particularly the discussions for how to categorize and label these genome modified crops.

12.2 The CRISPR Technology Applications in Maize

12.2.1 *Introduction to CRISPR/Cas System*

In 1987, Japanese scientists discovered the CRISPR/Cas (Clustered Regularly Interspaced Short Palindromic Repeats/CRISPR-associated proteins) system in *Escherichia coli* has now became a novel genetic tool based on bacterial adaptive immunity (Kozovska et al. 2021). The initially known as dismissed repetitive sequences CRISPR arrays in prokaryotic genomes, later related to immune function when researchers such as Philippe Horvath and Rodolphe Barrangou revealed that the spacer sequences in them matched to those of previous invading plasmids and viruses, showing a form of genetic memory (Ali et al. 2024). These arrays cause bacteria to develop an immunity by integrating segments of foreign DNA into their genome that are transcribed into guide RNAs to help Cas nucleases to target and cleave the same sequences during later infections (Agarwal and Gupta 2021). Because of site-specific double-stranded breaks in targeted DNA Cas9 and Cas12 have gained specific interest as they allow precise genome editing via repair mechanisms like nonhomologous end joining (NHEJ) or homology-directed repair (HDR; Fig. 12.1; Asmamaw and Zawdie 2021; Tariq et al. 2023). The CRISPR/Cas approach, especially the Class II types (Cas9, Cas12, and Cas13) are being using in gene editing across various organisms due to their simplicity, accuracy, and adaptability, with significant applications in animal and plant genome modification, epigenetic regulation, and in the resistance against different diseases (Janik et al. 2020; Liu et al. 2021; Pramitha et al. 2021).

12.2.2 *Mechanism of CRISPR/Cas9*

CRISPR/Cas works by leveraging bacteria's immunity to identify and combat viruses that invade. Cas9 protein interacts with sgRNA to produce the ribonucleoprotein complex (RNP). After recognizing the PAM region next to the target sequence, this complex anneals to the genetic code with base pairing in order to

Table 12.1 List of genes targeted in maize crop through CRISPR/Cas system for phytic acid biosynthetic pathway, male sterility, haploid induction, synthetic apomixis, plant development, stress response, nutritional content, hybrid breeding, and de novo domestication

Gene-editing tool	Target Gene	Anatomical expression	Function	References
CRISPR/Cas	*IPE1*	Express in the tapetum and function in tapetum regeneration and pollen exine formation	Encodes a glucose-methanol-choline oxidoreductase that is required for pollen exine formation	Chen et al. (2017)
CRISPR/Cas	*lpa1, lpa2*	Kernels	Phytic acid biosynthesis, phosphorus availability, reduced phytic acid levels via knockout	Raboy et al. (2000), Singh et al., (2020), Pandey et al., (2021)
CRISPR/Cas9	*MIPS*	Seeds, developing seeds particularly endosperms, seeds maturation	Reduction in phytic acid content, accumulation in maize grains, plant breeding, crop improvement	Perera et al. (2021), Young et al., (2019)
CRISPR/Cas9	*ZmMRP4*	Roots, leaves, seeds,	Reduced phytic acid content, increased nutrient accumulation in grains, embryo, and corn grains	Frittelli et al. (2023)
CRISPR/Cas9	*ARGOS8*	Leaves, kernels, roots	Regulates plant growth, and development, potentially impacting yield	Shi et al. (2015)
CRISPR/Cas9	*lpa3*	Seeds development and maturation	Lowered phytic acid synthesis and storage	Khatoon et al., (2024)
CRISPR/Cas9	*crtRB1, o₂*	Kernel development, carotenoid composition	Reduced phytic acid level, O_2 regulates zein gene expression and kernel texture	Babu et al. (2013)
CRISPR/Cas9	*Zmzb7*	Albino phenotype	Marker gene for MEP pathway in chloroplasts	Feng et al. (2016)
CRISPR/Cas9	*LIG (liguleless1)*	Plant architecture and structure	Ligule development and growth	Svitashev et al. (2015)
CRISPR/Cas9	*SH2*	Endosperm	Modify nutritional content and supersweet maize development	Dong et al., (2019)

(continued)

Table 12.1 (continued)

Gene-editing tool	Target Gene	Anatomical expression	Function	References
CRISPR/Cas	*WX*	Endosperm, Pollen grain	Waxy maize development	Dong et al., (2019)
CRISPR/Cas	*ALS1 / ALS2*	Leaves, roots, kernels	Enhances maize resilience to environmental stresses, kernel quality, herbicide resistance	Svitashev et al. (2015)
CRISPR/Cas9	*ZmBHLH51, ZmBHLH122, ZmTGA genes (9-1/2/3, 10), ZmMYB genes (84, 33-1/2), ZmPHD11, ZmLBD genes (10/27)*	Transcription factors involved in anther development	Male sterility, fertility	Jiang et al. (2021)
CRISPR/Cas9	*MS26, MS45, MS8*	Transcription factors involved in anther development	Male sterility, fertility	Svitashev et al. (2015)
CRISPR/Cas9	*ZmGA20ox3*	Plant length and height	Improve plant wellness, gibberellin biosynthesis, semi-dwarf phenotype	Zhang et al. (2020a)
CRISPR/Cas9	*stiff1*	Stalk, stalk strength	Plant maturation, lodging resistance	Zhang et al. (2020a)
CRISPR/Cas9	*ZmBADH2a*	Seeds, grains	Improves grain quality	Wang et al. (2021)
CRISPR/Cas	*ZmBADH2b*	Seeds, kernels,	Improves grain quality	Wang et al. (2021)
CRISPR/Cas	*ZmPHYC1*	Flower, different tissues of plant	Plant development, Early flowering time	Li et al. (2020)
CRISPR/Cas	*ZmPHYC2*	Flower	Supports the functions of flowering, Early flowering time	Li et al. (2020)
CRISPR/Cas9	*ZmROP9*	Root, leaves	Cell signaling, plant development, altered root development via knockout	Facette et al., (2015)
CRISPR/Cas9	*ZmPL1*	Roots, shoots, overall plant	Improves drought stress tolerance in maize	Wang et al., (2025)

(continued)

Table 12.1 (continued)

Gene-editing tool	Target Gene	Anatomical expression	Function	References
CRISPR/Cas9	*gl2 (glossy2)*	Epicutilcar wax	Plant architecture and Epicutilcar wax formation	Lei et al. (2019)
CRISPR/Cas9	*ZmARGOS8*	Leaves, stem	Plant growth and development	Shi et al. (2015)
CRISPR/Cas9	*ZmGA3ox2*	Stems, leaves	Gibberellin biosynthesis, plant growth, and regulation enhanced via knock in	Yang et al., (2025)
CRISPR/Cas9	*ZmSWEET4c*	Kernels	Sugar transport, kernel development, impaired kernel development via knockout, improved yield, stress tolerance, and nutritional quality	Sosso et al. (2015)
CRISPR/Cas9	*ZmNAC78*	Kernels	Plant development, iron fortification, biofortification	Yan et al. (2023)
CRISPR/Cas9	*ZmACOS5-2*	anther and pollen	Anther and pollen development	Liu et al. (2022)
CRISPR/Cas9	*ZmDCL5*	Kernels	Temperature-sensitive male sterility.	Teng et al. (2020)
CRISPR/Cas9	*ZmBADH2a, ZmBADH2b*	Grain	Improved Grain quality, developing aromatic maize	Wang et al. (2021)
CRISPR/Cas9	*ZmPHYC1, ZmPHYC2*	Flower	Photoperiod sensitivity, flowering time, early flowering time	Li et al. (2020)
CRISPR/Cas9	*ZmMS45*	Anthers	Male fertility, hybrid breeding	Zhang et al. (2021)
CRISPR/Cas9	*ZmbHLH51*	Anthers	Plant development, Improved maize breeding techniques by introducing new genes (GMS) Genic Male Sterility of male sterility and fertility	Jiang et al. (2021)
CRISPR/Cas9	*ZmEREB92*	Roots, shoots	Ethylene response, plant development, gene mutagenesis	Fu et al. (2023)
CRISPR/Cas9	*ZmABCG26*	Anther	Anther development and male fertility	Jiang et al. (2021)
CRISPR/Cas9	*ZmCOI2a*		Male fertility, Jasmonate signaling receptor	Qi et al. (2022)
CRISPR/Cas9	*ZmPOD65*	Seed set, plant vigor, and fertility	More breeding possibilities	Wang et al. (2022c)

(continued)

Table 12.1 (continued)

Gene-editing tool	Target Gene	Anatomical expression	Function	References
CRISPR/Cas9	*ZmPLD3*	Seed set, plant vigor, and fertility	Creation of multiple haploid inducers	Wang et al. (2022c)
CRISPR/Cas9	*ZmPLA1*	Seed set, plant vigor, and fertility	Haploid induction	Wang et al. (2022c)
CRISPR/Cas9	*ZmDMP*	Seed set, plant vigor, and fertility	Maternal Haploid Induction, haploid inducer line development	Wang et al. (2022c)

compatibility, and selectively cuts double-stranded DNA. Double-strand breaks activate nonhomologous end joining (NHEJ) as well as homology-directed repair (HDR) mechanisms (Fig. 12.1). In the absence of a homologous repair template, NHEJ frequently results in random base pair inserts (insertions or deletions) that damage the target sequence. Alternatively, an exact genome edition can be accomplished by using a donor DNA template and a repair method. Cas proteins, including Cas9 and Cas12 have been used in this mechanism to behave as molecular cutters, cutting DNA at specified sites. This focused cutting allows for exact modification of genes such as deletion, addition, or alteration (Asmamaw and Zawdie 2021).

The Cas9-guide RNA ribonucleoprotein complexes identify comparable PAM sequences and unwind the desired dsDNA, forming an RNA-DNA heteroduplex. The process of CRISPR/Cas9 is presented in Fig. 12.1. The guide RNA (gRNA) spacer hybridizes with a complementary DNA target strand while displacing the non-target strand, creating a single-stranded 'R-loop' that triggers nuclease activity. Discrepancies among the guide RNA spacers and target strand impair the Cas9 protein complex's changes in conformation, limiting activation. Cas9 activates the nuclease and hydrolyzes the phosphodiesters of DNA backbone via two different domains. The RuvC enzyme region chops the nontarget helix of DNA with PAM, whereas the HNH nuclease domain cuts the desired DNA template tethered by gRNA (Tariq et al. 2023).

The Cas12 nucleases differ significantly in comparison to the type-2 CRISPR/Cas mechanism. They have just one RuvC-like domain, which cuts both chains of the targeted DNA. Cas12a, additionally referred to as Cpf1, was the initially extensively utilized Cas12 nuclease for genetic engineering purposes. Cas12a enables multiplex modification of genes by using a specific RNase domain to handle single crRNAs. Cas12 variations, notably Cas12a, recognize target sites, activate, and cleave ssDNA or RNA. Cas12 effectors, including Cas12g, may cleave either DNA or RNA when activated. Cas12a, unlike Cas9, detects the T-rich PAM pattern and matures its own guiding RNA. This results in asymmetrical 5′ and 3′ end breaches

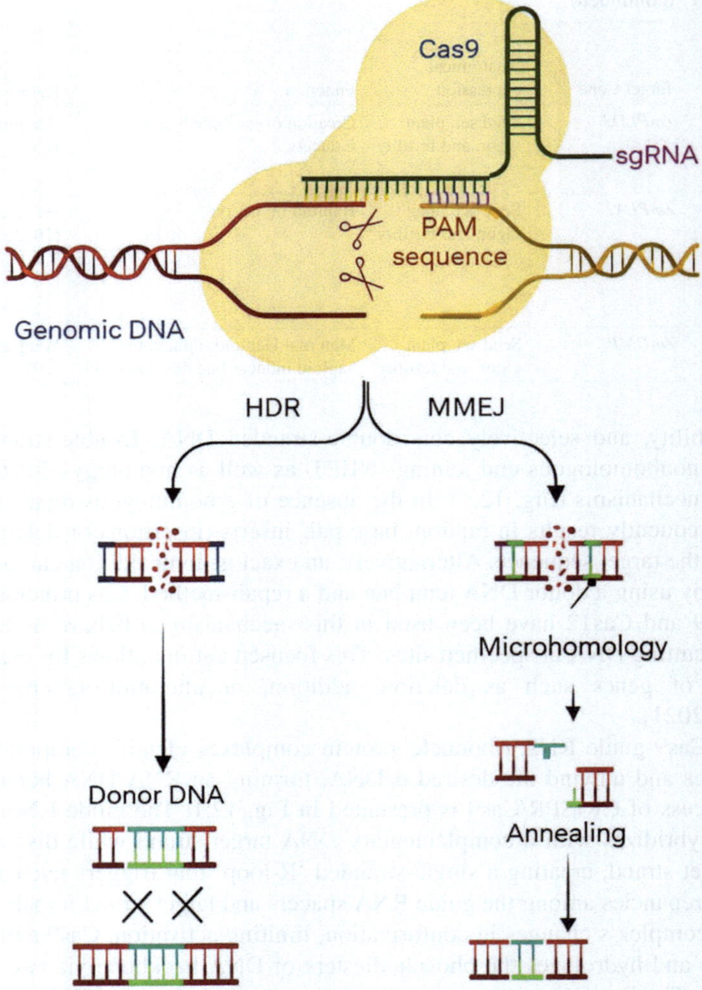

Fig. 12.1 Mechanism of gene editing. DNA double strand break can be repaired by HDR (Homology Directed Repair), and MMEJ (Micro-homology mediated end joining) mechanisms. MMEJ relies on short regions of microhomology for repair. It uses microhomologous sequence (2–20 bs) for repair after end resection but NHEJ doesn't need homology to ligate broken DNA ends. HDR occurs in the presence of a homologous donor template

in PAM distant dsDNA. Cas12b modification of the *GhCLA* gene in the cotton genomes caused in high-temperature tolerance. The LbCpf1 or LbCas12a genomic modification tool was employed to create DSB with 5′ overhangs (sticky ends) distal to the PAM site and replaced wild-type ALS along with mutant ALS to achieve resistance to herbicides in rice (*Oryza* sp.). LbCas12a is employed in rice to induce particular mutations in *Xa13* gene, resulting in resistance to bacterial leaf disease. Cas12-based GE mechanisms have the ability to increase crop resiliency (Janik et al. 2020).

12.2.3 CRISPR/Cas9 for Genome Editing in Maize

CRISPR/Cas9 technology has been identified as an effective method for improving maize nutrient value, digestibility, and taste. The increasing number of transgene-free plant types contribute significantly to the public's views, criticism of gene-edited crops, and food security regulatory issues. This is important to the widespread acceptance and success of GM crops. This approach shows significant promise for enhancing the quality and dietary value of sorghum, a key crop globally (Li et al. 2024). In maize, using CRISPR/Cas9 to knock out the Wx gene enabled the formation of 12 elite inbred lines of waxy maize variants (Gao et al. 2020), with an over 100% boost in amylopectin concentration in these maize grains. Waxy maize offers various benefits, notably greater palatability because it tastes excellent and is puffy; it is additionally used extensively as a thickener in a variety of cuisine dishes due to its peculiar texture. Another advantage of waxy maize is its early harvest, which might lead to food poverty. CRISPR/Cas9 can target and activate genes associated with the production of volatile compounds, which add to the aroma.

Such as aromatic maize was successfully produced by deactivating the betaine aldehyde dehydrogenase (*BADH2*) genes via genetic modification (Wang et al. 2021). One of study's primary goals is to increase crop production, and yield-related characteristics typically involve using parameters that include weight of grain, grain size, panicle size, and grain amount. Knocking down the *ZmCEP1* gene enhanced plant height, ear length, kernel size, and 100-kernel weight. The delivery of sugar and nitrite to the kernel is managed by ZmCEP1 gene (Xu et al. 2021a, b). Also, the angle of leaf orientation is a vital agronomic feature in cereals such as maize, rice, and sorghum, as the crop yields and plant number are regulated by it. The biotic and abiotic stress and also the climatic changes have a significant effect on crop yield, such as maize and sorghum (Njeru et al. 2023).

With the help of CRISPR/Cas, the genes for stress resilience can be modified, which include *ZmAbh4* (Blankenagel et al. 2022), *ZmGDIα* (Liu et al. 2022a, b, c, d), *ZmLOX3* (Pathi et al. 2020), *ZmHKT1*, *ZmALS1*, and *ZmALS2* to produce plants with high abiotic stress resistance such as salinity, drought, cold, and herbicides (Table 12.1). This potential of CRISPR-edited crops is highly beneficial as these changes are able to increase stress resistance with high yields when applied in properly established and high-producing cultivars. The genome of that specific variety remains intact as well. QTLs that are mainly made from multiple genes regulate abiotic stresses (Snehi et al. 2024). *ZMHKT1* mutants in corn were made by CRIPR/Cas, which increased root-to-shoot Na + transport and resistance to salt (Jiang et al. 2022a, b). ZmHKT1 is the main salt-resistant QTL that was successfully recognized as a sodium selective carrier (Wang et al. 2024a, b). The maize stiff1 gene expresses an F-box domain proteins, and the CRISPR/Cas9 knockdown plant (Cstiff1) has a stronger stalk than the unmodified standard. Several varieties of maize resistant to plant diseases such as *Ustilago maydis* and Fusarium graminearum have been created, resulting in increased biotic stress tolerance and maize yield. Studies exploited CRISPR/Cas9-mediated knockdown technique to develop maize genetic male-sterility (GMS) and sensitive to temperature male-sterility (MS) lines (Table 12.1; Farooq et al. 2024). Furthermore, viable male sterility maize mutants with Mendelian genetic principles were produced using CRISPR/Cas9 vectors (Liu et al. 2022a, b, c, d). Further RNA-seq approach combined with CRISPR/Cas9 mutation has successfully identified new maize GMS genes with redundant functions (Jiang et al. 2021a, b). Haploid plants have a single set of chromosomes, which when doubled can result in homozygous diploid plants (Yali 2022).

To speed up breeding actions, CRISPR/Cas9 has been used to knock down genes for functional genomics. In maize, some main genes are effective for haploid breeding activators which include *ZmPOD65*, *ZmPLD3*, *ZmDMP7*, and *ZmMTL* (Table 12.1). The way of efficient haploid identifier marker and development of double haploid crops can be possible by the removal of *ZmPLA1* gene in maize (Table 12.1; Jacquier et al. 2021). Also, the new research has suggested that haploid induction approach in maize can be applied to other crops like rice, sorghum, and fox tail millet. This is achieved by larger similarity of sequence between maize and sorghum, which is about 91% and 73% between maize and millet (Liu et al. 2022a, b, c, d). The combined effect of both *ZmPLA1* and *ZmDMP* genes was able to increase the rate of haploid inducers. Their similarities imply common tasks, which will aid future investigation into the molecular processes of haploid induction (Song et al. 2024).

12.2.4 Development of CRISPR/Cas9 Toolkits for Maize

The CRISPR-Cas toolbox has developed significantly since its introduction, with numerous components built for a variety of uses. In maize, popular techniques include using the *Streptococcus pyogenes* (SpCas9) enzyme to target coding areas and induce loss-of-function mutations. Efficiency of editing was evaluated by changing the promoters driving the nucleases and the gRNAs as well as the codon use of the nucleases. The maize Ubiquitin1 promoter has been one of the most widely used promoters to drive Cas9 expression (Gondalia et al. 2025) with reported mutational frequencies of monoallelic and biallelic mutations around 60–70%. Though highly efficient, constitutive expression of Cas9 can also lead to high somatic editing rates, which could complicate the identification of heritable mutations (Rahman et al. 2022). Thus, other tissue specific and/or inducible promoters have also been tested such as the promoter of the maize meiotic gene dmc1, pollen specific and callus specific promoters (Chamness 2023). The latter presenting high inheritability of induced mutations with low rate of somatic edits compared to maize Ubiquitin1 promoter. For sgRNAs, U3 and U6 promoters have been tested, with the former being the most efficient choice (Reza et al. 2024). Besides the use of distinct promoters to drive Cas9, also a Zea mays codon-optimized Cas9 (zCas9) has also been developed which displayed improved performance over the human optimized version (Aesaert et al. 2022). Also new nucleases were identified as alternatives to Cas9 such as Cas12a. Cas12a has a T-rich PAM (beneficial for targeting T-rich genomic regions) and creates double-strand breaks with staggered ends, which offers an advantage for gene insertion by NHEJ (Liu et al. 2024). Evaluation in maize has shown that, though not as efficient as Cas9, Cas12a can be effectively used for gene editing. Base editors are an amazing modification to Cas enzymes that allows for the selective substitution of one nucleotide for another (C-G to T-A and A-T to G-C) without the usage of a donor template or inducing double-strand breaks (Rallapalli and Komor 2023). Using an iterative improvement platform (ITER), extremely efficient Cas12a base editors were developed and analyzed in maize and wheat protoplasts (Gaillochet et al. 2023). Furthermore, nuclease deactivated Cas9 (dCas9) toolkits have been created for CRISPR-mediated transcription inhibition or activation (Cai et al. 2023).

One notable progress that is not focused on the system itself is the ability to reliably anticipate the efficiency of a gRNA at the start of studies. Multiple studies have successfully used machine learning models to anticipate the efficiency and even the most prevalent repair products of various gRNAs (Konstantakos et al. 2022). Because recognition of the protospacer adjacent motif (PAM) is needed for CRISPR-mediated gene editing, researchers have developed Cas9 variants with changed PAM detection (Zhao et al. 2023).

12.2.5 Early Applications of Cas9-gRNA Technology in Maize

CRISPR/Cas9 subsequently ruled the discipline of molecular biology. CRISPR, which is employed in a variety of plant and animal species, has evolved into an indispensable toolkit for guided manipulation of genes, either for gene interruption, substitution, or greater complexity. While CRISPR/Cas9 has numerous and stimulating applications, it is most commonly used in plants as a tool for editing genes to create functional loss knockout alleles. The emphasis will center on practical concerns for applying CRISPR in gene disruption as a resource for functional genetics in plants, specifically maize (Rasheed et al. 2021).

12.2.6 Targeted Knockout and Insertion

Targeted gene insertion has long been regarded as the complex method of modifying genomes in crops, owing to its dependency on numerous critical parameters. These include the efficacy of transforming and regenerating procedures, the efficient transmission of genetic editing elements, the copy count and access of the repair template at the double-strand break (DSB) site, the extent of the genetic code to be implemented the availability or lack of selectable marker genes, and the effectiveness of homologous directed repair (HDR) at a specific target site (Chen et al. 2022a, b, c). Furthermore, gene insertion experiments usually fail because of the integration process's intricacy, plant chimerism, and the challenge of separating intended modifications from helper genes like Cas9, gRNA, and selectable markers.

There are many methods developed to address these issues, which involve the use of particle bombardment and Agrobacterium-mediated transformations to improve the efficiency of genetic engineering. Also to reduce nonhomologous end joining (NHEJ) and effective delivery of editing elements as DNA, RNA, or RNP complexes with activating HDR pathways, the usage of viral replicons to increase the host copy number, and integration of cyclical and linear repaired templates are used. Still the effective, affordable and targeted trait insertion in crops is elusive.

Due to the requirement for a high donor template copy number, most targeted insertion experiments have used particle bombardment for delivery. While this method ensures a high copy number of templates, it frequently results in complex, often unusable integration patterns and can lead to chromosomal rearrangements, making it highly invasive (Wang et al. 2022a, b, c, d). In contrast, Agrobacterium-mediated transformation is chosen for its reproducibility, cost-effectiveness, broader species applicability, and lower attrition rates during regeneration (Thingnam et al. 2023). However, its primary limitation lies in the small number of T-DNA molecules delivered, resulting in reduced frequencies of targeted gene insertion. For example, studies, including unpublished data from Barone and Svitashev, reported targeted insertion frequencies of less than 0.1%, and other in planta experiments showed similarly low efficiencies, such as 0.15% in *Arabidopsis* and 0.085% in maize.

HDR occurs primarily in the late S, G2, and early M stages of cell cycle, promoting cell division through morphogenic genes supports an HDR-conducive environment. Further improvements in Agrobacterium-mediated transformation were achieved by employing a novel ternary vector containing an improved additional plasmid (pVIR9), which enhanced T-DNA transport and stable callus recovery, improving transformation efficiency by 6- to 7-fold compared to traditional systems (Tang et al. 2023). The integration of morphogenic genes and the optimized vector system significantly improved HDR-based insertion frequencies—up to 3% from less than 0.1%, thus making Agrobacterium-mediated targeted insertion not only viable but also acceptable for certain crop species (Wang et al. 2023).

Another innovative strategy includes surrounding the repair strand with desired regions in T-DNA vectors, which enhances the efficiency of HDR-based insertion. This design allows the linearization of the repair strand, facilitates template accessibility at the DSB region, and simplifies homologous recombination processes (Čermák 2021). This method also improved the quality of edits, evidenced by higher proportions of perfect insertions, similar efficiency with or without selectable markers, and stable Mendelian inheritance in T1 progeny(Sun et al. 2024).

12.2.7 Delivery of RNP Complexes for Precise Editing

Transferring ribonucleoproteins into a cell with no DNA is an alternate way for producing transgene-free targeted genome alterations (Kocsisova and Coneva 2023). DNA-free delivery is another objective in vegetatively propagated crops when crossing is not an option for breeding out transgenic insertion. Genetic modification using DNA-free administration with RNP might mitigate concerns about regulation regarding transgenic plants. Editing has been demonstrated in various plant species by transfecting protoplasts with built-in Cas9 RNP components (Yau et al. 2024). Despite the complexity and lengthy timeframe involved in regeneration crops from protoplasm, regular utilization of protoplasts alongside RNP transportation for the generation of steady altered lines is yet to be investigated in commercially major crops such as cotton, corn, soybeans, and the wheat (Zhang et al. 2021a, b, c). In contrast, several of these crops contain widely recognized extremely effective biolistic transforming mechanisms, and the efficient generation of transmissible changes is being performed using CRISPR-Cas9 in vitro transcription or RNPs provided with the biolistic technique (Chu and Agapito-Tenfen 2022).

12.2.8 Optimization of Promoters for CRISPR/Cas9

Promoters are vital in directing the expression of edited genes: in addition to the most often utilized 35S promoter of the constitutively cauliflower mosaic virus (CaMV), many other promoters are being utilized for expressing Cas9 and enhancing its modification performance (Khan et al. 2023). In 2018, Hashimoto et al. used the tomato elongation factor-1α (SlEF1α) promoter to activate Cas9 and enhance gene editing efficacy in the tomato genomes (Shao et al. 2024). In 2019, Bai et al. used the strong intrinsic promoter Progmscream M4 to activate the Cas9 protein (Freitas-Alves et al. 2024). In 2019, Kishi-Kaboshi et al. discovered that the parsley ubiquitin (PcUbi) promoter outperformed the CaMV35S promoter when employed in chrysanthemum breeding (Kishi-Kaboshi et al. 2023). In 2020, Wolabu et al. evaluated the influence of four separate Arabidopsis promoters and discovered that Cas9 activated by the Arabidopsis thaliana UBQ10 promoter increased gene editing efficacy by 95% (Gondalia et al. 2025). Four distinct RNA polymerase (Pol) III promoters (TaU3p, TaU6p, OsU3p, and OsU6p) in rice were revealed by (Li et al. 2020) based on a protoplasm transformation and found that the optimized sgRNA scaffolding stimulated by the TaU3 promoter was much more effective (Zhou et al. 2023). To activate Cas9, An et al. in 2021 exploited mannan synthase (MAS) promoter that accelerated the rates of mutation on edited side by approximately 75 percent (Pant et al. (2022). As compared to constitutive promoters, germline specific Cas9 promoters can spike the rate and reproducibility of crops while avoiding the somatic clone generation (Gondalia et al. 2025).

12.2.9 Multiplex Genome Editing in Maize

Multiplex genome editing method is used to express multiple sgRNAs packed with tRNA in polycistronic genes by only one RNA polymerase 3 (Pol3) promoter and Pol3 terminus (Xie et al. 2015). It enables a precise mutation in a genome at multiple DNA loci. CRISPR/Cas-based MGE technologies provide precise mutations at numerous loci in a gene. MGE is revolutionizing plant molecular sciences and crop breeding by allowing precise genome modification at single base-pair level, which was previously impossible with tools like zinc-finger nucleases and TALENs (Abdelrahman et al. 2021). By combining the proper alleles, genetic redundancy was reduced in corn. The capacity to do several gene manipulations in one plant is effective for creating intricate characteristics. The CRISPR/Cas9 has the benefit of multiplexing, which enables the simultaneous mutations of many genes. Cas9 protein expression can be coupled with several guide RNAs (gRNAs) with different spacer sequences to target various genes and gene families (Impens et al. 2023).

The Cas12a-based multiplex technology offers significant benefits over existing DNA editing techniques. In Cas12a system, crRNAs use their ribonuclease activity, eliminating the need for multiple Pol III promoters or specific sequences. Cas12a's qualities make it an appropriate system for MGE. Cas12a cleaves the DSB at cohesive ends away from the PAM site, allowing for continued DNA cleavage. Swarts and Jinek conducted a detailed comparison of Cas9 and Cas12a structure-function connections and their implications for genome editing. Cas12a was discovered to be very particular in plant MGE via whole-genome sequencing. Lee et al. (2019) observed similar results with CIRCLE-seq (Abdelrahman et al. 2021).

Base and prime editors enable an accurate gene editing in a living cell while avoiding double-strand cuts. DNA base editors (nCas9, D10A) employ deaminases for effectively transforming base pairs, such as C•G to T•A by CBEs1, A•T to G•C by ABEs2, C•G to G•C by CGBEs4, and dual-deaminase base editors. To accomplish targeted insertions, deletions, and point mutations, prime editors use nCas9 (H840A) and an edited reverse transcriptase (RT) regulated by prime editing gRNAs (pegRNAs). Base editing (BE) and prime editing (PE) are both employed in research and in many cell types and models of living organisms.

Editing many genomic regions with BEs or PEs enables complex functional genomics research and polygenic illness therapy. However, BE and PE have limited multiplex editing capabilities. Existing multiplex techniques need huge expression structures, leading to delivery costs. Pooling numerous gRNAs is a simple approach; however, it is not suitable for viral delivery in vivo. Cas12a is capable of maturing gRNAs, making it useful for gene knockouts, transcriptional control, and genetic modifications. Cas12a, unlike Cas9, does not have a nickase variant that only breaks the non-base-edited portion of DNA to maximize BE efficiency. However, Cas12a-based PEs are still being effectively constructed. Cas9 protein, often utilized in BE and PE, are designed to locate protospacer (PAM) adjacent patterns, edit more efficiently, and have excellent specificity.

Four lycopene metabolism genes were targeted using a multiplexed CRISPR/Cas9 technique, including SGR1, LCY-ε, BLC, LCY-B1, and LCY-B2 are lycopene β-cyclases. The triple-deleted mutant variants had a 2.5 times rise in lycopene and β-carotene contents in tomato fruit compared to the wild-type varieties. Zhang et al. (2020a, b) altered three essential genes in soybeans that encode isoflavonoid biosynthesis enzymes, including flavanone-3-hydroxylase 1 (GmF3H1 and GmF3H2) and flavone synthase II (GmFNS II), increasing isoflavone content by twofold compared to wild-type soybeans.

Thus, the multiplex CRISPR/Cas9 tool allows simultaneous modification of many genes on the same or other chromosomes, as well as the deletion of huge chromosomal segments including gene clusters. As a result, given the presence of metabolite gene clusters across a wide range of plant species, this method is an excellent choice for manipulating metabolic pathways. Dong et al. (2019) demonstrated that numerous genes may be inserted to increase the generation of beneficial metabolites without compromising other crop attributes. However, due to the poor effectiveness of effective insertion for many genes, this method has had minimal effect in plants. Leveraging the conserved PAM sequence of homologs, sgRNA-based CRISPR/Cas9-mediated genome alteration might target several homologs of the desired gene, additionally in polyploid varieties of plants (Bhambhani et al. 2021). Multiplex genetic modification with the CRISPR/Cas system is extremely beneficial since it is expected to significantly improve agricultural genome engineering and precision breeding (Saeed et al. 2025). Multiplexing is carried out primarily by inserting multiple number of guides RNAs (gRNAs) in vivo through multigene cassettes gRNA production, Csy4-based removal, ribozyme-flanked gRNAs, or tRNA-based gRNA fragmentation. Amongst this, the tRNA-based technique was extensively analyzed in crops using just one transcript and showed high editing efficacy (Gong et al. 2021). But its overlong tRNA-based units, difficulties in vector creation, and inadequate effectiveness of plant transformations hindered its capacity to do multiplex genome editing.

This allows the quick creation of genetically homogenous lines with multiple homozygous mutations in three generations (Impens et al. 2023). The identification of 36 genes that increase the growth of leaves, the study has made various varieties with a consistent 10% raise in size of the leaves (Vanhevel et al. 2024). This method is important for functional studies and trait improvement in corn that significantly reduces the time and effort need to generate high-order mutation. The BREEDIT technology uses multiplex CRISPR-based editing and selective crossing strategies to analyze and improve complicate traits in crops. By using this approach, the study has developed various mutant variations by recognizing 48 growth-regulating genes with the help of 12 guide RNA constructs (Lorenzo et al. 2022). Successful cross of lines containing variety of mutations allowed for the discovery of effective combinations of alleles influencing characteristics such as plant architecture and yield. This technology bridges gaps between conventional breeding and modern genome editing, giving a useful tool for sustainable agriculture (Buch et al. 2023).

12.2.10 Challenges with Stable Integration and Transgene-Free Techniques

The advanced genome editing methods have come to the point where transgene-free GE crops can be produced in just one generation. This is efficient for producing transgene-free mutants in vegetative propagative crops such as corn. Also, it removed the need for meiotic recombination-based identification of transgene-free plants in future generations (Qamar et al. 2024). Virus-induced genome editing (VIGE) is one method to develop transgene-free GE plants. Both DNA and RNA viruses have been used to deliver gRNA and Cas endonucleases into the tissues of plant, which can be carried out in a couple of ways: by inserting the gRNA into a plant cell that already has a Cas9 in its genome, or by simultaneously inserting the gRNA and Cas9 into the plant genome (Mahmood et al. 2023). Virus-induced genome editing (VIGE) has already been utilized in crops such as maize, wheat, rice, soybean, and tomato, deploying both DNA (e.g., BeYDV, WDV) and RNA-based viruses (e.g., TRV, PVX, TMV). VIGE offers high gRNA expression, avoids tissue culture, reduces off-target effects, and can generate transgene-free plants (Gentzel et al. 2022). However, achieving heritable edits is difficult due to poor virus movement to meristems—this can be improved using mobile RNA sequences like FT. Biolistics (particle bombardment) is also used to deliver plasmids or RNPs directly into plant cells. While plasmid delivery may create GMOs, using RNPs avoids foreign DNA integration and has been successful in maize, wheat, and rice. Agrobacterium-mediated transient expression is another method to produce transgenic free edits in plants like tobacco and tomato (Yan et al. 2022).

Despite transgene-free methods for editing genomes have various benefits over transgenes, multiple variables, such as a long lifespan, complicated genome with fluctuating chromosome counts from 80 to 130, large polyploidy with numerous alleles of only one gene, variability, and an enormous genomic size, make it challenging to design an effective genetic engineering system in sugarcane (Zhang and Yang 2024). The accessibility of well-annotated entire genome sequence databases permits the implementation of gRNA designing softwares, which include CHOP-CHOPv3, CRISPR-P 2.0, and numerous others to choose gRNAs that target specific sections of the genomes with high editing efficacy and minimal off-target site interactions (Surya Krishna et al. 2023). Though reference genetic sequences are accessible, they should not be utilized as a direct reference for building gRNAs. Before designing gRNA, clone and analyze the desired gene sequence from the plant selected for genomic modification. This technique assists to eliminate nucleotide incompatibilities within gRNA and the desired sequence while also ensuring proper gRNA binding at the target location. Maize has less successful effectiveness in transformation when compared to other plants, prompting the development of efficient maize transformation processes (Chen et al. 2024). Despite the many benefits of transgene-free techniques, the limited editing efficacy and lack of pressure to select for detecting transgene-free GM crops initially complicate and raise the expense of the procedure. Yet, these constraints could be solved by utilizing visual tags, such as pigments and fluorescent protein markers, to identify and eliminate transgene-containing sugarcane crops at a young stage, while also limiting the amount of plants required for subsequent sequencing and evaluation (Nivya and Shah 2023).

12.2.11 Improvement of Hybrid Breeding Techniques

In the past few years, there has been significant focus on reducing the duration of breeding while increasing cost efficiency. The DH process is an important method for rapidly producing homogenous and homogeneous offspring. It is better than traditional breeding because it addresses the problem of slow heterogeneity decline during the initial segregated generation (Lamichhane and Thapa 2022). DH progenies are homozygous lines, making them appropriate for selecting and checking within a wide range of environmental settings. DH has a benefit over F2 as it produces homozygous populations faster and backcross population because of repeated data across time and periods (Isnaini et al. 2023).

DH is an important technique for mapping populations and generating genetic maps for different crops (Qu et al. 2024). The HIR in maize haploid inducers can be increased by systematic QTL analysis of four main qualitative trait loci such as qhir1, qhir8, qhir11, and qhir12 (Khammona et al. 2024). The purpose of DH populations is to map different traits such as ease of digestion, agronomic properties, and stress of biotic and abiotic factors. In 2019, Zhang et al. identified three major QTLs, qSD6–2, qSD8–2, and qSBS1–2 that contributed to over 10% of the phenotypical variation in stalk lodging-related characteristics. The mapping investigation has been improved by using SNP markers due to discovery of affordable sequencers as mentioned in recent studies.

Marker-assisted cross-breed (MABC) inserts desirable genes into an agronomically important dominant line by using a molecular marker-based background and bypassing selection methods (Gupta et al. 2022). To maintain or set the intended gene in the repeated parent's DNA, a large number of recipients are needed. In the last phase of MABC, the stabilization of the targeted gene requires a reduced DH populations. Chaikam and his colleagues in 2018 used marker-aided insertion (flaking markers bnlg1811 and umc1917) to create 2GTAILs by inserting qhir1 via TAILs into exceptional tropical maize inbreds. One of these introduced lines, 2GTAIL006, emerged to be agronomically better to TAILs, with an overall HIR of 13.1% of the total (nearly 50% higher than the TAIL baseline).

As a result, MABC in combination with DH technique might help in rapid gene pyramiding (Khammona et al. 2024). The DUS evaluation, which measures distinctness, consistency, and rigidity, was carried out utilizing The International Union for the Protection of New Plant Varieties' (UPOV) PVP system. The result of this experiment reveals the unique traits of one variety over others that will remain intact following multiplication (Kock 2022). Smith et al. (2008) examined the genetic variables of both real and synthetic populations created using the Single Seed

Descent (SSD) and DH procedures during the first and following generations. Experimental data showed that DH progenies acquired more sections of intact parental chromosomes compared to SSD progenies. They also discovered that it is doable to choose DH progenies that are over 90% similar to both primary maternal lines (Santos et al. 2022).

12.2.12 Genome Editing for Linkage Drag Reduction

Trans genome modification techniques such as the desired-target mutator (DTM) (Yassitepe et al. 2021), haploid induction (HI) editing technological advances (HI-Edit; Li et al. 2025a, b), and haploid-inducer mediated genome editing system (IMGE; Hooghvorst and Nogués 2021) have the potential to speed up the creation of modified maize variations. Such strategies foster the direct transfer of the CRISPR/Cas system to transformation-recalcitrant lines by pairing superior inbred varieties against a stable transformed type carrying a CRISPR/Cas make up, enabling CRISPR/Cas to sustain active and create new mutant alleles.

DTM was first used to generate new edited variants of the LIGULELESS1 (*LG1*) gene in corn (Qin et al. 2023). Breeding T1 plants having CRISPR components along with six distinct maize strains produced progeny with around 20% changed alleles. As a way to breed, this procedure requires a number of backcrosses using marker-assisted selection to assimilate the new alleles into an elite line suitable for hybrid production. But this strategy overcomes the linkage drag effects and uses smaller population number for recovering the recurring genome, which increases the level of accuracy and duration necessary to insert a novel allele within elite strains (Singh et al. 2022). DTM was also employed to induce ZmWX knockout mutations, resulting in hybrids with high endosperm amylopectin content (Singh et al. 2022). Although the DTM strategy enhanced genetic modification methods for maize cultivation, particularly through minimizing the linkage drag impact, it keeps requiring several backcrosses to incorporate the new modified allele into an elite strain (Zhou et al. 2022).

12.2.13 Doubled Haploid (DH) Technology in Maize Breeding

Maize is one of the top three grains for global food security, with a significant presence in Africa and America and widely used for feed, food, biofuel, and industrial applications around the globe. Employing doubled haploid technique in maize requires immediate action for addressing the growing requirements of maize items. In compared to classical inbred lines development techniques, this technology significantly shortens the breeding cycles through swiftly creating completely genetically homogenous lines of breeding in 2–3 generations (Gupta et al. 2022). Haploid (n) cell double their chromosomes, either artificially or spontaneously, resulting in two entirely identical gene sets. Artificial DHs are able to generated both in vivo and in vitro.

The haploid induction-based DH approach for breeding of corn is significantly regarded a valuable toolkit for improving breeding effectiveness. The importance of DHs in genetic study soon became apparent and research in different methods causes of increasing haploid generation rate initiated (Grüning et al. 2025). In maize cultivation, various haploid approaches for producing and doubling haploids are being developed. When haploid (n) cells go through synthetic or natural chromosomal doubling, they produce two distinct gene sets, yielding the Doubled Haploid (DH) genotype (Mitiku 2022). In accordance with Barnabás et al. and Maluszynski et al., DHs can be intentionally created at varying frequency using in vivo or in vitro methods. Repetitive selection (RS) is an important aspect in the development of hybrids. A superior line selection could be achieved and utilized to generate hybrids that outperform prior rounds (Thakur et al. 2024). The goal of recurring selection in hybridization breeding aims to increase a heterotic category's General Combining Ability (GCA) with comparable to others, known by the term "opposite" pool (s). Recurrent selection based Doubled Haploid (DH) lines may prove particularly effective in this context. Adopting DH lines provides several benefits, namely enhanced variation in genetics among testing units and more consistency for estimating the genotypic significance of DH lineages and testcrosses. Doubling haploids are capable of being used in a recurrent breeding mechanism, with the population improving after several rounds of crossing, DH generation, and selection (Hossain et al. 2022). Mutation breeding is another application of DH technology. Homozygosity allows the fixation of changes in the initial generation. Purely haploid crops, as opposed to DH technology, are suitable for cultivation and study. It is related to their potential to produce healthy kernel after diploid fertilization (Qu et al. 2024). In microspore cultivation, chemical mutagenesis of microspores within the uninucleate phase is a highly successful method of making mutations, yielding pure elite mutant inbred lines. Recycling DH lines has been shown to speed up haploid rate and maintain fertilization. A comparative study concluded that firm stalk synthetic substances developed 0.13 percent haploid, whereas haploid-derived DHs created 0.43 percent haploids (Auinger 2022). The preliminary stiff stalk synthetic showed a 9.4% haploid fertility recovery, but the rate of DH-derived haploids went up to 33 percent.

Selections are associated with hereditary or germline concentration in order to create haploids and restore fertility. Random mating of high-producing DH lines as a resource for the next haploid selection cycle can increase the crop's germplasm in terms of productivity, overall health, and agronomic traits. Recurrent selection, or other types of breeding strategy, is often employed to improve inbred lines by introducing improved alleles for production, insecticide resistance, resilience to stress, and several other agricultural traits (El-Esawi 2024).

12.2.14 Haploid-Inducer Mediated Genome-Editing System (IMGE)

CRISPR-Cas has tremendous potential for improving crops, but it confronts hurdles in efficient delivery and regeneration, particularly in monocots like corn (Chen et al. 2022a, b, c). Conventional techniques, such as backcrossing, might produce linkage drag and be time consuming. New technologies for multiplex editing, such as the BREEDIT pipeline's interaction with haploid induction, did show significant improvement (Impens et al. 2023), although their use in elite lines remains limited. To overcome this, innovative approaches such as HI-Edit and IMGE combine gene editing and haploid induction to increase crop efficiency in maize, *Arabidopsis*, and wheat (Li et al. 2025a, b). IMGE technique uses haploid-inducing strains and CRISPR-Cas gene edits to modify genes of interest in target crops, resulting in improved crop yields. This technique increases the introduction of newly gene-edited qualities inside the its intended variety by using swift haploid induction accompanied by a doubled haploid (DH) procedure, eliminating the need for stable gene editing and diverse backcross generations, and putting the concept of "targeted mutations through pollination" into practice (Cao et al. 2025). Maize, the world's major commercially available grain crop, provides an essential concern for plant breeding development. Advances in genetic engineering and hybridization practices have the potential to significantly increase global food production. Despite these hypothetical benefits, there has been little advancement in establishing HI-Edit/IMGE for seed market purposes, notably for critical grains such as maize (Hernandes-Lopes et al. 2023a, b). To create an HI-Edit/IMGE system, either directly modify the inducer or employ a backcrossing technique to combine the necessary components.

Direct transformation, which HI-Edit/IMGE aims to avoid, has only recently demonstrated success in maize (Tian et al. 2024). Delzer et al. (2024) observed on the screening of haploid inducer lines with good genetic modification efficiency for HI-Edit/IMGE construction in maize. Yet, many researchers continue to encounter significant difficulties with direct transformations.

12.2.15 HI-Edit Technology for Elite Lines and Commercial Crop Varieties

HI-Edit integrates two revolutionary plant breeding techniques: haploid induction (HI) with gene editing (GE). In HI-Edit, cross-pollination between any line and a haploid inducer line expressing a GE transgene produces a modified haploid plant with just the non-inducer genome. The potential benefit of HI-Edit is the rapid transmission of novel GE features directly to commercially available varieties via just one cross and the double haploid (DH) technique, which eliminates generations of cross-breed and marker selections. A separate study has presented an analogous technique called inducer-mediated genome editing [IMGE](Hooghvorst and Nogués 2021).

The first HI-Edit study emphasized around NP222RS, which showed a moderate transform frequency (TF), lower haploid induction rates (HIR), and inadequate seed setting. The genetic conditions needed for effective mutation and haploid inducement were never combined into one line (Delzer et al. 2024). To generate a highly desirable HI-Edit line, biparental cross strategies have been developed that integrated the HIR of a current haploid inducer along with the transformability of top stiff stalk and non-stiff stalk variants. In modern maize cultivation, inbreeding and the hybridization are frequently used to preserve good traits and increase hybrid productivity (Resende et al. 2022). However, inbreeding can promote inbreeding depression by revealing recessive deleterious genes, until these alleles are removed by negative selection.

Hybridization can help reduce inbreeding linked to depression through balancing undesirable alleles and diminishing their influence. Surprisingly, the changing development of adverse mutations during breeding cycles is determined by the mutational impacts. Both ILs and individual hybrids acquired some negative mutations, which had been compensated by neutrality or valuable alleles whenever their rates became insufficient in the hybrid. Negative selection eliminated very detrimental mutations, whereas inbreeding made recessive deleterious variations susceptible to negative selection in ILs. The asymmetric abundance of jSFS among distinct heterotic groups of maize ILs, as well as its elevated rates in variants that are heterozygous in hybrids, facilitated the integration of a little adverse modifications, that contributed to the constant enhancement of major traits of agriculture in the commercial division varieties (Sun et al. 2023). HI-Edit/IMGE approaches involve inserting the CRISPR/Cas components inside a haploid inducer line, which is subsequently utilized for pollinating maternal top lineages. The paternal genomes, which contains the CRISPR/Cas9 transgenic material, is removed from the haploid progeny. But in just a few of such haploids, the maternal genome is changed in trans. The haploid progeny are then treated by using a chromosome-doubling agent to create DH strains. The lines are then checked for the presence of subsequent changes, making them instantly monoallelic (homozygous) (Li et al. 2025a, b). For example, genome-edited deletions of the *ZmLG1* and *UB2* genes have been created and traversed with a haploid inductor line, resulting in changed haploid generation with a 3 to 4 percent efficacy (Zargar et al. 2022). In a separate instance, a haploid inductor altered line for the MTL pollen-specific phospholipase gene was developed and used for pollinating additional maize lines, that produced GE double haploids (Impens et al. 2023). Additionally for being a transformation-free technique, the Hi-Edit/IMGE method generates transgene-free improved crops, which is important for marketplaces that are not GM-specific. Additional developments in the Hi-Edit methods are being described, largely to enhance the haploid induction rates and detect seeds with haploid embryo through modifying visual traits like as anthocyanin production (Hernandes-Lopes et al. 2023a, b) or add clear transgene marker to inducer strains (Yan et al. 2021; Xu et al. 2021a, b). Thus, trans gene editing technologies can precisely insert desirable features into the genome of elite line, bypassing the tedious conventional introgression efforts. Therefore, they are able to improve allele variation, that may be advantageous for further breeding actions.

12.3 Applications of Gene Editing in Trait Improvement

With the expected rise in population, the effects of climate shifts, and declining of accessible agricultural place, there should be an urgent need to genetically enhance crops to meet the increasing need of food. Current yield trends are inadequate to feed the global population by 2050, especially in light of persistent climate-related stress, shrinking farmland, and accelerating population growth (Ghosh et al. 2024). Maize, a key staple crop consumed by humans for diet, livestock feed, and a variety of industrial applications. To address the growing global need for food, high-quality cultivars must be developed to increase production (Wang et al. 2023). The purpose can be met by crop's genome improvement, a multidisciplinary strategy which incorporates plant physiology, genetics, as well as biotechnology (Gupta and Salgotra 2022). Modern corn breeding techniques, such as hybrid breeding, mutagenesis, marker-assisted selection, and gene engineering, rely on gene variation to boost agronomic qualities. While these technologies have substantially improved maize improvement, there are worries about the reduction of the genetic basis and the loss of genetic variety. This restriction could prevent future development in enhancing a variety of yield-related traits (Dalakouras and Vlachostergios 2021). Therefore, there is an immediate requirement to discover alternative methods for considerably improving maize yield-related characteristics. Despite genetic diversity, a greater knowledge of how epigenetic alterations influence plant phenotypes has aided agricultural growth (Idehen et al. 2023). Yield and resilience to abiotic stress are complex traits that are extensively impacted by environmental conditions and may include many genetic loci with little effects. This intricacy makes it difficult to understand the molecular mechanisms underlying gene function and properly assess phenotypes, limiting the efficient employment of gene editing technologies to generate superior cultivars for specific traits. While transgene maize varieties with increased insect and herbicide resistance have been commercially marketed for decades, only a few have been created to target complex features (Yassitepe et al. 2021).

The challenge of using genetically engineered approaches to improve complicated features that remain persistent across various settings has hindered the general advancement of synthetic crops (Simmons et al. 2021). Interestingly, new developments in genome modification, notably CRISPR/Cas technology, now enable the simultaneous targeting of many genes. Furthermore, combining editing of the genome with haploid induction expands the potential of genetically engineering challenging traits within a diverse spectrum of germplasms used in plant breeding initiatives. A great number of maize gene editing studies have employed genetic transformation methods applying Agrobacterium as well as biolistic transmission, in particular for a few moderate genotypes that are amenable to Agrobacterium infection and tissue repair. The absence of a universally applicable transformation approach, as well as the relatively poor transformation effectiveness, has impeded the widespread adoption of gene editing in maize. To address these challenges, new advancements have been proposed, such as the use of morphogenic regulators (MRs) and trans genome editing. These advancements hold potential for widespread usage in maize breeding programs, particularly those that incorporate tropical germplasm.

12.3.1 Agronomic Traits Improvement

Key agronomic factors, like yield, are influenced by the tiny effects of multiple genes, making improvement challenging. Optimizing these complex quantitative traits often involves significant and labor-intensive screening and multi-crossing procedures, which can take a long time to generate new top cultivars, even with modern approaches like Genome Prediction. On the other hand, CRISPR/Cas-based multiplex genome editing (MGE) is able to generate new lines rapidly with multiple genomic changes at once. This causes the improvement of better desirable traits in crops and as well as a step forward to more advanced transgene free techniques (Lorenzo et al. 2023).

The editing of both homozygous and heterozygous sequences can be done through multiplex genome engineering (MGE). For instance, one sgRNA can target many conserved regions at once, while multiple sgRNAs can edit different genes or multiple loci of same gene. MGE has become the most finest and powerful genome editing tool for tackling complex characteristics. The simultaneous removal of the *GW2*, *GW5*, and *TGW6* genes reduced the grain weight, yielding new rice lines with 30% greater grain weight is a perfect example of MGE compatibility (Yang et al. 2024a, b). Furthermore, classical crossing to remove the CRISPR machinery produced transgene-free changed offspring, demonstrating the technique's potential for developing innovative agricultural cultivars.

In maize, various MGE approaches are implemented. For example, Qi et al. introduced an standard method based on a tRNA-processing system, enabling the integration of up to four sgRNAs in a single pattern. The intrinsic tRNA-processing mechanism not just effectively handles main transcripts with small expression cassettes, it also appears to increase edited effectiveness, achieving rates between 85.7% and 100%. This improvement can be attributed to the A- and B-box elements within the tRNA sequences, which help insert transcriptional factors (Zhang et al. 2024a, b). This finding is particularly significant for genome editing in maize, where transformation efficiency is often low—making high editing efficiency essential to overcome this major limitation. Additionally, MGE in maize is able to ramped up from a significant amount sgRNAs. For example, vectors including up to 12 different sgRNA expression cassettes are effectively employed to alter Cas9-expressing "editor" corn varieties (Lorenzo et al. 2022).

Researchers analyzed both Cas12a and Cas9 systems for MGE specific to the Opaque2 (O2) bZIP transcriptional factor in maize. Although Cas12a showed poorer editing efficacy in both T0 and T1 generations as compared to Cas9, however, it was able to make a broader spectrum of changes in the T2 generation. Moreover, Cas12a-based editing efficiency was positively linked with nuclease expression levels to indicate the best alternative of multiplex editing in maize (Gong et al. 2021).

BREEDIT is a revolutionary technique in maize that is recently introduced. This technique is also intended to improve complex qualities controlled by various genes such as drought resistance and yield. The BREEDIT approach can transform two populations with multiple sgRNAs sequences, each contains 12 distinct sgRNAs, which are then crossed to gradually produce mutations across an increasing number of genes. BREEDIT helps in insertion of new genes for desired traits as well (Lorenzo et al. 2022). The complex phenotypes in short can be improved by multiplex genome editing (MGE) especially that are controlled by multiple genes. However, the limitations of this technique should be kept in mind. For example, the creating of sgRNA sets to target desired genes inside conserved gene families can be challenging. Furthermore, genotyping several genes may necessitate complex sequencing techniques, like, amplicon sequencing. Integrating multiple sgRNAs into a single vector is able to reduce transformation accuracy. To address the difficulties associated with handling large plasmids, an alternate technique of CRISPR-RNPs, which is generated in vitro by mixing Cas9 protein is useful.

Transgene-free gene editing introduces no foreign DNA into the plant's genome. Maize developing embryos were co-bombarded with sgRNA:Cas RNPs, morphogenic regulators that promote transformation, and a selectable marker (MoPAT-DsRED) to demonstrate effective genome editing using RNP complexes. The positive findings highlight the potential of RNP-based MGE in corn, paving the path for its use in plants that are difficult to regenerate from protoplasts. Other CRISPR/Cas approaches can also create transgene-free cultivars, although they usually need crossing to eradicate the exogenous editing components. Conversely, RNP-based editing is more likely to avoid regulatory hurdles without needing of further breeding stages.

Although biotech varieties of maize show resilience to pests and herbicides have greatly boosted global maize production, improving intricate quantitative traits including drought and heat resistance, absorption of nutrients and efficacy, and productivity is still critical to meet the rising needs of maize-based items. Today, maize breeding systems have access to an unprecedented array of advanced tools that accelerate and enhance the precision of developing elite lines. Such as high-throughput phenotyping, sophisticated statistical and computational strategies, crop modeling, cutting-edge gene sequencing and prediction technologies, as well as emerging genome editing.

12.3.2 Disease-Resistant Maize

Throughout the long history of maize cultivation, the prevalence of diseases has shifted dynamically, with some diseases becoming more or less significant compared to others. The occurrence of maize diseases is closely linked to factors such as pathogen availability, the varieties grown, weather conditions, farming practices, and agricultural ecosystems. Such as the 1970 outbreak of southern corn leaf blight (SCLB) in the United States occurred by the widespread usage of varieties that were susceptible with T-type male-sterile cytoplasm (cms-T), which proved vulnerable to the novel race T of Bipolaris maydis (Zhu et al. 2021). The over cultivation of susceptible XianYu335 variety recently caused the outbreak of northern maize leaf blight (NCLB) in North China (Gu et al. 2025). The stalk rot and ear rot diseases were developed due to the humid and warm conditions of the Huang-Huai-Hai plain. Also, maize plants are placed in field for longer time to dry when harvested by machines accelerating the diseases intensity. So, the failing to remove rotten ears during mechanical harvesting lowers kernel quality (Holland et al. 2020). The plant immune response is a complex, highly controlled, and multilayered system that may be classified into two types: qualitative resistance to diseases and quantitative disease resilience (QDR; Schurack 2021). Resistance-related genes (R-genes) are linked to qualitative resistance usually give complete or nearly full protection and are regarded as major genes (Quade 2023). While QDR gives partial or no resistance and is controlled by many genes with low impacts (Lipps 2021). Isolated R-genes are commonly short lived, but when coupled with QDR genes, their effectiveness improves for better agricultural disease resistance (Million et al. 2023). Consequently, integrating several R-genes and QDR genes into the same genome is the most successful way to create disease-resistant cultivars.

Conventional breeding mostly relies on professionalism of breeder's selecting phenotypes for disease-resistant varieties. This method is time-consuming, ineffective, and heavily impacted by external variables. Nevertheless, by the discovery of superior genes that are resilient alongside molecular markers, combining conventional breeding with marker-assisted selection (MAS) has shown an extremely successful strategy for creating elite resistant maize lines. Furthermore, genome selection (GS), transgenic methods, and gene mutations present interesting options. When combined with doubled haploid (DH) technology, these technologies can considerably increase the procedure of breeding in corn. Transgenic techniques, in particular, are still useful tools for producing disease-resistant maize.

These strategies use biotechnology to directly introduce or modify particular target genes (Limenie and Alehegn 2025). Nearly all genes that are disease resistant are functional in the produced transgenic maize lines, suggesting the high promise of these approaches for maize resistance breeding efforts. Majorly, transgene approaches are able to overcome reproductive barriers among crop species, enabling the transfer of resistant genes from different species. For example, expressing the enduring wheat resistant to disease gene *Lr34* in maize proved to give resistance to common rust and northern corn leaf blight (NCLB; Vanacore et al. 2023).

Functional studies of resistance genes taken from corn reveal that their effectiveness is often precisely linked to their expression levels. Therefore, breeding disease-resistant maize by either overexpressing or silencing specific genes is a plausible approach. However, enhanced expression of certain pleiotropic resistance genes can negatively impact other traits. For instance, while overexpressing *ZmCCT* enhances stalk rot resistance, it also delays flowering time in maize (Tang et al. 2022). These negative effects could be mitigated by using pathogen-inducible promoters to control resistance gene expression. Although this strategy has yet to be explored in maize, it has proven highly innovative in rice (Zingore et al. 2022).

The CRISPR/Cas9 technology revolutionized gene editing due to its precision, adaptability, stability, and efficacy, making it one of the most powerful tools for genetic research and crop improvement (Yang et al. 2022). CRISPR/Cas is successfully being used to increase the disease resistance of crops. After genome editing, the transient cassettes are removed by self-pollination or hybridization. However, removing the CRISPR-Cas9 cassette often takes multiple generations to delete (Das et al. 2023). Double-haploid (DH) technology is an efficient way to improve breeding effectiveness by reducing the need for several generations of selection. Integrating CRISPR/Cas9 with DH technology is an effective method for speeding maize breeding. Using the same approaches, two different teams developed systems called HI-EDIT (Li et al. 2025a, b) and IMGE, which yield transformed haploids free of the CRISPR/Cas9 cassette in a single phase. Employing the HI-Edit approach, 4.8–8.8% of the crops generated lacked the CRISPR/Cas9 cassette but contained changes to the inducer gene matl. Similarly, the IMGE approach achieved an editing efficiency of roughly 4.1% for the ZmLG1 gene in haploids, and all zmlg1-edited haploids are Cas9-free (Zhu et al. 2021).

12.3.3 Herbicide-Tolerant Maize

Modern agriculture faces a dual challenge: the growing human population requires a 25–70% increase in food production to meet nutritional demands, while weeds cause great yield losses in nearly each cropping procedure, posing a serious risk to global food safety (Khatri et al. 2024). With weeds developing resistance to herbicides, effective weed management has become increasingly difficult. Although numerous approaches have been used in agriculture, herbicide-based control of weeds remains the most common approach. Conventional and transgenic breeding for herbicide-resistant (HR) crops are frequently labor-intensive and time-consuming. On the contrary, CRISPR-Cas9 genome editing provides the potential to induce precise point mutations at target loci for insertions and deletions modification, making it a strong tool for desired trait enhancement in plants (Sultana et al. 2024).

Genome editing provides a new, very effective reliable technique for creating herbicide-resistant (HR) plants, addressing the rising weed challenge facing the agricultural sector. These genome-edited HR crops are expected to boost agricultural yield, simplify and improve weed control, and result in considerable cost savings (Kuang et al. 2024). To investigate the genetic heritage and phenotypic integrity of CRISPR-edited plants, the CRISPR components have to be removed, which is a prerequisite for commercial licensing of genome-edited crops. Studies employed the CRISPR technique to combine precisely manufactured toxin proteins with the male sterility gene, resulting in TKC (Transgene Killer CRISPR). Besides, genome editing enables the simultaneous alteration of many genes, allowing multiple traits to be modified within a single plant (Hussain et al. 2021). The study employed CRISPR/Cas9-mediated repair using homology to accurately insert amino acid modifications into the ZmEPSPS gene, which encodes the enzyme 5-enolpyruvylshikimate-3-phosphate synthase (EPSPS; Yu et al. 2023). The exact modifications, G172A, T173I, and P177S, allow for considerable glyphosate resistance while maintaining plant growth and production. Particularly, the produced maize lines are transgene-free, eliminating regulatory impediments and social concerns about genetically engineered plants (Kaul et al. 2024). Furthermore along with EPSPS, the acetolactate synthase (ALS) gene was also proven to be a key target for conferring resistance to many herbicide groups, involving as sulfonylureas and imidazolinones. Through CRISPR/Cas9 base editing, targeted point mutations were inserted into the ALS gene, producing maize mutants with increased herbicide resistance. These genome-edited plants show agronomic efficiency similar to their wild-type counterparts, indicating a long-term solution for weed management (Wang et al. 2022a, b, c, d). CRISPR/Cas9 technologies' accuracy and effectiveness emphasize their enormous promise for producing herbicide-resistant crops, resulting in increased agricultural production and a sustainable environment (Sony et al. 2023).

12.3.4 Abiotic Stress Tolerant Maize

To fulfill the nutritional needs of an increasing worldwide population, it is critical to dramatically expand food crop output. However, plant sensitivity to numerous abiotic stressors is a significant barrier to achieving this goal (Chaffai et al. 2024). Several abiotic stress-tolerant crops have previously been produced using standard marker-assisted breeding, but this process takes about a decade to produce resilient cultivars due to the significant screening and backcrossing required (Singh et al. 2024).

Although genetically altered stress-tolerant plants have shown promise, commercialization faces various difficulties. In this scenario, genome editing appears as a sophisticated technique for creating abiotic stress-resistant crops, allowing for precise change of numerous gene loci in a shorter timescale and lowering crop development program expenditures (Tyagi et al. 2024). Gene editing is already used to successfully improve plant attributes like nutritional value and yield (Gao 2021).

12.3.5 Drought Stress Tolerant Maize

A significant achievement in genome editing is the modification of the *ARGOS8* gene, which negatively controls ethylene signaling pathways. Using CRISPR/Cas9 to modify the regulation of this gene led to substantial drought tolerance and enhanced yield in water-stressed fields (Kiani 2023). Similarly, the removal of *ZmHDT103*, a histone deacetylase gene, showed that its absence improves drought resistance by affecting abscisic acid signaling and the activation of stress-responsive genes (Table 12.1; Wang et al. 2025). Amid water shortages with the help of these precise gene modifications, the plant growth and production can be supported to offer useful ways of building climate-resilient crops.

12.3.6 Salt Stress Tolerant Maize

The genes responsible for ionic and osmotic stresses can be regulated by CRISPR-based genome editing. For example, when the gene *ZmASR6* expressed in abscisic acid signaling, it has showed the increased maize salt tolerance by accelerating the antioxidant enzyme activity and decreasing oxidative damage (Li et al. 2025a, b). Similarly, ion transporter genes such as *ZmNHX1* have demonstrated the importance of maintaining ion balance under salt stress can lead to better maize growth and survival in salty soils (Cao et al. 2023).

12.3.7 Editing Cis-Regulatory Elements for Quantitative Trait Improvement.

Cis-regulatory elements (CREs) are noncoding DNA sequences that regulate gene expression. The better understanding of transcriptional regulation of genes that mainly control plant growth and development can be possible by identifying Cis-regulatory elements. They are important to develop gene regulation links to show the regulatory relationships between transcription factors and target genes.

CREs are transcription factor (TF)-binding sites (TFBSs) in enhancers, promoters, and silencers. They, like transposable elements (TEs), are linked to genome development and intergenic space (Zhang et al. 2022). In addition, chromatin status and connections among gene areas influence the expression of gene. The CRISPR/Cas9 approach speeds up crop development as versus conventional breeding and mutational breeding methods, and it enables transgene-free genome editing. It has been used to improve agricultural properties including quality, crop yield, and disease resistance due to its effectiveness and precision in targeted gene alteration. However, in plant genetic modification, most of the work has been confined to the coding region, with CRISPR/Cas9 typically employed for knocking out a number of genes. Particularly, the wheat Mildew Resistance Locus (TaMLO) allele TaMLO-A1 was initial gene edited in wheat using CRISPR/Cas9, and TaMLO-A1-knockout plants were more resistant to powdery mildew (PM) than native plants (Ingvardsen et al. 2023). Furthermore, using CRISPR/Cas9, rice p genes which include *GW2*, *GW5*, and Grains LENGTH AND WIDTH2 (*GLW2*) are believed to considerably increase grain yields, despite the fact that genome editing of regulatory areas is rather uncommon in plants. Because many plants have a conserved CLV-WUS system, changing the promoters of genes in this pathway, such as maize CLAVATA3/EMBRYO SURROUNDING REGION-RELATED7 (*ZmCLE7*) and Zm FON2-LIKE CLE PROTEIN1 (ZmFCP1), improves maize meristem development and numerous yield-related measures (Liu et al. 2021). Gene knockdown entailed introducing the natural maize GOS2 promoter into the 5′ UTR of the Auxin-Regulated Gene Involved in Organ Size8 (*ARGOS8*) or replacing the native *ARGOS8* promoter with HDR. This alteration improved *ARGOS8* expression and grain production under drought stress.

Site-directed integration of large chunks is a typical challenge in gene knocking development. But with the effective application of CRISPR/Cas9 engineering, a synthetically generated 5.2 kb carotenoid biosynthetic expressed gene factor was precisely inserted at a specified location in the rice genetic code, leading to the generation of golden rice having improved carotenoid synthesis (Dong et al. 2019). Base editing techniques, such as cytosine base editing (CBE) and adenine base editing (ABE), are widely employed to cause permanent nucleotide changes in plants. CBE generates C:G to T:A mutations by simply inserting the desired DNA area using a single-guide RNA (sgRNA); it was modified and enhanced in a number of plant species, including maize, where CENTROMERE SPECIFIC HISTONE H3 (*ZmCENH3*) has been successfully edited (Dawe et al. 2023). Furthermore, the single-base variation number for *OsCDC48* within the pnCas9-PBE system was 43.48%, and the modified genotypes had no unexpected InDel mutations (Cui et al. 2023).

12.3.8 Modern Approaches: Whole-Genome Mutant Libraries

Whole-genome mutant libraries are critical modern tools in functional genomics, allowing high-throughput analysis of gene function in crop species. Several mutagenesis techniques have been employed to develop these libraries, including transposon insertion (e.g., Mutator/Mu system), chemical mutagenesis like ethyl methanesulfonate (EMS), targeted CRISPR/Cas9 gene editing, and TILLING (Targeting Induced Local Lesions in Genomes; Kirov, 2023). The UniformMu is a transposon-based mutation library that is a prominent reverse genetics database to study how genes function in maize as well as it includes a searchable database of inserts across the genome. EMS mutant populations in sites such as MaizeGDB are used to detect SNP-based mutations that cause loss of function and generate allelic series (Wang et al. 2022a, 2022b, 2022c, 2022d). Recently, new libraries of CRISPR/

Cas9-based whole genome deletion are being used for precise gene disruptions to improve the efficiency of functional research in elite maize varieties (He et al. 2024). These mutant resources and high-throughput phenotyping technologies such as RNA-seq, qPCR, metabolomics, and automated imaging systems are giving powerful gene trait association research. This approach accelerates the gene discovery while also improving complex molecular breeding activities for strengthening agronomic traits like as drought tolerance, nutrient efficacy, and climatic stress (Zhang et al. 2021a, b, c).

12.3.9 High-Throughput Gene Targeting Platforms in Maize

Maize (Zea mays L.) is a widely grown crop in the whole world which is fulfilling the dietary requirements of humans and serving as feed for animals, and also producing biofuel. Many factors such as climatic, soil, and abiotic stresses which include drought, salt, extreme temperatures, and cold weather have a major impact on maize yield that reduce its output. Under light of these challenges, crop breeding has changed owing to high-throughput sequencing technologies (HSTs). These advancements have made it simpler to find beneficial quantitative trait loci (QTL), marker-trait associations (MTAs), genotypes, and alleles that contribute to crop development.

Well-established omics approaches such as genomics, transcriptomics, proteomics, and metabolomics are now being included into maize breeding research. These strategies have revealed novel biological markers that are being used to improve maize's ability to tolerate a variety of stressors (Farooqi et al. 2022). Although substantial developments are made in understanding how corn responds to various external factors in developmental stages, many abiotic stress resilient QTLs or SNPs have shown to be associated with fertility problems (Salika and Riffat 2021). Technological breakthroughs in sequencing, visualization, and machine learning can allow a deeper understanding of response to abiotic stress at the tissues, cellular or subcellular molecular level.

Despite high-throughput phenotyping technologies were used to collect phenotypic information associated with drought in maize, the collection and analysis of such pictures need additional validation to ensure phenotypic preciseness. Significantly, the link between phenotype and genotype has to be more clearly identified and measured for multiple factors examined within a range of external conditions. This will strengthen our potential to anticipate phenotypical efficiency using genetic data (Gill et al. 2022). Improvements in molecular and analytical methods, as well as machine learning, will help predict the inherited factors that cause a tolerant phenotype. As an example, a mGWAS experiment combined with machine learning has discovered multiple genes (such as *ZmCS3*, *ZmUGT*, and *ZmCYP709B2*) associated with osmotic stresses caused by salt in corn (Wang et al. 2022a, b, c, d), expanding our knowledge of the complex processes underlying resistance to salt. These findings would not just give an in-depth awareness of the signaling networks and cellular mechanisms that drive crop stress tolerance, but additionally allows in developing breeding methods for resilient to stress crops (Lin 2024).

12.3.10 Current Challenges in Maize Genome Editing

As the global population continues to rise, there is a growing need for sustainable food production on a large scale. By 2050, the world's population is going to reach 10 billion, insisting an corresponding increase in food output. However, modern agriculture is facing serious challenges due to the escalating effects of climate shift. Maize (Zea mays L.), a major crop worldwide, plays a key role in various sectors as a food source, feed for animal, biofuel, and industrial raw material (Amanjyoti et al. 2024). Notably, fluctuations in temperature, rainfall, and their combined effects have historically played a major role in influencing worldwide productivity of maize and other key plants (Ahmad et al. 2024). For instance, Brazil, the third-largest maize producer globally saw significant declines in maize output, with reductions of around 18 million tons in the 2015/16 season and 23 million tons in 2020/21. These numbers reflect around 21% losses over the 2014/15 and 2019/20 crop seasons, accordingly (CONAB 2021).

According to standard prices of maize between 2018 to 2021 (CEPEA-ESALQ 2021), yield losses are estimated to cost the economy between USD 3 billion and USD 5 billion. These shortages in production occurred during times of severe drought (INMET 2021), resulting in decreased yields across several important maize-producing regions. Similarly, the 2012 droughts in the United States, the globe's top maize producer, caused significant output declines and dramatic hikes in prices (Zhang et al. 2022). As a result, continuous research of maize cultivars with increased genetic resilience, combined with improved farming techniques, is critical to mitigating future yield losses caused by the rising frequency, intensity, and length of climate-related stresses (IPCC 2022).

Yield and tolerance to abiotic stresses are complicated parameters that are heavily influenced by environmental conditions and numerous genomic loci with small effects, that cause it challenging to reveal the specific molecular mechanisms for a gene function and precisely assess related phenotypes so complicate the use of genomic engineering tools for improving these traits. Although, many transgene maize varieties are available that are highly resistant to insects and herbicides but there are only a few examples of targeting more complex traits (Yassitepe et al. 2021).

Maintaining the complex traits of crops that remain stable in various conditions has limited the use of transgene technologies caused the development of more advanced biotech crops (Simmons et al. 2021). New techniques like CRIPR/Cas are revolutionizing the agriculture by targeting multiple genes at once. The integration of gene editing and haploid induction improve the potential of creating complex traits of different desired genes used in plant breeding.

A great deal of maize genome editing studies has focused on genetic transformation approaches including *Agrobacterium* or biolistic approaches, especially in a small number of moderate genotypes that are susceptible to *Agrobacterium* infection and tissue regeneration (Kausch et al. 2021). But the absence of transformation methodology that works effectively over a wider variety of genotypes, along with relatively low transformation efficiencies, has hampered the widespread adoption of gene alteration in maize. For addressing these issues, new solutions have recently been proposed, like, the use of morphogenic regulators (MRs) as well as trans genome editing. These advancements have an opportunity for greater implementation in maize breeding programs around the globe, especially those that use tropical germplasm.

12.4 Prospects

12.4.1 The BREEDIT Initiative for Multiplex Gene Editing in Maize

The fundamental issue for farming in today's world is to provide sufficient food to a growing global population in addition to adapting to climate change. Although revolutionary methods have already been employed to solve the problem, we are approaching a turning point in crop development through traditional breeding. Latest developments in genetic modification using the CRISPR/Cas system have the potential to accelerate breeding for plants to meet this expanding need. Over the last century, conventional breeding has been crucial in adapting crops to local ecosystems and enhancing production under stress conditions (Snowdon et al. 2021). Genome-assisted selection has significantly contributed to the production of novel crops by incorporating haplotype traits into breeding procedures (Bhat et al. 2021). CRISPR/Cas approach emerged became a powerful tool for multiplex-targeting several GRGs at once, resulting in variation in genetic makeup in a large number of sites and allowing studying of a broad range of combinatorial mutations. Several studies have shown the way CRISPR can be used to modify plant structures and address intricate traits across a number of plants, including maize (Doll et al. 2019). Large-scale CRISPR screenings in maize were carried out as part of a wider application. Small contributions from multiple genes influence important agronomic factors such as yield. MGE is capable of targeting both similar and different sequences. For instance, a single sgRNA may choose several genes with homologous segments, while many sgRNAs can address different genes or possibly different loci within one gene. Improving these complex quantitative characteristics involves arduous and laborious selection and multi-crossing techniques, which might take a while to generate new top cultivars, even when using current technologies such as Genome Predictions. CRISPR/Cas-based multiplexed editing of genomes (MGE) enables the creation of novel lines with a variety of genomic changes in a few generations. This potential represents a considerable leap over conventional breeding of plants and mutations, allowing for the accumulation of advantageous characteristics.

MGE is one of the most successful GE methodologies for enhancing complicated characteristics. Maize has been grown using multiple MGE ways. For example, Qi et al. proposed an enhanced tRNA-processing system-based approach in maize that allowed for the insertion of as many as four sgRNAs in a series. The indigenous tRNA-processing machinery not only successfully processes the primary transcript from an increased smaller expressing cassette, yet it also appears to increase editing efficacy (85.7–100%). The A- and B-boxes in tRNA sequences, which attract transcriptional factors, might be responsible for the increased editing efficiency (Rafique et al. 2024). This discovery is important for maize GE since higher editing performance is required whenever crop transformation rate is low, therefore reducing this is essential barrier.

MGE in maize can also be done using a larger number of sgRNAs. For example, vectors containing up to 12 separate sgRNA expression cassettes have been successfully employed to transform a "editor" (i.e., Cas9-expressing) maize line (Lorenzo et al. 2022). Gong et al. (2021) investigated the CRISPR/Cas12a and CRISPR/Cas9 systems for MGE against the maize bZIP transcription factor Opaque2 (O2). CRISPR/Cas12a had poorer editing effectiveness than CRISPR/Cas9 in the T0 and T1 generations, but it resulted in a wider range of mutations in T2. Furthermore, the Cas12a-based system's editing efficiency was found to be positively connected with nuclease expression levels, indicating that it is a viable alternative to MGE in maize. The approach, also known as BREEDIT, tries to improve complex qualities regulated by several genes, including resistance to drought and productivity.

In BREEDIT, produced population transformed using multiple sets of sgRNAs (12 distinct sgRNAs per set) undergo crossing, resulting in alterations in a growing quantity of genes. This technique is beneficial when it comes to trait development and identifying additional genes that contribute to a certain trait of interest (Lorenzo et al. 2022). Put in common these results demonstrate that MGE has the ability to rapidly modify complex phenotypes, particularly those governed by several genes. It is critical to understand the limitations of this strategy. For example, creating a collection of sgRNAs for individual genes in homologous families may prove challenging. Many genes may be genotyped using more advanced sequencing techniques, such as amplicon sequencing. Combining a large number of sgRNAs into a single vector may also limit transformation efficiency.

Direct administration of ribonucleoprotein complexes is an alternate strategy that avoids the difficulties associated with managing overly large plasmid sizes. In vitro, Cas9 is combined with mature sgRNA molecules to produce such RNPs (Lin et al. 2022). This approach does not introduce any foreign DNA into the plant, making it transgenic. Maize premature embryos co-bombarded along with sgRNA:Cas viral nuclear proteins, coupled with genes for improving transformation (morphogenic regulator) and a selectable marker (MoPAT-DsRED), have also been shown to successfully modify their genomes (Hernandes-Lopes et al. 2023a, b). This promising finding shows that RNPs have the potential to be exploited for MGE in maize, opening the way for their usage in fields that are not susceptible to protoplasm regeneration. Other CRISPR/Cas techniques can result in transgenic breeds, but crosses are necessary to eliminate the foreign editing process. In the example of RNPs, modified events will most likely escape restrictions with little effort.

12.4.2 Advancements in Genome Editing for Precise Maize Improvement

Genome engineering using CRISPR/Cas is an effective method for generation of better crops. The accuracy with which this technology targets specific regions, the ability to simultaneously alter many genes, and the potential of separating the CRISPR components through cross-breed while retaining the modified region is transforming the breeding programs for plants (Hernandes-Lopes et al. 2023a, b).

Furthermore, CRISPR/Cas has benefits over traditional breeding and transient methods, which include the ability to prevent GMO rules by producing alleles that are distinct from the ones generated naturally, the ability to quickly create stable lines with homozygous mutations mutated at specific loci, along with the quite simple stacking of various desirable characteristics using multiplex methods. Although the initial CRISPR/Cas approach evolved into a powerful toolset that transformed plant cultivation and the breeding process, drawbacks related to the random nature of the generated mutations precluded adaptation to certain situations.

Nonetheless, innovative CRISPR/Cas-based techniques continue to be discovered, such as the usage of different Cas nucleases, which include Cas12a (also named Cpf1), improved Cas nucleases with greater efficacy, inactivated Cas (dCas) or Cas nickases (nCas), synthetic sgRNAs, and combining of nCas or dCas to the functional areas of various proteins (Gao 2021). Furthermore, certain CRISPR-based methods, including base and prime editing, enable selective and precise alterations at the desired position, reducing the unpredictability of inserts produced by the nonhomologous end joining (NHEJ) pathway (Molla et al. 2021).

Furthermore, improving CRISPR/Cas systems requires developing novel methods of modifying the genome and widening their application to crop breeding, which will enable the enhancement of characteristics governed by several genes or perhaps the creation of multiple traits at once.

Multiplex genome editing (MGE) is an important advancement of genome editing which is being used to improve complex characteristics in crops such as rice and maize. There are many examples for maize are present where MGE technique has used such as tRNA-based system that increases editing efficiency and constructs that can express up to 12 sgRNAs (Lorenzo et al. 2022). Cas12a-based MGE has demonstrated great potential of gene editing due to its mutant variety (Gong et al. 2021. In one example, the semi dwarf maize was produced by knocking out the ZmGA20ox3 gene and in another example the finding male sterility related genes (Liu et al. 2022a, b, c, d), and also accelerating mutations for complex traits using BREEDIT (Lorenzo et al. 2022). While there are challenges in sgRNA designing, genotyping, and vector size, RNP delivery provides a transgene free and efficient possibilities that make MGE a breakthrough crop improvement tool. Gene editing (GE) is not restricted to coding areas; it can also target regulatory elements such as promoters to control gene expression. Cis-regulatory elements (CREs) arose during domestication to maintain expression patterns (Marand et al. 2023). Editing promoters with CRISPR/Cas can produce a variety of phenotypes by modifying enhancers or repressors, even without affecting coding sequences (Ganguly et al. 2022).

Multiplex genome editing can produce various promoter variations, resulting in varying expression levels and symptoms. However, predicting these effects is difficult due to enhancer-repressor interactions, CRE distance, and chromatin structure (Schmitz et al. 2022).

Despite challenges, promoter editing is promising for breeding. It avoids exogenous DNA and transgene-related regulatory issues, which can help with public acceptance. For example, editing the *ZmCLE7* edits in maize affected inflorescence meristem size (Liu et al. 2021). Editing the *ZmACO2* promoter improved grain yield (Ning et al. 2021). In drought-tolerant maize, the *ARGOS8* promoter was swapped with a stronger *GOS2* promoter, mimicking beneficial transgene effects (Varotto 2023).

Base and prime editing are advanced CRISPR/Cas strategies enabling precise genome manipulations without introducing double-strand breaks (DSBs). Base editing fuses a catalytically inactive Cas9 with a deaminase to enable targeted nucleotide conversions, such as C-to-T or A-to-G (Saber et al. 2023). Base editors have been implemented in various crops including rice, maize, wheat, and cotton, such as creating herbicide-tolerant maize via editing *ZmALS1/2* (Qiaoet al. 2023).

Prime editing employs a Cas9 nickase coupled to reverse transcriptase and a pegRNA to enable all 12 base conversions, and also deletions insertions, without DSBs or donor templates. Though editing efficiency remains low, improvements such as modifying reverse transcriptase domains have significantly increased success rates (Zong et al. 2022; Xu et al. 2022). These developments enable precision trait development, such as herbicide resistance in maize and soybean crops and sugar content modification in strawberries using uORF editing. While additional improvement is required, base and prime editors show great promise for accurate plant breeding (Molla et al. 2021).

12.4.3 Challenges in Field Testing and Regulatory Considerations for CRISPR-Edited Maize

Maize, an important crop for nutrition, livestock feed, and industry, benefits significantly from the CRISPR-Cas9 system's specific genome-modifying capabilities. CRISPR has been used for targeted mutations in maize to get desirable such as resistant to diseases, increased yield, and high nutritional content. The long-term impacts of using CRISPR technology to change maize crop performance and the environment must be extensively examined. Consistent surveillance is necessary to assess the sustainability of CRISPR-edited maize varieties. In general, the latest advances in maize crop transformations with CRISPR technology have resulted in significant improvements in crop resilience, efficiency, and dietary value.

Effective maize applications, together with case studies, demonstrate the actual advantages of CRISPR technology in agriculture (Ammar et al. 2023). Yet, resolving the accompanying technical and ethical considerations is critical to ensuring the responsible and long-term implementation of CRISPR-edited maize cultivars. There should be proper management, open debates, and constant research is required to understand the potential of this editing technique for improving the maize crops across the globe. We can insert different targeted traits in a single maize line with CRISPR/Cas. In this regard, a single maize line can have disease resistance, have higher nutritional value, and can be more resilient to different environmental conditions (Iqbal et al. 2024).

12.5 Limitations and Strategies for the Application of Genome Editing in Maize Breeding

Since its introduction, plant biotechnology has made great progress. For example, in 1996, Bt genes were inserted into maize to give immunity against the European corn pests (García et al. 2023). It was a historic progress as it caused the selling of the first transgenic cereal crop. The maize cultivars with multiple or stacked features, such as insecticide and resistant to herbicides, was a great development in genetic engineering that allowed for improved weed and pesticide resilient crops (Limenie and Alehegn 2025). Similarly, the genetically modified maize crops that showed resistance to drought is an important step in addressing one of agriculture's major issues. Genetic engineering has enabled the integration of genes encoding proteins that confer resistance to insects and diseases such as the insertion of Bt genes in maize and cotton genome have combat lepidopteran pests (Ali et al. 2023; Li et al. 2021a, b). Abiotic stresses, for example drought, salt, and high or low temperatures, have an impact on crop yields. Genetic engineering has revolutionized agriculture, giving an opportunity to improve cereal crop output and resistance to diseases As we look forward, several emerging technologies are predicted to have an influence on the cereal crop improvement. CRISPR-Cas9 technology allows for precision, targeted mutation of genomes, for example, the targeted insertion of disease-resistant genes, removing undesirable traits, and increase of yield-related characteristics (Wang et al. 2023). GWAS can examine variations across a large populations to identify genes and genetic markers associated with pertinent traits. Which, later on can be used to design crops with improved yields and resistance to infectious diseases. Such field influences the developing and engineering biological characteristics that are not present in the wild. In the agriculture sector, gene editing has a great importance as it can create crops with desired properties such as higher nutrient value, resilience to stress, and productivity. The integration of genetic engineering and precise agricultural technologies can improve resource conservation and crops integrity. For example, genetically modified crops can engineered to respond to certain environmental conditions, removing the need for manual intervention (Srivastav et al. 2024).

The new doors can be opened for growth and development of crops by understanding the mechanism of how external factors impact the gene expression without the change of DNA sequence. The modification of epigenetic markers can improve stress and disease resistance. Microbiome has a great influence on plant growth and production. So, genetic engineering can potentially be used to incorporate beneficial bacteria into cereal crops, increasing resistance to pathogens and environmental stresses (Bigini et al. 2021). Climate change gives new challenges to agriculture, genetic engineering can be used to develop crops that are more resistant to drought, heat, and other environmental stresses. This technology use genetic markers to indicate the growth of individual plants and also breeders to choose the best plants for breeding without waiting for them to mature. It is able to improve the breeding procedures and increase the efficacy of genetic engineering programs (Getahun et al. 2025). The advancement of biological computation and machine learning can enable more exact predictions of how genetic mutations will affect plant traits. This can guide to proper design the genetic engineering research and increase the crop improvement programs. As genetic engineering technologies advance, improved laws and regulations and public education will be necessary to ensure its responsible and ethical use (Bigini et al. 2021).

12.5.1 Commercialization of Transgenic Maize Hybrids

Genetic engineering may be used to modify the genetic features of maize, and several other transformation techniques are commonly employed. Currently, the most often studied features are pesticide and herbicide susceptibility. In 1996, the US allowed the marketing of Cry1Ab GM maize ("Bt176," "MON810," and "Bt11"), which has been promoted and used for 27 years (García et al. 2023). According to this, 244 GM maize varieties were licensed for commercialization (Yi et al. 2022). The study from the International Service for the Application of Agricultural Biotechnology (ISAAA), genetically modified crops are grown in nearly thirty nations in 2019.

In 2019, GM maize had a planting area of 6.09 × 107 hm2, making about 32% of the overall GM crop acreage. The largest planting area was in the United States (3.317 × 107 hm2), followed by Brazil and Argentina. Globally, genetically modified maize is predominantly utilized as livestock feed (Anderson et al. 2021) as well as an industrialized raw material for producing alcohol. Just a slight fraction is consumed entirely by human beings. Genetically modified maize utilized in food is mainly processed for extracting maize oil (Boccia and Punzo 2021), for producing maize syrup, corn flour, or different maize substances (Zhang et al. 2021a, b, c), specifically corn starch, that is commonly utilized as a thickening agent, gel-forming agent, a filler, and water-retention agent in the culinary industry (Hashemzadeh et al. 2022).

GM maize is also utilized as a raw material for food manufacturing and processing, such as tortillas in Mexico (Delgado-Valerio et al. 2022); white maize in South Africa, where it is a staple diet for most people; and maize flakes, popcorn, and maize-related snacks.

There's little debate that modified maize has become an essential part of people's lives all around the world, so it is necessary to effectively control its cultivation and manufacture. The European Union (EU) has created an accountability control system that requires tagging of GM products on the marketplace. The EU Regulation 1830/2003/EC, which went into action in April of 2004, mandates that goods having over 0.9% (mass fraction, % m/m) be marked with the terms "genetic improvement" or "processed from a GM crop."

Furthermore, the expression technique should be based on the DNA copy number ratio (Deng 2024). The measured value of the typical substance determined by the percentage of mass will govern, whereas the measuring outcome of the replica number percentages (% cpT/cpE) has to be translated to mass percent. There are laws and regulations in China, the US, Canada, Australia, New Zealand, India, and many other nations that outline a labels system of management for genetically modified goods (Ahire et al. 2024). This guarantees that enterprises and customers have accessibility to correct data, allowing them to use their rights to decide while regulating and validating label claims. The latter scenario needs the employment of precise detection tools to identify the possibility of GM organisms. Depending on the objective, GM crop detection may be divided into three categories: nucleic acids, proteins, and metabolites. DNA is more persistent in live cells than protein, and it is difficult to remove regardless of the processing of crops; thus, trace or identifiable quantities of segments of DNA can be found in the final product. Thus, identifying DNA is typically the preferred approach for detecting GM components in cereals (Luo et al. 2023).

12.5.2 *Limitations of Transgenic Approaches for Complex Traits*

When transgenic plants are planted in a garden, they will undoubtedly come into contact with a diverse range of species that collaborate to perform a variety of ecological roles in agricultural settings. Despite strong scientific consensus that GE crops are safe to ingest, many individuals remain skeptical about this technology.

The development of transgenic crops into the regular food production system has sparked public outrage due to ethical and risk concerns. Five regulatory issues concerning genetically modified crops have been raised: possible adverse effects on the health of humans, damages to the environment, detrimental effects on conventional agricultural techniques, unnecessary management monitoring, and the process's "artificiality."

The introduction of transgenic plants into the worldwide situation, where food supply is straining to keep up with an ever-increasing population, has aroused outrage (Sumanth et al. 2022). In context of human health, there is no conclusive evidence that genetically modified foods are more likely than ordinary meals to cause allergic reactions. Genetic transformation does not cause allergies. However, the sort of genes selected for introduction into the transgenic host plant may result in allergic reactions. Microbial transgenes are most likely responsible for horizontal gene transfer from genetically modified crops to the gut microbiome. A transgenic plant has a part of foreign DNA that is not naturally present in the plant. These DNA fragments are commonly acquired from trans (similar plants) or cis (totally different species), such as viruses and bacteria. There is growing concern about whether eating "alien" DNA offers a risk to human or animal systems (Koul 2022). Despite genetically modified crops have the capacity to ease global food shortages, their viability for cultivation remains uncertain. The growing cultivation of genetically modified foods to combat hunger has unforeseen repercussions for the environment and human health. There are several future uses of genetically modified crops (i) enhanced and more efficient weed management; (ii) reduced losses due to pests like insects and viruses, and reduced demand for insecticide; (iii) lower post harvest losses because of greater shelf life and marketing adaptability (tomatoes due to storage pest resistance); (iv) increased nutritional value (canola oil); and (v) more efficient hybrid seed production. Transgenic crops will be used in the future not only for improved agronomic properties, but also for food processing, medicine (including edible immunizations), and unique chemicals (Benavente and Giménez 2021).

12.6 "Transformation-Free" Methods for Genome Editing

Conventional plant transformation methods have relied heavily on transgene-dependent transport of CRISPR/Cas elements (Gao 2021). Many plant viral plasmids have recently became as viable toolkits for transitory CRISPR/Cas transfer via infections of intact plants (Oh et al. 2021). The benefits of viral distribution systems are simplicity, rapidity, and durability. Negative-stranded viral genomes, such as Sonchus yellow netting virus and tomato spotting wilt virus (TSWV), stand out among these vectors due to their large carrying capacity. These capabilities allow the transfer of whole CRISPR/Cas elements as well as larger base editors, reducing the need for stable transformations (Liu and Dawe 2023). CRISPR-Cas systems in plants are frequently supplied via transformation methods, notably *Agrobacterium tumefaciens*, due to their simplicity and low cost. This technique enables the steady insertion of CRISPR-Cas DNA, resulting in exceptional modifying efficacy in crops including Arabidopsis, rice, tomatoes, corn, and grapes (Ali et al. 2023). Particle bombardment, used mainly in monocots, also delivers CRISPR components via metal microprojectiles, enabling targeted edits (Gordon et al. 2021). Despite their efficiency, transformation methods often cause random foreign DNA integration, leading to off-target mutations and GMO regulatory concerns. Removing these transgenes requires traditional breeding. Transient expression methods, which avoid stable DNA integration and selection markers, offer a way to produce transgene-free plants, though some DNA fragments may still integrate (Qamar et al. 2024). Viral-induced genome editing (VIGE) enables CRISPR/Cas delivery by expressing Cas protein and guide RNAs (gRNAs), offering a transgene-free alternative to stable transformation (Mikhaylova et al. 2022). So far, over 22 plant viruses have been modified for this purpose, but a universally effective vector for agricultural use is yet to emerge due to challenges in balancing editing efficiency, virus replication, and plant development. Initially, viruses delivered only gRNAs into plants already expressing Cas9 (Shen et al. 2024). Limited cargo capacity in commonly used vectors like TRV restricted their ability to carry large genes like Cas9, prompting development of systems for transient, full CRISPR/Cas expression without transformation (Kang et al. 2025). Scientists now use unarmed or deconstructed viruses lacking transmission functions, which increases cargo space but reduces mobility, often necessitating tissue culture. Most studies have been limited to *Nicotiana benthamiana*, and methods are not easily transferable to other species. Moreover, without transgenes, traditional selection methods (e.g., antibiotics or fluorescence) cannot be used, making detection of edited events difficult. This limitation often drives researchers to target marker genes like *PDS*. High CRISPR/Cas expression is needed for efficient modification but can trigger plant immune responses. Active research is underway to resolve these interconnected limitations, and various solutions have been proposed (Shen et al. 2024). Transgenic-free techniques provide an exciting new horizon in current crop cultivation, with enormous economic possibilities. It is critical to move beyond traditional stable transformations and develop temporary expression methods that do not rely on standard selection markers, which often result in transmissible editing. Scientists are still trying to deliver the entire CRISPR-Cas system within a single virus-derived plasmid while ensuring transition to the meristem. On the other hand, in the past few years, a variety of approaches were developed to achieve this goal. Such distinct strategies involve the integration of transmissible components, RNAi inhibitors, innovative small Cas proteins, and borne by seeds viruses to achieve the desired impact (Mikhaylova 2025).

The most promising inherited and tissue-culture-free genome editing methods employ a TRSV-TRV-ALSV triple plasmid complex including Cas9 and FT:gRNAs, as well as an RNAi inhibitor. But this technique was only used for the N. benthamiana PDS gene (Yoshida et al. 2024). Most CRISPR/Cas researchers have yet to make a decision among simple selection processes and transgene-free methods, temporary and stable expression, and in vitro culture vs agroinfiltration. Before VIGE may be extensively implemented, numerous critical concerns must be overcome (Zhao et al. 2023).

References

Abdelrahman M, Wei Z, Rohila JS, Zhao K (2021) Multiplex genome-editing technologies for revolutionizing plant biology and crop improvement. Front Plant Sci 12:721203

Ammar A, Khalid MN, Amjad I (2023) Maize crop transformation with CRISPR: state-of-the-art and future directions. Trends Biotech Plant Sci 1(1):16–28

Anzalone AV, Randolph PB, Davis JR, Sousa AA, Koblan LW, Levy JM, Chen PJ, Wilson C, Newby GA, Raguram A, Liu DR (2019) Search-and-replace genome editing without double-strand breaks or donor DNA. Nature 576(7785):149–157

Auinger HJ (2022) Genomic prediction in advanced cycle breeding programs, Doctoral dissertation, Technische Universität München

Aesaert S, Impens L, Coussens G, Van Lerberge E, Vanderhaeghen R, Desmet L, Vanhevel Y, Bossuyt S, Wambua AN, Van Lijsebettens M, Inzé D (2022) Optimized transformation and gene editing of the B104 public maize inbred by improved tissue culture and use of morphogenic regulators. Front Plant Sci 13:883847

Agarwal N, Gupta R (2021) History, evolution and classification of CRISPR-Cas associated systems. Prog Mol Biol Transl Sci 179:11–76

Ahire JJ, Rohilla A, Kumar V, Tiwari A (2024) Quality management of probiotics: ensuring safety and maximizing health benefits. Curr Microbiol 81(1):1

Ahmad U, Hussain M, Ahmad W, Javed J, Arshad Z, Akram Z (2024) Impact of global climate change on maize (Zea mays): physiological responses and modern breeding techniques. Trends Biotech Plant Sci 2(1):62–77

Alam MJ, Chowdhury MAH, Nahar Q, Rashid MHO, Ahmed KS (2020) Comparative efficacy of some insecticides against maize aphid (Rhopalosiphum maidis) and its influence on natural enemies in maize ecosystem. Trop Agrobiodivers 1(2):95–101

Ali A, Zafar MM, Farooq Z, Ahmed SR, Ijaz A, Anwar Z, Abbas H, Tariq MS, Tariq H, Mustafa M, Bajwa MH (2023) Breakthrough in CRISPR/Cas system: current and future directions and challenges. Biotechnol J 18(8):2200642

Ali F, Hameed A, Rehman A, Sarfraz S, Rajput NA, Atiq M (2024) CRISPR system discovery, history, and future perspective. In: OMICs-based techniques for global food security, pp 159–170

Alsharairi, N.A., 2022. Dietary antioxidants and lung cancer risk in smokers and non-smokers. In Healthcare (Vol. 10, No. 12, p. 2501). MDPI

Amanjyoti, Singh J, Sowdhanya D, Rasane P, Singh J, Ercisli S, Verma H, Ullah R (2024) Maize. In: Cereals and nutraceuticals. Springer Nature Singapore, Singapore, pp 47–80

Anderson JA, Herman RA, Carlson A, Mathesius C, Maxwell C, Mirsky H, Roper J, Smith B, Walker C, Wu J (2021) Hypothesis-based food, feed, and environmental safety assessment of GM crops: a case study using maize event DP-202216-6. GM Crops Food 12(1):282–291

Asmamaw M, Zawdie B (2021) Mechanism and applications of CRISPR/Cas-9-mediated genome editing. Biologics 15:353–361

Babu R, Palacios N, Prasanna BM (2013) Biofortified maize–a genetic avenue for nutritional security. Trans Gen Crop Breed Abiotic Stress Yield Quality 2:161–176

Bhambhani S, Kondhare KR, Giri AP (2021) Diversity in chemical structures and biological properties of plant alkaloids. Molecules 26(11):3374

Bhat MA, Mir RA, Kumar V, Shah AA, Zargar SM, Rahman S, Jan AT (2021) Mechanistic insights of CRISPR/Cas-mediated genome editing towards enhancing abiotic stress tolerance in plants. Physiol Plant 172(2):1255–1268

Bigini V, Camerlengo F, Botticella E, Sestili F, Savatin DV (2021) Biotechnological resources to increase disease-resistance by improving plant immunity: a sustainable approach to save cereal crop production. Plants 10(6):1146

Benavente E, Giménez E (2021) Modern approaches for the genetic improvement of rice, wheat and maize for abiotic constraints-related traits: a comparative overview. Agronomy 11(2):376

Bhushan B, Jat BS, Dagla MC, Aggarwal SK, Rakshit S (2021) Anthocyanins and Proanthocyanidins as anticancer agents. In: Exploring plant cells for the production of compounds of interest, pp 95–124

Blankenagel S, Eggels S, Frey M, Grill E, Bauer E, Dawid C, Fernie AR, Haberer G, Hammerl R, Barbosa Medeiros D, Ouzunova M (2022) Natural alleles of the abscisic acid catabolism gene ZmAbh4 modulate water use efficiency and carbon isotope discrimination in maize. Plant Cell 34(10):3860–3872

Boccia F, Punzo G (2021) A choice experiment on consumer perceptions of three generations of genetically modified foods. Appetite 161:105158

Buch K, Kaushik A, Mishra U, Beese S, Samanta S, Singh R (2023) Unravelling the complexity of plant breeding through modern genetic techniques and tools: a review. Int J Plant Soil Sci 35(21):97–105

Burlini I, Grandini A, Tacchini M, Maresca I, Guerrini A, Sacchetti G (2020) Different strategies to obtain corn (Zea mays L.) germ extracts with enhanced antioxidant properties. Nat Prod Commun 15(1):1934578X20903562

Casas MI, Duarte S, Doseff AI, Grotewold E (2014) Flavone-rich maize: an opportunity to improve the nutritional value of an important commodity crop. Front Plant Sci 5:440

Čermák T (2021) Sequence modification on demand: search and replace tools for precise gene editing in plants. Transgenic Res 30(4):353–379

Chen L, Liu Y, Wu G, Veronican Njeri K, Shen Q, Zhang N, Zhang R (2016) Induced maize salt tolerance by rhizosphere inoculation of bacillus amyloliquefaciens SQR9. Physiol Plant 158(1):34–44

Chen X, Lu X, Shu N, Wang S, Wang J, Wang D, Guo L, Ye W (2017) Targeted mutagenesis in cotton (Gossypium hirsutum L.) using the CRISPR/Cas9 system. Sci Rep 7(1):44304

Chu P, Agapito-Tenfen SZ (2022) Unintended genomic outcomes in current and next generation gm techniques: a systematic review. Plants 11(21):2997

Cigan AM, Unger E, Xu RJ, Kendall T, Fox TW (2001) Phenotypic complementation of ms45 maize requires tapetal expression of MS45. Sex Plant Reprod 14(3):135–142

Cui Y, Cao Q, Li Y, He M, Liu X (2023) Advances in cis-element-and natural variation-mediated transcriptional regulation and applications in gene editing of major crops. J Exp Bot 74(18):5441–5457

Cai R, Lv R, Shi XE, Yang G, Jin J (2023) CRISPR/dCas9 tools: epigenetic mechanism and application in gene transcriptional regulation. Int J Mol Sci 24(19):14865

Cao Y, Zhou X, Song H, Zhang M, Jiang C (2023) Advances in deciphering salt tolerance mechanism in maize. Crop J 11(4):1001–1010

Cao Y, Yu H, Guo X, Lu Y, Li W, Fu F (2025) Marker-assisted backcross breeding of drought-tolerant maize lines transformed by vacuolar H+-pyrophosphatase gene (AnVP1) from Ammopiptanthus nanus. Plants 14(6):926

CEPEA-ESALQ - Centro de Estudos Avocado's em Economia Aplicada ESALQ Indicador do Milho Esalq/BM&FBovespa (2021) [12 Oct 2021]. CEPEA-ESALQ - Centro de Estudos Avançados em Economia Aplicada ESALQ (2021) Indicador do Milho Esalq/BM&FBovespa, https://www.cepea.esalq.usp.br/br/indicador/milho.aspx

Chaffai R, Ganesan M, Cherif A (2024) Abiotic stress in plants: challenges and strategies for enhancing plant growth and development. In: Plant adaptation to abiotic stress: from signaling pathways and microbiomes to molecular mechanisms. Springer Nature Singapore, Singapore, pp 1–30

Chamness JC (2023) Improved methods of transgenic and viral reagent delivery for genome engineering in plants, Doctoral dissertation. University of Minnesota

Chen C, Liu X, Li S, Liu C, Zhang Y, Luo L, Miao L, Yang W, Xiao Z, Zhong Y, Li J (2022a) Co-expression of transcription factors ZmC1 and ZmR2 establishes an efficient and accurate haploid embryo identification system in maize. Plant J 111(5):1296–1307

Chen J, Li S, He Y, Li J, Xia L (2022b) An update on precision genome editing by homology-directed repair in plants. Plant Physiol 188(4):1780–1794

Chen Z, Debernardi JM, Dubcovsky J, Gallavotti A (2022c) Recent advances in crop transformation technologies. Nat Plants 8(12):1343–1351
Chen F, Chen L, Yan Z, Xu J, Feng L, He N, Guo M, Zhao J, Chen Z, Chen H, Yao G (2024) Recent advances of CRISPR-based genome editing for enhancing staple crops. Front Plant Sci 15:1478398
CONAB - Companhia Nacional de Abastecimento (2021) Acompanhamento de safra brasileira [12 Oct 2021]. CONAB - Companhia Nacional de Abastecimento (2021) Acompanhamento de safra brasileira, https://www.conab.gov.br/info-agro/safras/graos
Dawe RK, Gent JI, Zeng Y, Zhang H, Fu FF, Swentowsky KW, Kim DW, Wang N, Liu J, Piri RD (2023) Synthetic maize centromeres transmit chromosomes across generations. Nat Plants 9(3):433–441
Doll NM, Gilles LM, Gérentes MF, Richard C, Just J, Fierlej Y, Borrelli VM, Gendrot G, Ingram GC, Rogowsky PM, Widiez T (2019) Single and multiple gene knockouts by CRISPR–Cas9 in maize. Plant Cell Rep 38(4):487–501
Dong L, Qi X, Zhu J, Liu C, Zhang X, Cheng B, Xie C (2019) Supersweet and waxy: meeting the diverse demands for specialty maize by genome editing. Plant Biotechnol J 17(10):1853
Dalakouras A, Vlachostergios D (2021) Epigenetic approaches to crop breeding: current status and perspectives. J Exp Bot 72(15):5356–5371
Das K, Ayim BY, Borodynko-Filas N, Das SC, Aminuzzaman FM (2023) Genome editing (CRISPR/Cas9) in plant disease management: challenges and future prospects. J Plant Prot Res:159–172
Delgado-Valerio P, Ramón-Amado A, Piñeyro-Nelson A, Álvarez-Buylla ER, Ayala-Angulo N, Molina-Sánchez A (2022) Presencia de secuenciastransgénicasen masa para tortillas de pobladosurbanos y rurales de la mesetapurépecha, michoacán, México. Revistafitotecniamexicana 45(3):283–292
Delzer B, Liang D, Szwerdszarf D, Rodriguez I, Mardones G, Elumalai S, Johnson F, Nalapalli S, Egger R, Burch E, Meier K (2024) Elite, transformable haploid inducers in maize. Crop J 12(1):314–319
Deng Y (2024) Study on international genetically modified food labeling system via the perspective of consumers' right to know. Trans Econ Bus Manage Res 6:292–298
Ding XM, Mu YD, Zhang KY, Wang JP, Bai SP, Zeng QF, Peng HW (2020) Vitamin E improves antioxidant status but not lipid metabolism in laying hens fed a aged corn-containing diet. Anim Biosci 34(2):276
El-Esawi MA (ed) (2024) Recent trends in plant breeding and genetic improvement. BoD–Books on Demand
Facette MR, Park Y, Sutimantanapi D, Luo A, Cartwright HN, Yang B, ... Smith, L. G. (2015). The SCAR/WAVE complex polarizes PAN receptors and promotes division asymmetry in maize. Nat Plants 1(2), 1–8
Farooq M, Wahid A, Zahra N, Hafeez MB, Siddique KH (2024) Recent advances in plant drought tolerance. J Plant Growth Regul:1–33
Farooqi MQU, Nawaz G, Wani SH, Choudhary JR, Rana M, Sah RP, Afzal M, Zahra Z, Ganie SA, Razzaq A, Siddique KH (2022) Recent developments in multi-omics and breeding strategies for abiotic stress tolerance in maize (Zea mays L.). Front Plant Sci 13:965878
Feng C, Yuan J, Wang R, Liu Y, Birchler JA, Han F (2016) Efficient targeted genome modification in maize using CRISPR/Cas9 system. J Genet Genomics 43(1):37–43
Freitas-Alves NS, Moreira-Pinto CE, Távora FT, Paes-de-Melo B, Arraes FB, Lourenço-Tessutti IT, Moura SM, Oliveira AC, Morgante CV, Qi Y, Grossi-de-Sa MF (2024) CRISPR/Cas genome editing in soybean: challenges and new insights to overcome existing bottlenecks. J Adv Res
Frittelli A, Botticella E, Palombieri S, Masci S, Celletti S, Fontanella MC, Astolfi S, De Vita P, Volpato M, Sestili F (2023) The suppression of TdMRP3 genes reduces the phytic acid and increases the nutrient accumulation in durum wheat grain. Front Plant Sci 14:1079559

Fu J, Pei W, He L, Ma B, Tang C, Zhu L, Wang Q (2023) ZmEREB92 plays a negative role in seed germination by regulating ethylene signaling and starch mobilization in maize. PLoS Genet 19(11):e1011052
Gallavotti A, Barazesh S, Malcomber S, Hall D, Jackson D, Schmidt RJ, McSteen P (2008) Sparse inflorescence1 encodes a monocot-specific YUCCA-like gene required for vegetative and reproductive development in maize. Proc Natl Acad Sci 105(39):15196–15201
Ganguly DR, Hickey LT, Crisp PA (2022) Harnessing genetic variation at regulatory regions to fine-tune traits for climate-resilient crops. Mol Plant 15(2):222–224
García M, García-Benítez C, Ortego F, Farinós GP (2023) Monitoring insect resistance to Bt maize in the European Union: update, challenges, and future prospects. J Econ Entomol 116(2):275–288
Getahun A, Alemu A, Nida H, Woldesemayat AA (2025) Multi-locus genome-wide association mapping for major agronomic and yield-related traits in sorghum (Sorghum bicolor (L.) moench) landraces. BMC Genomics 26(1):304
Gill HS, Halder J, Zhang J, Rana A, Kleinjan J, Amand PS, Bernardo A, Bai G, Sehgal SK (2022) Whole-genome analysis of hard winter wheat germplasm identifies genomic regions associated with spike and kernel traits. Theor Appl Genet 135(9):2953–2967
Gordon-Kamm W, Barone P, Svitashev S, Sander JD, Kumar S, Jones T (2021) Strategies for CRISPR/Cas9-mediated genome editing: from delivery to production of modified plants
Grüning VK, Lübberstedt T, Frei UK (2025) Doubled haploid technology: generation of doubled haploid maize lines using haploid inducers. Cold Spring Harb Protoc 2025(3):pdb-prot108624
Gupta HS, Hossain F, Muthusamy V (2015) Biofortification of maize: An Indian perspective. Indian J Gen Plant Breed 75(01):1–22
Gaillochet C, Peña Fernández A, Goossens V, D'Halluin K, Drozdzecki A, Shafie M, Van Duyse J, Van Isterdael G, Gonzalez C, Vermeersch M, De Saeger J (2023) Systematic optimization of Cas12a base editors in wheat and maize using the ITER platform. Genome Biol 24(1):6
Gao C (2021) Genome engineering for crop improvement and future agriculture. Cell 184(6):1621–1635
Gao H, Mutti J, Young JK, Yang M, Schroder M, Lenderts B, Wang L, Peterson D, St. Clair G, Jones S, Feigenbutz L (2020) Complex trait loci in maize enabled by CRISPR-Cas9 mediated gene insertion. Front Plant Sci 11:535
Gao L, Yang G, Li Y, Sun Y, Xu R, Chen Y, Wang Z, Xing J, Zhang Y (2021) A kelch-repeat superfamily gene, ZmNL4, controls leaf width in maize (Zea mays L.). Plant J 107(3):817–830
Gentzel IN, Ohlson EW, Redinbaugh MG, Wang GL (2022) VIGE: virus-induced genome editing for improving abiotic and biotic stress traits in plants. Stress Biol 2(1):2
Ghosh A, Kumar A, Biswas G (2024) Exponential population growth and global food security: challenges and alternatives. In: Bioremediation of emerging contaminants from soils. Elsevier, pp 1–20
Gondalia N, Quiroz LF, Lai L, Singh AK, Khan M, Brychkova G, McKeown PC, Chatterjee M, Spillane C (2025) Harnessing promoter elements to enhance gene editing in plants: perspectives and advances. Plant Biotechnol J 23:1375
Gong C, Huang S, Song R, Qi W (2021) Comparative study between the CRISPR/Cpf1 (Cas12a) and CRISPR/Cas9 systems for multiplex gene editing in maize. Agriculture 11(5):429
Gu Y, Yan B, Yang Y, Huang Y, Liu X, Liu S (2025) Metabolomic analysis of maize response to northern corn leaf blight. Metabolites 15(2):113
Gulati A, Singh J, Rasane P, Kaur S, Kaur J, Nanda V (2023) Anti-cancerous effect of corn silk: a critical review on its mechanism of action and safety evaluation. 3 Biotech 13(7):246
Guo J, Tan L, Kong L (2022a) Multiple levels of health benefits from resistant starch. J Agric Food Res 10:100380
Guo Z, Kuang Z, Tao Y, Wang H, Wan M, Hao C, Shen F, Yang X, Li L (2022b) Miniature inverted-repeat transposable elements drive rapid microRNA diversification in angiosperms. Mol Biol Evol 39(11):msac224

Gupta C, Salgotra RK (2022) Epigenetics and its role in effecting agronomical traits. Front Plant Sci 13:925688

Gupta M, Choudhary M, Kumar H, Kaswan V, Kaur Y, Choudhary JR, Yadav S (2022) Doubled haploid technology in maize (Zea mays): status and applications. Indian J Agric Sci 92(3):283–291

Hashemzadeh H, Karbasi A, Mohammadi H, Firoozzare A, Boccia F (2022) Investigating the effect of nudges on consumers' willingness to pay for genetically modified corn oil. Sustainability 14(19):12705

Hernandes-Lopes J, Pinto MS, Vieira LR, Monteiro PB, Gerasimova SV, Nonato JVA, Bruno MHF, Vikhorev A, Rausch-Fernandes F, Gerhardt IR, Pauwels L (2023a) Enabling genome editing in tropical maize lines through an improved, morphogenic regulator-assisted transformation protocol. Front Genome Ed 5:1241035

Hernandes-Lopes J, Yassitepe JEDCT, Koltun A, Pauwels L, Silva VCHD, Dante RA, Gerhardt IR, Arruda P (2023b) Genome editing in maize: improving complex traits in a global crop. Genet Mol Biol 46:e20220217

Holland JB, Marino TP, Manching HC, Wisser RJ (2020) Genomic prediction for resistance to Fusarium ear rot and fumonisin contamination in maize. Crop Sci 60(4):1863–1875

Hossain F, Muthusamy V, Bhat JS, Zunjare RU, Kumar S, Prakash NR, Mehta BK (2022) Maize breeding. In: Fundamentals of field crop breeding. Springer Nature Singapore, Singapore, pp 221–258

Hussain A, Ding X, Alariqi M, Manghwar H, Hui F, Li Y, Cheng J, Wu C, Cao J, Jin S (2021) Herbicide resistance: another hot agronomic trait for plant genome editing. Plants 10(4):621

He B, Pan S, Zhao J, Zou X, Liu X, Wu S (2024) Maize improvement based on modern breeding strategies: progress and perspective. ACS Agri Sci Technol 4(3):274–282

Hooghvorst I, Nogués S (2021) Chromosome doubling methods in doubled haploid and haploid inducer-mediated genome-editing systems in major crops. Plant Cell Rep 40(2):255–270

Hung HY, Browne C, Guill K, Coles N, Eller M, Garcia A, Lepak N, Melia-Hancock S, Oropeza-Rosas M, Salvo S, Holland JB (2012) The relationship between parental genetic or phenotypic divergence and progeny variation in the maize nested association mapping population. Heredity 108(5):490–499

Ingvardsen CR, Brinch-Pedersen H (2023) Challenges and potentials of new breeding techniques in Cannabis sativa. Front Plant Sci 14:1154332

Iqbal J, Kamran M, Shafique MM, Jawad M, Arshad ZN, Raza F, Nawaz MS (2024) Harnessing genetic diversity for sustainable maize production: a contemporary perspective. Trends Animal Plant Sci 3:10–19

Idehen EO, Abegunde IO, Kehinde OB (2023) Genetic variability of yield and yield-related traits among 27 drought-tolerant maize genotypes. Agric Trop Subtrop 56(1):81–91

Impens L, Lorenzo CD, Vandeputte W, Wytynck P, Debray K, Haeghebaert J, Herwegh D, Jacobs TB, Ruttink T, Nelissen H, Inzé D (2023) Combining multiplex gene editing and doubled haploid technology in maize. New Phytol 239(4):1521–1532

INMET - Instituto Nacional de Meteorologia (2021) SPI - Índice de Precipitação Padronizada [12 Oct 2021]. INMET - Instituto Nacional de Meteorologia (2021) SPI - Índice de Precipitação Padronizada. https://portal.inmet.gov.br/servicos/spi-índice-de-precipitação-padronizada

IPCC - The Intergovernmental Panel on Climate Change Climate Change 2022: Impacts, adaptation, and vulnerability. Sixth Assessment Report 2022 [12 Oct 2021]. Sixth Assessment Report. https://www.ipcc.ch/report/ar6/wg2/

ISAAA (2019) ISAAA Brief 55-2019: executive summary biotech crops drive socio-economic development and sustainable environment in the new frontier. International Service for the Acquisition of Agri-Biotech Applications

Isnaini I, Nugraha Y, Baisakh N, Carsono N (2023) Toward food security in 2050: gene pyramiding for climate-smart rice. Sustainability 15(19):14253

Jacquier NM, Gilles LM, Martinant JP, Rogowsky PM, Widiez T (2021) Maize in planta haploid inducer lines: a cornerstone for doubled haploid technology. In: Doubled haploid technology: volume 2: hot topics, *Apiaceae, Brassicaceae, Solanaceae*, pp 25–48

Janik E, Niemcewicz M, Ceremuga M, Krzowski L, Saluk-Bijak J, Bijak M (2020) Various aspects of a gene editing system—crispr–cas9. Int J Mol Sci 21:9600–9604
Jiang Y, An X, Li Z, Yan T, Zhu T, Xie K, Liu S, Hou Q, Zhao L, Wu S, Liu X (2021a) CRISPR/Cas9-based discovery of maize transcription factors regulating male sterility and their functional conservation in plants. Plant Biotechnol J 19(9):1769–1784
Jiang Y, Li Z, Liu X, Zhu T, Xie K, Hou Q, Yan T, Niu C, Zhang S, Yang M, Xie R (2021b) ZmFAR1 and ZmABCG26 regulated by microRNA are essential for lipid metabolism in maize anther. Int J Mol Sci 22(15):7916
Jiang C, Sun J, Li R, Yan S, Chen W, Guo L, Qin G, Wang P, Luo C, Huang W, Zhang Q (2022a) A reactive oxygen species burst causes haploid induction in maize. Mol Plant 15(6):943–955
Jiang Y, Sun K, An X (2022b) CRISPR/Cas system: applications and prospects for maize improvement. ACS Agric Sci Technol 2(2):174–183
Jiang Y, An X, Li Z, Yan T, Zhu T, Xie K, Wan X (2021) CRISPR/Cas9-based discovery of maize transcription factors regulating male sterility and their functional conservation in plants. Plant Biotechnol J 19(9):1769–1784
Kang M, Azanu MK, Lee K, Wang K (2025) A rapid agrobacterium-mediated transformation method using maize B104 immature embryos. Cold Spring Harb Protoc 2025(11):pdb-prot108595
Khatoon G, Lal SK, Kumar S, Mehta S, Kumar M (2024) Advances in genetic strategies for reducing Phytic acid content in maize: from traditional mutants to CRISPR/Cas9 innovations
Kirov I (2023) Toward transgene-free transposon-mediated biological mutagenesis for plant breeding. Int J Mol Sci 24(23):17054
Kuzma J, Grieger K (2020) Community-led governance for gene-edited crops. Science 370(6519):916–918
Kaul T, Thangaraj A, Jain R, Bharti J, Kaul R, Verma R, Sony SK, Motelb KFA, Yadav P, Agrawal PK (2024) CRISPR/Cas9-mediated homology donor repair base editing system to confer herbicide resistance in maize (Zea mays L.). Plant Physiol Biochem 207:108374
Kaur R, Singh A (2022) Coloured maize and its unique features. In: Maize. CRC Press, pp 247–284
Kaur R, Pahwa AA, Bhise SH (2021) Maize: a potential grain for functional and nutritional properties. In: Cereals and cereal-based foods. Apple Academic Press, pp 65–80
Kaur N, Kumar R, Singh A, Shobha D, Das AK, Chaudhary D, Kaur Y, Kumar P, Sharma P, Singh B (2022) Improvement in nutritional quality of traditional unleavened flat bread using quality protein maize. Front Nutr 9:963368
Kausch AP, Wang K, Kaeppler HF, Gordon-Kamm W (2021) Maize transformation: history, progress, and perspectives. Mol Breed 41:1–36
Khammona K, Dermail A, Suriharn K, Lübberstedt T, Wanchana S, Thunnom B, Poncheewin W, Toojinda T, Ruanjaichon V, Arikit S (2024) Accelerating haploid induction rate and haploid validation through marker-assisted selection for qhir1 and qhir8 in maize. Front Plant Sci 15:1337463
Khan A, Nasim N, Pudhuvai B, Koul B, Upadhyay SK, Sethi L, Dey N (2023) Plant synthetic promoters: advancement and prospective. Agriculture 13(2):298
Khatri P, Kumar P, Shakya KS, Kirlas MC, Tiwari KK (2024) Understanding the intertwined nature of rising multiple risks in modern agriculture and food system. Environ Dev Sustain 26(9):24107–24150
Kiani BH (2023) Genome editing of maize with CRISPR/Cas systems: past, present, and future. In: CRISPR/Cas-mediated genome editing in plants. Apple Academic Press, pp 173–190
Kim HY, Lee KY, Kim M, Hong M, Deepa P, Kim S (2023) A review of the biological properties of purple corn (Zea mays L.). Sci Pharm 91(1):6
Kishi-Kaboshi M, Nishizawa-Yokoi A, Mitsuhara I, Toki S, Sasaki K (2023) Excision of DNA fragments with the piggyBac system in Chrysanthemum morifolium. Plant Biotechnol 40(2):157–165
Kock MA (2022) A new holistic system for plant innovations: UPOV 2030. In: Intellectual property protection for plant related innovation: fit for future? Springer International Publishing, Cham, pp 305–362

Kocsisova Z, Coneva V (2023) Strategies for delivery of CRISPR/Cas-mediated genome editing to obtain edited plants directly without transgene integration. Front Genome Ed 5:1209586
Konstantakos V, Nentidis A, Krithara A, Paliouras G (2022) CRISPR–Cas9 gRNA efficiency prediction: an overview of predictive tools and the role of deep learning. Nucleic Acids Res 50(7):3616–3637
Koul B (2022) Transgenics and crop improvement. In: Cisgenics and transgenics: strategies for sustainable crop development and food security. Springer Nature Singapore, Singapore, pp 131–347
Kozovska Z, Rajcaniova S, Munteanu P, Dzacovska S, Demkova L (2021) CRISPR: history and perspectives to the future. Biomed Pharmacother 141:1–9
Kuang Y, Yu H, Qi F, Zhou X, Li X, Zhou H (2024) Developing herbicide-resistant crops through genome editing technologies: a review. Crop Prot 183:106745
Lamichhane S, Thapa S (2022) Advances from conventional to modern plant breeding methodologies. Plant Breed Biotechnol 10(1):1–4
Li Q, Wu G, Zhao Y, Wang B, Zhao B, Kong D, Wei H, Chen C, Wang H (2020a) CRISPR/Cas9-mediated knockout and overexpression studies reveal a role of maize phytochrome C in regulating flowering time and plant height. Plant Biotechnol J 18(12):2520–2532
Li Y, Zhu J, Wu H, Liu C, Huang C, Lan J, Zhao Y, Xie C (2020b) Precise base editing of non-allelic acetolactate synthase genes confers sulfonylurea herbicide resistance in maize. Crop J 8(3):449–456
Li P, Li Z, Xie G, Zhang J (2021a) Trihelix transcription factor ZmThx20 is required for kernel development in maize. Int J Mol Sci 22(22):12137
Li Y, Lin Z, Yue Y, Zhao H, Fei X, E L, Liu C, Chen S, Lai J, Song W (2021b) Loss-of-function alleles of ZmPLD3 cause haploid induction in maize. Nat Plants 7(12):1579–1588
Li Q, Feng Q, Snouffer A, Zhang B, Rodríguez GR, van der Knaap E (2022) Increasing fruit weight by editing a cis-regulatory element in tomato KLUH promoter using CRISPR/Cas9. Front Plant Sci 13:879642
Li X, Liu W, Wang G, Sun SSM, Yuan L, Wang J (2024) Improving digestibility of sorghum proteins by CRISPR/Cas9-based genome editing. Food Energy Secur 13(1):e506
Li A, Yang Y, Guo Y, Li Q, Zhou A, Wang J, Lu R, Shelden MC, Wu C, Wu J (2025a) ZmASR6 positively regulates salt stress tolerance in maize. New Crops 2:100067
Li L, Fu X, Qi X, Xiao B, Liu C, Wu Q, Zhu J, Xie C (2025b) Harnessing haploid-inducer mediated genome editing for accelerated maize variety development. Plant Biotechnol J 23:1604
Limenie AD, Alehegn M (2025) Genetic engineering for cereal crop yield improvement and disease resistant breeding. Sci World J 2025(1):6743917
Lipps S (2021) Identifying loci conferring quantitative disease resistance in sorghum and maize against fungal foliar pathogens, Doctoral dissertation. University of Illinois at Urbana-Champaign
Liu J, Dawe RK (2023) Large haplotypes highlight a complex age structure within the maize pan-genome. Genome Res 33(3):359–370
Liu S, Li C, Wang H, Wang S, Yang S, Liu X, Yan J, Li B, Beatty M, Zastrow-Hayes G, Song S (2020) Mapping regulatory variants controlling gene expression in drought response and tolerance in maize. Genome Biol 21:1–22
Liu Q, Yang F, Zhang J, Liu H, Rahman S, Islam S, Ma W, She M (2021) Application of CRISPR/Cas9 in crop quality improvement. Int J Mol Sci 22:4200–4206
Liu C, Kong M, Yang F, Zhu J, Qi X, Weng J, Di D, Xie C (2022a) Targeted generation of Null Mutants in ZmGDIα confers resistance against maize rough dwarf disease without agronomic penalty. Plant Biotechnol J 20(5):803
Liu C, Kong M, Zhu J, Qi X, Duan C, Xie C (2022b) Engineering null mutants in ZmFER1 confers resistance to ear rot caused by Fusarium verticillioides in maize. Plant Biotechnol J 20(11):2045

Liu X, Zhang S, Jiang Y, Yan T, Fang C, Hou Q, Wu S, Xie K, An X, Wan X (2022c) Use of CRISPR/Cas9-based gene editing to simultaneously mutate multiple homologous genes required for pollen development and male fertility in maize. Cells 11(3):439
Liu Z, Fan H, Ma Z (2022d) Comparison of SWEET gene family between maize and foxtail millet through genomic, transcriptomic, and proteomic analyses. Plant Genome 15(3):e20226
Liu Z, Liu H, Huang C, Zhou Q, Luo Y (2024) Hybrid Cas12a variants with relaxed PAM requirements expand genome editing compatibility. ACS Synth Biol 13(6):1809–1819
Luo T, Li L, Wang S, Cheng N (2023) Research progress of nucleic acid detection technology for genetically modified maize. Int J Mol Sci 24(15):12247
Lee K, Zhang Y, Kleinstiver BP, Guo JA, Aryee MJ, Miller J, Malzahn A, Zarecor S, Lawrence-Dill CJ, Joung JK, Wang K (2019) Activities and specificities of CRISPR/Cas9 and Cas12a nucleases for targeted mutagenesis in maize. Plant Biotechnol J 17(2):362–372
Li F, Chung T, Pennington JG, Federico ML, Kaeppler HF, Kaeppler SM et al (2015) Autophagic recycling plays a central role in maize nitrogen remobilization. Plant Cell 27(5):1389–1408
Li K, Liu Y, Cao H, Zhang Y, Gu Z, Liu X, Yu A, Kaphle P, Dickerson KE, Ni M, Xu J (2020a) Interrogation of enhancer function by enhancer-targeting CRISPR epigenetic editing. Nat Commun 11(1):485
Li N, Lin B, Wang H, Li X, Yang F, Ding X, Yan J, Chu Z (2019) Natural variation in Zm FBL41 confers banded leaf and sheath blight resistance in maize. Nat Genet 51(10):1540–1548
Li Q, Wu G, Zhao Y, Wang B, Zhao B, Kong D, Wang H (2020b) CRISPR/Cas9-mediated knockout and overexpression studies reveal a role of maize phytochrome C in regulating flowering time and plant height. Plant Biotechnol J 18(12):2520–2532
Li Z, Zhang X, Zhao Y, Li Y, Zhang G, Peng Z, Zhang J (2018) Enhancing auxin accumulation in maize root tips improves root growth and dwarfs plant height. Plant Biotechnol J 16(1):86–99
Liang Y, Liu Q, Wang X, Huang C, Xu G, Hey S, Lin HY, Li C, Xu D, Wu L, Tian F (2019) Zm MADS 69 functions as a flowering activator through the ZmRap2. 7-ZCN 8 regulatory module and contributes to maize flowering time adaptation. New Phytol 221(4):2335–2347
Lin K, Williams DV, Zeng M, Ahmed IM, Dai H, Cao F, Wu F (2022) Identification of low grain cadmium accumulation genotypes and its physiological mechanism in maize (Zea mays L.). Environ Sci Pollut Res 29(14):20721–20730
Liu H, Ding Y, Zhou Y, Jin W, Xie K, Chen LL (2017) CRISPR-P 2.0: an improved CRISPR-Cas9 tool for genome editing in plants. Mol Plant 10(3):530–532
Liu X, Zhang S, Jiang Y, Yan T, Fang C, Hou Q, Wan X (2022) Use of CRISPR/Cas9-based gene editing to simultaneously mutate multiple homologous genes required for pollen development and male fertility in maize. Cells 11(3):439
Lorenzo CD, Debray K, Herwegh D, Develtere W, Impens L, Schaumont D, Vandeputte W, Aesaert S, Coussens G, De Boe Y, Inzé D (2023) BREEDIT: a multiplex genome editing strategy to improve complex quantitative traits in maize. Plant Cell 35(1):218–238
Lorenzo P, Alvarez-Iglesias L, González L, Revilla P (2022) Assessment of Acacia dealbata as green manure and weed control for maize crop. Renew Agri Food Sys 37(4):322–336
Mao Y, Yang X, Zhou Y, Zhang Z, Botella JR, Zhu JK (2018) Manipulating plant RNA-silencing pathways to improve the gene editing efficiency of CRISPR/Cas9 systems. Genome Biol 19(1):149
Marand AP, Eveland AL, Kaufmann K, Springer NM (2023) Cis-regulatory elements in plant development, adaptation, and evolution. Annu Rev Plant Biol 74(1):111–137
Mikhaylova E (2025) Virus-induced genome editing (VIGE): one step away from an agricultural revolution. Int J Mol Sci 26(10):4599
Mikhaylova EV, Khusnutdinov EA, Chemeris AV, Kuluev BR (2022) Available toolkits for CRISPR/Cas genome editing in plants. Russ J Plant Physiol 69(1):3
Molla KA, Sretenovic S, Bansal KC, Qi Y (2021) Precise plant genome editing using base editors and prime editors. Nat Plants 7(9):1166–1187
Mahmood MA, Naqvi RZ, Rahman SU, Amin I, Mansoor S (2023) Plant virus-derived vectors for plant genome engineering. Viruses 15(2):531

Mallakpour S, Hussain CM, Banerjee S, Prakash PC, Kadeppagari RK (2021) Zein-based nanoproducts in nutrition and food sectors. In: Handbook of consumer nanoproducts, pp 1–16
Marangoni F, Agostoni C, Borghi C, Catapano AL, Cena H, Ghiselli A, La Vecchia C, Lercker G, Manzato E, Pirillo A, Riccardi G (2020) Dietary linoleic acid and human health: focus on cardiovascular and cardiometabolic effects. Atherosclerosis 292:90–98
Million CR, Wijeratne S, Karhoff S, Cassone BJ, McHale LK, Dorrance AE (2023) Molecular mechanisms underpinning quantitative resistance to Phytophthora sojae in Glycine max using a systems genomics approach. Front Plant Sci 14:1277585
Mitiku T (2022) Review on haploid and double haploid maize (Zea mays) breeding technology. Int J Agric Sci Food Technol 8(1):52–58
Mrowicka M, Mrowicki J, Kucharska E, Majsterek I (2022) Lutein and zeaxanthin and their roles in age-related macular degeneration—neurodegenerative disease. Nutrients 14(4):827
Ning Q, Jian Y, Du Y, Li Y, Shen X, Jia H, Zhao R, Zhan J, Yang F, Jackson D, Liu L (2021) An ethylene biosynthesis enzyme controls quantitative variation in maize ear length and kernel yield. Nat Commun 12(1):5832
Nivya VM, Shah JM (2023) Recalcitrance to transformation, a hindrance for genome editing of legumes. Front Genome Ed 5:1247815
Njeru F, Wambua A, Muge E, Haesaert G, Gettemans J, Misinzo G (2023) Major biotic stresses affecting maize production in Kenya and their implications for food security. PeerJ 11:e15685
Oh Y, Kim H, Kim SG (2021) Virus-induced plant genome editing. Curr Opin Plant Biol 60:101992
Pandey AK, Aggarwal S, Meena V, Kumar A (2021) Phytic acid reduction in cereal grains by genome engineering: potential targets to achieve low phytate wheat. Gen Eng Crop Improve:146–156
Perera OP, LIttle NS, Abdelgaffar H, Jurat-Fuentes JL, Reddy GV (2021) Genetic knockouts indicate that the ABCC2 gene in the bollworm Helicoverpa Zea is not a major receptor for the Cry1Ac insecticidal protein
Pant S, Nag P, Ghati A, Chakraborty D, Maximiano MR, Franco OL, Mandal AK, Kuila A (2022) Employment of the CRISPR/Cas9 system to improve cellulase production in Trichoderma reesei. Biotechnol Adv 60:108022
Pathi KM, Rink P, Budhagatapalli N, Betz R, Saado I, Hiekel S, Becker M, Djamei A, Kumlehn J (2020) Engineering smut resistance in maize by site-directed mutagenesis of LIPOXYGENASE 3. Front Plant Sci 11:543895
Pramitha JL, Rana S, Aggarwal PR, Ravikesavan R, Joel AJ, Muthamilarasan M (2021) Diverse role of phytic acid in plants and approaches to develop low-phytate grains to enhance bioavailability of micronutrients. Adv Genet 107:89–120
Qamar Z, Rizwan M, Rani R, Shahzad R, Jamil S, Faheem M, Abro S, Shimelis H (2024) Trends towards the production of biologically safe marker free transgenic plants. Appl Ecol Environ Res 22(1)
Qin L, Wu X, Zhao H (2023) Molecular and functional dissection of LIGULELESS1 (LG1) in plants. Front Plant Sci 14:1190004
Qu Y, Fernie AR, Liu J, Yan J (2024) Doubled haploid technology and synthetic apomixis: recent advances and applications in future crop breeding. Mol Plant 17:1005
Quade A (2023) Breeding for durable resistance to Puccinia sorghi: status and strategies
Qi X, Guo S, Wang D, Zhong Y, Chen M, Chen C, Liu C (2022) ZmCOI2a and ZmCOI2b redundantly regulate anther dehiscence and gametophytic male fertility in maize. Plant J 110(3):849–862
Qin H, Xiao H, Zou G, Zhou Z, Zhong JJ (2017) CRISPR-Cas9 assisted gene disruption in the higher fungus Ganoderma species. Process Biochem 56:57–61
Raboy V, Gerbasi PF, Young KA, Stoneberg SD, Pickett SG, Bauman AT, Murthy PP, Sheridan WF, Ertl DS (2000) Origin and seed phenotype of maize low phytic acid 1-1 and low phytic acid 2-1. Plant Physiol 124(1):355–368

Rafique AA, Zu UM, Khan N, Hussain A, Rauf A (2024) Enhancing wheat (Triticum Aestivum l.) yield through genetic modification. Biomed Clin Res 3(4)

Rahman F, Mishra A, Gupta A, Sharma R (2022) Spatiotemporal regulation of CRISPR/Cas9 enables efficient, precise, and heritable edits in plant genomes. Front Genome Ed 4:870108

Rallapalli KL, Komor AC (2023) The design and application of DNA-editing enzymes as base editors. Annu Rev Biochem 92(1):43–79

Rani R, Singh PK, Tewari H, Agarwal P (2021) Growth and instability in maize production and export in India

Rasheed A, Gill RA, Hassan MU, Mahmood A, Qari S, Zaman QU, Ilyas M, Aamer M, Batool M, Li H, Wu Z (2021) A critical review: recent advancements in the use of CRISPR/Cas9 technology to enhance crops and alleviate global food crises. Curr Issues Mol Biol 43(3):1950–1976

Resende MPM, Filho AJC, Antunes AM, de Oliveira BM, de Oliveira RG (2022) Population genomics of maize. In: Population genomics: crop plants. Springer International Publishing, Cham, pp 377–455

Reza MAN, Harvey TN, Regmi A, Torgersen JS, Sandvik GK (2024) Exploring the use of alternative promoters for enhanced transgene and sgRNA expression in Atlantic salmon cells. Mar Biotechnol 26(6):1143–1154

Saeed F, Manzoor H, Bukhat S, Azeem F, Rasul S (2025) Multiplex genome editing in plants. In: Genome editing for crop improvement: theory and methodology. CABI, pp 68–87

Sangare DB, Shetti P, Yankanchi S, Kadirimangalam SR, Baguda R, Xingming F, Fan J, Singh S, Jadhav RA, Ashrutha MA, Aditi K (2023) Maize Nutraceutomics: genomics, biotechnology, and nanotechnology. In: Compendium of crop genome designing for nutraceuticals. Springer Nature Singapore, Singapore, pp 85–113

Santos IGD, Verzegnazzi AL, Edwards J, Frei UK, Boerman N, Tonello Zuffo L, Pires LP, de La Fuente G, Lübberstedt T (2022) Usefulness of temperate-adapted maize lines developed by doubled haploid and single-seed descent methods. Theor Appl Genet 135(6):1829–1841

Saritha A, Ramanjaneyulu AV, Sainath N, Umarani E (2020) Nutritional importance and value addition in maize. Biotica Res Today 2(9):974–977

Schurack S (2021) Mechanisms of quantitative disease resistance in the maize-Ustilago maydis interaction, Doctoral dissertation. Universität zu Köln

Shao J, Peng B, Zhang Y, Yan X, Yao X, Hu X, Li L, Fu X, Zheng H, Tang K (2024) A high-efficient protoplast transient system for screening gene editing elements in salvia miltiorrhiza. Plant Cell Rep 43(2):45

Simmons CR, Lafitte HR, Reimann KS, Brugière N, Roesler K, Albertsen MC, Greene TW, Habben JE (2021) Successes and insights of an industry biotech program to enhance maize agronomic traits. Plant Sci 307:110899

Singh M, Nara U, Kumar A, Thapa S, Jaswal C, Singh H (2022) Enhancing genetic gains through marker-assisted recurrent selection: from phenotyping to genotyping. Cereal Res Commun 50(3):523–538

Singh SB, Kumar B, Singh A, Kumar S (2024) Conventional and advance breeding approaches for developing abiotic stress tolerant maize. In: Adapting to climate change in agriculture-theories and practices: approaches for adapting to climate change in agriculture in India. Springer Nature Switzerland, Cham, pp 281–302

Snehi S, Choudhary M, Kumar S, Jayaswal D, Kumar S, Prakash NR (2024) Mapping of quantitative traits loci: harnessing genomics revolution for dissecting complex traits. In: Genomics data analysis for crop improvement. Springer Nature Singapore, Singapore, pp 125–157

Song J, Datla R, Zou J, Xiang D (2024) Haploid induction: an overview of parental factor manipulation during seed formation. Front Plant Sci 15:1439350

Sony SK, Kaul T, Motelb KFA, Thangaraj A, Bharti J, Kaul R, Verma R, Nehra M (2023) CRISPR/Cas9-mediated homology donor repair base editing confers glyphosate resistance to rice (Oryza sativa L.). Front Plant Sci 14:1122926

Sultana SN, Zubayer M, Islam MS, Ashab KR, Shanta SH, Kayesh E (2024) Sustainable and eco-friendly approach to weed management. In: Weed management-global strategies, p 83

Sumanth KS, Anusha T, Prakash S (2022) Transgenic crops: present status, problems and future prospects. Pharma Innov J 11:753–775
Sun S, Wang B, Li C, Xu G, Yang J, Hufford MB, Ross-Ibarra J, Wang H, Wang L (2023) Unraveling prevalence and effects of deleterious mutations in maize elite lines across decades of modern breeding. Mol Biol Evol 40(8):msad170
Sun L, Lai M, Ghouri F, Nawaz MA, Ali F, Baloch FS, Nadeem MA, Aasim M, Shahid MQ (2024) Modern plant breeding techniques in crop improvement and genetic diversity: from molecular markers and gene editing to artificial intelligence—a critical review. Plants 13(19):2676
Surya Krishna S, Harish Chandar SR, Ravi M, Valarmathi R, Lakshmi K, Prathima PT et al (2023) Transgene-free genome editing for biotic and abiotic stress resistance in sugarcane: prospects and challenges. Agronomy 13(4):1000
Saber Sichani A, Ranjbar M, Baneshi M, Torabi Zadeh F, Fallahi J (2023) A review on advanced CRISPR-based genome-editing tools: base editing and prime editing. Mol Biotechnol 65(6):849–860
Sahu A, Verma R, Gupta U, Kashyap S, Sanyal I (2024) An overview of targeted genome editing strategies for reducing the biosynthesis of phytic acid: An anti-nutrient in crop plants. Mol Biotechnol 66(1):11–25
Salika R, Riffat J (2021) Abiotic stress responses in maize: a review. Acta Physiol Plant 43(9):130
Schmitz RJ, Grotewold E, Stam M (2022) Cis-regulatory sequences in plants: their importance, discovery, and future challenges. Plant Cell 34(2):718–741
Shen X, Liu L, Tran T, Ning Q, Li M, Huang L, Zhao R, Li Y, Qing X, Jackson D, Zhao H (2024) KRN5b regulates maize kernel row number through mediating phosphoinositol signalling. Plant Biotechnol J 22(12):3427–3441
Shi J, Habben JE, Archibald RL, Drummond BJ, Chamberlin MA, Williams RW, Lafitte HR, Weers BP (2015) Overexpression of ARGOS genes modifies plant sensitivity to ethylene, leading to improved drought tolerance in both Arabidopsis and maize. Plant Physiol 169(1):266–282
Singh A, Karjagi C, Rakshit S (2020) Minimally altering a critical kinase for low-phytate maize. Sci Rep 10(1):6324
Smith JSC, Hussain T, Jones ES, Graham G, Podlich D, Wall S, Williams M (2008) Use of doubled haploids in maize breeding: implications for intellectual property protection and genetic diversity in hybrid crops. Mol Breed 22(1):51–59
Snowdon RJ, Wittkop B, Chen TW, Stahl A (2021) Crop adaptation to climate change as a consequence of long-term breeding. Theor Appl Genet 134(6):1613–1623
Sosso D, Luo D, Li QB, Sasse J, Yang J, Gendrot G, Suzuki M, Koch KE, McCarty DR, Chourey PS, Frommer WB (2015) Seed filling in domesticated maize and rice depends on SWEET-mediated hexose transport. Nat Genet 47(12):1489–1493
Srivastava AK, Suresh Kumar J, Suprasanna P (2021) Seed 'primeomics': plants memorize their germination under stress. Biol Rev 96(5):1723–1743
Svitashev S, Young JK, Schwartz C, Gao H, Falco SC, Cigan AM (2015) Targeted mutagenesis, precise gene editing, and site-specific gene insertion in maize using Cas9 and guide RNA. Plant Physiol 169(2):931–945
Tang B, Zhang Z, Zhao X, Xu Y, Wang L, Chen XL, Wang W (2022) Multi-omics analysis reveals a regulatory network of ZmCCT during maize resistance to Gibberella stalk rot at the early stage. Front Plant Sci 13:917493
Tang Y, Zhang Z, Yang Z, Wu J (2023) CRISPR/Cas9 and Agrobacterium tumefaciens virulence proteins synergistically increase efficiency of precise genome editing via homology directed repair in plants. J Exp Bot 74(12):3518–3530
Tariq A, Mushtaq M, Yaqoob H, Bhat BA, Zargar SM, Raza A, Ali S, Charagh S, Mubarik MS, Zaman QU, Prasad PV (2023) Putting CRISPR-Cas system in action: a golden window for efficient and precise genome editing for crop improvement. GM Crops & Food 14(1):1–27
Teng C, Zhang H, Hammond R, Huang K, Meyers BC, Walbot V (2020) Dicer-like 5 deficiency confers temperature-sensitive male sterility in maize. Nat Commun 11(1):2912

Thakur P, Kumari N, Kumar A, Sharma P, Chadha S (2024) Recent advances in development and utilization of double haploids (DHs) in economically important vegetable crops. Plant Cell Tissue Organ Cult 156(1):15

Thingnam SS, Lourembam DS, Tongbram PS, Lokya V, Tiwari S, Khan MK, Pandey A, Hamurcu M, Thangjam R (2023) A perspective review on understanding drought stress tolerance in wild banana genetic resources of Northeast India. Genes 14(2):370

Tian J, Wang C, Chen F, Qin W, Yang H, Zhao S, Xia J, Du X, Zhu Y, Wu L, Cao Y (2024) Maize smart-canopy architecture enhances yield at high densities. Nature 632(8025):576–584

Tyagi A, Mir ZA, Almalki MA, Deshmukh R, Ali S (2024) Genomics-assisted breeding: a powerful breeding approach for improving plant growth and stress resilience. Agronomy 14(6):1128

Vanhevel Y, De Moor A, Muylle H, Vanholme R, Boerjan W (2024) Breeding for improved digestibility and processing of lignocellulosic biomass in Zea mays. Front Plant Sci 15:1419796

Varotto S (2023) Current status and future prospective of genome editing application in maize. In: A roadmap for plant genome editing. Springer Nature Switzerland, Cham, pp 165–182

Vanacore MFG, Sartori M, Giordanino F, Barros G, Nesci A, García D (2023) Physiological effects of microbial biocontrol agents in the maize Phyllosphere. Plants 12(24):4082

Voytas DF, Gao C (2014) Precision genome engineering and agriculture: opportunities and regulatory challenges. PLoS Biol 12(6):e1001877

Wang B, Lin Z, Li X, Zhao Y, Zhao B, Wu G, Ma X, Wang H, Xie Y, Li Q, Song G (2020) Genome-wide selection and genetic improvement during modern maize breeding. Nat Genet 52(6):565–571

Wang L, Kaya HB, Zhang N, Rai R, Willmann MR, Carpenter SC, Read AC, Martin F, Fei Z, Leach JE, Martin GB (2021) Spelling changes and fluorescent tagging with prime editing vectors for plants. Front Genome Ed 3:617553

Wang J, He Z, Wang G, Zhang R, Duan J, Gao P, Lei X, Qiu H, Zhang C, Zhang Y, Yin H (2022a) Efficient targeted insertion of large DNA fragments without DNA donors. Nat Methods 19(3):331–340

Wang P, Pu Y, Abid MA, Kang L, Ye Y, Zhang M, Liang C, Wei Y, Zhang R, Meng Z (2022b) A rapid and efficient method for isolation and transformation of cotton callus protoplast. Int J Mol Sci 23(15):8368

Wang Y, Tang Q, Pu L, Zhang H, Li X (2022c) CRISPR-Cas technology opens a new era for the creation of novel maize germplasms. Front Plant Sci 13:1049803

Wang Y, Zafar N, Ali Q, Manghwar H, Wang G, Yu L, Ding X, Ding F, Hong N, Wang G, Jin S (2022d) CRISPR/Cas genome editing technologies for plant improvement against biotic and abiotic stresses: advances, limitations, and future perspectives. Cells 11(23):3928

Wang N, Ryan L, Sardesai N, Wu E, Lenderts B, Lowe K, Che P, Anand A, Worden A, Van Dyk D, Barone P (2023) Leaf transformation for efficient random integration and targeted genome modification in maize and sorghum. Nat Plants 9(2):255–270

Wang W, Javed T, Shen L, Sun T, Yang B, Zhang S (2024a) Establishment of an efficient sugarcane transformation system via herbicide-resistant CP4-EPSPS gene selection. Plants 13(6):852

Wang WY, Ren ZJ, Li PQ, Huang J, Chai WW, Shi YQ, Bao AK, Hou CC, Li LG, Ma Q, Wang SM (2024b) HKT1; 1 modulated by vacuolar Na+ compartmentation functions in maintaining the salt-accumulating trait in a xerophyte. Environ Exp Bot 226:105921

Wang C, Zhou Y, Wang Y, Jiao P, Liu S, Guan S, Ma Y (2025) CRISPR-Cas9-mediated editing of ZmPL1 gene improves tolerance to drought stress in maize. GM Crops Food 16(1):1–16

Wang CT, Ru JN, Liu YW, Li M, Zhao D, Yang JF, Fu JD, Xu ZS (2018) Maize WRKY transcription factor ZmWRKY106 confers drought and heat tolerance in transgenic plants. Int J Mol Sci 19(10):3046

Wang C, Zhou Y, Wang Y, Jiao P, Liu S, Guan S, Ma Y (2025) CRISPR-Cas9-mediated editing of *ZmPL1* gene improves tolerance to drought stress in maize. GM Crop Food 16(1):1–16

Wang X, Wang H, Liu S, Ferjani A, Li J, Yan J, Yang X, Qin F (2016) Genetic variation in ZmVPP1 contributes to drought tolerance in maize seedlings. Nat Genet 48(10):1233–1241

Wang Y, Liu X, Zheng X, Wang W, Yin X, Liu H, Wang F (2021) Creation of aromatic maize by CRISPR/Cas. J Integr Plant Biol 63(9):1664–1670
Wu Y, Wang S, Du W, Ding Y, Li W, Chen Y, Wang Y (2023) Sugar transporter *ZmSWEET1b* is responsible for assimilate allocation and salt stress response in maize. Funct Integr Genomics 23(2):137
Widomska J, SanGiovanni JP, Subczynski WK (2020) Why is zeaxanthin the most concentrated xanthophyll in the central fovea? Nutrients 12(5):1333
Xie K, Minkenberg B, Yang Y (2015) Boosting CRISPR/Cas9 multiplex editing capability with the endogenous tRNA-processing system. Proc Natl Acad Sci 112(11):3570–3575
Xu G, Zhang X, Chen W, Zhang R, Li Z, Wen W, Warburton ML, Li J, Li H, Yang X (2022) Population genomics of Zea species identifies selection signatures during maize domestication and adaptation. BMC Plant Biol 22(1):72
Xu JH, Messing J (2008) Organization of the prolamin gene family provides insight into the evolution of the maize genome and gene duplications in grass species. Proc Natl Acad Sci 105(38):14330–14335
Xu J, Yin Y, Jian L, Han B, Chen Z, Yan J, Liu X (2021a) Seeing is believing: a visualization toolbox to enhance selection efficiency in maize genome editing. Plant Biotechnol J 19(5):872
Xu R, Li Y, Sui Z, Lan T, Song W, Zhang M, Zhang Y, Xing J (2021b) A C-terminal encoded peptide, ZmCEP1, is essential for kernel development in maize. J Exp Bot 72(15):5390–5406
Yali W (2022) Haploids and doubled haploid technology application in modern plant breeding. J Plant Sci 10(2):71–75
Yan Y, Zhu J, Qi X, Cheng B, Liu C, Xie C (2021) Establishment of an efficient seed fluorescence reporter-assisted CRISPR/Cas9 gene editing in maize. J Integr Plant Biol 63(9):1671–1680
Yan Y, Zhu X, Yu Y, Li C, Zhang Z, Wang F (2022) Nanotechnology strategies for plant genetic engineering. Adv Mater 34(7):2106945
Yang Y, Xu C, Shen Z, Yan C (2022) Crop quality improvement through genome editing strategy. Front Genome Edit 3:819687
Yang T, Dong J, Xiong X, Zhang L, Wang J, Hu H, Zhou L, Yang W, Ma Y, Fu H, Chen J (2024a) A novel function of GW5 on controlling the early growth vigor and its haplotype effect on shoot dry weight and grain size in Rice (Oryza sativa L.). Rice 17(1):49
Yang W, Zhai H, Wu F, Deng L, Chao Y, Meng X, Chen Q, Liu C, Bie X, Sun C, Yu Y (2024b) Peptide REF1 is a local wound signal promoting plant regeneration. Cell 187(12):3024–3038
Yassitepe JEDCT, da Silva VCH, Hernandes-Lopes J, Dante RA, Gerhardt IR, Fernandes FR, da Silva PA, Vieira LR, Bonatti V, Arruda P (2021) Maize transformation: from plant material to the release of genetically modified and edited varieties. Front Plant Sci 12:766702
Yau YY, Easterling M, Kumar A (2024) Advances in delivery of CRISPR–Cas reagents for precise genome editing in plants. In: Gene editing in plants: CRISPR-Cas and its applications. Springer Nature Singapore, Singapore, pp 543–569
Yan J, Kandianis CB, Harjes CE, Bai L, Kim EH, Yang X, Skinner DJ, Fu Z, Mitchell S, Li Q, Rocheford T (2010) Rare genetic variation at Zea mays crtRB1 increases β-carotene in maize grain. Nat Genet 42(4):322–327
Yan P, Du Q, Chen H, Guo Z, Wang Z, Tang J, Li WX (2023) Biofortification of iron content by regulating a NAC transcription factor in maize. Science 382(6675):1159–1165
Yang J, Xu G, Zhang M, Xue W, Wu J, Li Y, Wang H (2025) Dual role of Glossy15 in regulating flowering by modulating gibberellins and floral organ gene expression in maize. New Phytol 248(4):2143–2159
Yoshida K, Sakamoto S, Mitsuda N (2024) Synthetic-biology approach for plant lignocellulose engineering. Plant Biotechnol 41(3):213–230
Young J, Zastrow-Hayes G, Deschamps S, Svitashev S, Zaremba M, Acharya A, Paulraj S, Peterson-Burch B, Schwartz C, Djukanovic V, Kumar S (2019) CRISPR-Cas9 editing in maize: systematic evaluation of off-target activity and its relevance in crop improvement. Sci Rep 9(1):6729

Yi H, Liang Z, Ge J, Zhang H, Liu F, Ren X, Ren J, Wang H, Ren J, Ren X, Zhang Y (2022) A multiplex PCR system for the screening of genetically modified (GM) maize and the detection of 29 GM maize events based on capillary electrophoresis. Agriculture 12(3):413
Yu X, Huo G, Yu J, Li H, Li J (2023) Prime editing: its systematic optimization and current applications in disease treatment and agricultural breeding. Int J Biol Macromol 253:127025
Zargar M, Zavarykina T, Voronov S, Pronina I, Bayat M (2022) The recent development in technologies for attaining doubled haploid plants in vivo. Agriculture 12(10):1595
Zhang L, Yang X (2024) Advances in sugarcane genomics: navigating through complex polyploid genomes. Mol Plant Breed 15
Zhang J, Feng C, Su H, Liu Y, Liu Y, Han F (2020a) The cohesin complex subunit ZmSMC3 participates in meiotic centromere pairing in maize. Plant Cell 32(4):1323–1336
Zhang Z, Zhang X, Lin Z, Wang J, Liu H, Zhou L, Zhong S, Li Y, Zhu C, Lai J, Li X (2020b) A large transposon insertion in the stiff1 promoter increases stalk strength in maize. Plant Cell 32(1):152–165
Zhang J, Augustine RC, Suzuki M, Feng J, Char SN, Yang B, McCarty DR, Vierstra RD (2021a) The SUMO ligase MMS21 profoundly influences maize development through its impact on genome activity and stability. PLoS Genet 17(10):e1009830
Zhang R, Ma S, Li L, Zhang M, Tian S, Wang D, Liu K, Liu H, Zhu W, Wang X (2021b) Comprehensive utilization of corn starch processing by-products: a review. Grain Oil Sci Technol 4(3):89–107
Zhang Y, Iaffaldano B, Qi Y (2021c) CRISPR ribonucleoprotein-mediated genetic engineering in plants. Plant Commun 2(2):100168
Zhang T, van der Wiel K, Wei T, Screen J, Yue X, Zheng B, Selten F, Bintanja R, Anderson W, Blackport R, Glomsrød S (2022) Increased wheat price spikes and larger economic inequality with 2 C global warming. One Earth 5(8):907–916
Zhang J, Li K, Sun Y, Yao C, Liu W, Liu H, Zhong Y (2024a) An efficient CRISPR/Cas9 genome editing system based on a multiple sgRNA processing platform in Trichoderma reesei for strain improvement and enzyme production. Biotechnol Biofuels Bioprod 17(1):22
Zhang Y, Mo Y, Ren H, Wu X, Han L, Sun Z, Xu W (2024b) Improving sedum plumbizincicola genetic transformation with the SpGRF4–SpGIF1 gene and the self-excision CRE/LoxP system. Planta 259(5):1–10
Zhang J, Zhang X, Chen R, Yang L, Fan K, Liu Y, Liu Y (2020a) Generation of transgene-free semidwarf maize plants by gene editing of gibberellin-oxidase20-3 using CRISPR/Cas9. Front Plant Sci 11:1048
Zhang S, Wu S, Niu C, Liu D, Yan T, Tian Y, Wan X (2021) ZmMs25 encoding a plastid-localized fatty acyl reductase is critical for anther and pollen development in maize. J Exp Bot 72(12):4298–4318
Zhang X, Huang C, Wu D, Qiao F, Li W, Duan L, Wang K, Xiao Y, Chen G, Liu Q, Yan J (2017) High-throughput phenotyping and QTL mapping reveals the genetic architecture of maize plant growth. Plant Physiol 173(3):1554–1564
Zhang X, Qiao Y, Meng F, Fan C, Zhang M (2018) Identification of maize leaf diseases using improved deep convolutional neural networks. IEEE Access 6:30370–30377
Zhang Z, Dong J, Ji C, Wu Y, Messing J (2019) NAC-type transcription factors regulate accumulation of starch and protein in maize seeds. Proc Natl Acad Sci 116(23):11223–11228
Zhang Z, Zhang X, Lin Z, Wang J, Liu H, Zhou L, Lin Z (2020b) A large transposon insertion in the stiff1 promoter increases stalk strength in maize. Plant Cell 32(1):152–165
Zhang Z, Zheng X, Yang J, Messing J, Wu Y (2016) Maize endosperm-specific transcription factors O2 and PBF network the regulation of protein and starch synthesis. Proc Natl Acad Sci 113(39):10842–10847
Zhou L, Jiang L (2024) Genomics-assisted breeding in maize: techniques and outcomes. Maize Gen Gen 15
Zong Y, Liu Y, Xue C, Li B, Li X, Wang Y, Li J, Liu G, Huang X, Cao X, Gao C (2022) An engineered prime editor with enhanced editing efficiency in plants. Nat Biotechnol 40(9):1394–1402

Zhao L, Koseki SR, Silverstein RA, Amrani N, Peng C, Kramme C, Savic N, Pacesa M, Rodríguez TC, Stan T, Tysinger E (2023) PAM-flexible genome editing with an engineered chimeric Cas9. Nat Commun 14(1):6175

Zhou L, Zhang C, Zhang Y, Wang C, Liu Q (2022) Genetic manipulation of endosperm amylose for designing superior quality rice to meet the demands in the 21st century. J Cereal Sci 105:103481

Zhou J, Luan X, Liu Y, Wang L, Wang J, Yang S, Liu S, Zhang J, Liu H, Yao D (2023) Strategies and methods for improving the efficiency of CRISPR/Cas9 gene editing in plant molecular breeding. Plants 12(7):1478

Zhu M, Tong L, Xu M, Zhong T (2021) Genetic dissection of maize disease resistance and its applications in molecular breeding. Mol Breed 41(5):32

Zingore S, Adolwa IS, Njoroge S, Johnson JM, Saito K, Phillips S, Kihara J, Mutegi J, Murell S, Dutta S, Chivenge P (2022) Novel insights into factors associated with yield response and nutrient use efficiency of maize and rice in sub-Saharan Africa. A review. Agron Sustain Dev 42(5):82

Chapter 13
Applications of CRISPR-Cas for Genetic Improvement of Triticeae

Nayla Munawar, Annena Ijaz, Iram Zovia, Paulo Zaini, Muhammad Naeem Khan, Ahmd Al-Harrasi, and Aftab Ahmad

Abstract The Triticeae tribe has major crops including wheat, rye, and barley, which comprise a major part of the global food source. However, enhancing the quality of these crops is challenging due to their limited genetic diversity, polyploid genomes, and vulnerability to biotic and abiotic stresses. Genetic improvement in these crops has been modified by the CRISPR/Cas system, which allows precise and effective genome editing. The applications of CRISPR/Cas in Triticeae enhance different characteristics including disease resistance, improved grain quality with enhanced yield, and stress tolerance. The potential of CRISPR/Cas9 for targeted genetic modifications has increased with recent developments like base editing, prime editing, and multiplex genome editing. Moreover, the use of CRISPR-based transcriptional regulation offers new possibilities for trait improvement without inducing permanent genomic changes. Despite regulatory obstacles and technical constraints, CRISPR/Cas technology offers a potent method for expediting Triticeae breeding. Incorporating this approach with omics and genomic techniques would increase genetic advantages even further, providing opportunities for the production of sustainable cereal crops to enhance food security.

Nayla Munawar and Annena Ijaz contributed equally with all other contributors.

N. Munawar · I. Zovia
Department of Chemistry, College of Science, United Arab Emirates University, Al-Ain, United Arab Emirates

A. Ijaz
Department of Biochemistry, Government College University, Faisalabad, Pakistan

P. Zaini
Department of Plant Sciences, University of California, Davis, USA

M. N. Khan
School of Pharmaceutical Science and Technology, Tianjin University, Tianjin, China

A. Al-Harrasi · A. Ahmad (✉)
Natural and Medical Sciences Research Center (NMSRC), University of Nizwa, Nizwa, Oman
e-mail: aftab.ahmad@unizwa.edu.om

K. A. Abd-Elsalam, A. Ahmad (eds.), *Gene Editing in Cereals*, Sustainability Sciences in Asia and Africa, https://doi.org/10.1007/978-981-95-6722-5_13

Keywords Triticeae · CRISPR/Cas · Crop Resilience · Genetic Improvement · Food Security

13.1 Introduction

Since the inception of agriculture, cereals have been consumed worldwide due to their nutritional value and beneficial effects on human health. Moreover, the production cost of cereal crops is low, and production strategies are relatively simple (Poutanen et al. 2022). Cereals, such as the Triticeae (wheat, barley, rye, and triticale) species, are important sources for brewing, human consumption, and animal feed. Wheat is the second most important cereal crop as it occupies a central place in human nutrition providing 30% of the daily proteins and calories (Li et al. 2021a, b). Durum wheat (tetraploid) and bread wheat (hexaploid) have similar characteristics but have unique properties that make their use in different bakery products. Durum wheat, due to its high protein and gluten content, has been used in pasta production, whereas the bread wheat is mainly used for bread production due to its soft texture (Hickey et al. 2019).

Barley is the fourth-largest cereal crop in terms of cultivation and global production. Due to its significant importance in agriculture, it ranks after wheat, maize, and rice. Barely is also known as functional grain as it is high in dietary fiber, β-glucans, minerals, and vitamins essential for human diet (Kirschner et al. 2021). Rye is one of the important cereal crops with significant nutritional value mainly micronutrients as well as a potent source of fiber, protein, starch, and antioxidants. Rye comprises the gluten protein, which makes it a valuable addition for bakery products. Rye is considered an important dietary source due to the unique blend of amino acids, minerals (calcium, iron, zinc), and vitamins (tocopherol, folate, B6, riboflavin, thiamin). Plant scientists considered the rye a material to improve targeted mutation induction (Bucsella et al. 2016). Modern varieties of the cereal crops have contributed to the food security from the past half-century. Triticale is one of the modern varieties of cereal crops produced by artificial crossing of durum wheat and rye. The contribution of both crops in various triticale genotypes mainly depends on the proportions of wheat and rye chromatin incorporated. Modern varieties have a greater contribution from the D genome of the wheat (Kuluev et al. 2022).

Cereal crops have been significantly enhanced through different agricultural techniques such as chemical and radiation-mediated mutation breeding, molecular marker-assisted breeding, and conventional breeding. Recent development in genetic engineering have expedited the genetic research and developed new potential to modify the unique traits of the cereal crops for better quality and yield (Jinek et al. 2012). Site-specific nucleases such as zinc finger nucleases (ZFNs) and transcriptional activator-effector nucleases (TALENs) were used for genome editing that induce the double-stranded breaks in the genome thereby altering the target location of the gene. These breaks are subsequently repaired by either by homologous recombination (HR) or nonhomologous end joining (NHEJ). But these systems have few drawbacks as they are time-consuming, expensive, and labor extensive, and the construction protocols are difficult (Murovec et al. 2017).

CRISPR/Cas (Clustered Regularly Interspaced Short Palindromic Repeats) has revolutionized the genome editing in plant biology. It comprises DNA endonucleases known as Cas9 protein and a small fragment called guide RNA. CRISPR-derived RNA (crRNA) and trans-activating RNA (tracrRNA) are the two components that make up the gRNA framework. Since the crRNA contains a tiny homology area, it can bind the crRNA to tracrRNA and precisely target the double-stranded DNA cutting site. TracrRNA mediates the stem loop formation by integrating with the Cas9 protein. The crRNA and tracrRNA are combined to form a single guide RNA chimera (sgRNA), which also coordinates the sequence-dependent Cas9 dsDNA break, in the genome editing method mediated by CRISPR/Cas9. The Cas9-sgRNA complex forms a double-strand break (DSB) by attaching to the target site where the sgRNA matches the homologous sequence.

CRISPR/Cas9 has been widely utilized in crops that have an agricultural importance mainly in cereals crops due to its enhanced and focused targeting, cost-effectiveness, and acceptability. CRISPR/Cas9-based genome editing in cereals has been employed to improve specific traits such as improved quality and yield and make them disease resistant from bacterial, viral, and fungal pathogens. In cereal crops, the major concern is the improvement in nutritional quality that can be achieved by targeting the single or multiple genes to achieve the desired trait.

13.2 CRISPR-Cas Systems for Genetic Improvement in Cereal Crops

13.2.1 CRISPR-Cas9

CRISPR/Cas9, a revolutionary advancement in genetic engineering, has become a prime choice to meet the increasing demand of cereal crops. More than 90% of the global food production is obtained from cereal crops due to the presence of dietary nutrients that play a vital role in growth and development. Genome editing with CRISPR/Cas9 has been successfully demonstrated in cereal crops to enhance disease resistance and improve agronomic traits such as improved yield, drought tolerance, plant height, and flowering time. CRISPR/Cas has been employed for the silencing of *TaEDR1* gene in wheat plant to develop the resistance against powdery mildew with the mutation frequency resulted in up to 28.50% (Char and Yang 2020). Similarly, CRISPR/Cas9 has been used to edit multiple genes (*TaDEP1*, *TaNAC2*, *TaGW2*, and *TaGASR7*) in wheat, thereby resulting in edited plants with improved grain length, area, weight, and length as compared to the wild type plants (Wang et al. 2018a, b). Moreover, CRISPR/Cas9 has potential to edit the *HvOPM1* and *HvOPM2* gene, *OVULE PECTIN MODIFIER1* and two pectin methylesterase genes present in barley to improve the plant yield, thus the mutant lines displayed increased grain size, increased length of ovaries and ovules (Yang et al. 2023). Cheng et al. (2023) successfully knockout the Gibberellin 3-oxidase1 (*HvGA3ox1*) in barley to produce the semi dwarf mutant lines with increased coleoptile length. Consequently, seed dormancy (SD) has been adjusted to the ideal level, and crop yield has been maintained under climate change. Similarly in wheat, CRISPR/Cas9-based deletion of *ZnF-B* gene that codes for the brassinosteroid signaling resulted in semi dwarf wheat mutant plants with increased grain yield by 15.2%. Moreover, CRISPR/Cas9-based knockout of *PDIL5–1* gene in barley resulted in broad-spectrum resistance against viral pathogens such as barley yellow and mild mosaic virus (Basu et al. 2023). This accumulating body of evidence highlights the potential of CRISR-based editing to alter the architecture of cereal crops that may launch the modern green revolution.

13.2.2 CRISPR-Cas12

Type V CRISPR system includes the versatile protein Cas12a for genome editing. CRISPR-Cas12a has become a viable effector RNA-guided DNA endonuclease substitute for the Cas9 protein. To effectively cut ssDNA and dsDNA, Cas12a only needs crRNAs, not tracrRNA or RNase III, in contrast to Cas9. The RuvC and nuclease lobe (NUC) domains for cleavage activity have been identified in Cas12a (Fonfara et al. 2016). Since the Cas12 protein produces staggered DSBs and facilitates the HDR repair mechanism rather than both NHEJ and HDR, it was recognized to be superior to the Cas9 protein, thus exhibiting promise in plant editing (Hillary and Ceasar 2022).

The rapid advancement of CRISPR-Cas12 genome editing has proved a revolution in various crops. CRISPR/Cas12 along with 17 gRNA has been successfully employed to target the eight genes *TaAn-1, TaGASR7, TaGS3, TaGSE5, TaGW2, TaGW7, TaPDS,* and *TaSPL16* of wheat, thus increasing the grain size, weight, and number with high editing efficiency (Wang et al. 2021). Moreover, CRISPR-Cas12 has been used to produce Wheat Dwarf Virus (WDV) resistant wheat plants by targeting the coat protein (CP) gene using two gRNAs, achieving 90–100% editing efficiency, thus promising the defense against crop diseases (Baran et al. 2024).

13.3 Application of Genetic Improvement of Triticae

Human solely depends on the agriculture for food production, but climate change, abiotic stress (such as drought), and pathogens pose a significant threat to the production of crops, therefore affecting food security. Plant scientists have used several cutting-edge molecular techniques to address these issues to improve plant quality and enhance yield. The CRISPR/Cas9 genome editing is one of the groundbreaking techniques that have been applied to a variety of crops to boost production and

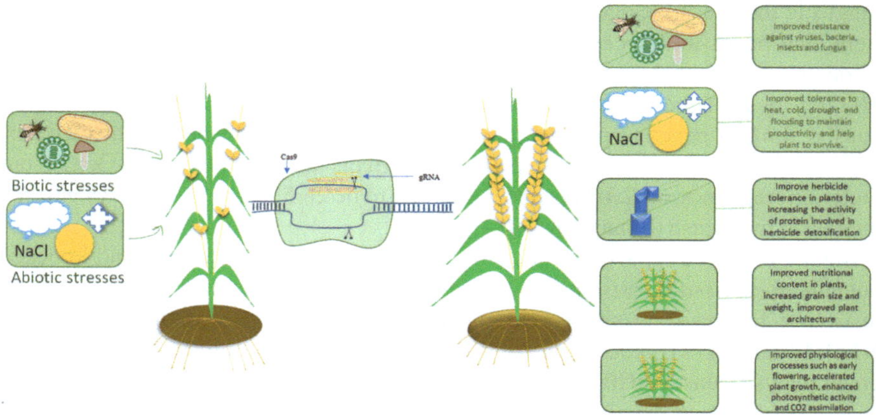

Fig. 13.1 Diagrammatic illustration of Successful Application of CRISPR/Cas Genome Editing in Triticaea. CRISPR/Cas9 genome editing allows the development of biotic and abiotic stress-resistant crops with improved agronomical traits such as enhanced resistance to viruses, bacteria, and fungi, making plants tolerant to herbicides, increased nutritional content, and improved physiological processes to create more germplasm resources that fulfill the need for global food security

enhance disease resistance (Hillary and Ceasar 2019). These tools either enhance the function or expression of the gene that has agronomical importance to understand the genetic architecture of traits variation (Song et al. 2022) (Fig. 13.1).

13.3.1 Pathogen Defense and Durable Resistance

Pest and pathogens such as fungi, viruses, and bacteria are the major threats that cause diseases in many plants including the major crops, i.e., cotton and wheat. Biotic factors not only influence the plant growth and development but also reduce crop yield and quality, and even lead to dramatic crop loss in certain cases. Many genes are associated with the complicated mechanism by which plant response to multiple disease caused by pathogens (Zhang et al. 2021a, b). During plant development, S genes are involved in different biological processes, and the frequent knockout of these S genes can lead to the major pleiotropic consequences. Susceptibility (S) and resistant (R) genes could be targeted using CRISPR/Cas to enhance the plant-pathogen interaction and develop transgenic plants having resistance to pathogens. Instead of single gene, knockout of multiple genes resulting high tolerance against these diseases (Zaidi et al. 2020).

Brauer et al. (2020) introduced mutation in *TaNFXL1* gene of *Triticeae* that codes for the transcriptional factor induced by the mycotoxin deoxynivalenol that leads to the wheat resistance against the fungus *Fusarium graminearum*, which causes the Fusarium head blight. The knockout mutants were resistant to *F. graminearum,* reducing the need for fungicides (Brauer et al. 2020). Phyto pathogenic fungus is the causative agent of powdery mildew disease that affects wheat leaves, stem, and heads in *Triticeae* family substantially reducing the crop yield, and the respective fungus disrupting the normal functioning of *TaMLO* gene, located on the *MLOB1* locus of the wheat plant. Li et al. (2022) used the CRISPR/Cas9 to knock out the *TaMLO* genes to produce powdery mildew-resistant wheat plant varieties, which also minimize the negative pleiotropic effects in edited plants. The knock out mutation frequency of *TaMLO* gene was found to be 28.5% which improves wheat's resilience to disease (Shan et al. 2014).

Zhang et al. (2017) also introduced the mutations into *EDR1* gene and knocked out *enhanced disease resistance 1* (*TaEDR1*) gene resulting in the improved wheat's resistance to powdery mildew. Similarly Chinese group successfully altered the *MLO-A1* and *EDR1* gene using CRISPR/Cas9 to develop mildew-resistant wheat bread plants (Hensel 2020).

Accumulation of reactive oxygen species and expression of pathogenesis-relative genes are the major aspects associated with the Overexpression of *TaCIPK14* genes. There by decreasing the defense response of pathogenic interactions with wheat plants. Research studies have demonstrated CRISPR/Cas9 gene editing toward disease resistance in wheat caused by fungus *Puccinia striiformis f.* sp. *Tritici* in which plant undergoes various disabilities.

Fungus that causes stripe rust damages plants by growing in the leaves and releasing spores that burst through the leaf surface. The resistance to wheat stripe rust may be enhanced by editing the genes linked to the Calcineurin B-like protein (CBL)-Interacting Protein Kinases (CIPK14) (He et al. 2023). Genes like *Ta-eIF4E* and *Ta-eIF(iso)4E*, which code for highly conserved translation initiation and susceptibility (S) factors, are found in viruses of the Potyviridae family. These genes are essential for the completion of the virus's life cycle. The plant genes whose products are important in viral infection may be altered by CRISPR/Cas technology. Virus infection has been the major concern during Triticeae cultivation. Kan et al. (2023) introduced the mutation in *PDIL5–1* and *eIF4E* genes to prevent the wheat plant from wheat yellow mosaic virus (WYMV) (Kan et al. 2023). Similarly, CRISPR/Cas9-based mutation in *Ta-eIF4E* and *Ta-eIF(iso)4E* genes of bread wheat resulted in broad-spectrum resistance against wheat yellow mosaic virus and wheat spindle streak mosaic virus (Hahn et al. 2021).

13.3.2 Abiotic Stress Tolerance

Abiotic stress on crop growth and development is getting worse due to global warming. Extreme temperatures, salinity, drought, and pollution are the most prevalent abiotic stressors. They all have a major impact on plant growth and development, which in turn influences plant biomass, yield, and quality, therefore threatening global food security, which drastically lower agricultural yields, especially for important crops like wheat and triticale (Rey et al. 2021).

Plants respond to various environmental challenges via a large number of structural and regulatory genes, including noncoding RNAs (Zhang 2015). Both conventional transgenic technology and advanced genome editing technologies can be used to target all of these genes to increase crop resistance to abiotic stressors. CRISPR/Cas technology has advanced in producing genome editing mutants to improve plant resistance to abiotic stressors because a single stress event may be associated with thousands of genes, and no single gene is dominant during this process, unlike biotic stresses (Zafar et al. 2020). More prominently CRISPR/Cas9 system has emerged as a highly efficient technology with the potential to develop new wheat and especially triticale varieties with high yield (Bansal et al. 2021). Hypothermia and drought are the major abiotic stress factors caused by deficiency of water, and they affect crops cultivation. A bread wheat, an important, widely grown agricultural crop providing 20% of food calories per day and has been widely used for bread production. It has complex genome structure with equal contribution from A, B, and D sub genomes (Choulet et al. 2014).

Wheat ethylene-responsive factor 3 (TaERF3) and wheat dehydration-responsive element binding protein 2 (TaDREB2) in wheat protoplasts are two abiotic stress-responsive transcription factor genes that are edited using the CRISPR/Cas9 genome editing system. This allows plant tissues to withstand drought without affecting growth parameters. Moreover, CRISPR/Cas9 constructs were developed for targeted mutations in two stress-responsive genes (*DREB2* and *ERF3*) of wheat to enhance the drought tolerance in edited plants. Edited plants were found to have a high survival rate relative to the control plants under stress conditions (Kim et al. 2018). Another study demonstrated the use of CRISPR/Cas9 to introduce mutations in 3′(2′), 5′-bisphosphate nucleotidase (Sal1), Multiprotein Binding Factor 1, and high-affinity Pi transporters (*PHT1*) genes of wheat plants, which results in drought and stress tolerance in the edited plants (Abdallah et al. 2022). Similarly, when the histone acetyltransferase *HAG1* gene in Triticae family were knockout using Cas9 protein, the plants were able to survive the salt tolerance (Zheng et al. 2021).

CRISPR/Cas has been employed in cereal crops such as barley to develop a mutation in the *HvITPK1* gene to elucidate the role of this gene in stress signaling and IP6 synthesis. Inositol hexaphosphate (IP6), sometimes referred to as phytic acid, is a crucial phosphate storage form found in cereal grains and mainly produced by phosphorylating inositol phosphate. Inositol triphosphate kinase (ITPKS), an essential set of enzymes, phosphorylates wheat seeds via a lipid-independent mechanism of phytic acid production. Development of mutant lines with reduced IP6 content increase or enhanced the phosphate availability in food and forage. So, the mutant lines have an altered level of phosphate in mature grains ranging from 65% to 174% when compared with wild-type grains. Moreover, insertion mutants have higher salt resistance relative to the deletion mutants and wild-type plants. These findings demonstrated that *ITPK* genes are involved in phosphate accumulation as well as the signaling pathways for abiotic stress responses (Vlcko and Ohnoutkova 2020).

13.3.3 Improving Yield and Biomass

Enhancing agricultural production through conventional breeding and transgenic technology is always a long-term objective. However, scientists and breeders have not made significant progress using gene cloning and transgenic approaches because yield traits are complex traits controlled by several genes and gene networks. Numerous studies demonstrate that the expression of specific genes has a significant impact on agricultural output, even though it is practically impossible to find a single gene that significantly controls yield. Crop yield may be directly increased if these genes were suppressed or even eliminated. The rapidly evolving CRISPR/Cas system offers an ideal tool to knock out these genes that adversely affect plant yield.

Preharvest sprouting (PHS) poses a severe danger to crop productivity globally especially in the case of triticale. Few labs perform genome editing in triticale because it is time-consuming and expensive, due to its complex genomic structure that has not been studied in detail yet. This plant also has low regeneration ability (Debernardi et al. 2020). The agronomic traits of plants are mostly determined by seed dormancy and germination, and the degree of dormancy potentially influences crop yield and quality in agricultural production. Abe et al. (2019) used CRISPR/Cas to create mutations in the three *Qsd1* homologs to alter seed dormancy. Further crossings resulted in edited plants, whose seeds are less likely to emerge on spikes and have a longer dormant period (Abe et al. 2019).

The *ABA 8'-HYDROXYLASE 1* gene is responsible for the grain sprouting on spikes; therefore, Michalski et al. developed a CRISPR/Cas9 construct to knock out *ABA 8'-HYDROXYLASE 1* gene in triticale protoplast. The mutant lines were found to be unaffected to seed dormancy (Michalski et al. 2021). The CRISPR/Cas-based modification in wheat and barley influences the plant hormone by producing alterations in the cytokinin gene leading to the development of additional spikelets and increased yield while not affecting other agronomic parameters (Hensel 2020). *TaGW2* and *TaGW7* homologs in the B and D genomes of the tribe Triticeae were knocked out using CRISPR/Cas9. The phenotypes of the knockout mutants increased in weight and grain size, which may result in increased plant yield overall (Wang et al. 2019).

Enhancing the efficacy of nitrogen nutrition is a key objective in Triticeae crop breeding. Three *TaARE1* homeologs were recently discovered and described by Zhang et al. (2021a, b). CRISPR/Cas technology was used to create mutant wheat lines with three *TaARE* null alleles. All mutant lines exhibited a greater yield, slower aging, and greater tolerance to nitrogen deficiency in fields (Zhang et al. 2021a, b). Similarly, Zhang et al. (2019) modify all the homologs of *CKX2.1* genes that encode the cytokinin oxidase/dehydrogenase, resulting in a higher spike yield (Zhang et al. 2019).

Moreover, CRISPR/Cas9 has potential to edit the *HvOPM1* and *HvOPM2* gene, *OVULE PECTIN MODIFIER1,* and two pectin methylesterase genes present in barley to improve the plant yield, thus the mutant lines displayed increased grain size, increased length of ovaries and ovules (Yang et al. 2023). Semi-dwarf barley mutant lines have longer coleoptiles as a result of the CRISPR/Cas9 knockout of a gene called Gibberellin 3-oxidase1 (*HvGA3ox1*). However, the yield of the knockout mutants increased by 20% when the temperature was changed, and the seed dormancy (SD) was adjusted to the optimal level (Cheng et al. 2023). CRISPR/Cas9 method was used to knock out the *ZnF-B* gene in wheat, which encodes a positive regulator of brassino-steroid signaling. This resulted in semi-dwarf height, with a higher yield of 15.2% of the mutant plants (Song et al. 2023). The growing amount of research suggests that the modern green revolution could be sparked by the use of CRISPR-based editing to alter the architecture of cereal plants.

MPK genes have a functional role in the development of plants, but their role in Triticeae remains unknown. MPK genes mainly encode the mitogen-activated protein kinases, and their functional role is well studied in crop plants as well as in model plants (*Arabidopsis thaliana* and *Oryza sativa*). Křenek et al. (2021) developed the CRISPR/Cas construct to knock out the HvMPK6 gene in barley plants, but the results were found to be surprising, as the mutant lines had no phenotypic difference relative to the wild type. All mutants were identified as heterozygous, with the second allele invariably being native *HvMPK6*. The results suggest lethality for the biallelic mutation. The mutants were then used to produce seeds; however, the sprouts were damaged, had fewer roots, and lacked branches. HvMPK6 was found to be crucial for the development of barley embryos and early sprouts based on the findings of the research on the *MPK6* orthologs in *A. thaliana* and rice (Křenek et al. 2021).

13.3.4 Nutritional Content

The prevalence of malnutrition has a long-term impact on healthcare, society, the economy, and development at the individual, local, and national levels. Through the use of CRISPR-Cas-based crop genome editing, quantitative improvements in starch, protein, gamma-aminobutyric acid (GABA), oleic acid, anthocyanin, phytic acid, gluten, and steroidal glycoalkaloid contents have been made, as well as qualitative improvements in aroma, shelf life, and sweetness. Starch is a key nutritional component of cereal endosperm that significantly influences grain end-of-use quality, such as cooking properties and digestibility (Chen et al. 2021). Four genes, the granule-bound starch synthase (*GBSSI*), the starch synthase (*SSIIa*), the starch debranching enzyme (isoamylase, *ISAI*), and starch regulator 1 (*RSR1*), were selected for CRISPR/Cas9-mediated editing because simultaneous knockout of genes with opposing functions can yield alternative combinations of starch types, achieving 46% gene editing efficacy. Edited plants demonstrated an amylose-deficient phenotype with potential agricultural applications such as improved yield and lodging resistance (Miroshnichenko et al. 2024).

The MPK genes are found to encode for a starch branching enzyme; thus, the resulting plants after CRISPR/Cas knockout were produced, i.e., spring and winter wheat lines that produce with the increased amylose levels and decreased amylopectin content (Li et al. 2021a, b, c). Asparagine, a precursor of acrylamide that is formed during the processing of wheat when the temperature is high. Elevated levels of asparagine content cause carcinogenicity in plants. Therefore, CRISPR/Cas9 has the potential to obtain the free asparagine content in bread wheat grains. Raffan et al. (2021) investigated gene editing in bread wheat by targeting the six alleles of the *TaASN2* gene, which codes for asparagine. The high editing efficiency was achieved, and the mutant plants have decreased asparagine content and low carcinogenicity.

One of the studies knocked out the lipoxygenase genes (*TaLpx1* and *TaLox2*) and *TaGASR7*, *TaNAC2*, *TaGW2*, and *TaDEP1* genes in wheat to increase grain yield and size. The mutant wheat plant results in increased grain area, length, width, and weight when compared with wild-type plants, thus achieving edited efficiency of grain weight (27.7%), grain width (10.9%), grain area (17.0%), and grain length (6.1%) (Wang et al. 2018a, b). These studies demonstrate the use of genome editing using CRISPR/ Cas9 in wheat to improve yield and to overcome adverse conditions. Two essential quality parameters such as flour protein content and gluten strength have been enhanced by targeting *TaGW2* in the B1 and D1 genomes using CRISPR/Cas9 (Zhang et al. 2018). Similarly, Sánchez-León et al. (2018) used the CRISPR/Cas9 approach to knockout the 35 members of α-Gliadin gene to generate the low-gluten bread wheat.

Moreover, CRISPR/Cas9 genes involved in the vitamin E biosynthesis to improve the nutritional content in barley. CRISPR/Cas targeted knockout of *HvHPT* and *HvHGGT* genes produces the mutant barley plants that have reduced tocopherol and tocotrienol content in barley grains. Moreover, HPLC results demonstrated that tocopherol and tocotrienol contents reduced substantially due to the mutation in *HvHPT* gene, and tocotrienol biosynthesis has been blocked completely due to HvHGGT mutation. The findings reveal that the HvHGGT product functions as a crucial tocotrienol biosynthesis enzyme (Zhou et al. 2020; Zeng et al. 2020)

13.3.5 Improving Crop Quality

Genome editing can lead to highly efficient improvements in yield, quality, and other agronomic traits of crop plants as well as reduce the genome complexity and enhance the phenotypic features of triticale crop species. The CRISPR/Cas9 system is highly significant in increasing the overall yield of wheat plants. Negative regulators of wheat plants such as *GW7, GW2, LOX2*, and *GASR7* are knocked out by using CRISPR/ Cas technology, resulting in the formation of longer grains, which improved the overall yield of wheat plants (Wang et al. 2020).

A CRISPR/Cas9-generated knockout mutation of the SPL13 gene led to enhanced grain size and quantity, decreased flowering time, and shorter tillers and plant height (Gupta et al. 2023). Increased spikelet number and delayed heading were the outcomes of a CRISPR/Cas9-induced mutation in the wheat *FT-D1* gene, which eventually improved yield (Chen et al. 2022). In another study, four genes associated to wheat quality named *pinb, waxy, ppo, and psy* were knocked out using CRISPR/Cas9, which enhanced the dough's color, starch quality, and wheat grain hardness (Zhang et al. 2021a, b).

Moreover, CRISPR/Cas9-mediated editing of the *SBEIIa* gene led to the development of high-amylose wheat lines with significantly increased levels of resistant starch. These edited lines exhibited improved efficiency and superior grain quality compared to conventional wheat cultivars (Li et al. 2021b). Barley strains possessing the *HVARE1* mutation exhibited increased plant height, a greater number of shoots, higher grain protein content, and improved yield. Additionally, these mutants demonstrated enhanced chlorophyll content, delayed senescence, and elevated nitrogen levels in shoots, even under low soil nitrogen conditions (Karunarathne et al. 2022). The CRISPR/Cas construct was developed to knock out the four genes *HvCslH1, HvCslF3, HvCslF6,* and *HvCslF9* that typically code for (1,3;1,4)-β-glycan synthase and GT2 glycosyltransferase content. A mutation in the *HvCslF6* gene leads to a reduced content of the particular enzyme in the mutant plants, while the mutation in other genes has no subsequent effect on the enzymes. So, the mutant plants have (1,3;1,4)-β-glycan in grains (Garcia-Gimenez et al. 2020).

13.3.6 Base Editing (BE) and Prime Editing (PE)

Gene knockouts have been the primary focus in wheat genome editing research, while base editing was given less attention. Consequently, it is imperative to enhance the effectiveness of genome editing technologies in cereal crops, mainly Triticae. Furthermore, the biosafety of modified plants has been significantly increased using the base editing approach because it modifies or alters specific base pair in a plant's DNA sequence. Therefore, the base-editing approach will play a more important role for the genetic modification in wheat breeding in future (Zhang et al. 2019).

Base editing (BE) and prime editing (PE) have also been successfully employed in most cereal crops allowing the targeted site editing without introducing double-strand breaks. Numerous BE systems, such as adenine base editors (ABEs), cytosine base editors (CBEs), C-to-G base editors (CGBEs), and dual base editors have been developed and are currently in use for editing or replacing single bases in rice, wheat, and maize (Molla et al. 2021). Base editing, a revolutionary method for precision genome editing, allows for irreversible base conversion. Cytosine base editors (CBEs) and adenine base editors (ABEs) are widely used for targeted base pair conversions, such as C/G to T/A and A/T to G/C, respectively. In wheat, precise C-to-T editing was demonstrated at the TaLOX2 locus (Zong et al. 2017). ABE-mediated modifications enabled efficient A-to-G conversion in genes like *TaDEP1* and *TaGW2*. This approach also facilitated the generation of transgene-free, herbicide-tolerant wheat by editing the TaALS-P174 gene (Li et al. 2019). Dual-site mutants at *TaALS-P174* and *TaALS-G631* exhibited enhanced resistance to nicosulfuron. Unexpected C-to-G edits were also observed in certain lines. Additionally, modifications in TaACCase-A1992 conferred tolerance to quizalofop herbicide (Zhang et al. 2019).

Prime editing technique has been effectively applied in plants, enabling precise insertions, deletions, and point mutations without inducing double-strand breaks or requiring DNA donor templates for repair. Prime editing (PE), being the precise genome editing technique, utilizes a catalytically modified Cas9 endonuclease fused to a specialized reverse transcriptase, guided by a prime editing RNA (pegRNA). This regulates the target site and installs the desired edits to generate new genetic alterations at a specific DNA site. Prime editing (PE) enables diverse genome modifications and holds significant potential for applications such as improving yield, enhancing product quality, and increasing resistance to various abiotic and biotic stresses (Table 13.1).

13.3.7 CRISPR Reagents and Cargos

CRISPR cargos can be delivered as DNA, RNA, or ribonucleoproteins (RNPs). Three approaches mainly employed for CRISPR/Cas9 cargos including DNA plasmid containing Cas9 and gRNA, mRNA for Cas9 translation along with sgRNA and purified Cas9 protein along with transcribed gRNA (Ribonucleoprotein complex). Transgene-free editing, nontoxicity, high specificity, and high efficiency being the ideal factors for successful delivery of CRISPR cargos (Gordon-Kamm et al. 2021). Different methods such as Agrobacterium, bombardment or biolistic method, floral-dip, and PEG-mediated protoplast have been employed for the successful delivery of CRISPR cargos in plants. One of the most widely used delivery method for achieving genetically modified plants is plasmid delivery through *Agrobacterium*-mediated genetic transformation, being the simple and robust mainly for *Arabidopsis* plant. Increased off-target effects due to the stable integration of plasmid cassettes represent a major limitation associated with DNA cargo based delivery systems (Lino et al. 2018). Presence of transgene in the final product allows the use of alternative method such as Cas9 RNPs into regenerative cells, enables the transgene-free editing, reduces off target effects and transient expression, but it affects the editing efficiency of CRISPR/Cas due to the large size of Cas9. Achievement of high editing efficiency attains only when delivery of CRISPR/Cas9 construct occurs into rigid plant cell (Chuang et al. 2021). Public acceptance and regulatory status of GMO crops are highly influenced by choice of cargo for efficient delivery. Direct and Indirect methods have been employed for the delivery of CRISPR Cargo. Direct methods involve physical, i.e., Electroporation and chemical, i.e., polyethylene glycol (PEG), while indirect methods involve the use of bacteria as vector to direct the CRISPR cargo in the plant cell (Kocsisova and Coneva 2023). Electroporation allows the electric pulses to create transient pores in the plasma membrane promptly destroying it, allowing water molecules to enter the lipid bilayer and generate hydrophobic pores leading to the entry of nucleic acid into the protoplast of the cells. Electroporation can deliver a CRISPR cargo containing CRISPR/Cas9. Scientists utilized Cas9 along with sgRNA being employed in wheat to introduce edit in expressed DsRed gene and 2 wheat genes (TaLox2 and TaUbiL1). But the

Table 13.1 Significant studies demonstrating the CRISPR/Cas modification in Triticaea for improved crop production

Sr. No.	Target gene	No of gRNA	Crop Species	Genome editing Approach	Gene function	Improved trait	Transformation strategy	Mechanism	References
1	*Rht-B1, ZnF-B*	Single gRNA	Wheat *(Triticum aestivum)*	CRISPR/ Cas9	Balance the level of phytohormones brassinolide and gibberellin	Transgene-free semi-dwarf plant, high yield, and efficient nitrogen use	Agrobacterium-mediated transformation	NHEJ	Song et al. (2023)
2	*TaINOX, TaPDS*	Single gRNA	Wheat *(Triticum aestivum)*	CRISPR/ Cas9	Inositol metabolism and carotenoid biosynthesis	Albinism expression of plants	Agrobacterium-mediated transformation	NHEJ	Wolt et al. (2016)
3	*TaLpx-1*	Single gRNA	Wheat *(Triticum aestivum)*		Higher fusarium headblight resistance	Disease-resistant plants	Bombardment of immatureembryos with gold particles	N/A	Nalam et al. (2015)
4	*WDV1, WDV2, WDV3* and *WDV4*	Four gRNA	Barley *(Hordeum vulgare)*	CRISPR/ Cas9	Wheat dwarf virus resistance	Disease-resistant plants	Agrobacterium-mediated transformation	NHEJ	Kis et al. (2019)
5	*PDIL5–1*	Single gRNA	Barley *(Hordeum vulgare)*	CRISPR/ Cas9	Resistance against Barley yellow mosaic virus and barley mild mosaic virus.	Disease-resistant plants	Agrobacterium-mediated transformation	NHEJ	Holubová et al. (2018)
6	*HvCKX1* and *HvCKX3*	Single gRNA	Barley *(Hordeum vulgare)*	CRISPR/ Cas9	Cytokinin Metabolism	Improved yield, increase root length of plant	Agrobacterium-mediated transformation	NHEJ	Holubová et al. (2018)

(continued)

Table 13.1 (continued)

Sr. No.	Target gene	No of gRNA	Crop Species	Genome editing Approach	Gene function	Improved trait	Transformation strategy	Mechanism	References
7	*TaAn-1, TaGASR7, TaGS3, TaGSE5, TaGW2, TaGW7, TaPDS* and *TaSPL16*	Seventeen gRNA	Wheat (*Triticum aestivum*)	CRISPR/ Cas12	Improved plant quality	Increased grain size, weight and number	Agrobacterium-mediated transformation	NHEJ	Wang et al. (2021)
8	Coat Protein of WDV	Two gRNA	Wheat (*Triticum aestivum*)	CRISPR/ Cas12	Defence against Wheat plant diseases	Wheat Dwarf Virus-resistant plants	Agrobacterium-mediated transformation	NHEJ	Baran et al. (2024)
9	*TaNFXL1*	Single gRNA	Wheat (*Triticum aestivum*)	CRISPR/ Cas9	Codes for the transcriptional factor induced by the mycotoxin deoxynivalenol	*Fusarium graminearum* resistance in wheat plant	Agrobacterium-mediated transformation	NHEJ	Brauer et al. (2020)
10	*TaML, TaEDR1*	Single gRNA	Wheat (*Triticum aestivum*)	CRISPR/ Cas9	Powdery mildew resistance in wheat plant, minimize the negative pleiotropic effects	Powdery Mildew-resistant wheat plant	Agrobacterium-mediated transformation	NHEJ	Shan et al. (2014)
11	Calcineurin B-like protein (CBL)-Interacting Protein Kinases *(TaCIPK1)*	Single gRNA	Wheat (*Triticum aestivum*)	CRISPR/ Cas9	Improve the resistance against wheat stripe rust	*Puccinia striiformis*Fungus resistance wheat plant	Agrobacterium-mediated transformation	NHEJ	He et al. (2023)

12	*Ta-eIF4, Ta-eIF(iso)4E, PDIL5–1*	Single gRNA	Bread Wheat	CRISPR/Cas9	Code for highly conserved translation initiation as well as susceptibility (S) factors	Wheat spindle streak mosaic virus (WSSMV) and wheat yellow mosaic virus (WYMV) resistant breadwheat plant	Agrobacterium-mediated transformation	NHEJ	Hahn et al. (2021)
13	*TaERF3, TaDREB2*	Single gRNA	Wheat (*Triticum aestivum*)	CRISPR/Cas9	Abiotic stress-responsive transcription factor genes that withstand drought	High survival of plants against drought stress	Agrobacterium-mediated transformation	NHEJ	Kim et al. (2018)
14	Histone acetyltransferase *HAG1 gene*	Single gRNA	Triticae family	CRISPR/Cas9	Provides stress tolerance	Edited plants having resistance to salt stress	Agrobacterium-mediated transformation	NHEJ	Zheng et al. (2021)
15	*PHT1*	Single gRNA	Wheat (*Triticum aestivum*)	CRISPR/Cas9	Provides heat and drought stresses, prevent phosphate deprivation	Edited plants having drought and heat resistance	Agrobacterium-mediated transformation	NHEJ	Abdallah et al. (2022)
16	*HvITPK1*	Single gRNA	Barley	CRISPR/Cas9	Phosphate accumulation, Transduction of abiotic stress signals	Edited plant having abiotic stress resistance	Agrobacterium-mediated transformation	NHEJ	Vlcko and Ohnoutkova (2020)
17	*ABA 8'-HYDROXYLASE 1*	Single gRNA	Triticale	CRISPR/Cas9	Responsible for grain sprouting on spikes	Edited lines were found to be unaffective to seed dormancy	Agrobacterium-mediated transformation	NHEJ	Michalski et al. (2021)

(continued)

Table 13.1 (continued)

Sr. No.	Target gene	No of gRNA	Crop Species	Genome editing Approach	Gene function	Improved trait	Transformation strategy	Mechanism	References
18	*TaGW2, TaGW7*	Single gRNA	Triticea family	CRISPR/Cas9	Enhanced the plant yield	Edited plants have increased size and weight	Agrobacterium-mediated transformation	NHEJ	Wang et al. (2019)
19	*CKX2., DUO-B1*	Single gRNA	Wheat (*Triticum aestivum*)	CRISPR/Cas9	Production of additional spikelets	Higher spike yield	Agrobacterium-mediated transformation	NHEJ	Zhang et al. (2019), Wang et al. (2022)
20	*HvOPM1* and *HvOPM2*	Single gRNA	Barley (*Hordeum vulgare)*	CRISPR/Cas9	Involved in enhanced plant yeild	Increase in ovarian, grain size, and length of ovule	Agrobacterium-mediated transformation	NHEJ	Yang et al. (2023)
21	Gibberellin 3-oxidase1 (*HvGA3ox1*)	Single gRNA	Barley (*Hordeum vulgare*)	CRISPR/Cas9	Adjusting seed dormancy at optimum level	Increased coleoptile length of the semi-dwarf mutants of barley, Yield increase of 20%	Agrobacterium-mediated transformation	NHEJ	Cheng et al. (2023)
22	*ZnF-B*	Single gRNA	Wheat (*Triticum aestivum*)	CRISPR/Cas9	Encodes a positive regulator of brassino-steroid signaling	Mutants plants having greater yield	Agrobacterium-mediated transformation	NHEJ	Song et al. (2023)
24	*GBSSI, SSIIa, ISAI, RSR1*	Four gRNA	Cereals	CRISPR/Cas9	Involved in increased yeild and lodging resistance	Edited plants showed amylose deficient phenotype	Agrobacterium-mediated transformation	NHEJ	Miroshnichenko et al. (2024)

25	*TaSBEIIa*	Single gRNA	Bread wheat lines	CRISPR/Cas9	Codes for starch branching enzyme	Plants having increased amylose levels and decreased amylopectin	Agrobacterium-mediated transformation	NHEJ	Li et al. (2021a, b, c)
26	*TaASN2*	Single gRNA	Bread wheat	CRISPR/Cas9	Involved in obtaining free asparagine content	Edited lines showed lower carcinogenicity	Agrobacterium-mediated transformation	NHEJ	Raffan et al. (2021)
27	*TaLpx1, TaLox2, TaGASR7, TaNAC2, TaGW2, TaDEP1*	Multiple gRNA	Wheat (*Triticum aestivum*)	CRISPR/Cas9	Involved in increasing plant size, weight, and yield	Edited plants having improved grain size, length and area	Agrobacterium-mediated transformation	NHEJ	Wang et al. (2018a, b)
28	*HvHPT, HvHGGT*	Two gRNA	Barley (*Hordeum vulgare*)	CRISPR/Cas9	Encode enzymes essential for vitamin E biosynthesis	Significantly reduced tocopherol and tocotrienol content in edited plants	Agrobacterium-mediated transformation	NHEJ	Zheng et al. (2021)
29	*GW7, GW2, LOX2, GASR7*	Multiple gRNA	Wheat (*Triticum aestivum*)	CRISPR/Cas9	Improved quality of plant	Edited plants exhibited longer grain size	Agrobacterium-mediated transformation	NHEJ	Wang et al. (2019)
30	*SPL1, FT-D1*	Single gRNA	Wheat (*Triticum aestivum*)	CRISPR/Cas9	Improved quality of plant	Edited plants having decreased flowering time, shorter tillers, and plant height	Agrobacterium-mediated transformation	NHEJ	Gupta et al. (2023), Chen et al. (2022)
31	*pinb, waxy, ppo, and psy*	Multiple gRNA	Wheat (*Triticum aestivum*)	CRISPR/Cas9	Improved quality of plant	Edited plants having improved dough's color and starch quality	Agrobacterium-mediated transformation	NHEJ	Zhang et al. (2021a, b)

(continued)

Table 13.1 (continued)

Sr. No.	Target gene	No of gRNA	Crop Species	Genome editing Approach	Gene function	Improved trait	Transformation strategy	Mechanism	References
32	*SBEIIa*	Single gRNA	Wheat (*Triticum aestivum*)	CRISPR/ Cas9	Improved plant quality	Edited plants having amylose content	Agrobacterium-mediated transformation	NHEJ	Li et al. (2021b)
33	*TaALS-P174*, *TaALS-G631*	Two gRNA	Wheat (*Triticum aestivum*)	CRISPR/ Cas9	Base editing (provide resistance to herbicides)	Transgene-free wheat variant carrying a herbicide tolerance	Agrobacterium-mediated transformation	NHEJ	Zhang et al. (2019)

mutation and electroporation efficiency were obtained to be low (Ghogare et al. 2021). The major drawbacks associated with this system are the use of strong currents that lead to the death of the cell. PEG is one the most important genetic transformation method. gRNA along with Cas9 construct incubate with protoplast in the presence of PEG. This method has been first reported in maize (Li et al. 2019). Cas9/gRNA RNPs have been delivered by PEG method in *Arabidopsis* and tobacco to make transgene-free edited lines. Protoplast isolation and establishment of suspension cells are the major challenges associated with this delivery method. Out of all delivery methods, *Agrobacterium* transformation has been preferred due to its high efficiency as CRISPR construct has been easily transformed into desired explants such as callus, leaf, and floral parts of plants. To date, many plants have been reported with high editing efficiency due to the employment of agrobacterium-mediated delivery (Sandhya et al. 2020).

13.4 Off-targets Effects

CRISPR/Cas9 technique has been successfully employed for altering the genome of numerous cereal crops, including barley, rice, wheat, and maize to achieve the desired traits with agronomical importance and provide new insights on large gene families, biochemical pathways, and functionally redundant genes. But the off-target effects remain the major problem associated with CRISPR/Cas9-mediated genome editing. These off-target effects are toxic for plants at cellular level as they induce the unwanted cleavage in the genome. Furthermore, plants may suffer from undesirable chromosomal rearrangements including translocations, deletions and insertions that are typically caused by the repair of these off-target double-stranded breaks (Brunet et al. 2009; Lee et al. 2012; Kim et al. 2013). Occasional off-targeting can be problematic as it alters the desired phenotype, which typically leads to the incorrect interpretation of results.

The specificity of Cas9 is determined mainly by a 20-base pair gRNA along with the PAM sequence. PAM sequence has a significant role as it contributes to Cas9 binding, whereas the 3′ end is essential for target recognition, formation of R-loop and also the activation of the nuclease activity of Cas9. Numerous studies reported that the single gRNA can tolerate up to five base pair mismatches (Li et al. 2019). Several approaches have been developed to minimize off-targeting. The first strategy involves the manipulation of the structure and composition of sgRNA to reduce the off-target effects because the composition and structure of gRNA are the two major factors that control the off-target and on-target cleavages (Tsai et al. 2015). A significantly low or high GC content reduces the effectiveness of sgRNA. To address this, an alternative strategy involves truncating the sgRNA at the 5′ or 3′ end using sgRNA nickase. This approach helps in selecting efficient gRNA based on the GC content of the desired sequence while minimizing predicted off-target effects (Sander and Joung 2014).

Another technique is to use the ribonucleoprotein (RNP) delivery mechanism to shorten the exposure period of the genomic DNA to the CRISPR reagents, hence minimizing off-targeting effects (Vakulskas and Behlke 2019). Another strategy to minimize the off-targets involves the use of high-fidelity Cas9 enzymes, including spCas-HF and eSpCas9 to enhance the specificity of the CRISPR/Cas system (Kleinstiver et al. 2016). Various Cas9 variants have been developed to minimize the off-target effects. Guilinger et al. (2014) developed fCas9 by fusing the dCas9 with FokI nuclease, and transformed cells produced from fCas9 exhibited a 140-fold improvement in specificity. Kleinstiver et al. (2016) made 3–4 amino acid modifications in Cas9 and found no detectable off-targets. These Cas9 variants can be substituted with Cas9 to eliminate the problem of off-target events. Several approaches have been described to investigate off-target events, including in silico prediction, T7E1 test, HTGTS, ChIP-seq, IDLV, fluorescence in situ hybridization, deep sequencing, and so on. Of these methods, Digenome-seq and GUIDE-seq were found to be the most accurate, with a sensitivity of 0.1% for off-target identification.

13.5 Regulatory Challenges

Genome editing shows great potential for crop improvement. CRISPR/Cas9 editing has several advantages over conventional methods such as the ability to produce multiple, precise alterations and comparatively being simpler and less expensive. CRISPR/Cas9 has been used to modify a variety of traits in cereal crops to achieve improved grain quality thus providing consumers with nutritional advantages. The continuous discovery of novel protein scissors with distinct proto-spacer adjacent motives, such as CRISPR-Cas12a in conjunction with Cas9, has expanded the technical possibilities and helped improve the efficiency of editing various genes and regulatory components. Different countries have different regulatory frameworks. Laws and regulations pertaining to gene-edited crops are controversial and unclear worldwide. The primary goal should be the commercialization of gene-edited crops, and in order to expedite this process, technological and policy barriers should be addressed. The regulatory frameworks around genetically modified organisms (GMOs) primarily govern the development and commercialization of novel gene-edited crops in different countries (Waltz 2016; Lema 2019).

Different countries impose regulations on genome editing crops so that they can be accepted worldwide. New Zealand (NZ) and the European Union have implemented strict process-based regulation for genetically edited crops because these crops undergo GM safety tests that are time-consuming and require high cost. Moreover, China also relies on a process-based regulatory framework to regulate GMOs that involves the multistage evaluation of the GM crops. In the USA, genetically modified crops undergo product-based regulation, and the product has been assessed by the existing laws; likewise, in Colombia, Chile, Argentina, and Brazil. Gene-edited products have been exempted from GMOs classification if the final product has no exogenous DNA (Du et al. 2016). GM crops should be closely monitored; till date, few genome-edited crops have been available commercially such as high GABA tomatoes, waxy corn and non-browning mushrooms. Meanwhile, the benefits of genome editing crops have been restricted due to ethical concerns and stringent regulations. To overcome these restrictions, several regulatory frameworks must be established its alignment with global biosafety standards allows us to achieve the benefit of the genome-edited crops, and it promotes the agricultural innovation while fulfilling the strict and complex regulatory challenges (Wolt et al. 2016).

National concerns about the growing use of novel and emerging gene editing techniques in plants raise questions about whether these developments (a) can be exempted from current GMO regulations and (b) whether current regulations need to be updated and modified to adequately monitor new products resulting from these techniques. Most of the developed countries could gain advantage economically and in terms of domestic agriculture if they amend their current biosafety laws and GMO regulatory frameworks to allow gene-edited crops and their products. Development of biosafety laws for gene-edited crops may progress if different GMO standards and regulations are harmonized globally (Ahmad et al. 2021).

13.6 Role of Artificial Intelligence and CRISPR/Cas in Cereal Crops

CRISPR/Cas9 has been widely used genome editing tool with the potential to alter the precise target in the genome. Despite having accuracy and high efficiency of the gRNA, off-targets still exist that lead to genome instability. One way to prevent the adverse effects of gene editing is to ensure accurate prediction of off-targets. Numerous software and web tools have been available that calculate the scores based on the position of mismatched base pairs in the gRNA, concentrating on the simple sequence properties of gRNA and neglecting the genomic context of the particular target sequence. Multiple machine learning and deep learning based algorithms have been available for accurately finding the divergent and convergent cleavage sites to predict the off-targets and on-targets in the crop genome (Liu et al. 2017). These models have significant importance in genetic engineering, thus demonstrating the characteristics of these models in genome-wide engineering in crops that incorporate trained data sets, which explain the variety of features such as sgRNA thermodynamics and complementarity between the target site and gRNA These models do not include the defined conditions or steps; rather they include the methods or programs to perform the task from the specialized experimental data sets to predict the best gRNA (Xie et al. 2017). Machine learning algorithms have been widely employed in plant sciences such as pathogen identification, plant identification and breeding and stress physiology (Minkenberg et al. 2018). Three machine learning models such as random forests (RFs), support vector machines

(SVM), and artificial neural networks (ANNs) were recently used to evaluate sgRNA, on targets and off-target sites in the genomes of 15 distinct plants, including maize, wheat, and rice. Comparatively, the ML model exceeded the other models in accuracy with 96.27%. Predicting off-target effects, choosing the best target sequences, and enhancing current systems are all possible with artificial intelligence in CRISPR/Cas systems (Das et al. 2023). For analyzing large-scale image data sets, identifying phenotypic trends, and linking certain genetic modifications to ensuing morphological changes, machine learning and deep learning models can be applied to the phenotyping of CRISPR mutants. Different ML algorithms, such as hyperspectral imaging data, successive projections algorithm (SPA), SVM, and extreme learning machine (ELM) have been employed in rice to achieve the best possible gRNA. Rice mutants were found to have an accuracy of 93%. Mini Cas proteins have a 3D structure that is 90% similar to their original counterparts. Surprisingly, these proteins were designed by AI and are smaller in size when compared with different Cas proteins, such as Cas9 and Cas12. These machine learning and deep learning models have datasets to select the best sgRNA expression formats and scaffold sequence, along with PAM compatibilities; therefore, these models predicted the activities of 17 different small Cas9s proteins (Feng et al. 2017). These findings suggest that novel CRISPR-based plant gene-editing tools could be developed using small Cas9 proteins.

13.7 Conclusion

CRISPR/Cas genome editing offers immense potential for both basic and applied research, allowing researchers to modify crop-specific traits with greater precision and efficiency. The CRISPR/Cas technique has been used to create several novel bread wheat and barley lines with improvements in features that are commercially useful. In cereal crops, the CRISPR-Cas9 and Cas12 systems have already been effectively used to improve agronomically significant traits such as grain yield, disease and pest resistance, grain and flour quality, gluten-free properties, nutritional value, and nitrogen use efficiency. However, different crops in the Triticeae tribe, including barley, einkorn wheat, and rye, have not been extensively investigated by genetic engineering But the recent developments in genetic engineering and the transformation of these species make it possible that new Triticeae species will soon be investigated using CRISPR/Cas genome editing. Gene-editing technology will become more popular in the future and will play a significant role in improving crop quality, even though there are few challenges that need to be resolved. Therefore, it is crucial to explore innovative delivery methods for CRISPR/Cas components, particularly targeting apical meristems or pollen grains, to generate genetically edited plants without relying on in vitro cultivation. Consequently, the development of novel carrier materials would be highly beneficial. In addition to these factors, another barrier to innovations in plant breeding is the government's stringent regulation of gene-editing technology and public concerns. Different strategies and several AI-based tools have been developed to optimize the CRISPR/Cas system that includes the designing of optimized gRNA, enhancing the precision of gene editing while significantly reducing the likelihood of off-target effects. These tools utilize advanced algorithms to predict and select the most effective gRNA sequences, ensuring high editing efficiency and improved accuracy in genome modifications.

References

Abdallah NA, Elsharawy H, Abulela HA, Thilmony R, Abdelhadi AA, Elarabi NI (2022) Multiplex CRISPR/Cas9-mediated genome editing to address drought tolerance in wheat. GM Crops Food 6:1–17

Abe F, Haque E, Hisano H, Tanaka T, Kamiya Y, Mikami M, Kawaura K, Endo M, Onishi K, Hayashi T, Sato K (2019) Genome-edited triple-recessive mutation alters seed dormancy in wheat. Cell Rep 28:1362–1369

Ahmad A, Munawar N, Khan Z, Qusmani AT, Khan SH, Jamil A et al (2021) An outlook on global regulatory landscape for genome-edited crops. Int J Mol Sci 22(21):11753

Bansal M, Jindal S, Wani SH, Ganie SA, Singh R (2021) Genome editing and trait improvement in wheat. In: Wani SH, Mohan A, Singh GP (eds) Physiological, molecular, and genetic perspectives of wheat improvement. Springer, Cham, pp 263–283

Baran B, Teymur T, Morca AF, Konuk M, Tastan C (2024) Harnessing CRISPR/Cas12a technology to combat wheat dwarf virus without genetic modification. *bioRxiv*, 2024-07. Biotechnol J 19(5):937–951

Basu, U., Riaz Ahmed, S., Bhat, B. A., Anwar, Z., Ali, A., Ijaz, A., & Mushtaq, M. (2023). A CRISPR way for accelerating cereal crop improvement: Progress and challenges. Frontiers in Genetics, 13, 866976.

Brauer EK, Balcerzak M, Rocheleau H, Leung W, Schernthaner J, Subramaniam R, Ouellet T (2020) Genome editing of a deoxynivalenol-induced transcription factor confers resistance to fusarium graminearum in wheat. Mol Plant-Microbe Interact 33(3):553–560

Brunet E, Simsek D, Tomishima M, DeKelver R, Choi VM, Gregory P, Jasin M (2009) Chromosomal translocations induced at specified loci in human stem cells. Proc Natl Acad Sci 106(26):10620–10625

Bucsella, B., Molnár, D., Harasztos, A. H., & Tömösközi, S. (2016). Comparison of the rheological and end-product properties of an industrial aleurone-rich wheat flour, whole grain wheat and rye flour. Journal of Cereal Science, 69, 40-48.

Char, S. N., & Yang, B. (2020). Genome editing in grass plants. Abiotech, 1(1), 41-57.

Chen J, Hawkins E, Seung D (2021) Towards targeted starch modifcation in plants. Curr Opin Plant Biol 60:102013

Chen Z, Ke W, He F, Chai L, Cheng X, Xu H, Wang X, Du D, Zhao Y, Chen X, Xing J (2022) A single nucleotide deletion in the third exon of FT-D1 increases the spikelet number and delays heading date in wheat (Triticum aestivum L.). Plant Biotechnol J 20(5):920–933

Cheng J et al (2023) New semi-dwarfing alleles with increased coleoptile length by gene editing of gibberellin 3-oxidase 1 using CRISPR-Cas9 in barley (Hordeumvulgare L.). Plant Biotechnol J 21:806–818

Chuang, Y. F., Phipps, A. J., Lin, F. L., Hecht, V., Hewitt, A. W., Wang, P. Y., & Liu, G. S. (2021). Approach for in vivo delivery of CRISPR/Cas system: a recent update and future prospect. Cellular and Molecular Life Sciences, 78, 2683-2708.

Das J et al (2023) Machine learning in the estimation of CRISPR-Cas9 cleavage sites for plant system. Front Genet 13:1085332

Debernardi JM, Tricoli DM, Ercoli MF, Hayta S, Ronald P, Palatnik JF, Dubcovsky J (2020) A GRF-GIF chimeric protein improves the regeneration efficiency of transgenic plants. Nat Biotechnol 38:1274–1279

Du H, Zeng X, Zhao M, Cui X, Wang Q, Yang H, Yu D (2016) Efficient targeted mutagenesis in soybean by TALENs and CRISPR/Cas9. J Biotechnol 217:90–97

Feng X, Peng C, Chen Y, Liu X, Feng X, He Y (2017) Discrimination of CRISPR/Cas9-induced mutants of rice seeds using near-infrared hyperspectral imaging. Sci Rep 7(1):15934

Fonfara I, Richter H, Bratovič M, Le Rhun A, Charpentier E (2016) The CRISPR-associated DNA-cleaving enzyme Cpf1 also processes precursor CRISPR RNA. Nature 532(7600):517–521

Garcia-Gimenez G, Barakate A, Smith P, Stephens J, Khor SF, Doblin MS, Houston K (2020) Targeted mutation of barley (1, 3; 1, 4)-β-glucan synthases reveals complex relationships between the storage and cell wall polysaccharide content. Plant J 104(4):1009–1022

Ghogare, R., Ludwig, Y., Bueno, G. M., Slamet-Loedin, I. H., & Dhingra, A. (2021). Genome editing reagent delivery in plants. Transgenic Research, 30(4), 321-335

Gordon-Kamm W, Barone P, Svitashev S, D Sander J, Kumar S, & Jones, T (2021) Strategies for CRISPR/Cas9-mediated genome editing: from delivery to production of modified plants

Guilinger JP, Thompson DB, Liu DR (2014) Fusion of catalytically inactive Cas9 to FokI nuclease improves the specificity of genome modification. Nat Biotechnol 32(6):577–582

Gupta A, Hua L, Zhang Z, Yang B, & Li W (2023) CRISPR-induced miRNA156-recognition element mutations in TaSPL13 improve multiple agronomic traits in wheat. Plant Biotechnology Journal, 21(3): 536–548.

Hahn F, Sanjurjo LL, Sparks CA, Kanyuka K, Nekrasov V (2021) Efficient CRISPR/Cas-mediated targeted mutagenesis in spring and winter wheat varieties. Plants (Basel) 10:1481

He F, Wang C, Sun H, Tian S, Zhao G, Liu C, Guo J (2023) Simultaneous editing of three homoeologues of TaCIPK14 confers broad-spectrum resistance to stripe rust in wheat. Plant Biotechnol J 21(2):354–368

Hensel G (2020) Genetic transformation of Triticeae cereals–summary of almost three-decade's development. Biotechnol Adv 40:107484

Hickey, L. T., N. Hafeez, A., Robinson, H., Jackson, S. A., Leal-Bertioli, S. C., Tester, M., & Wulff, B. B. (2019). Breeding crops to feed 10 billion. Nature biotechnology, 37(7), 744-754.

Hillary VE, Ceasar SA (2019) Application of CRISPR/Cas9 genome editing system in cereal crops. Open Biotechnol J 13(1)

Hillary VE, Ceasar SA (2022) Prime editing in plants and mammalian cells: mechanism, achievements, limitations, and future prospects. BioEssays 44(9):2200032

Holubová K, Hensel G, Vojta P, Tarkowski P, Bergougnoux V, Galuszka P (2018) Modification of barley plant productivity through regulation of cytokinin content by reverse-genetics approaches. Front Plant Sci 9:1676

Jinek M, Chylinski K, Fonfara I, Hauer M, Doudna JA, Charpentier E (2012) A programmable dual-RNA–guided DNA endonuclease in adaptive bacterial immunity. Science 337(6096):816–821

Kan J, Cai Y, Cheng C, Chen S, Jiang C, He Z, Yang P (2023) CRISPR/Cas9- guided knockout of *eIF4E* improves wheat yellow mosaic virus resistance without yield penalty. Plant Biotechnol J 21:893–895

Karunarathne SD, Han Y, Zhang XQ, Li C (2022) CRISPR/Cas9 gene editing and natural variation analysis demonstrate the potential for HvARE1 in improvement of nitrogen use efficiency in barley. J Integr Plant Biol 64(3):756–770

Kim Y, Kweon J, Kim A, Chon JK, Yoo JY, Kim HJ, Kim JS (2013) A library of TAL effector nucleases spanning the human genome. Nat Biotechnol 31(3):251–258

Kim D, Alptekin B, Budak H (2018) CRISPR/Cas9 genome editing in wheat. Funct Integr Genomics 18:31–41

Kirschner GK, Rosignoli S, Guo L, Vardanega I, Imani J, Altmüller J, Hochholdinger F (2021) ENHANCED GRAVITROPISM 2 encodes a STERILE ALPHA MOTIF–containing protein that controls root growth angle in barley and wheat. Proc Natl Acad Sci 118(35):e2101526118

Kis A, Hamar É, Tholt G, Bán R, Havelda Z (2019) Creating highly efficient resistance against wheat dwarf virus in barley by employing CRISPR/Cas9 system. Plant Biotechnol J 17(6):1004

Kleinstiver BP, Pattanayak V, Prew MS, Tsai SQ, Nguyen NT, Zheng Z, Joung JK (2016) High-fidelity CRISPR–Cas9 nucleases with no detectable genome-wide off-target effects. Nature 529(7587):490–495

Kocsisova, Z., & Coneva, V. (2023). Strategies for delivery of CRISPR/Cas-mediated genome editing to obtain edited plants directly without transgene integration. Frontiers in Genome Editing, 5, 1209586.

Křenek P, Chubar E, Vadovič P, Ohnoutková L, Vlčko T, Bergougnoux V, Šamaj J (2021) CRISPR/Cas9-induced loss-of-function mutation in the barley mitogen-activated protein kinase 6 gene causes abnormal embryo development leading to severely reduced grain germination and seedling shootless phenotype. Front Plant Sci 12:670302

Kuluev BR, Mikhailova EV, Kuluev AR, Galimova AA, Zaikina EA, Khlestkina EK (2022) Genome editing in species of the tribe Triticeae with the CRISPR/Cas system. Mol Biol 56(6):885–901

Lee HJ, Kweon J, Kim E, Kim S, Kim JS (2012) Targeted chromosomal duplications and inversions in the human genome using zinc finger nucleases. Genome Res 22(3):539–548

Lema MA (2019) Regulatory aspects of gene editing in Argentina. Transgenic Res 28(Suppl 2):147–150

Li J, Manghwar H, Sun L, Wang P, Wang G, Sheng H, Zhang X (2019) Whole genome sequencing reveals rare off-target mutations and considerable inherent genetic or/and somaclonal variations in CRISPR/Cas9-edited cotton plants. Plant Biotechnol J 17(5):858–868

Li J, Jiao G, Sun Y, Chen J, Zhong Y, Yan L, Jiang D, Ma Y, Xia L (2021a) Modification of starch composition, structure and properties through editing of *TaSBEIIa* in both winter and spring wheat varieties by CRISPR/Cas9. Plant Biotechnol J 19:937–951

Li J, Jiao G, Sun Y, Chen J, Zhong Y, Yan L, Xia L (2021b) Modification of starch composition, structure and properties through editing of TaSBEIIa in both winter and spring wheat varieties by CRISPR/Cas9. Plant Biotechnol J 19(5):937–951

Li G, Wang L, Yang J, He H, Jin H, Li X, Ren T, Ren Z, Li F, Han X, Zhao X (2021c) A high-quality genome assembly highlights rye genomic characteristics and ergonomically important genes. Nat Genet 53(4):574–584

Lino, C. A., Harper, J. C., Carney, J. P., & Timlin, J. A. (2018). Delivering CRISPR: a review of the challenges and approaches. Drug delivery, 25(1), 1234-1257.

Li S, Lin D, Zhang Y, Deng M, Chen Y, Lv B, Li B, Lei Y, Wang Y, Zhao L, Liang Y, Liu J, Chen K, Liu Z, Xiao J, Qiu JL, Gao C (2022) Genome-edited powdery mildew resistance in wheat without growth penalties. Nature 602:455–460

Liu H, Ding Y, Zhou Y, Jin W, Xie K, Chen LL (2017) CRISPR-P 2.0: an improved CRISPR-Cas9 tool for genome editing in plants. Mol Plant 10(3):530–532

Michalski K, Hertig C, Mankowski DR, Kumlehn J, Zimny J, Linkiewicz AM (2021) Functional validation of Cas9/guideRNA constructs for site-directed mutagenesis of triticale *ABA8'OH1* loci. Int J Mol Sci 22:7038

Minkenberg B, Zhang J, Xie K, Yang Y (2018) CRISPR-PLANT v2: An online resource for highly specific guide RNA spacers based on improved off-target analysis. Plant Biotechnol J 17(1):5

Miroshnichenko D, Timerbaev V, Divashuk M, Pushin A, Alekseeva V, Kroupin P, Dolgov S (2024) CRISPR/Cas9-mediated multiplexed multi-allelic mutagenesis of genes located on A, B and R subgenomes of hexaploid triticale. Plant Cell Rep 43(3):59

Molla KA, Sretenovic S, Bansal KC, Qi Y (2021) Precise plant genome editing using base editors and prime editors. Nat Plants 7:1166–1187

Murovec, J., Pirc, Ž., & Yang, B. (2017). New variants of CRISPR RNA-guided genome editing enzymes. Plant biotechnology journal, 15(8), 917-926.

Nalam VJ, Alam S, Keereetaweep J, Venables B, Burdan D, Lee H, Shah J (2015) Facilitation of Fusarium graminearum infection by 9-lipoxygenases in Arabidopsis and wheat. Mol Plant-Microbe Interact 28(10):1142–1152

Poutanen KS, Kårlund AO, Gómez-Gallego, C, Johansson DP, Scheers NM, Marklinder IM, & Landberg, R (2022) Grains–a major source of sustainable protein for health. Nutrition reviews, 80(6): 1648–1663.

Raffan S, Sparks C, Huttly A, Hyde L, Martignago D, Mead A, Hanley SJ, Wilkinson PA, Barker G, Edwards KJ, Curtis TY, Usher S, Kosik O, Halford NG (2021) Wheat with greatly reduced accumulation of free asparagine in the grain, produced by CRISPR/Cas9 editing of asparagine synthetase gene *TaASN2*. Plant Biotechnol J 19:1602–1613

Rey MD, Ramírez C, Martín AC (2021) Wheat, Rye, and Barley genomes can associate during meiosis in newly synthesized trigeneric hybrids. Plants 10:113

Sánchez-León S, Gil-Humanes J, Ozuna CV, Giménez MJ, Sousa C, Voytas DF, Barro F (2018) Low-gluten, nontransgenic wheat engineered with CRISPR/Cas9. Plant Biotechnol J 16(4):902–910

Sander JD, Joung JK (2014) CRISPR-Cas systems for genome editing, regulation and targeting. Nat Biotechnol 32(4):347

Sandhya, D., Jogam, P., Allini, V. R., Abbagani, S., & Alok, A. (2020). The present and potential future methods for delivering CRISPR/Cas9 components in plants. Journal of Genetic Engineering and Biotechnology, 18(1), 25.

Shan Q, Wang Y, Li J, Gao C (2014) Genome editing in rice and wheat using the CRISPR/Cas system. Nat Protoc 9(10):2395–2410

Song X et al (2022) Targeting a gene regulatory element enhances rice grain yield by decoupling panicle number and size. Nat Biotechnol 40:1403–1411

Song L, Liu J, Cao B, Liu B, Zhang X, Chen Z, Ni Z (2023) Reducing brassinosteroid signalling enhances grain yield in semi-dwarf wheat. Nature 617(7959):118–124

Tsai SQ, Zheng Z, Nguyen NT, Liebers M, Topkar VV, Thapar V, Joung JK (2015) GUIDE-seq enables genome-wide profiling of off-target cleavage by CRISPR-Cas nucleases. Nat Biotechnol 33(2):187–197

Vakulskas CA, Behlke MA (2019) Evaluation and reduction of CRISPR off-target cleavage events. Nucleic Acid Ther 29(4):167–174

Vlcko T, Ohnoutkova L (2020) Allelic variants of CRISPR/Cas9 induced mutation in an inositol trisphosphate 5/6 kinase gene manifest different phenotypes in barley. Plants 9(2):195

Waltz E (2016) Gene-edited CRISPR mushroom escapes US regulation. Nature 532(7599):293

Wang W, Pan Q, He F et al (2018a) Transgenerational CRISPR-Cas9 activity facilitates multiplex gene editing in allopolyploid wheat. CRISPR J 1(1):65–74

Wang W, Simmonds J, Pan Q, Davidson D, He F, Battal A, Akhunova A, Trick HN, Uauy C, Akhunov E (2018b) Gene editing and mutagenesis reveal inter-cultivar differences and additivity in the contribution of *TaGW2* homoeologues to grain size and weightin wheat. Theor Appl Genet 131:2463–2475

Wang W, Pan Q, Tian B, He F, Chen Y, Bai G, Akhunova A, Trick HN, Akhunov E (2019) Gene editing of the wheat homologs of TONNEAU1-recruiting motif encoding gene affects grain shape and weight in wheat. Plant J 100(2):251–264

Wang K, Gong Q, Ye X (2020) Recent developments and applications of genetic transformation and genome editing technologies in wheat. Theor Appl Genet 133(5):1603–1622

Wang W, Tian B, Pan Q, Chen Y, He F, Bai G, Akhunov E (2021) Expanding the range of editable targets in the wheat genome using the variants of the Cas12a and Cas9 nucleases. Plant Biotechnol J 19(12):2428–2441

Wang Y et al (2022) Improving bread wheat yield through modulating an unselected AP2/ERF gene. Nat Plants 8:930–939

Wolt JD, Wang K, Yang B (2016) The regulatory status of genome-edited crops. Plant Biotechnol J 14(2):510–518

Xie X, Ma X, Zhu Q, Zeng D, Li G, Liu YG (2017) CRISPR-GE: a convenient software toolkit for CRISPR-based genome editing. Mol Plant 10(9):1246–1249

Yang X et al (2023) Ovule cell wall composition is a maternal determinant of grain size in barley. New Phytol 237:2136–2147

Zafar SA, Zaidi SSEA, Gaba Y, Singla-Pareek SL, Dhankher OP, Li X, Pareek A (2020) Engineering abiotic stress tolerance via CRISPR/Cas-mediated genome editing. J Exp Bot 71(2):470–479

Zaidi SS et al (2020) Engineering crops of the future: CRISPR approaches to develop climate-resilient and disease-resistant plants. Genome Biol 21:289

Zeng Z, Han N, Liu C, Buerte B, Zhou C, Chen J, Wang M, Zhang Y, Tang Y, Zhu M., Wang J., Yang Y, Bian H. (2020). Functional dissection of HGGT and HPT in barley vitamin E biosynthesis via CRISPR/Cas9-enabled genome editing. Ann. Bot. 126:929–942. https://doi.org/10.1093/aob/mcaa115.

Zhang B (2015) MicroRNA: a new target for improving plant tolerance to abiotic stress. J Exp Bot 66(7):1749–1761

Zhang Y, Bai Y, Wu G et al (2017) Simultaneous modification of three homoeologs of TaEDR1 by genome editing enhances powdery mildew resistance in wheat. Plant J 91(4):714–724

Zhang Y, Li D, Zhang D et al (2018) Analysis of the functions of *TaGW2* homoeologs in wheat grain weight and protein content traits. Plant J Cell Mol Biol 94(5):857–866

Zhang Z, Hua L, Gupta A, Tricoli D, Edwards KJ, Yang B, Li W (2019) Development of an Agrobacterium-delivered CRISPR/Cas9 system for wheat genome editing. Plant Biotechnol J 17(8):1623–1635

Zhang J, Zhang H, Li S, Li J, Yan L, Xia L (2021a) Increasing yield potential through manipulating of an *ARE1* ortholog related to nitrogen use efficiency in wheat by CRISPR/Cas9. J Integr Plant Biol 63:1649–1663

Zhang D, Zhang Z, Unver T, Zhang B (2021b) CRISPR/Cas: a powerful tool for gene function study and crop improvement. J Adv Res 29:207–221

Zheng M, Lin J, Liu X, Chu W, Li J, Gao Y, An K, Song W, Xin M, Yao Y (2021) Histone acetyltransferase TaHAG1 acts as a crucial regulator to strengthen salt tolerance of hexaploid wheat. Plant Physiol 186(4):1951–1969

Zhou C, Chen J, Bian H (2020) Functional dissection of HGGT and HPT in barley vitamin E biosynthesis via CRISPR/Cas9-enabled genome editing. Ann Bot 126(5):929–942

Zong Y, Wang Y, Li C, Zhang R, Chen K, Ran Y, Qiu JL, Wang D, Gao C (2017) Precise base editing in rice, wheat and maize with a Cas9-cytidine deaminase fusion. Nat Biotechnol 35:438–440

Chapter 14
Applications of CRISPR-Cas for Genetic Improvement of Triticale

Komal Pujare and Siddhesh B. Ghag

Abstract Modernization of agriculture by introducing novel crop varieties with improved productivity, agronomic qualities, and climate resilience is essential to circumvent the food security issue. Triticale, a fertile hybrid developed from wheat and rye, offers tremendous potential as an alternative to wheat in marginal unfavorable production environments. Though several hybrids have been developed over years, genetic modification using recombinant DNA technology in triticale is limited due to its polyploidy, poor genetic information, and restricted regeneration protocols. CRISPR-Cas system for genome editing has revolutionized agriculture by developing genome-edited climate-resilient crops. Application of this technology for genetic improvement of triticale will help in unlocking its full potential in developing climate-resilient, nutritionally rich, and adaptable to harsh environmental changes. This chapter provides a comprehensive summary of the current status of research on conventional genetic manipulation technology and use of CRISPR-Cas system along with the challenges associated in manipulating triticale genomes to introduce beneficial agronomic traits.

Keywords *Triticale* · CRISPR-Cas system · Cereal genetic improvement · *Secale* · *Triticum*

14.1 Introduction

Agriculture is essential for ensuring a steady food supply for humans. Crops have always been exposed to various environmental challenges such as drought, heat, salinity, pest and pathogen attacks, and climate changes, all of which have led to reduced yields. According to the Food and Agriculture Organization (FAO) of the

K. Pujare · S. B. Ghag (✉)
School of Biological Sciences, UM-DAE Centre for Excellence in Basic Sciences, University of Mumbai campus, Kalina, Santacruz (East), Mumbai, India
e-mail: siddhesh.ghag@cbs.ac.in

K. A. Abd-Elsalam, A. Ahmad (eds.), *Gene Editing in Cereals*, Sustainability Sciences in Asia and Africa, https://doi.org/10.1007/978-981-95-6722-5_14

United Nations, global annual cereal consumption in 2024 increased by 0.6% compared to the previous year (FAO 2024). This highlights the importance of essential cereal crops such as rice, wheat, and maize for human population that is increasing every year. It is essential to maintain the quality of staple foods alongside increasing the quantity to feed the population. Therefore, there is an increasing demand for these staples that are nutritionally rich and also resist biotic and abiotic stresses. With this intention in mind, plant breeders have been utilizing existing germplasms to develop novel varieties. Triticale (X *Triticosecale* Wittmack) was the first successful "man-made" cereal grain. This crop is a hybrid resulting from the cross between the genera *Triticum* (wheat) and *Secale* (rye), the name reflecting its dual ancestry. The development of this crop was driven by the goal of combining the high yield and superior grain quality of wheat with resistance to biotic and abiotic stresses found in rye (Mergoum and Gómez Macpherson 2004).

In 1884, Carman published an illustration of a partial wheat-rye plant, but the first amphidiploid stable wheat-rye hybrids were discovered by Rimpau in 1888 (Sturtevant 1886; Rimpau 1891). This crop is distinct from other cereals because it can thrive in drought and nutritionally poor soil conditions. Moreover, it offers higher yield and superior nutritional qualities compared to wheat (O'mara 1953; Müntzing 2010). In 2017, a group of researchers published a study that compared the yield responses of wheat and triticale in response to nitrogen fertilizer on high-yielding soils. The results showed that triticale grain had lower protein content and specific weight as compared to wheat. However, it yielded higher biomass and straw, had a lower harvest index, and demonstrated greater total nitrogen uptake than wheat (Roques et al. 2017). Triticale has a wide range of applications, from being used as feed and food grain to rotating it with other staple crops for reducing soil pests and pathogens. Presently, it is a popular livestock feed as compared to bread making (Mergoum et al. 2009; Malherbe et al. 2023). Grain from modern triticale varieties has been found to have an energy value similar to that of other cereal grains (Zakir and Belete 2019). Triticale has higher protein content than wheat and a better balance of essential amino acids, which are advantageous for the livestock industries (Salmon 1984). Depending on the harvesting time, triticale varieties can offer highly digestible forage and grain feed for ruminants and the amount of phytic acid is also same as other cereals (Evci et al. 2024). Moreover, recent studies have shown that triticale can serve as a raw material for bioethanol production through bioprocessing with a conversion efficiency of 96% (Malherbe et al. 2023).

Majority of the triticales produced originated either by crossing wheat (*Triticum aestivum* L., $2n = 42 = AABBDD$) or durum wheat (*Triticum durum* L., $2n = 28 = AABB$) as the seed parent and rye (*Secale cereale* L., $2n = 14 = RR$) as a pollen parent to produce hexaploid ($2n = 42 = AABBRR$) or octoploid ($2n = 56 = AABBDDRR$) triticale (Fig. 14.1). Vast research has been carried out to improve the hexaploid triticale due to its reproductive stability and great breeding strength as compared to octoploid triticale (Mergoum et al. 2009). It has high potential for grain and forage production, but its improvement is hindered by several agricultural challenges, one of them being limited genetic diversity. As a relatively new crop with minimal natural evolution compared to wheat and barley, developing diverse germplasm for triticale poses a challenge for breeders (Mergoum and Gómez Macpherson 2004). Although triticale germplasm shows strong resistance to diseases and pests, it is susceptible to the emerging races of pathogens and insects (Arseniuk 2015). Abiotic stresses, including acidic and alkaline soils, micronutrient deficiencies, and moisture stress, also present significant problems in triticale cultivation (Blum 2014). Additionally, least preferred by farmers, restricted utilization, and partial acceptance in human diets are challenges that must be addressed by the scientific and political community to utilize triticale to its full potential that can contribute effectively to agricultural production (Mergoum and Gómez Macpherson 2004). Proof of principle studies have been carried out in the past demonstrating development of transgenic triticale expressing antibiotic resistance, herbicide-tolerant marker gene, and reporter gene such as GFP and GUS (Bińka et al. 2012;

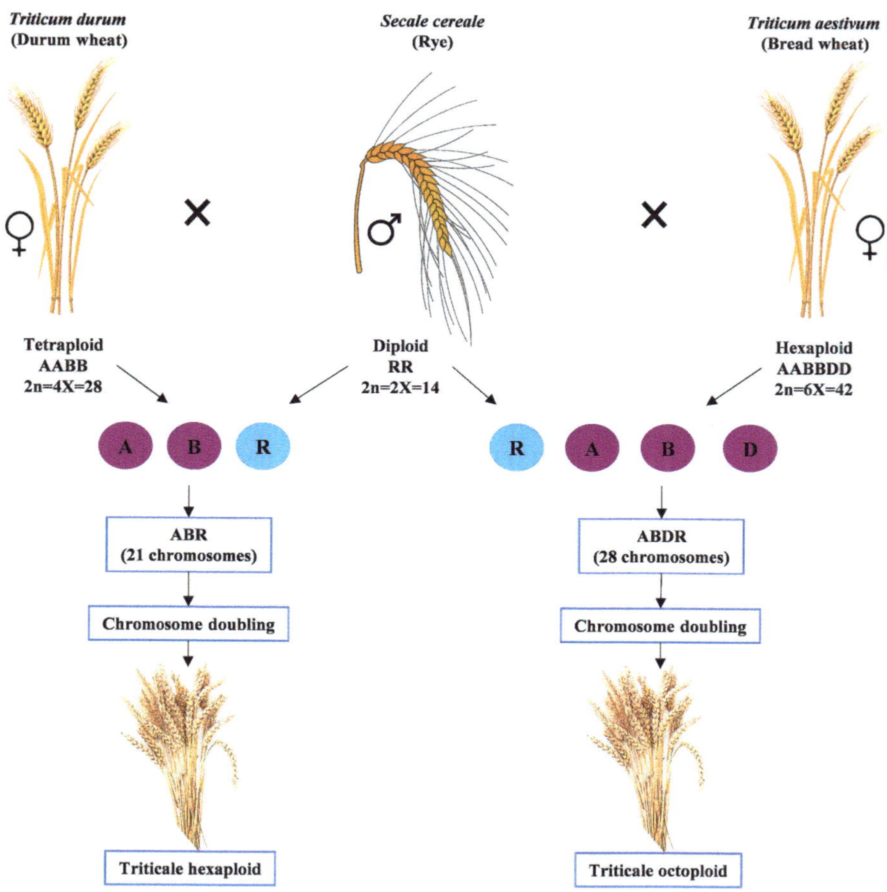

Fig. 14.1 Breeding strategy utilized for developing hexaploid and octoploid triticale from *Triticum* and *Secale*

Hensel et al. 2012). However, research on developing transgenic triticale is at a very naive stage as compared to other cereals such as rice, wheat, and maize. A query search in National Library of Medicine (Pubmed, NCBI accessed on 24th January, 2025) database with keyword "transgenic triticale" retrieved only 24 entries, whereas for transgenic wheat, rice, and maize gave 2087, 6123, and 3710 entries, respectively. Genome editing has become the modern revolutionizing technology for improving crop plants in terms of agricultural productivity, abiotic stress tolerance, and biotic resistance. This technology can be very well employed in triticale for overcoming their agricultural challenges and providing alternative sources to wheat and rice (Mengstie and Wondimu 2021). Traditional genome modification techniques such as Sequence-Specific Nucleases (SSNs), Zinc Finger Nucleases (ZFNs), and Transcription Activator-Like Effector Nucleases (TALENs) have been used to manipulate the genome of cereals. These approaches rely on the specific DNA–protein interactions to locate target sequences, making them specific, complicated, and time-consuming. However, the clustered regularly interspaced short palindromic repeats (CRISPR)/CRISPR-associated protein (Cas) system utilizes DNA-RNA interaction to target specific DNA sequences and manipulate them in order to introduce mutation (Gaj et al. 2013). This chapter entails the CRISPR-Cas9 system and its potential application for improvement of triticale.

14.2 CRISPR-Cas9 Technology for Crop Improvement

Owing to the global food security issue, there is a primary goal to improve crop plants for better agricultural productivity and climate resilience. These issues can be essentially addressed using modern molecular tools such as the CRISPR-Cas9 genome editing technique that offers significant potential to enhance crop yield. Bacteria and Archaea utilize CRISPR-Cas as an adaptive immune defence response against viruses by incorporating phage-derived sequences (spacers) into their genome, forming CRISPR arrays. Upon subsequent viral attack, a double-stranded RNA structure directs the Cas9 protein (endonuclease) to induce a double-stranded break in the cognate DNA molecule. This RNA structure forms when CRISPR RNA (crRNA) pairs with the transactivating crRNA (tracrRNA) and cleaves the target DNA strands with the help of Cas9. Following this binding, the two Cas9 domains, RuvC-like and HNH-like, cleave the complementary and noncomplementary strands, respectively (Gasiunas et al. 2012; Jinek et al. 2012). The role of CRISPR-Cas in lactic acid bacterium *Streptococcus thermophilus* was evaluated by inserting the phage sequence into spacer region of CRISPR that conferred resistance to the phage. However, this resistance was lost when the corresponding protospacer sequence was deleted from the phage genome. This technology of using guide RNA (gRNA) for targeting specific DNA sequences has been efficiently used in targeted genome modification techniques (Barrangou et al. 2007; Deveau et al. 2008). CRISPR-Cas system has been used in several crop plants such as rice, wheat, maize, banana, melons, tomatoes, etc. to introduce mutations (insertion or deletion) in the genes rendering them nonfunctional. The upgraded CRISPR system can be used to edit nucleotide sequences within the genome through base-editing or prime-editing methods (Mengstie and Wondimu 2021; Ahmar et al. 2024; Tiwari and Ghag 2024). CRISPR-Cas system has been used to enhance abiotic stress tolerance, disease resistance, increase nutritional content, grain quality, and other agronomic or breeding traits in crop plants (Basu et al. 2023). For instance, CRISPR-mediated knockout of *OsANN1* (Rice Annexin1) in rice plants imparted resistance to rice blast disease caused by *Magnaporthe oryzae* (Zhao et al. 2023). Additionally, one of the parents of triticale, *T. aestivum*, has been genetically modified by targeting *TaATX4* gene by CRISPR-Cas9 to enhance drought tolerance (Wang et al. 2023). In maize, *Zmzb7* marker gene essential for methyl-D-erythritol-4-phosphate (MEP) pathway was successfully mutated using CRISPR-Cas9 system giving albino phenotype (Feng et al. 2016).

14.3 Applications of CRISPR-Cas9 Technology in Triticale

In today's world, with the ever-increasing population, changing climatic conditions, and limited resources, triticale offers an excellent alternative to other cereals due to its ability to thrive in harsh environments while requiring minimal inputs and supporting sustainable farming systems. Furthermore, it yields higher biomass and grain production across a range of soil types and climatic conditions (Mergoum et al. 2009). To increase triticale production and broaden its applications, genetic modifications are essential. Researchers have employed various methods to achieve this, including the development of transgenic triticale. Hansel and co-workers successfully developed transgenic triticale cv. Bogo containing the selectable marker gene hygromycin phosphotransferase (*HPT*) and a synthetic green fluorescent protein gene (*gfp*) and analyzed their T-DNA integration, copy number, integrity, and inheritance (Hensel et al. 2009, 2012). Recently, the CRISPR system was employed for genome modification in triticale, leading to the enhancement of various traits. As triticale is a relatively new crop in the field of genome editing, there are only a few studies demonstrating the successful application of CRISPR-Cas9 (Table 14.1).

Multiplexed CRISPR-Cas9 editing was carried out in Triticale cv. Dublet and Triticale cv. Tikhon cultivars to modulate starch biosynthesis. Sub-genome independent introduction of panel of novel mutations in genes associated with starch biosynthesis such as *GBSSI*, *SSIIa*, *ISA1*, and *RSR1* was carried out. The CRISPR-Cas9 harboring vectors (with wheat optimized Cas9) p*GCB* and p*SGU3–34* were delivered into the embryogenic callus cultures of triticale using the microprojectile bombardment method. Transgenic triticale plants were selected using a dual selection strategy. The transgenic lines (harboring the *bar* gene) were selected on a phosphinothricin-supplemented medium and further checked for the expression of green fluorescent protein (GFP). The editing efficiency of sgRNAs varied with mutagenesis rate ranging from 2% to 27%. The authors successfully generated 7 single mutants, 2 double mutants, 1 triple mutant, and 2 quadruple mutants, all of

Table 14.1 Summary of CRISPR-Cas system used in Triticale

Sr. No.	Name of plant	Gene name	Method of delivery	Gene	Gene function	CAS9 promoter	SgRNA Promoter	Type of mutagenesis	References
1	Triticale cv. Dublet (spring type) and Triticale cv. Tikhon (winter type)	Starch synthase	Microprojectile bombardment	*GBSSI*	Elongation of linear amylose chain	*ZmUbi*	*U3*	Multi-allelic mutagenesis	Miroshnichenko et al. (2024)
2		Starch synthase	Microprojectile bombardment	*SSIIa*	Elongation of short amylopectin chains to intermediate length	*ZmUbi*	*U3*	Multi-allelic mutagenesis	
3		Starch debranching enzyme (Isoamylase)	Microprojectile bombardment	*ISAI*	Debranching	*ZmUbi*	*U3*	Multi-allelic mutagenesis	
4		Starch regulator 1	Microprojectile bombardment	*RSR1*	Regulation of starch synthesis	*ZmUbi*	*U3*	Multi-allelic mutagenesis	
5	Triticale cv. Bogo	ABA 8′-hydroxylase	Protoplast transformation	*TsABA 8′OH1*	Controls ABA catabolic degradation in grains	*ZmUBI-1*	*U6*	Site-directed mutagenesis	Michalski et al. (2021)

them containing indels in the A, B, and R sub-genomes of triticale (Miroshnichenko et al. 2024).

Michalski et al. (2021) pre-validated Cas9/gRNA constructs in a highly regenerable hexaploid Triticale cv. Bogo for the use of CRISPR-Cas9 technology in triticale. Given the complexity of genome editing in triticale, efforts were made to enhance efficiency of mutagenesis by incorporating a non-processive 3′-5′ exonuclease, Three Prime Repair Exonuclease (*TREX2*), into the plasmid constructs. This modified plasmid, combined with two different gRNA constructs, was used to target the *TsABA8′OH1* single-copy gene on each of the 6 chromosomes in triticale, which is known to regulate catabolic degradation of abscisic acid in grains. Protoplast transfection method was employed to deliver the vector constructs into the cells. There was an increase in the indel induction frequency in triticale due to the increased TREX2 nuclease activity as the gRNA construct without TREX2 showed poor editing. The presence of the *TREX2* gene showed maximum editing efficiency of the triticale A genome followed by B genome and lastly the R genome (Michalski et al. 2021).

Current research on genome editing in triticale is in a naive stage, but it has high potential to genetically improve triticale. Genome editing has been successfully used to introduce novel genetic modifications in cereal crops, such as wheat, barley, and rice to enhance their quality, resistance, and agronomic value (Matres et al. 2021). Since triticale shares a genome from one of its parents, the durum wheat, CRISPR-Cas system targeting wheat genes for improving agronomic traits can be employed in triticale. For instance, in Coeliac disease, ingestion of gluten triggers an autoimmune disorder primarily caused due to the presence of α-gliadin (a highly stimulatory peptide). The CRISPR-Cas9 technique was employed to mutate α-gliadin in the hexaploid wheat by designing two gRNAs targeting its conserved regions, resulting in transgene-free, low-gluten wheat lines. The mutant wheat lines showed 85% reduction in the immunoreactivity, making it a potential source for low-gluten food products (Jouanin et al. 2020). The same genes can be targeted in triticale, nevertheless the toxicity of α-gliadin in triticale is least (Qi et al. 2013). Thus, adapting methodologies from similar crops could make genome editing more effective in a relatively new crop, such as triticale.

14.3.1 Challenges Associated with Gene Editing of Triticale

Triticale is the best alternative to other cereal crops as it can grow in stressed environments; however, improving its quality is associated with challenges in its genome editing (Fig. 14.2). One of the major reasons is relatively limited germplasm. Unlike wheat, barley, and other cereals, triticale is new and hence has a very less genetic diversity, genome data availability, and therefore it is difficult to produce diverse germplasm important for genome editing (Al-Khayri et al. 2019; Michalski et al. 2021; Subedi et al. 2024). Since the wheat genome is well-annotated, numerous CRISPR-Cas-based studies have been conducted in wheat to enhance its agronomic value. Recently, multiplexed CRISPR-Cas9-mediated genome editing was employed in wheat to increase drought tolerance by generating knockouts of 3′(2′), 5′-bisphosphate nucleotidase (*TaSal1*) gene (Abdallah et al. 2025). It has been more cumbersome to generate genome-edited triticale due to its polyploid nature and larger genome size making it difficult for the Cas9/gRNA complexes to find all the targets and modify them (Kumar et al. 2019). Another common challenge faced by all monocotyledon plants is the genetic transformation method and regeneration of transformants (Sood et al. 2011; Hensel 2020). Various transformation techniques such as polyethylene glycol (PEG), electroporation, microinjection, silicon carbide whiskers, and biolistic method have been optimized over years in monocots to produce transgenic lines (Sood et al. 2011). The age and physiological conditions of the explant, along with its capability of proliferation and regeneration post-transformation, are crucial for obtaining successful transgenic lines (Mohammed et al. 2019). Moreover, recovery of transformed explants in monocots remains difficult, posing a challenge for effective genome editing. To increase plant regeneration and transformation efficiency, a chimeric protein complex comprising Growth-Regulating Factor 4 (GRF4) and its cofactor Grf-Interacting Factor1 (GIF1) were used. The results demonstrated a regeneration efficiency of 9–19%, although no transgenic lines were generated in this study. To address the challenge of monocot transformation, several studies were carried out, leading to the successful delivery of transgenes and production of transgenic plants (Debernardi et al. 2020). In one such study, *in vitro* nano complex consisting of transferred DNA, virulence protein D2, and recombination protein A were delivered into the microspores of triticale using a Tat_2 Cell-Penetrating Peptide (CPP) as a carrier. This delivery

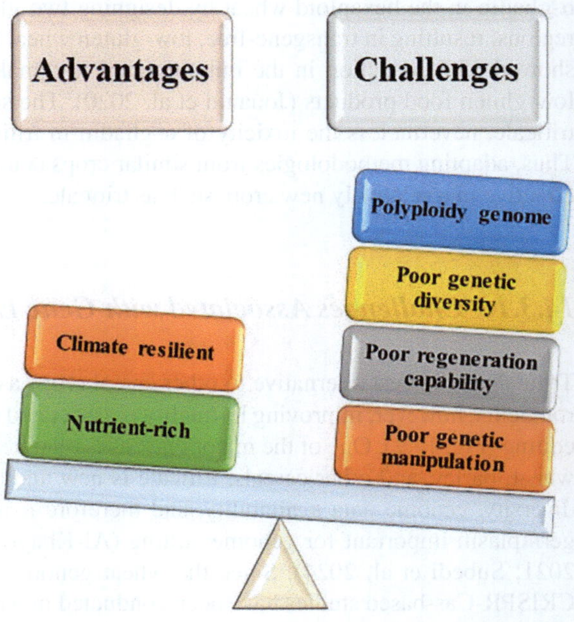

Fig. 14.2 Advantages to be exploited and challenges to overcome for triticale as an ideal cereal crop

method was able to successfully transfer a single copy of *uidA* (*GUS*) gene without degradation into plants by using *Agrobacterium tumefaciens*-mediated transformation (Ziemienowicz et al. 2012). Moreover, the *Agrobacterium*-mediated transformation method was optimized by adjusting parameters such as nutrient concentration, temperature, pH, presence and varying concentration of acetosyringone and antioxidants, and time of cultivation to achieve successful transgenic lines. The results showed that the efficiency of obtaining transgenic plants was highest in barley (86%), followed by maize (24%), wheat (10%), and triticale (4%). These methods can serve as an alternative transformation approach for the efficient delivery of transgenes into monocot plants such as triticale (Hensel et al. 2009).

14.4 Other Genome Editing Techniques Employed in Triticale

In addition to CRISPR-Cas, there are various other molecular breeding techniques and chromosomal manipulation strategies that have been used to genetically improve triticale varieties (Subedi et al. 2024). Triticale is a key focus for molecular breeders because of its self-pollinating nature, presence of a male fertility restorer gene, and cytoplasmic male sterility (Warzecha et al. 2014). To make hybrid cultivars, male sterility is essential. Researchers developed male sterile lines through the conventional technique of recombining triticale with rye maintainers. Results obtained showed that, out of 180 offspring, 7.8% of the individuals maintained male sterility, and their progeny from subsequent crosses maintained the same trait. However, these traditional methods are labor-intensive, less efficient, time-consuming, and result in limited genetic variations (Warzecha et al. 2014). Another method that has been successfully employed to improve triticale is chromosomal manipulations, such as insertions, deletions, substitutions, and translocations (Subedi et al. 2024). Hexaploid triticale and bread wheat (the female parent of triticale) differ in their R and D genomes, leading to variations in both agronomic and end-use quality traits. Budak et al. (2004) examined the introgression of individual chromosomes from the D-genome of wheat that influenced the characteristics of the winter triticale cv. Presto. Their findings showed that substituting chromosomes 2D, 4D, and 6D with 2R, 4B, and 6R, respectively, reduced plant height. Additionally, substituting chromosome 5D with 5R increased kernel hardness. Introgression of chromosomes 1D and 6D enhanced the end-use quality characteristics of cv. Presto (Budak et al. 2004). Chromosomal addition and substitution techniques have been used in triticale to enhance its tolerance to powdery mildew infection caused by fungus *Blumeria graminis*. The *Pm13* gene, which confers resistance to powdery mildew, was found to be evolutionarily transferred from *Aegilops longissima* to *Aegilops variabilis* Eig. (UUSS). This resistance gene was then transferred into hexaploid triticale using *A. variabilis* × *S. cereale* amphiploid bridge forms. The addition or substitution lines of hybrids carrying chromosome 3Sv, which contains the powdery mildew resistant loci, were screened by cytogenetic techniques, visual disease symptoms of powdery mildew and through marker-assisted selection using *Pm13* as marker (Kwiatek et al. 2016).

14.5 Conclusion

Triticale provides significant advantages over other crops due to its nutritional qualities and tolerance to biotic and abiotic stresses. Due to its relatively low bread-making property, it has now been explored primarily for bioethanol production, as forage, rotational crop, and in poultry farming (Gaviley et al. 2024). In order to maximize its potential for human consumption as an alternate to wheat and enhance specific agronomic traits, genetic modifications are deemed essential. There are a few conventional genetic manipulation methods available for triticale, but CRISPR-Cas9 offers great advantages as it requires less time, results in fewer off-target effects, greater customization potential, and cost-effectiveness. Currently, the application of CRISPR-Cas9 system in triticale is limited due to challenges such as low genomic diversity, polyploid genome, restricted germplasm, and a poorly annotated genome. However, CRISPR-Cas-based genome editing technology is well-established for one of its parents, i.e., the hexaploid wheat. Similar methodologies can be employed and further optimized in triticale to enhance its agronomic value and unlock its full potential to be a beneficial crop for farmers compared to other cereal crops.

References

Abdallah NA, Elsharawy H, Abulela HA, Thilmony R, Abdelhadi AA, Elarabi NI (2025) Multiplex CRISPR/Cas9-mediated genome editing to address drought tolerance in wheat. GM Crops Food 16(1):1–17. https://doi.org/10.1080/21645698.2022.2120313

Ahmar S, Usman B, Hensel G, Jung KH, Gruszka D (2024) CRISPR enables sustainable cereal production for a greener future. Trends Plant Sci 29(2):179–195. Elsevier Ltd. https://doi.org/10.1016/j.tplants.2023.10.016

Al-Khayri JM, Jain SM, Johnson DV (2019) Advances in plant breeding strategies: cereals. In: Advances in plant breeding strategies: cereals, vol 5. Springer International Publishing. https://doi.org/10.1007/978-3-030-23108-8

Arseniuk E (2015) Triticale biotic stresses—known and novel foes. In: Triticale. Springer International Publishing, pp 83–110. https://doi.org/10.1007/978-3-319-22551-7_4

Barrangou R, Fremaux C, Deveau H, Richards M, Boyaval P, Moineau S, Romero DA, Horvath P (2007) CRISPR provides acquired resistance against viruses in prokaryotes. Science 315(5819):1709–1712. https://doi.org/10.1126/science.1138140

Basu U, Riaz Ahmed S, Bhat BA, Anwar Z, Ali A, Ijaz A, Gulzar A, Bibi A, Tyagi A, Nebapure SM, Goud CA, Ahanger SA, Ali S, Mushtaq M (2023) A CRISPR way for accelerating cereal crop improvement: Progress and challenges. Front Genet 13. https://doi.org/10.3389/fgene.2022.866976

Bińka A, Orczyk W, Nadolska-Orczyk A (2012) The agrobacterium-mediated transformation of common wheat (Triticum aestivum L.) and triticale (x Triticosecale Wittmack): role of the binary vector system and selection cassettes. J Appl Genet 53(1):1–8. https://doi.org/10.1007/s13353-011-0064-y

Blum A (2014) The abiotic stress response and adaptation of triticale—a review. Cereal Res Commun 42(3):359–375. https://doi.org/10.1556/CRC.42.2014.3.1

Budak H, Stephen Baenziger P, Beecher BS, Graybosch RA, Campbell BT, Shipman MJ, Erayman M, Eskridge KM (2004) The effect of introgressions of wheat D-genome chromosomes into "Presto" triticale. In: Euphytica, vol 137. Kluwer Academic Publishers

Debernardi JM, Tricoli DM, Ercoli MF, Hayta S, Ronald P, Palatnik JF, Dubcovsky J (2020) A GRF–GIF chimeric protein improves the regeneration efficiency of transgenic plants. Nat Biotechnol 38(11):1274–1279. https://doi.org/10.1038/s41587-020-0703-0

Deveau H, Barrangou R, Garneau JE, Labonté J, Fremaux C, Boyaval P, Romero DA, Horvath P, Moineau S (2008) Phage response to CRISPR-encoded resistance in *Streptococcus thermophilus*. J Bacteriol 190(4):1390–1400. https://doi.org/10.1128/JB.01412-07

Evci Ş, Kara K, Demirci M (2024) Determination of yield nutrient contents and in vitro digestibility of some triticale varieties harvested at different maturity stages. Turk J Vet Anim Sci 48(4):174–179. https://doi.org/10.55730/1300-0128.4352

FAO (2024) FAO Cereal Supply and Demand Brief | Food and Agriculture Organization of the United Nations. (n.d.). Retrieved January 28, 2025, from https://www.fao.org/worldfoodsituation/csdb/en

Feng C, Yuan J, Wang R, Liu Y, Birchler JA, Han F (2016) Efficient targeted genome modification in maize using CRISPR/Cas9 system. J Genet Genomics 43(1):37–43. https://doi.org/10.1016/j.jgg.2015.10.002

Gaj T, Gersbach CA, Barbas CF (2013) ZFN, TALEN, and CRISPR/Cas-based methods for genome engineering. Trends Biotechnol 31(7):397–405. https://doi.org/10.1016/j.tibtech.2013.04.004

Gasiunas G, Barrangou R, Horvath P, Siksnys V (2012) Cas9–crRNA ribonucleoprotein complex mediates specific DNA cleavage for adaptive immunity in bacteria. Proc Natl Acad Sci 109(39). https://doi.org/10.1073/pnas.1208507109

Gaviley OV, Katerynych OO, Ionov IA, Dekhtiarova OO, Griffin DK, Romanov MN (2024) Triticale: a general overview of its use in poultry production. Encyclopedia 4(1):395–414. https://doi.org/10.3390/encyclopedia4010027

Hensel G (2020) Genetic transformation of Triticeae cereals—summary of almost three-decade's development. In: Biotechnology advances, vol 40. Elsevier Inc. https://doi.org/10.1016/j.biotechadv.2019.107484

Hensel G, Kastner C, Oleszczuk S, Riechen J, Kumlehn J (2009) *Agrobacterium*-mediated gene transfer to cereal crop plants: current protocols for barley, wheat, triticale, and maize. Int J Plant Genomic 2009(1). https://doi.org/10.1155/2009/835608

Hensel G, Oleszczuk S, Daghma DES, Zimny J, Melzer M, Kumlehn J (2012) Analysis of T-DNA integration and generative segregation in transgenic winter triticale (x Triticosecale Wittmack). BMC Plant Biol 12(1):171. https://doi.org/10.1186/1471-2229-12-171

Jinek M, Chylinski K, Fonfara I, Hauer M, Doudna JA, Charpentier E (2012) A programmable dual-RNA–guided DNA endonuclease in adaptive bacterial immunity. Science 337(6096):816–821. https://doi.org/10.1126/science.1225829

Jouanin A, Gilissen LJWJ, Schaart JG, Leigh FJ, Cockram J, Wallington EJ, Boyd LA, van den Broeck HC, van der Meer IM, America AHP, Visser RGF, Smulders MJM (2020) CRISPR/Cas9 gene editing of gluten in wheat to reduce gluten content and exposure—reviewing methods to screen for coeliac safety. Front Nutr 7. https://doi.org/10.3389/fnut.2020.00051

Kumar R, Kaur A, Pandey A, Mamrutha HM, Singh GP (2019) CRISPR-based genome editing in wheat: a comprehensive review and future prospects. Mol Biol Rep 46(3):3557–3569. https://doi.org/10.1007/s11033-019-04761-3

Kwiatek M, Belter J, Majka M, Wiśniewska H (2016) Allocation of the S-genome chromosomes of Aegilops variabilis Eig. Carrying powdery mildew resistance in triticale (× Triticosecale Wittmack). Protoplasma 253(2):329–343. https://doi.org/10.1007/s00709-015-0813-6

Malherbe SJM, Cripwell RA, Favaro L, van Zyl WH, Viljoen-Bloom M (2023) Triticale and sorghum as feedstock for bioethanol production via consolidated bioprocessing. Renew Energy 206:498–505. https://doi.org/10.1016/j.renene.2023.02.047

Matres JM, Hilscher J, Datta A, Armario-Nájera V, Baysal C, He W, Huang X, Zhu C, Valizadeh-Kamran R, Trijatmiko KR, Capell T, Christou P, Stoger E, Slamet-Loedin IH (2021) Genome editing in cereal crops: an overview. Transgenic Res 30(4):461–498. https://doi.org/10.1007/s11248-021-00259-6

Mengstie MA, Wondimu BZ (2021) Mechanism and applications of crispr/ cas-9-mediated genome editing. Biologics Targets Ther 15:353–361). Dove Medical Press Ltd. https://doi.org/10.2147/BTT.S326422

Mergoum M, Gómez Macpherson H (2004) Triticale improvement and production. Food and Agriculture Organization of the United Nations

Mergoum M, Singh PK, Peña RJ, Lozano-del Río AJ, Cooper KV, Salmon DF, Gómez Macpherson H (2009) Triticale: a "new" crop with old challenges. In: Cereals. Springer, pp 267–287. https://doi.org/10.1007/978-0-387-72297-9_9

Michalski K, Hertig C, Mańkowski DR, Kumlehn J, Zimny J, Linkiewicz AM (2021) Functional validation of cas9/guideRNA constructs for site-directed mutagenesis of triticale ABA8′OH1 loci. Int J Mol Sci 22(13):7038. https://doi.org/10.3390/ijms22137038

Miroshnichenko D, Timerbaev V, Divashuk M, Pushin A, Alekseeva V, Kroupin P, Bazhenov M, Samarina M, Ermolaev A, Karlov G, Dolgov S (2024) CRISPR/Cas9-mediated мultiplexed multi-allelic mutagenesis of genes located on A, B and R subgenomes of hexaploid triticale. Plant Cell Rep 43(3):59. https://doi.org/10.1007/s00299-023-03139-x

Mohammed S, Samad AA, Rahmat Z (2019) Agrobacterium-mediated transformation of rice: constraints and possible solutions. Rice Sci 26(3):133–146. https://doi.org/10.1016/j.rsci.2019.04.001

Müntzing A (2010) Studies on the properties and the ways of production of rye-wheat amphidiploids. Hereditas 25(4):387–430. https://doi.org/10.1111/j.1601-5223.1939.tb02704.x

O'mara JG (1953) The cytogenetics of triticale. Bot Rev 19(10):587–605. https://doi.org/10.1007/BF02861827

Qi P-F, Chen Q, Ouellet T, Wang Z, Le C-X, Wei Y-M, Lan X-J, Zheng Y-L (2013) The molecular diversity of α-gliadin genes in the tribe Triticeae. Genetica 141(7–9):303–310. https://doi.org/10.1007/s10709-013-9729-2

Rimpau W (1891) Crossing products of agricultural crops. Agricultural yearbooks, Berlin 20:335–371

Roques SE, Kindred DR, Clarke S (2017) Triticale out-performs wheat on range of UK soils with a similar nitrogen requirement. J Agric Sci 155(2):261–281. https://doi.org/10.1017/S0021859616000356

Salmon RE (1984) True metabolizable energy and amino acid composition of wheat and triticale and their comparative performance in Turkey starter diets. Poult Sci 63(8):1664–1666. https://doi.org/10.3382/ps.0631664

Sood P, Bhattacharya A, Sood A (2011) Problems and possibilities of monocot transformation. Biol Plant 55(1):1–15. https://doi.org/10.1007/s10535-011-0001-2

Sturtevant EL (1886) The claimed wheat and Rye hybrid. Science 7(154):56–56. http://www.jstor.org/stable/1760402

Subedi M, Sandal SS, Savindra P, Kulbhushan, Paudel P, Walia P, Gopera S, Mane OP (2024) Advances in triticale (X Triticosecale) improvement: chromosome manipulation and biotechnological approaches. J Exp Agric Int 46(10):817–829. https://doi.org/10.9734/jeai/2024/v46i103007

Tiwari A, Ghag SB (2024) Base editing and prime editing in horticulture crops: potential applications, challenges, and prospects. In: CRISPRized horticulture crops. Elsevier, pp 105–126. https://doi.org/10.1016/B978-0-443-13229-2.00002-8
Wang Z, Zhang Y, Kang Z, Mao H (2023) Improvement of wheat drought tolerance through editing of TaATX4 by CRISPR/Cas9. J Genet Genomics 50(11):913–916. https://doi.org/10.1016/j.jgg.2023.10.001
Warzecha T, Sutkowska A, Góral H (2014) Male sterility of triticale lines generated through recombination of triticale and rye maintainers. Span J Agric Res 12(4):1124–1130. https://doi.org/10.5424/sjar/2014124-6144
Zakir M, Belete T (2019) Review on history and achievements of triticale breeding status in Ethiopia. Int J Res Stud Agric Sci 5(9). https://doi.org/10.20431/2454-6224.0509002
Zhao Q, Liu R, Zhou Q, Ye J, Meng F, Liu J, Yang C (2023) Calcium-binding protein OsANN1 regulates rice blast disease resistance by inactivating jasmonic acid signaling. Plant Physiol 192(2):1621–1637. https://doi.org/10.1093/plphys/kiad174
Ziemienowicz A, Shim Y-S, Matsuoka A, Eudes F, Kovalchuk I (2012) A novel method of transgene delivery into triticale plants using the *Agrobacterium* transferred DNA-derived Nanocomplex. Plant Physiol 158(4):1503–1513. https://doi.org/10.1104/pp.111.192856

Chapter 15
CRISPR Editing for Quality and Nutritional Improvement of Cereals

Andres Saldaña-Padilla, Luis Fernando Muñoz-Mateo, Luis Uriel Gonzalez-Avila, Luis Alberto Castillo-Díaz, Juan Manuel Bello-López, Cecilia Hernández-Cortez, and Graciela Castro-Escarpulli

Abstract As essential foods in many cultures' diets, cereals collectively account for about half of global caloric intake. The world's food supply relies on a small number of species, including rye, oats, barley, sorghum, rice, wheat, and maize, because they are all members of the same family. This poses a risk to global food safety, given the extreme weather conditions caused by climate change and/or geopolitical conflicts. Therefore, improving both nutritional characteristics and those that enable them to survive increasingly adverse weather conditions through technologies such as CRISPR-Cas gene editing is essential. The components of the CRISPR-Cas system allow genes to be edited specifically to silence or overexpress them, without the need to introduce genetic material other than that belonging to the cereal in question. This, in turn, could improve consumer acceptance and facilitate the procedures that different countries' regulations impose on genetically modified products. This technique has already been used to fortify cereals that are deficient in certain vitamins or amino acids, for example, by editing the genes (overexpression

A. Saldaña-Padilla
Laboratorio de Investigación Clínica y Ambiental, Departamento de Microbiología, Escuela Nacional de Ciencias Biológicas, Instituto Politécnico Nacional, Ciudad de México, Mexico

Laboratorio de Bioquímica Microbiana, Departamento de Microbiología, Escuela Nacional de Ciencias Biológicas, Instituto Politécnico Nacional, Ciudad de México, Mexico

L. F. Muñoz-Mateo · L. U. Gonzalez-Avila · G. Castro-Escarpulli (✉)
Laboratorio de Investigación Clínica y Ambiental, Departamento de Microbiología, Escuela Nacional de Ciencias Biológicas, Instituto Politécnico Nacional, Ciudad de México, Mexico

L. A. Castillo-Díaz
Departamento de Medicina y Ciencias de la Salud, Universidad de Sonora, Hermosillo, Mexico

J. M. Bello-López
División de Investigación, Hospital Juárez de México, Ciudad de México, Mexico

C. Hernández-Cortez
Laboratorio de Bioquímica Microbiana, Departamento de Microbiología, Escuela Nacional de Ciencias Biológicas, Instituto Politécnico Nacional, Ciudad de México, Mexico

K. A. Abd-Elsalam, A. Ahmad (eds.), *Gene Editing in Cereals*, Sustainability Sciences in Asia and Africa, https://doi.org/10.1007/978-981-95-6722-5_15

of proteins) that code for a metabolic pathway that synthesizes them, thereby increasing their content. Similarly, genes have been edited so that cereals capture and retain more minerals such as zinc and iron. Likewise, editing with CRISPR-Cas has made it possible to reduce cultivation times, plant size, and productivity, as well as to improve their organoleptic and cooking characteristics, ensuring greater production and increased cereal reserves. In turn, it has been possible to generate cereals capable of resisting abiotic factors such as drought, heat, and salinity, among others, and to resist biotic factors such as diseases caused by microorganisms. Finally, editing cereal genomes without inserting exogenous genes poses a lower biosafety risk than other techniques and could be more morally acceptable than their genetically modified counterparts.

Keywords CRISPR-Cas editing · Cereals · Food safety · Biofortification · Cereal quality

15.1 Introduction

Although "cereal" and "grain" are frequently used interchangeably, cereal is a more specific term because it solely refers to grasses. Cereal grains are monocotyledonous fruits consisting of bran, endosperm, and germ, according to botanical theory. Pseudocereals, like quinoa or amaranth, are classified as dicotyledonous plants in botany (Skeie and Fadnes 2024; Sá and House 2024).

With 12 subfamilies and over 11,000 species, the grass family *Poaceae*, also known as *Gramineae*, is the sixth-largest plant family and is found around the world. In many areas, humans have domesticated members of this family to serve as the principal source of food such as wheat, barley, or maize; as feed for farm animals, and as industrial materials such as sugar cane, bamboo, and sedges. Cereals can also be processed industrially to produce oils, malts, alcoholic beverages, and other products. One of the primary objectives of agriculture since the dawn of civilization has been the domestication and cultivation of these species (Huang et al. 2022a, b; Chaudhary et al. 2023).

Ten millennia ago, major cereal varieties were domesticated and improved from their wild ancestors through human selection, cultivation practices, and agricultural environments. Strong selection was applied to the wild cereal ancestors during domestication in order to make them high-yielding and easy to grow. The 1960s experienced the emergence of the Green Revolution with increased yields of the two main cereals, wheat and rice, which led to a reduction in food grain shortages. In order to increase yield and other agronomic characteristics and provide sustainable yields that could satisfy rising nutritional needs, the leading cultivars were then further optimized. Currently, 90% of the world's cereal production comes from rice, wheat, and maize, with the other 10% coming from minor cereals including barley, oats, rye, and millet (Chaudhary et al. 2023; Liu et al. 2024).

Cereals have been and continue to be the principal food group in many cultures worldwide. Since the economy has improved, other food groups that were formerly taken in smaller amounts now account for a larger share of energy and nutritional consumption across a range of people, frequently at the expense of cereals. Despite this, cereals continue to make an important contribution to carbohydrate, protein, and energy intake, being key sources of dietary fiber, selenium, iron, B vitamins, and vitamin E (Skeie and Fadnes 2024).

Advances and improvements in the production of large quantities of cereals have not been fully reflected in their nutritional composition. Approximately one in three individuals globally suffers from malnutrition as a result of this long-standing disregard for the nutrients included in cereals. Therefore, it is necessary to use advanced genetic resources to fortify cereals in order to close this nutritional gap and guarantee sustained food security (Chaudhary et al. 2023).

To meet the challenge of global food security in the context of climate change, it is essential to improve the nutritional quality of cereals as well as their resistance to biotic stresses such as insects, pests, and diseases; and abiotic stresses such as heat, cold, salinity, and drought. This is because the negative effects of climate change, such as rising temperatures, altered rainfall patterns, and an increase in extreme weather events, have seriously threatened global agricultural systems. Developing a desired characteristic through crop improvement is still a multifaceted process. Currently, a significant portion of the processes are genetic in nature and entail altering, removing, or adding one or more genes that code for the desired cereal feature. Currently, a significant portion of the processes used are genetic and involve changing, deleting, or adding one or more genes that code for the desired cereal feature (Inam et al. 2024).

Cereals are among the crops that have benefited most from transgenic technology. However, because of population increase and declining agricultural resources, these solutions are still unable to supply world demand. Genomic techniques, such as gene editing, which helps to manipulate genes more precisely without inserting exogenous DNA, are still a viable option to meet this demand and increase the nutritional value of cereals. Unlike previous genetic modification techniques, these edited cereals may gain consumer acceptance more readily than transgenic cereals (Das et al. 2023).

15.1.1 Cereals

A large proportion of the world's food and feed production is generated from a single plant family called *Poaceae*. This family includes cereal crops, wheat, rice, corn, barley, sorghum, and millet, which provide more than 50% of the calories consumed worldwide. Therefore, a large part of the global food supply, i.e., food security, depends on a few closely related species (Bettgenhaeuser and Krattinger 2019). This means that the area planted with cereals represents more than 73% of the total harvested area in the world, making them the main crop and the most common staple food for all populations (Guo et al. 2022). The main cereals with significant quantities of global production are maize (*Zea mays*), rice (*Oryza sativa*), wheat (*Triticum* spp.), barley (*Hordeum vulgare*), and sorghum (*Sorghum bicolor*). Minor cereals with great local significance include millet (*Eleusine coracana*), oats (*Avena sativa*), rye (*Secale cereale*), and others (Zhu 2018). Of the cereals with the greatest consumption worldwide, some of their particularities are highlighted below.

15.1.1.1 Maize

Domesticated about 9000 years ago from its wild relative, *teosintle*, in southwestern Mexico, maize has become the most cultivated grain in the world. Despite the loss of genetic diversity following domestication, modern maize remains relatively rich in genetic variation, facilitating its cultivation under diverse environmental conditions. The popularity of corn is due not only to an increase in direct human consumption as food but also to the increasing production of corn bioethanol, feed for farm animals, and additional products such as corn syrup, oil, and flour (Madeiros et al. 2021).

15.1.1.2 Rice

Rice is, along with wheat and corn, one of the most cultivated and consumed cereals in the world. According to data from the Food and Agriculture Organization, global rice production has increased notably in recent years, with more than 80% used for food (Habibpourmehraban et al. 2022). Cultivated rice includes two species, *Oryza sativa* and *Oryza glaberrima*, which were domesticated independently in Asia and Africa, respectively. Rice cultivated in Asia was domesticated from *Oryza rufipogon* 9000 years ago and includes two subspecies, japonica and indica, while rice cultivated in Africa was domesticated from *Oryza barthii* in West Africa 3000 years ago (Chen et al. 2021).

15.1.1.3 Wheat

Wheat began to be cultivated approximately 10,000 years ago during the Neolithic Revolution. The genus name of wheat, i.e., *Triticum*, is derived from the Latin word *tero* (trillo). The modern name, *Triticum aestivum*, represents hexaploid bread wheat, differentiating it from tetraploid semolina wheat, which is *Triticum durum* and is predominantly grown for pasta production. Today, bread wheat (*Triticum aestivum*) is the most cultivated wheat. Wheat is the source of half of the food calories consumed worldwide and is rich in protein (gluten) (Khalid et al. 2023).

15.1.1.4 Barley

Cultivated barley was domesticated about 10,000 years ago in the Fertile Crescent from *Hordeum vulgare* subsp. *spontaneum*, the genus comprises almost 33 annual and perennial species (Patial et al. 2023). It is the fourth most important cereal in terms of production and has high value in terms of animal feed, human food, and the brewing industry (Yuan et al. 2022).

15.1.1.5 Sorghum

Sorghum was domesticated 8000 years ago in northeast Africa and later spread to other parts of the world (Dahlberg 2019). As a commercial crop, it ranks fifth among cereals by production level. Due to its high resistance to drought conditions, it is an ideal alternative to face the challenges posed by global warming and the drastic decline in corn cultivation caused by fallows formed by climate changes (Thilakarathna et al. 2022).

15.1.2 Structure of Cereal Grains

Cereal grains have a complex structure characterized by different layers of cells. These parts may differ significantly in terms of composition and size in each cereal; however, the overall structure remains largely the same. Three main parts have been established: the germ, the endosperm, and the bran. The starchy endosperm represents 80–85% of the grain and is composed of starch and protein. The remainder of the dry weight of the grain is divided between the bran (12–18%) and the germ (2–3%). The germ has the highest content of lipids and fat-soluble vitamins in cereal. The endosperm has the greatest economic importance, since it contains starch and some proteins, as a source of energy for the developing plant after germination. The outermost layer of the endosperm is called aleurone and may have pigmentation. The bran consists of several different layers that can be distinguished from each other, such as external pericarp, internal pericarp, testa, and nucellar epidermis. The inner and outer pericarp are rich in highly cross-linked polysaccharides, such as cellulose, lignin, and heteroxylan (Nirmala Prasadi and Joye 2020). Whole grains, that is, those that have the parts, are sources of vitamins, minerals, dietary fiber, and phytochemicals. Nevertheless, these are lost when the grain is ground and cereal flour is produced (Francavilla and Joye 2020).

Table 15.1 Nutritional characteristics of cereals per 100 g

	Cereals				
Composition	Maize	Rice	Wheat	Barley	Sorghum
Carbohydrates (g)	72.2	76	71	77.2	70.7
Protein (g)	8.9	7.9	11.6	9.9	10.4
Lipids (g)	3.9	2.7	2	1.2	3.1
Fiber (g)	2	1	2	0.6	2
Ca (mg)	22	33	30	29	25
Fe (mg)	2.1	1.8	3.5	2.5	3.9
P (mg)	268	221	372	–	287
K (mg)	284	400	370	–	350
Mg (mg)	147	140	113	–	–
Thiamine (mg)	0.37	0.41	0.41	0.19	0.38
Riboflavin (mg)	0.12	0.04	0.10	0.11	0.
Niacin (mg)	2.2	4.3	5.1	4.6	4.3

Modified from Guo et al. (2022)

15.1.3 Nutritional Characteristics

Cereals provide between 70% and 80% carbohydrates, between 7.5–15% proteins, between 1% and 4% lipids, and minerals between 1.5% and 3%, which are mainly phosphorus and calcium, in the form of phytate, and a minimum of vitamins (Guo et al. 2022). Certainly, these percentages vary depending on the cereal and whether it is processed or not, as shown in Table 15.1.

15.1.3.1 Starch

It is a polysaccharide composed of two molecules: amylose and amylopectin. Amylose is a linear chain of D-glucose units with an α-(1, 4) bond, while amylopectin comprises D-glucose branches from an α-(1, 4) bond on the branches to α-(1.6) link at branch points. This polysaccharide is found in plant tissues in granular form, in different granular shapes and sizes, where it serves as the main carbohydrate storage reserve (Mathobo et al. 2021).

15.1.3.2 Proteins

Seed proteins are classified into four categories according to their solubility by the Osborne protein fractionation method. Albumins are soluble in water. Globulins are soluble in saline solutions. Prolamins are soluble in alcohol solutions and are rich in proline and glutamine. In almost all cereals, prolamins are the most abundant class of seed storage proteins, followed by glutelins. Most cereals are usually rich in glutamine, proline, leucine, methionine, and cysteine, although limited in the essential amino acids: lysine, tryptophan, threonine, and valine. The high content of prolamin is responsible for the low content of essential amino acids (Bouchard et al. 2022).

15.1.3.3 Dietary Fiber

It is defined as carbohydrates with a degree of polymerization of 3 or more that are found naturally in foods of plant origin and that are not digested or absorbed by the small intestine. It can be classified according to its solubility in water into insoluble dietary fiber (cellulose, hemicellulose, and lignin) and soluble dietary fiber (pectins, β-glucans, and gums) (Nirmala Prasadi and Joye 2020).

15.1.4 Food Safety

Agri-food systems remain highly vulnerable to shocks and disruptions arising from conflict, climate variability and extreme weather events, and economic contraction. These factors, combined with growing inequalities, continue to challenge the ability of agri-food systems to provide nutritious, safe, and affordable diets for all (FAO, 2023). These factors generate the opportunity to apply molecular biology techniques, especially those that offer genetic editing as opposed to genetic modification, whose impact on the environment or humans can be harmful. These technologies are Meganucleases or Homing Endonucleases, Zinc Finger Nucleases, or the CRISPR-Cas system, with the aim of obtaining foods with a greater amount of nutrients (biofortification) or that can better withstand the challenges associated with climate change (Kumar et al. 2022).

15.2 CRISPR-Cas Cereal Editing

15.2.1 CRISPR-Cas System

The CRISPR system (Clustered Regularly Interspaced Short Palindromic Repeats) can be found in the DNA of bacteria and archaea, and its function is to eliminate foreign DNA that may come from viruses and other genetic invaders. Cas proteins (CRISPR-Associated Proteins), of which the best known is Cas9, are used in genome editing. These proteins act as "molecular scissors" that can cut DNA at specific locations and with unprecedented high precision (Mojica et al. 2000).

The general operation of the CRISPR system consists of three events: (1) Spacer acquisition, the adaptation phase in which new DNA sequences are incorporated and recognized as foreign, detected by a sequence called Protospacer Adjacent Motif (PAM). Proteins associated with the CRISPR system, such as Cas1 and Cas2, capture a fragment of the invading DNA. The spacers are sandwiched between palindromic repeats at the CRISPR locus and serve as a "record" of previous infections (Hille et al. 2018). This integration is specific, generally occurring at the 5′ end, making it the newest spacer and thus, closest to the leading repeat. (2) CRISPR RNA biogenesis (crRNA) guides the Cas proteins to the invading DNA that needs to be cut. This locus is transcribed as a single long RNA precursor, called pre-crRNA, which needs to be processed to generate individual crRNAs. This processing is carried out by different Cas proteins. The crRNA biogenesis involves an additional RNA molecule called tracrRNA (trans-activating crRNA). It is paired with the pre-crRNA in the repetitive regions, and this duplex is recognized and processed by the RNase III enzyme, with the help of the Cas9 protein, to produce mature crRNAs. The crRNAs may still require additional maturation to reach their final form. This step ensures the crRNA is of the correct length and sequence to accurately guide the Cas protein to its target. (3) Interference Complex Formation: In its mature form, the crRNA associates with the corresponding Cas protein to form an interference complex, which allows the CRISPR-Cas system to recognize and degrade the invading DNA (Hille et al. 2018).

15.2.2 CRISPR-Cas Editing Technique

The CRISPR/Cas technique is a molecular tool used to "edit" or "correct" the genome of any cell, allowing the modification of specific DNA sequences in a precise and efficient manner (Tao et al. 2023). The main components of the CRISPR/Cas technique include the guide RNA (gRNA) and the Cas enzyme, usually Cas9. The guide RNA is a short RNA sequence that is complementary to the target DNA segment, allowing it to bind to the specific DNA sequence to be modified. The Cas9 enzyme acts like molecular scissors, cutting the DNA at the site determined by the guide RNA. The process of the CRISPR-Cas technique can be divided into three main stages: recognition, cutting, and repair. In the recognition stage, the guide RNA binds to the target DNA at a specific location in the genome. Then, in the cutting stage, the Cas9 enzyme cuts both strands of DNA at the target site. Finally, in the repair stage, the cell attempts to repair the DNA cut, which may result in the inactivation of a gene, the deletion of a DNA segment, or the insertion of a new segment, depending on the type of repair induced. This process allows precise modifications to the genome, opening a wide range of possibilities in genetic research and biotechnology (Anzalone et al. 2020) (Fig. 15.1).

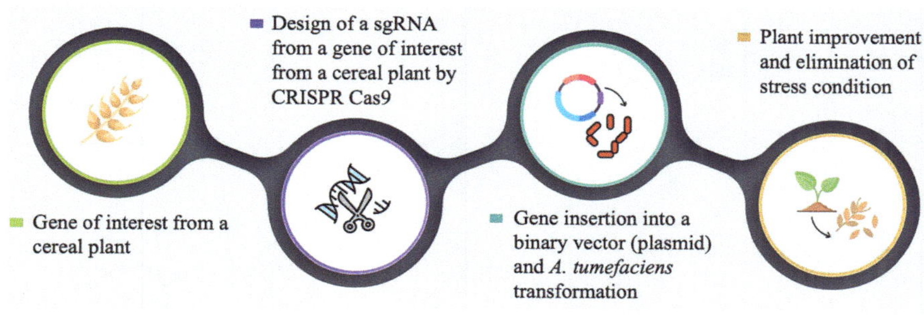

Fig. 15.1 Transformation process of a cereal plant through CRISPR Cas9. A gene of interest from the plant is selected, an sgRNA is generated using CRISPR Cas9 to insert it into a vector, and then the bacterium *Agrobacterium tumefaciens* is transformed and grown with the plant cells to generate seedlings and obtain the genetically modified plant. Generated from this work

15.2.3 Advantages and Disadvantages of Its Use in Cereals

Selective breeding in plants and cereals is an age-old agricultural practice that aims to improve the characteristics of cultivated plants through the selection and crossing of individuals with desirable traits. This technique involves the careful selection of plants that exhibit favorable attributes, such as higher yields, disease resistance, tolerance to adverse climatic conditions, and better nutritional quality. In the case of cereals, selective breeding has been instrumental in increasing production. By selecting and crossing varieties of wheat, maize, and rice, more resistant and productive crops have been developed that can thrive in diverse environmental conditions (Rahim et al. 2020).

With the advancement of biotechnology and genetics, modern techniques such as gene editing and molecular markers have complemented and improved traditional selective breeding practices, enabling faster and more accurate breeding of cultivated cereals. CRISPR-Cas has revolutionized gene editing in the case of cereals, for example, it has been used to reduce the gliadin content in wheat, which could benefit people with celiac disease. In addition, this technology has enabled the development of maize varieties that are more resistant to pests and drought (Li et al. 2021a, b) (Fig. 15.2).

The use of CRISPR-Cas in gene editing of cereals does not involve the insertion of transgenes, which could increase consumer acceptance of these products. However, genetic modification is still a matter of debate and is regulated in many countries. CRISPR-Cas technology offers numerous advantages in the gene editing of cereals. Its precision allows accurate modifications to be made to specific genes, while its efficiency makes it a fast and relatively easy-to-implement technique compared to previous methods. In agriculture, CRISPR-Cas is used to improve disease resistance, increase yields, and improve the nutritional quality of cereals. In addition, this technology facilitates the creation of cereal varieties that can better adapt to adverse climatic conditions, thus contributing to food security in a context of

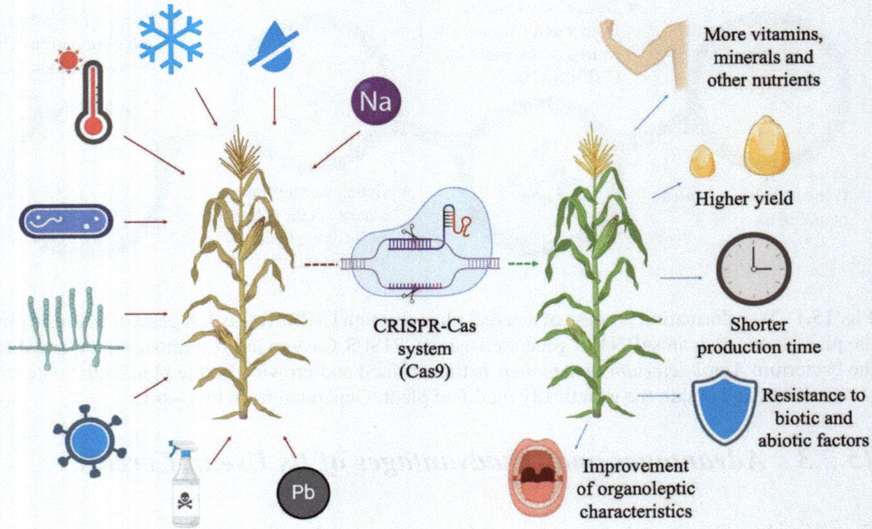

Fig. 15.2 CRISPR-Cas editing targets in cereals. CRISPR-Cas gene editing can be used to improve the nutritional quality of cereals and provide stable food security by making them resistant to biotic and abiotic factors, improving their pre-consumption quality. Generated from this work

climate change. The use of CRISPR-Cas in gene editing presents several significant challenges and some disadvantages. One of the main ones is the risk of generating unintended cuts in DNA that may result in unwanted mutations. In addition, CRISPR-Cas-modified organisms face regulatory and public acceptance challenges like transgenics. The deletion of certain genes may decrease genetic diversity, which could compromise the ability of crop populations to adapt and survive in the long term, examples of which may include maize and wheat. Finally, the use of CRISPR in food modification raises important ethical debates, especially in relation to genetic manipulation and sustainability (Kumar et al. 2021; Basu et al. 2023) (Fig. 15.3).

15.3 Biofortification with CRISPR-Cas

Fortification is a method of food enrichment where fortifiers are added directly to foods during processing; and biofortification involves fortification at the production level of crops through the deliberate use of specific mineral fertilizers to increase their concentration especially in the edible parts, making the latter more cost-effective, with a one-time investment and low recurring costs, offering a brighter and more promising future to address nutritional challenges by providing

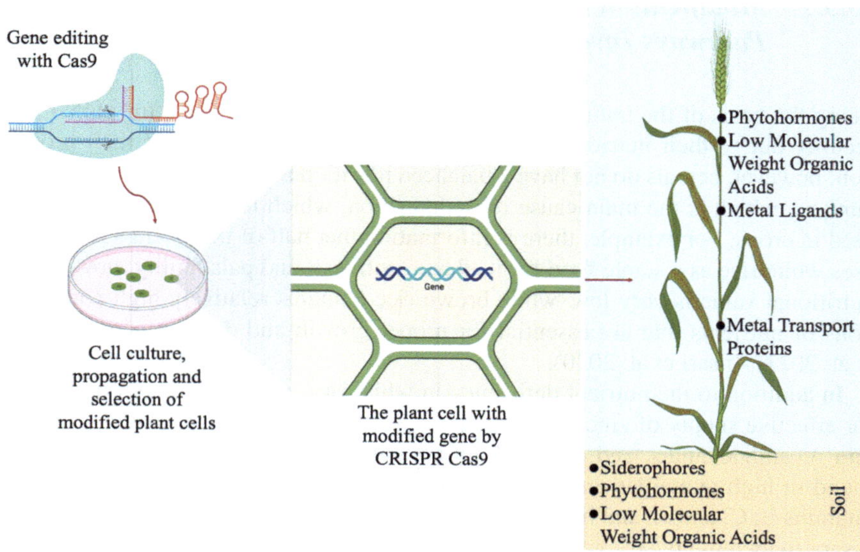

Fig. 15.3 Advantages obtained by genetic modification of cereal plants. Genetic modification of cereal plants provides several advantages, including environmental resistance, nutritional improvements, food safety, bioremediation, and selection against pests. Generated from this work

micronutrient-rich crops to people deficient in them, since malnutrition is a serious health problem worldwide and is rapidly advancing as the population increases. Therefore, biofortification could prevent, control or even avoid malnutrition, vitamin and mineral insufficiency, obesity and diet-related noncommunicable disorders, since micronutrient deficiency for a short period may not cause much harm, but if it is for a long time it causes diseases such as iron deficiency anemia, beriberi due to vitamin B deficiency, pellagra due to niacin deficiency, rickets due to vitamin D deficiency, and scurvy due to vitamin C deficiency (Bhattacharya et al. 2024; Kumar et al. 2022; Shahzad et al. 2021).

The CRISPR-Cas9 technique has been effectively used in important crops, mainly in cereal crops due to its simple design, acceptability, cost-effectiveness, high efficiency, good reproducibility, shorter time required, high engineering feasibility, ability to create knockout genes, and RNA editing. This genome editing technique, like others, has been used to improve cereal crops by developing new varieties that provide better yield, better nutritional quality, and stress tolerance. It has been described that as of September 2023, CRISPR-Cas technology has been applied to 120 plant species. Among cereals, this system has been used in rice (*Oryza sativa*), wheat (*Triticum aestivum*), and maize (*Zea mays*) (Ahmar et al. 2024; Kumar et al. 2022; Ansari et al. 2020).

15.3.1 Modification of Nutritional Values and Metabolic Pathways Involved

Being the basis of the human diet, cereals are of great importance for global food security due to their nutritional value and health benefits, as well as their production; however, cereals do not have a balanced nutritional profile and are required by humans, which is the main cause of malnutrition, which is why biofortification is used in crops. For example, there is information that half of the world's population uses white rice as a staple food for its flavor, softness, and palatability; however, its nutritional value is very low, while brown rice contains relatively high concentrations of nutrients that are essential for proper growth and development (Shahzad et al. 2021; Ansari et al. 2020).

In addition to the nutrient deficiency in white rice, wheat needs biofortification for effective supply of zinc, iron, selenium, yellow grain pigment content, provitamin A, anthocyanins, and essential amino acids, as well as phytic acid, which is found in high quantities and should therefore be controlled. Maize is deficient in vitamins E, C, and A, and has the presence of antinutrient components, so it is necessary to increase the tryptophan and lysine content to increase the nutritional quality of zein proteins. Barley is also known to have deficiencies of micronutrients such as zinc, iron, γ-linolenic acid, polyunsaturated fatty acids, and stearidonic acid. Conversely, sorghum is deficient in iron and zinc along with iodine, selenium, magnesium, and calcium. For this reason, various cereal crops have been subjected to biofortification processes using CRISPR-Cas9 technology (Bhattacharya et al. 2024; Shahzad et al. 2021).

15.3.1.1 Vitamins

Given that carotenoids are precursors of vitamin A, it is necessary to enrich the so-called "golden crops" with these isoprenoid pigments in order to develop them so that humans can acquire this vitamin and combat its deficiency, with CRISPR being the most effective, fast, and widespread method, without DNA or transgenes, and with multiple targeted genetic modifications of organisms. For this reason, several genome-edited golden crops have been produced. As an example, there is the golden rice variety Kitaake, in which the genes for phytoene desaturase (*CrtI*) and phytoene synthase (*PSY*) from maize were introduced into the rice genome through a 5.2 kb cassette for carotenoid biosynthesis, determining that this variety contains 7.9 μg/g dry weight of β-carotene in the endosperm (Kumar et al. 2022; Shahzad et al. 2021; Dong et al. 2020).

Vitamin E or α-tocopherol is also essential in the human diet, as cardiovascular diseases and certain types of cancer are associated with insufficient intake of this vitamin, with the daily requirement ranging from 15 to 30 mg. CRISPR-Cas9 technology has significantly increased the content of tocopherols and tocotrienols in barley through targeted overexpression of the enzyme homogentisate phytyltransferase (*HvHPT*) and the enzyme homogentisate geranylgeranyltransferase (*HvHGGT*) from *H. vulgare*. Thus, the *HvHPT* and *HvHGGT* genes can be used to improve vitamin E content in other crops (Kumar et al. 2022).

15.3.1.2 Minerals

The reference concentrations of iron and zinc in wheat and corn grains are 30 μg/g and 25 μg/g, respectively. In rice grains, however, the concentrations are 2 μg/g of Fe and 16 μg/g of Zn. Nonetheless, wheat grains should contain 59 μg of Fe/g and 38 μg of Zn/g to meet 30–40% of the average adult dietary requirement. Therefore, studies have been conducted to improve the accumulation of Fe and Zn in grains. The need for iron in humans can be met by biofortification of crops using CRISPR-Cas9, altering the *IPK1* gene that encodes the enzyme Inositol pentakisphosphate 2-kinase 1, in wheat. In the case of zinc deficiency in bread wheat (*T. aestivum*) crops, the CRISPR-Cas system alters the enzyme Inositol pentakisphosphate 2-kinase 1 (*TaIPK1*), which reduces phytic acid, leading to improved zinc accumulation in grains. Some wheat varieties such as BHU1, BHU-6, and Zincol-2016, with 35–42 μg Zn/g have been developed through conventional breeding; however, there has been little success with iron concentration (Kumar et al. 2022; Ibrahim et al. 2021; Shahzad et al. 2021).

CRISPR-Cas9 gene editing technology will allow the specific modification of genes responsible for nutrient content in cereal crops, which will make the biofortification process more efficient and controlled. Table 15.2 shows the possible candidate genes associated with micronutrients such as Fe, Zn, and vitamins present in cereals (Bhattacharya et al. 2024).

15.3.1.3 Biofortification Through Cytokinin Metabolism

Plants can absorb essential minerals for their growth from the soil, so another way to perform biofortification is through the metabolism of cytokinins, phytohormones that function as negative regulators of root elongation and branching, and shape the architecture of the plant root system, which decreases crop yield. Cytokinin levels in plants are regulated by isopentenyl transferase (*IPT*) enzymes, which are then destroyed by cytokinin oxidase/dehydrogenase (*CKX*) and inactivated by glucosylation through cytokinin glucosyl transferases (*CGTs*), so low cytokinin levels increase root growth and the absorption of mineral nutrients such as Zn and Fe, increasing crop yield by overexpressing *CKX* and *CGTs* genes in the root zone. In a study using the Cas9 system, the *CKX* genes were eliminated by RNA guidance to generate ckx1 and ckx3 mutant lines with knockout mutations in the *HvCKX1* and *HvCKX3* genes, respectively, to increase the grain yield of barley, since it was determined that the reduced *CKX* activity in the ckx1 lines induced longer roots, greater surface area and a greater number of root hairs, while in the ckx3 mutants opposite results were obtained. Another study in brown rice crops to increase the

Table 15.2 Genes associated with Fe, Zn, and vitamins as possible candidates for biofortification in cereals

Cereal	Candidate genes	Micronutrient
Barley	*HvHGGT*	Vitamin E
	HvMTP1, HVPT1, HVPT2	Zn
	HvYS1, NAS, NAAT, DMAS, IDS2, IDS3	Fe
Maize	*crtRB1, LcyE,* Y1/Psy1	Vitamin A
	ZmFer1, ZmYS1	Fe
	ZmZIP	Fe and Zn
Rice	Crt1	Vitamin A
	HvNAS1, MxIRT1	Fe and Zn
	OsDMAS1, OsFer2, OsIRO2, OsIRT1, OsNAAT1, OsVIT1, OsVIT2, OsYSL13, OsYSL, OsYSL16	Fe
	PSY	Vitamin A
Wheat	*Crt1, CrtB*	Vitamin A
	Gpc-B1	Zn
	OsNAS2, PvFERRITIN	Fe and Zn
	TaFer1, TaFer2	Fe

Modified from Bhattacharya et al. (2024)

concentration of Zn using CRISPR-Cas9 technology was inactivating the *CYP735A* gene involved in the formation of trans-zeatin cytokinins (tZ type). Other studies have focused on developing mutants of a repressor of the abnormal response to cytokinins in barley, HvARE1, with high nitrogen content in shoots from soils deficient in this mineral through the genetic transformation of immature embryos with Agrobacterium-mediated CRISPR-Cas with sgRNAs targeting HvARE1. Therefore, these crops have efficiency in the use of nitrogen, reducing the contribution of fertilizers in the soil, thus preventing environmental pollution due to the use of these substances (Kumar et al. 2022).

15.4 Modification of Cereal Quality with CRISPR-Cas

15.4.1 Cultivation Time

Conventional crossbreeding of grains is a long-term process, which involves time-consuming steps that delay crop yielding and production (Ahmar et al. 2023; Basu et al. 2023). Therefore, using a CRISPR-Cas9 *Wus2* construct, transformation of *S. bicolor*-generated embryonic sorghum occurred only 14 days after insertion of genes for morphogenetic proteins, *Zm-Wus2* and Z (*Zm-Bbm*). Subsequently, genes remotion prior germination, yielded mutants without altering plant traits, making this strategy an encouraging alternative to accelerate somatic embryo formation and the regeneration of sorghum crops (Che et al. 2022).

Dormancy, which lasts months, allows crops survival throughout dry and hot summers. Thus, to extend dormancy of wheat, CRISPR-Cas9 was used to target the *Qsd1* gene in cv. Haruyoki, Yumechikara, and Kitanokaori lines. This favored seed germination and extended dormancy, which is in good agreement with previous work where the deletion of *Qsd1* with CRISPR-Cas9, led to a longer germination in wheat (> 30 days) without impairing plant traits (Liu et al. 2021; Abe et al. 2019).

Likewise, CRISPR-Cas9 gene edition allowed the transformation and regeneration of several Australian malting barley varieties only within 6–8 weeks. Strategy allows to avoid tedious steps, which usually last over 35 weeks during typical barley culture (Han et al. 2021). The *grain size 3* (*GS3*) gene was deleted in the rice variety *indica* maintainer Mei1B with CRISPR-Cas9. After several crossings, *GS3* KO lines displayed an increased length (79%) and length/width (4.19%) ratio. Mei2B lines exhibited an increased thousand-grain weight (6.7%), grain yield per plant (14–9%), and their seed appearance addressed standards from traditional Simiao type cultivars (Huang et al. 2022a, b). Previously, after targeting *GS3* in the *indica* rice Bac Thom 7 with CIRPR-Cas9, edited strains rendered long grains and an increased thousand-grain weight without plant traits alterations (Nguyen et al. 2021).

15.4.2 Organoleptic Features

Nowadays, providing fragrance in cereals represents a major goal for agricultural industry. Particularly, 2-acetyl-1-pyrroline (2-AP), derived from the conversion of g-aminobutyraldehyde (GABald) in the absence of betaine aldehyde dehydrogenase 2 (OsBADH2), is associated with aroma in rice. CRISPR-Cas9 genome editing allowed specific editions in *OsBADH2* gene that led to the accumulation of 2-AP (from 600 to 800 ng/g) in the SH313 rice line; however, it also caused changes in plant traits, which could be tackled after crossing the KO lines with a nonaromatic rice line (Ashokkumar et al. 2020; Liao et al. 2024). Previously, the deletion of *OsBADH2* using CRISPR-Cas9 rendered aroma in two nonaromatic rice varieties, *japonica* Ningjing 1 and *indica* Huang Huazhan, without impairing rice quality (amylose rate, gel consistency, starch viscosity, etc.) (Hui et al. 2022). Strains of maize with aroma were also generated after deletion of two *BADH2* homologous genes in maize (*ZmBADH2a* and *ZmBADH2b*) from various *Z. mays* lines with CRISPR-Cas9 system. Seeds of mutants were able to produce AP-2, which exhibited a popcorn-like fragrance. Overall, the literature discussed supports the suitability of CRISPR/Cas9 to provide organoleptic features in some cereals. Noteworthy, further research is needed to refine protocols to avoid non-desirable traits alterations in plants and grains (Wang et al. 2021).

15.4.3 Cooking and Consumption Quality

Currently, there is an urge to generate cereals with low glutelin and amylose content due to grains with high carbohydrates content, and low rates of gel consistence are related to poor eating and cooking quality (Wang et al. 2024; Yang et al. 2010). Similarly, the deletion of *OsAPP6* and *OsAPP10* in *japonica* rice lines using CRISPR-Cas9 generated rice lines exhibiting significant reduction in glutelin and amylose content (Wang et al. 2020). CRISPR-Cas9 was also successful in reducing acrylamide content in the wheat line *T. aesticum* cv Cadenza. Acrylamide is a carcinogenic compound derived from asparagine, which forms during wheat cooking. After CRISPR-Cas9 and a 4-gRNA polycistronic gene construct were introduced into wheat embryos through particle bombardment, a reduction of asparagine synthetase gene expression was recorded. Moreover, mutants with edits in six *TaASN* alleles showed low asparagine total content in their seeds mainly at T2 generation (Raffan et al. 2021).

High amylose content and resistant starch (RS) are desirable features for wheat consumption because RS is not digested and absorbed by the body, leading to a decrease in blood glucose levels. Therefore, the mutagenesis of *TaSBEIIa* gene, which is involved in the regulation of the amount of starch in plants, was carried out in *cv* ZM and *cv* Bobwhite wheat strains using CRISPR-Cas9. As a result, total starch content significantly decreased, whereas nutritional properties and RS contents increased in mutants. However, decrements in height, tiller number, length, and width of grains were observed (Li et al. 2021a, b).

A-amylase/trypsin inhibitors (ATI) are associated with coeliac disease and IgE-mediated wheat allergies. Using CRISPR/Cas9 multiplex edition, the deletion of *ATI* genes, *WTAI-CM3,* and *WTAI-CM16* (major allergens) was finely conducted, although the deletion of *CM16* could not be confirmed. Therefore, further research must be carried out to confirm *CM16* deletion for the potential consumption of wheat edited lines in a reliable manner (Camarlengo et al. 2020).

15.5 Resistance to Biotic and Abiotic Factors

CRISPR-Cas9 technology has been used to generate crops resistant to climate and to pathogens such as bacteria, fungi, insects, and viruses (Zenda et al. 2021; Maharajan et al. 2022). These changes can be beneficial by reducing the use of pesticides, insecticides, and antibiotics, reducing their traces in cereals that feed most of the population that cannot access quality products that do not contain this type of residue, giving access to quality food to all sectors of the population, but also contributing to reducing the undesirable effects on the environment.

15.5.1 Resistance to Bacterial Infections

Bacteria cause various diseases in cereals, reducing their growth and yield. Sucrose family transporters (*SWEET*) have been reported to improve rice growth and yield when plants are affected by bacterial blight. The *OsSWEET13* gene encodes a sucrose transporter that plays an important role in the interaction between plant and pathogen. *PthXo2*, an effector protein produced by *Xanthomonas oryzae*, causes *OsSWEET13* expression in the host and, as a result, susceptibility. Deletion of the *OsSWEET13* gene using CRISPR-Cas9 improved rice growth against this disease (Maharajan et al. 2022; Li et al. 2022).

15.5.2 Resistance to Fungal Infections

These types of diseases are devastating to crops, among which powdery mildew seriously affects crop productivity. The *mildew resistance locus* (*MLO*) is a susceptibility gene, and its loss of function confers long-lasting, broad-spectrum resistance to powdery mildew in various plants. Editing *MLO* genes (*TaMLO-A1*, *B1,* and *D1*) with CRISPR-Cas9 has allowed the development of wheat plants resistant to powdery mildew. In addition to these genes, the *improved disease resistance 1* (*EDR1*) gene has also been used, with the help of the CRISPR-Cas9 system, wheat resistant to powdery mildew was generated (Li et al. 2022; Tripathi et al. 2022; Maharajan et al. 2022). Likewise, genes associated with the calcineurin B-like protein (*CBL*)-Interacting protein kinases (*CIPK14*) have been silenced, which could improve resistance against wheat rust caused by *Puccinia striiformis* f. sp. *Tritici* (Ahmar et al. 2023).

Rice blast is a fungal disease caused by *Magnaporthe oryzae*. Ethylene response factors (*ERF*) participate in the response to resistant diseases against *M. oryzae*. *OsERF922* has been reported as a negative regulator of blast resistance and is positively regulated by virulent and avirulent strains of *M. oryzae* (Matres et al. 2021). CRISPR-Cas9 has been used to improve resistance to this disease by disrupting the *OsERF922* and *OsSEC3A* genes in rice. CRISPR-Cas9-induced rice *Bsr-d1* and *Pi21* mutations could also cause partial resistance to rice blast, but their effect has not been as strong as that of *oserf922* (Li et al. 2022). Regarding maize, CRISPR-Cas9 has been used to delete the *lox3* gene, which improved resistance to *Ustilago maydis,* and similarly, deletion of *COI1a* provided resistance against *Gibberella* stem rot (Ahmar et al. 2023).

15.5.3 Resistance to Viral Infections

One of the viral diseases affecting rice is the rice tungro disease (RTD), caused by a mixed infection of rice tungro bacilliform virus (RTBV) and rice tungro spherical virus (RTSV). For this infection, targeted mutagenesis with CRISPR-Cas9 was used to induce mutations upstream of the YVV residues in *eIF4G* of *IR64*, an indica variety susceptible to RTSV. The mutation was inherited in the next generation. Developed mutant lines showing resistance to RTSV could be used as starting material for RTD resistance (Matres et al. 2021). To combat wheat yellow mosaic viral disease, wheat orthologs of the barley (*Hordeum vulgare*) susceptibility gene known as *PDIL5-1* were used, which have been eliminated to combat wheat yellow mosaic virus (WYMV). The *PDIL5-1* gene (protein disulfide isomerase-like 5-1) encodes a chaperone located in the endoplasmic reticulum, involved in correct protein folding, and the loss-of-function mutation provides resistance against bimoviruses such as WYMV. In the experiments, the homozygous triple knockout mutant generated by CRISPR-Cas9 (*Tapdil5–1-4a*, *Tapdil5–1-4b,* and *Tapdil5–1-4d*) has shown significant resistance against WYMV compared to the wild-type plant without any penalty in agronomic performance (Awan et al. 2022). Regarding maize, the *GDIα* genes have been eliminated to increase resistance against viruses with CRISPR-Cas9 (Ahmar et al. 2023).

15.5.4 Resistance to Herbicides and Pesticides

The elimination with CRISPR-Cas9 of the *ALS1/ALS2* gene from corn allows greater resistance to the herbicide chlorsulfuron (Ahmar et al. 2023). For rice, the use of CRISPR improved herbicide tolerance by editing *BEL* and *ALS* for sulfonylurea herbicide resistance and *EPSPS* for glyphosate resistance (Chaudhary et al. 2022). The CRISPR-Cas9-induced knockout mutation in the α-6 nicotinic acetylcholine receptor (nAchR) enhances the resistance of rice plants to Spinosad, a natural substance produced by a soil bacterium that can be toxic to insects (Ahmar et al. 2023). Also, for rice, modifications have been made with CRISPR-Cas in the *OsCYP71A1* gene for resistance to the stem borer and plant hoppers (Li et al. 2022).

15.5.5 Drought Stress Resistance

When applying the CRISPR-Cas9 technique in the wheat protoplast, the *DREB2* and *ERF3* genes have been edited. The mutant plants obtained showed greater resistance to drought stress. Meanwhile, in maize, the expression of the *ARGOS8* gene was increased using CRISPR-Cas9, and field tests showed that *ARGOS8* mutations increased grain yield under drought stress conditions (Ahmar et al. 2023).

Knockdown of the *OsDST* gene in rice using CRISPR-Cas9 showed increased leaf growth, water retention, and reduced stomatal density under drought stress. Knockdown of the *OsERA1* gene with CRISPR-Cas improved drought tolerance in rice. Deletion of *OsEPFL9* gene in rice using CRISPR/Cas12a system reduced stomatal count and increased water use efficiency under drought stress conditions (Maharajan et al. 2022). Similarly, *SlMAPK3*, *AtMIR169a*, *AtOST2*, and *OsSAPK2* genes have been targeted for drought resistance in rice, while *OsMPK2* and *ZmARGOSS8* are targeted for increasing grain yield under drought stress (Chaudhary et al. 2022).

15.5.6 Resistance to Salinity Stress

Like drought stress, the CRISPR/Cas system identified the functions of some genes in rice that respond to salt concentration. Deletion of the *OsRR22* gene by using CRISPR-Cas9 improved salinity resistance in rice. Mutations have also been generated with this technology in the *OsRR9* gene, which revealed the role it plays in salinity resistance by regulating HKT family transporters. Similar results have been obtained with the *OsOTS1* gene in rice (Maharajan et al. 2022). Salinity resistance in rice is also increased by a knockout mutation of a negative regulator *bHLH024*, as the mutant showed an increase in shoot weight, chlorophyll content, and chlorophyll fluorescence. As for maize, the *HKT1* gene, related to resistance to salinity, was modified by using CRISPR-Cas9. The mutants obtained showed an increase in the supply of sodium ions from root to shoot, which shows a new line of research to improve salt resistance in maize. A knockout mutation of the histone acetyltransferase gene *HAG1* in wheat, via CRISPR/Cas9, resulted in resistance against salt stress (Ahmar et al. 2023).

15.5.7 Temperature Stress Resistance

The *heat-sensitive albino1* gene (*HSA1*) has been characterized in rice, and its deletion with CRISPR-Cas9 significantly delayed chloroplast activity at high temperatures compared to wild-type rice (Maharajan et al. 2022). Similarly, with CRISPR, knockout mutants of several cold resistance genes (*WRKY63*, *PRP1*, and *ANN3*) have been generated. Mutants of the *WRKY63* and *PRP1* genes showed better resistance to cold, while mutants of the *OsANN3* gene showed lower resistance to cold (Ahmar et al. 2023). Deletion of annexin 3 (*OsAnn3*) gene in rice improved rice growth under cold stress. Deletion of *OsPRP1* in rice reduced the survival rate under cold stress. Rice genes have also been modified with CRISPR-Cas9: *OsPIN5b* (a panicle length gene), *OsGS3* (a grain size gene), and *OsMYB30* (a cold tolerance gene), obtaining improvements in panicle length and grain size under cold stress (Maharajan et al. 2022).

Among the CRISPR-Cas variants, the CRISPR-Cas12b system has potential applications in the development of cereals resistant to temperature changes. Therefore, targeting genes that respond to heat stress using the CRISPR-Cas12 system may enable the development of heat-tolerant cereals compared to other CRISPR-Cas variants (Maharajan et al. 2022).

15.5.8 Resistance to Heavy Metals

Knockdown of *OsNramp5* and *OsLCT1* by CRISPR-Cas9 reduced cadmium accumulation in rice; similarly, knockdown of *OsNRAMP1* resulted in lower levels of cadmium and lead in rice grains. Likewise, this technique has been used to inhibit the absorption and transport of arsenic in rice by eliminating an R2R3 MYB transcription factor from the *OsARM1* gene. To create low cesium rice plants, CRISPR-Cas has been used to inactivate the K^+ transporter *OsHAK1* (Li et al. 2022).

15.6 Effect of Genetic Modification on Biodiversity and Use in Food

The research on genetic modifications based on CRISPR Cas technology is a reality. Due to this situation, the scientific community must be guided by the four principles of bioethics, in addition to knowing whether there are rules or laws in their country regarding genetic modification. Historically in this field, scientists have been committed to the advancement of science, but they have set limits; these limits are based on ethics and the preservation of species, with the intention of maintaining the balance of biodiversity and the nature of evolution (Gonzalez-Avila et al. 2021; Lorenzo et al. 2022).

Despite the effectiveness of CRISPR Cas for genetic modification, the scientific community has raised different concerns, that the editing is carried out outside the target gene, illegal experimentation, and therefore, the lack of legal regulation. Thus, editing outside the target gene would in some cases cause chromosomal rearrangements that in turn cause mutations and DNA mismatches. This is different in the modification of plant cells, which would give an opportunity to use this technique for food improvement (Gonzalez-Avila et al. 2021; Lorenzo et al. 2022; Kolkur et al. 2024).

The natural process of obtaining food from agricultural areas has been modified over time to optimize and multiply the harvest. In this context, the need to meet the world's demand for food has been emphasized and, in contrast, attempts have been made to improve food nutritionally. Without forgetting, for economic reasons, the improvement of food includes increasing shelf life. These elements generate a natural question: is it correct to genetically modify food to ensure food? (Kolkur et al. 2024).

The possibility of genetically improving food has important implications for humans, the society, and all biological species. Bioethics is the discipline that attempts to respond to the moral dilemmas that arise from scientific advances in medicine and science, without altering the balance of society and value systems. This is a current problem, so bioethics must offer a vision of the future because of the uncertainty raised by these issues and the lack of knowledge of the developments that will result from the use of these techniques. In the same way, this position does not try to limit or cancel progress, but to avoid possible negative effects. In the context of food, the cultivation of new breeds of food that have been genetically edited in the laboratory has generated the classic concerns, including moral, regulatory and biosecurity questions, given that they can affect the moral living conditions of present and future generations, not only of humans, but also of plants, animals, and insect species (Basu et al. 2023).

The discussion about the application of these techniques questions the efficiency of the technique and the difficulty of knowing in time whether there will be a consequence. Nevertheless, other scientific positions suggest that it should be used to improve food and support the eradication of hunger in the world, from several points of view: on the one hand, the improvement in terms of increasing resistance to putrefaction in perishable foods; on the other hand, the improvement of food in terms of nutrient enrichment or nutrient enhancement, and finally, that foods susceptible to pests should be improved to avoid crop losses. These three points focus the debate on whether it is feasible to implement and enforce them, in addition to food safety and the effects on consumers. Some studies have shown that off-target changes in plants are rare, the mutation rate for these reasons is low. The above forces the scientific community to pay attention to the timely detection of unwanted effects before any application of CRISPR-Cas (Bartkowski et al. 2018).

The main argument for rejecting genetic engineering is the risk to human health from an anthropocentric point of view; however, the probable damage to the environment must be considered. If the arguments are presented from a utilitarian vision, genetic modification by using CRISPR-Cas without the insertion of a gene is less risky compared to other techniques, so it could become morally more acceptable, because until now no harmful effect on humans has been demonstrated after consuming genetically modified foods, in addition to the fact that genetically modified foods are practically identical to the original foods (Singh et al. 2024; Lorenzo et al. 2022).

Based on the above, the debate in the scientific community and some social organizations focuses on the discussion of ethics for the use of CRISPR-Cas in food improvement in modern agriculture together with ecological and social acceptance, due to the benefits it brings to humanity. Thus, the applications of CRISPR-Cas9 present some fundamental questions about the principle of justice. Alternatively, there are concerns about how the issue of equity in the distribution of these foods will be addressed, ensuring that there is no social segregation that does not have access to these foods that are beneficial to the consumer. This puts on the stage the need to carry out a social and political debate on legal regulation on the development and use of any novel technology that modifies foods using CRISPR Cas (Kumar et al. 2022).

Due to the ethical and regulatory concerns for allowing the use and application of gene editing technology, it is considered important to establish guidelines, laws, and principles to proceed with genome editing applications, including the CRISPR-Cas system, but these principles and guidelines are not equivalent to the techniques being safe due to some technical problems inherent to the techniques. In addition, biotechnology policies in the world are obsolete, not universal, little discussed, and limited, so it is required that the arguments have scientific and technical rigor for the genome editing of food using the CRISPR-Cas system. It is important to raise awareness in nations and their governments that they must have and comply with strict genome editing laws with internal and external observation before proceeding with any application of gene editing through institutions and the establishment of bioethics committees, which promote the safety of humanity, other animals and the nonliving environment and safe agriculture (Lorenzo et al. 2022; Brokowski and Adli 2019).

Acknowledgments A.S.P and LFMM received grant-aided support from "Secretaría de Ciencia, Humanidades, Tecnología e Innovación" (SECIHTI, Mexico), while L.A.C-D, C.H.-C, J.M. B-L, and G.C.-E. received support from "Sistema Nacional de Investigadores (SNII)" from SECIHTI, Mexico. G.C.-E. received support from "Estímulos al Desempeño de los Investigadores, Comisión de Operación y Fomento de Actividades Académicas (Instituto Politécnico Nacional)". This study was approved by "Secretaría de Investigación y Posgrado del Instituto Politécnico Nacional" (SIP 20232102,20240986). L.U.G.-A. designed figures for this manuscript. The authors are also grateful to Sofia Mulia for correcting the style of the manuscript. All authors have read and agreed to the published version of the manuscript.

References

Abe F, Haque E, Hisano H, Tanaka T, Kamiya Y, Mikami M, Kawaura K, Endo M, Onishi K, Hayashi T, Sato K (2019) Genome-edited triple-recessive mutation alters seed dormancy in wheat. Cell Rep 28(5):1362–1369.e4. https://doi.org/10.1016/j.celrep.2019.06.090

Ahmar S, Hensel G, Gruszka D (2023) CRISPR/Cas9-mediated genome editing techniques and new breeding strategies in cereals—status, improvements, and perspectives. Biotechnol Adv 69:108248. https://doi.org/10.1016/j.biotechadv.2023.108248

Ahmar S, Usman B, Hensel G, Jung KH, Gruszka D (2024) CRISPR enables sustainable cereal production for a greener future. Trends Plant Sci 29(2):179–195. https://doi.org/10.1016/j.tplants.2023.10.016

Ansari WA, Chandanshive SU, Bhatt V, Nadaf AB, Vats S, Katara JL, Sonah H, Deshmukh R (2020) Genome editing in cereals: approaches, applications and challenges. Int J Mol Sci 21(11):4040. https://doi.org/10.3390/ijms21114040

Anzalone AV, Koblan LW, Liu DR (2020) Genome editing with CRISPR-Cas nucleases, base editors, transposases and prime editors. Nat Biotechnol 38(7):824–844. https://doi.org/10.1038/s487-020-0561-9

Ashokkumar S, Jaganathan D, Ramanathan V, Rahman H, Palaniswamy R, Kambale R, Muthurajan R (2020) Creation of novel alleles of fragrance gene OsBADH2 in rice through CRISPR/Cas9 mediated gene editing. PLoS One (8):e0237018. https://doi.org/10.1371/journal.pone.0237018

Awan MJA, Pervaiz K, Rasheed A, Amin I, Saeed NA, Dhugga KS, Mansoor S (2022) Genome edited wheat-current advances for the second green revolution. Biotechnol Adv 60:108006. https://doi.org/10.1016/j.biotechadv.2022.108006

Bartkowski B, Theesfeld I, Pirscher F, Timaeus J (2018) Snipping around for food: economic, ethical and policy implications of CRISPR/Cas genome editing. *Geoforum 96*:172–180. https://doi.org/10.1016/j.geoforum.2018.07.017

Basu U, Riaz Ahmed S, Bhat BA, Anwar Z, Ali A, Ijaz A, Gulzar A, Bibi A, Tyagi A, Nebapure SM, Goud CA, Ahanger SA, Ali S, Mushtaq M (2023) A CRISPR way for accelerating cereal crop improvement: Progress and challenges. Front Genet 13:866976. https://doi.org/10.3389/fgene.2022.866976

Bettgenhaeuser J, Krattinger SG (2019) Rapid gene cloning in cereals. Theor Appl Genet 132(3):699–711. https://doi.org/10.1007/s00122-018-3210-7

Bhattacharya S, Sathiyabalan A, Amir M (2024) Advancing nutritional quality through Genomic approaches for biofortification in Cereal Crops: A review. Plant Cell Biotechnol Mol Biol 25(5–6):110–123. https://doi.org/10.56557/pcbmb/2024/v25i5-68713

Bouchard J, Malalgoda M, Storsley J, Malunga L, Netticadan T, Thandapilly SJ (2022) Health benefits of cereal grain- and pulse-derived proteins. Molecules (Basel) 27(12):3746. https://doi.org/10.3390/molecules27123746

Brokowski C, Adli M (2019) CRISPR ethics: moral considerations for applications of a powerful tool. J Mol Biol 431(1):88–101. https://doi.org/10.1016/j.jmb.2018.05.044

Camerlengo F, Frittelli A, Sparks C, Doherty A, Martignago D, Larré C et al (2020) CRISPR-Cas9 multiplex editing of the α-amylase/trypsin inhibitor genes to reduce allergen proteins in durum wheat [original research]. Front Sustain Food Syst 4. https://doi.org/10.3389/fsufs.2020.00104

Chaudhary M, Mukherjee TK, Singh R, Gupta M, Goyal S, Singhal P, Kumar R, Bhusal N, Sharma P (2022) CRISPR/Cas technology for improving nutritional values in the agricultural sector: an update. Mol Biol Rep 49(7):7101–7110. https://doi.org/10.1007/s11033-022-07523-w

Chaudhary N, Salgotra RK, Chauhan BS (2023) Genetic enhancement of cereals using Genomic resources for nutritional food security. Genes 14(9):1770. https://doi.org/10.3390/genes14091770

Che P, Wu E, Simon MK, Anand A, Lowe K, Gao H, Sigmund AL, Yang M, Albertsen MC, Gordon-Kamm W, Jones TJ (2022) Wuschel2 enables highly efficient CRISPR/Cas-targeted genome editing during rapid de novo shoot regeneration in sorghum. Commun Biol 5(1):344. https://doi.org/10.1038/s42003-022-03308-w

Chen Q, Li W, Tan L, Tian F (2021) Harnessing knowledge from maize and rice domestication for new crop breeding. Mol Plant 14(1):9–26. https://doi.org/10.1016/j.molp.2020.12.006

Dahlberg J (2019) The role of sorghum in renewables and biofuels. Method Mol Biol (Clifton) 1931:269–277. https://doi.org/10.1007/978-1-4939-9039-9_19

Das N, Ghosh Dhar D, Dhar P (2023) Editing the genome of common cereals (rice and wheat): techniques, applications, and industrial aspects. Mol Biol Rep 50(1):739–747. https://doi.org/10.1007/s11033-022-07664-y

Dong OX, Yu S, Jain R, Zhang N, Duong PQ, Butler C, Li Y, Lipzen A, Martin JA, Barry KW, Schmutz J, Tian L, Ronald PC (2020) Marker-free carotenoid-enriched rice generated through targeted gene insertion using CRISPR-Cas9. Nat Commun 11(1):1178. https://doi.org/10.1038/s41467-020-14981-y

FAO, FIDA, OMS, PMA y UNICEF (2023) El estado de la seguridad alimentaria y la nutrición en el mundo 2023. Urbanización, transformación de los sistemas agroalimentarios y dietas saludables a lo largo del continuo rural-urbano. FAO, Roma. https://doi.org/10.4060/cc3017es

Francavilla A, Joye IJ (2020) Anthocyanins in whole grain cereals and their potential effect on health. Nutrients 12(10):2922. https://doi.org/10.3390/nu12102922

Gonzalez-Avila LU, Vega-López JM, Pelcastre-Rodríguez LI, Cabrero-Martínez OA, Hernández-Cortez C, Castro-Escarpulli G (2021) The challenge of CRISPR-Cas toward bioethics. Front Microbiol 12:657981. https://doi.org/10.3389/fmicb.2021.657981

Guo H, Wu H, Sajid A, Li Z (2022) Whole grain cereals: the potential roles of functional components in human health. Crit Rev Food Sci Nutr 62(30):8388–8402. https://doi.org/10.1080/10408398.2021.1928596

Habibpourmehraban F, Atwell BJ, Haynes PA (2022) Unique and shared proteome responses of rice plants (*Oryza sativa*) to individual abiotic stresses. Int J Mol Sci 23(24):552. https://doi.org/10.3390/ijms2324552

Han Y, Broughton S, Liu L, Zhang XQ, Zeng J, He X, Li C (2021) Highly efficient and genotype-independent barley gene editing based on another culture. Plant Commun 2(2):100082. https://doi.org/10.1016/j.xplc.2020.100082

Hille F, Richter H, Wong SP, Bratovič M, Ressel S, Charpentier E (2018) The biology of CRISPR-Cas: backward and forward. Cell 172(6):1239–1259. https://doi.org/10.1016/j.cell.2011.032

Huang J, Gao L, Luo S, Liu K, Qing D, Pan Y, Dai G, Deng G, Zhu C (2022a) The genetic editing of GS3 via CRISPR/Cas9 accelerates the breeding of three-line hybrid rice with superior yield and grain quality. Mol Breed 42(4):22. https://doi.org/10.1007/s11032-022-01290-z

Huang W, Zhang L, Columbus JT, Hu Y, Zhao Y, Tang L, Guo Z, Chen W, McKain M, Bartlett M, Huang CH, Li DZ, Ge S, Ma H (2022b) A well-supported nuclear phylogeny of Poaceae and implications for the evolution of C4 photosynthesis. Mol Plant 4:755–777. https://doi.org/10.1016/j.molp.2022.01.0

Ibrahim S, Saleem B, Naeem MK, Arain SM, Khan MR (2021) Next-generation technologies for iron and zinc biofortification and bioavailability in cereal grains. Crop Pasture Sci. https://doi.org/10.1071/CP20498

Inam S, Muhammad A, Irum S, Rehman N, Riaz A, Uzair M, Khan MR (2024) Genome editing for improvement of biotic and abiotic stress tolerance in cereals. FPB 51:FP24092. https://doi.org/10.1071/FP24092

Khalid A, Hameed A, Tahir MF (2023) Wheat quality: a review on chemical composition, nutritional attributes, grain anatomy, types, classification, and function of seed storage proteins in bread making quality. Front Nutr 10:1053196. https://doi.org/10.3389/fnut.2023.1053196

Kolkur SU, Sharma A, Gouda MR, Praveen KV, Singh A (2024) CRISPR in agriculture and it's ethical implications: A bibliometric analysis. Food Humanit:100322. https://doi.org/10.1016/j.foohum.2024.100322

Kumar S, Rymarquis LA, Ezura H, Nekrasov V (2021) Editorial: CRISPR-Cas in agriculture: opportunities and challenges. Front Plant Sci 12:672329. https://doi.org/10.3389/fpls.2021.672329

Kumar D, Yadav A, Ahmad R, Dwivedi UN, Yadav K (2022) CRISPR-based genome editing for nutrient enrichment in crops: a promising approach toward global food security. Front Genet 13:932859. https://doi.org/10.3389/fgene.2022.932859

Li J, Jiao G, Sun Y, Chen J, Zhong Y, Yan L, Jiang D, Ma Y, Xia L (2021a) Modification of starch composition, structure and properties through editing of TaSBEIIa in both winter and spring wheat varieties by CRISPR/Cas9. Plant Biotechnol J 19(5):937–951. https://doi.org/10.1111/pbi.13519

Li J, Li Y, Ma L (2021b) Recent advances in CRISPR/Cas9 and applications for wheat functional genomics and breeding. aBIOTECH 2:375–385. https://doi.org/10.1007/s42994-021-00042-5

Li Y, Wu X, Zhang Y, Zhang Q (2022) CRISPR/Cas genome editing improves abiotic and biotic stress tolerance of crops. Front Genome Ed 4:9878. https://doi.org/10.3389/fgeed.2022.987817

Liao Y, Li M, Wu H, Liao Y, Xin J, Yuan X, Li Y, Wei A, Zou X, Guo D, Xue Z, Zhu G, Wang Z, Xu P, Zhang H, Chen X, Du K, Zhou H, Xia D, Ali A et al (2024) Generation of aroma in three-line hybrid rice through CRISPR/Cas9 editing of Betaine Aldehyde Dehydrogenase2 (OsBADH2). Physiol Plant 176(1):e14206. https://doi.org/10.1111/ppl.14206

Liu Y, Luo W, Linghu Q, Abe F, Hisano H, Sato K, Kamiya Y, Kawaura K, Onishi K, Endo M, Toki S, Hamada H, Nagira Y, Taoka N, Imai R (2021) *In planta* genome editing in commercial wheat varieties. Front Plant Sci 12:648841. https://doi.org/10.3389/fpls.2021.648841

Liu L, Zhan J, Yan J (2024) Engineering the future cereal crops with big biological data: toward intelligence-driven breeding by design. J Genetic Genomic = Yi chuan xue bao 51(8):781–789. https://doi.org/10.1016/j.jgg.2024.03.005

Lorenzo D, Esquerda M, Palau F, Cambra FJ (2022) Ethics and genomic editing using the crispr-cas9 technique: challenges and conflicts. NanoEthics 16(3):313–321. https://doi.org/10.1007/s169-022-00425-y

Maharajan T, Krishna TPA, Rakkammal K, Ceasar SA, Ramesh M (2022) Application of CRISPR/Cas system in cereal improvement for biotic and abiotic stress tolerance. Planta 256(6):106. https://doi.org/10.1007/s00425-022-04023-w

Mathobo VM, Silungwe H, Ramashia SE, Anyasi TA (2021) Effects of heat-moisture treatment on the thermal, functional properties and composition of cereal, legume and tuber starches—a review. J Food Sci Technol 58(2):412–426. https://doi.org/10.1007/s13197-020-04520-4

Matres JM, Hilscher J, Datta A, Armario-Nájera V, Baysal C, He W, Huang X, Zhu C, Valizadeh-Kamran R, Trijatmiko KR, Capell T, Christou P, Stoger E, Slamet-Loedin IH (2021) Genome editing in cereal crops: an overview. Transgenic Res 30(4):461–498. https://doi.org/10.1007/s11248-021-00259-6

Medeiros DB, Brotman Y, Fernie AR (2021) The utility of metabolomics as a tool to inform maize biology. Plant Commun 2(4):100187. https://doi.org/10.1016/j.xplc.2021.100187

Mojica FJ, Díez-Villaseñor C, Soria E, Juez G (2000) Biological significance of a family of regularly spaced repeats in the genomes of Archaea, Bacteria and mitochondria. Mol Microbiol 36(1):244–246. https://doi.org/10.1046/j.1365-2958.2000.01838.x

Nguyen T-D, Nguyen X-V, La V-H, Bui T-T, Ngo X-B, Cao L-Q, Park S-K (2021) Increasing grain size and weight by CRISPR/CAS9-mediated GS3 gene editing in BT7 rice variety. Plant Cell Biotechnol Mol Biol 22:96–104

Nirmala Prasadi PV, Joye IJ (2020) Dietary fibre from whole grains and their benefits on metabolic health. Nutrients 12(10):3045. https://doi.org/10.3390/nu12103045

Patial M, Chauhan R, Chaudhary HK, Pramanick KK, Shukla AK, Kumar V, Verma RPS (2023) Au-courant and novel technologies for efficient doubled haploid development in barley (*Hordeum vulgare* L.). Crit Rev Biotechnol 43(4):575–593. https://doi.org/10.1080/07388551.2022.2050181

Raffan S, Sparks C, Huttly A, Hyde L, Martignago D, Mead A, Hanley SJ, Wilkinson PA, Barker G, Edwards KJ, Curtis TY, Usher S, Kosik O, Halford NG (2021) Wheat with greatly reduced accumulation of free asparagine in the grain, produced by CRISPR/Cas9 editing of asparagine synthetase gene TaASN2. Plant Biotechnol J 19(8):1602–1613. https://doi.org/10.1111/pbi.13573

Rahim, M. S., Bhandawat, A., Rana, N., Sharma, H., Parveen, A., Kumar, P., Madhawan, A., Bisht, A., Sonah, H., Sharma, T. R., & Roy, J. (2020). Genomic selection in cereal crops: methods and applications. Accelerated Plant Breeding, Volume 1: Cereal Crops, 51-88

Sá AGA, House JD (2024) Protein quality of cereals: digestibility determination and processing impacts. J Cereal Sci 117:103892. https://doi.org/10.1016/j.jcs.2024.103892

Shahzad R, Jamil S, Ahmad S, Nisar A, Khan S, Amina Z, Kanwal S, Aslam HMU, Gill RA, Zhou W (2021) Biofortification of cereals and pulses using new breeding techniques: current and future perspectives. Front Nutr 8:721728. https://doi.org/10.3389/fnut.2021.721728

Singh PK, Devanna BN, Dubey H, Singh P, Joshi G, Kumar R (2024) The potential of genome editing to create novel alleles of resistance genes in rice. Front Genome Ed 6:14244. https://doi.org/10.3389/fgeed.2024.14244

Skeie G, Fadnes LT (2024) Cereals and cereal products—a scoping review for Nordic nutrition recommendations 2023. Food Nutr Res 14:68. https://doi.org/10.29219/fnr.v68.10457

Tao J, Bauer DE, Chiarle R (2023) Assessing and advancing the safety of CRISPR-Cas tools: from DNA to RNA editing. Nat Commun 14(1):212. https://doi.org/10.1038/s41467-023-35886-6

Thilakarathna RCN, Madhusankha GDMP, Navaratne SB (2022) Potential food applications of sorghum (Sorghum bicolor) and rapid screening methods of nutritional traits by spectroscopic platforms. J Food Sci 87(1):36–51. https://doi.org/10.1111/1750-3841.16008

Tripathi L, Dhugga KS, Ntui VO, Runo S, Syombua ED, Muiruri S, Wen Z, Tripathi JN (2022) Genome editing for sustainable agriculture in Africa. Front Genome Ed 4:876697. https://doi.org/10.3389/fgeed.2022.876697

Wang S, Yang Y, Guo M, Zhong C, Yan C, Sun S (2020) Targeted mutagenesis of amino acid transporter genes for rice quality improvement using the CRISPR/Cas9 system. Author links open overlay panel. Crop J 8:457–464. https://doi.org/10.1016/j.cj.2020.02.005

Wang Y, Liu X, Zheng X, Wang W, Yin X, Liu H, Ma C, Niu X, Zhu JK, Wang F (2021) Creation of aromatic maize by CRISPR/Cas. J Integr Plant Biol 63(9):1664–1670. https://doi.org/10.1111/jipb.13105

Wang Q, Gao H, Liu K, Wang H, Zhang F, Wei L, Lu K, Li M, Shi Y, Zhao J, Zhou W, Peng B, Yuan H (2024) CRISPR/Cas9-mediated enhancement of semi-dwarf glutinous traits in elite Xiangdaowan rice (*Oryza sativa* L.): targeting *SD1*and *Wx* genes for yield and quality improvement. Front Plant Sci:1333191. https://doi.org/10.3389/fpls.2024.1333191

Yuan Z, Yang H, Pan L, Zhao W, Liang L, Gatera A, Tucker MR, Xu D (2022) Systematic identification and expression profiles of the BAHD superfamily acyltransferases in barley (Hordeum vulgare). Sci Rep 12(1):5063. https://doi.org/10.1038/s498-022-08983-7

Zenda T, Liu S, Dong A, Duan H (2021) Advances in cereal crop genomics for resilience under climate change. Life (Basel) 11(6):502. https://doi.org/10.3390/life11060502

Zhu F (2018) Anthocyanins in cereals: composition and health effects. Food Res Int (Ottawa) 109:232–249. https://doi.org/10.1016/j.foodres.2018.04.0

MIX
Papier aus verantwortungsvollen Quellen
Paper from responsible sources
FSC® C105338

If you have any concerns about our products,
you can contact us on
ProductSafety@springernature.com

In case Publisher is established outside the EU,
the EU authorized representative is:
Springer Nature Customer Service Center GmbH
Europaplatz 3, 69115 Heidelberg, Germany

Printed by Libri Plureos GmbH
in Hamburg, Germany